Lecture Notes in Energy

Volume 20

For further volumes:
http://www.springer.com/series/8874

Ranjan Vepa

Dynamic Modeling, Simulation and Control of Energy Generation

Springer

Ranjan Vepa
University of London
London
UK

ISSN 2195-1284 ISSN 2195-1292 (electronic)
ISBN 978-1-4471-5399-3 ISBN 978-1-4471-5400-6 (eBook)
DOI 10.1007/978-1-4471-5400-6
Springer London Heidelberg New York Dordrecht

Library of Congress Control Number: 2013943467

Printed on acid-free paper

Springer is part of Springer Science+Business Media (www.springer.com)

To Sudha

Preface

In the last decade we have seen the gradual erosion of the world's fossil fuel resources and this in turn has led to the need increased efficiencies both in the generation of power and in the utilization of it. Moreover, the emphasis has gradually increased on the need for harnessing renewable energy sources that do not cause environmental pollution, such as wind, photovoltaic, hydraulic, among others. The need for improving the efficiency of energy generation has in turn led to the use for a range of automatic control-based techniques for regulating the conversion of the available energy into a usable form in the most optimum way. To be able to apply modern automatic controller design techniques to energy generation, it is often necessary to be able to effectively model the most significant aspects of the dynamics of energy generation and conversion. Such modeling will facilitate not only the design of automatic control systems for regulating the process of energy generation but also in ensuring that the system is operating with optimum efficiency. Moreover, it will facilitate the continuous monitoring of the relevant system variables by the use of modern estimation techniques.

This book addresses the core issues involved in the dynamic modeling, simulation, and control of selection of energy systems such as gas turbines, wind turbines, fuel cells, and batteries. The principles of modeling and control could be applied to other non-convention methods of energy generation such as solar energy and wave energy. A primary feature of this book is that it brings together a range of diverse topics relevant to the design of energy generation, its regulation, and control.

This book will help the reader understand the methods of modeling energy systems for controller design application as well as gain a basic understanding of the processes involved in the design of control systems and regulators. It will also be a useful guide for simulation of the dynamics of energy systems and for implementing monitoring systems based on the estimation of internal system variables from measurements of observable system variables.

A central feature of this book is that it brings together diverse topics in thermodynamics, fluid mechanics, heat transfer, electro-chemistry, electrical networks and electrical machines, and focuses on their applications in the field of energy generation, its control, and regulation. It will serve as a useful aid to designers of hybrid power generating systems involving advanced technology systems such as floating or offshore wind turbines and fuel cells. For the first time the book introduces case studies of the practical control laws for a variety of energy generation systems based on nonlinear dynamic models without relying on linearization. Also for the first time this book introduces the reader to the use nonlinear model-based estimation techniques and their application to energy systems.

Chapter 1 of the book covers the basic principle of electrical machines, fluid mechanics, thermodynamics, and heat transfer that are essential to Chaps. 3–8. The topic of Chap. 2 is principles of energy conversion to usable electrical form. All the basic principles of both motors and generators are covered in this chapter. An extensive section on motors is included, which are key elements of control system. Chapter 3 focuses on the dynamics and characteristics of alternating current machines, primarily induction and synchronous generators. Chapter 4 focuses on wind power generation, the components of a typical horizontal axis wind turbine, basics of wind power generation, the wind field, its characteristics and spectrum as well the dynamics of the rotor, tower, and floater or supporting structure. Chapter 5 deals with the dynamic modeling of gas turbines and compressors including combustion chambers, turbines and, nozzles and diffusers. Chapter 6 is about the electro-chemical modeling of fuel cells while Chap. 7 is devoted to high performance batteries. The last chapter is concerned about non-conventional methods of energy generation.

I would like to thank my colleagues, and present and former students of the School of Engineering and Material Science, at Queen Mary, University of London for their support in this endeavour.

I would like to thank my wife Sudha, for her love understanding and patience. Her encouragement was a principal factor that provided the motivation to complete the project. I would also like to thank our children Lullu, Satvi, and Abhinav for their understanding during the course of this project.

London, UK, 2013 R. Vepa

Contents

Abbreviations

AC	Alternating current
BEM	Blade element momentum
CCM	Catalyst-coated membrane
CS	Control surface
CV	Control volume
DC	Direct current
DFIG	Doubly fed induction generator
EMF	Electro motive force
EKF	Extended Kalman filter
EnKF	Ensemble Kalman filter
FACTS	Flexible AC transmission system
FADEC	Full authority digital engine controller
FC	Fuel cell
GDL	Gas diffusion layer
IGBT	Insulated-gate bipolar transistor
JONSWAP	Joint north sea wave project
KBM	Kinetic battery model
KF	Kalman filter
KVA	Kilovolt-amperes
KW	Kilowatt
LPV	Linear parameter varying
MCFC	Molten carbonate fuel cell
MMF	Magneto motive force
MPP	Maximum power point
MPPT	Maximum power point tracking
MVA	Megavolt-amperes
MW	Megawatt
NACA	National Advisory Committee on Aeronautics
NARMAX	Nonlinear autoregressive moving average models with exogenous inputs
PAFC	Phosphoric acid fuel cell
PEMFC	Proton exchange membrane fuel cell
PMSG	Permanent magnet synchronous generator
PV	Photo voltaic cells

RC	Resistive capacitive
SHE	Standard hydrogen electrode
SOC	State of charge
SOFC	Solid oxide fuel cell
STATCOM	Static synchronous compensators
SVC	Static var compensator
UKF	Unscented Kalman filter

Chapter 1
Introduction to Energy Generation Principles

1.1 Fluid Mechanics

Before embarking upon the detailed studies of the modeling of energy systems, the basic principles of fluid mechanics and thermodynamics as well as those of gas dynamic flows relevant to energy systems are discussed in this chapter. Further detailed expositions of these topics may be found in White (2003), Liepmann and Roshko (2002), and Babu (2008).

1.1.1 Mass, Momentum, and Energy Balance

A convenient and unified manner in which the governing equations of a flow can be developed is to consider a transported scalar or vector quantity, denoted by the variable Φ, associated with a mass of fluid within a predefined volume of fluid referred to as the *control volume* (CV). Given that the specific value of Φ per unit mass is ϕ and that ρ is the material density, one can establish a fundamental integral relationship between the quantities Φ and ϕ by considering an infinitesimal element of volume $d\mathbf{x}$ as

$$\Phi = \int_{CV} \rho\phi d\mathbf{x}.$$

If $\mathbf{u}$ is the velocity vector in the flow domain within CV and $\mathbf{n}$ is the outward drawn unit normal defined over the *control surface* (CS) enclosing the CV, the time rate of change of the variable Φ is

$$\frac{d\Phi}{dt} = \int_{CV} \frac{\partial(\rho\phi)}{\partial t} d\mathbf{x} + \int_{CS} \rho\phi(\mathbf{u}\cdot\mathbf{n})ds \qquad (1.1.1)$$

where $\mathbf{u}\cdot\mathbf{n}$ is the dot product of the vectors $\mathbf{u}$ and $\mathbf{n}$, i.e., the rate of generation of Φ is the sum of the rate of production of ϕ within the entire control volume and the

R. Vepa, *Dynamic Modeling, Simulation and Control of Energy Generation*, Lecture Notes in Energy 20,
DOI: 10.1007/978-1-4471-5400-6_1, © Springer-Verlag London 2013

net outflow of ϕ through the bounding surface. The variable Φ may be, in general, the total mass in which case $\phi = 1$, it may be the momentum in which case ϕ is the vector $\mathbf{u}$ or it may be the total energy in which case ϕ is the specific energy e.

Considering the time rate of change of the total mass within the CV, the Eq. (1.1.1) for the general transport dynamics of a flow variable may be expressed as

$$\frac{\mathrm{d}\Phi_m}{\mathrm{d}t} = \int_{\mathrm{CV}} \frac{\partial \rho}{\partial t} \mathrm{d}\mathbf{x} + \int_{\mathrm{CS}} \rho(\mathbf{u} \cdot \mathbf{n}) \mathrm{d}s = 0. \tag{1.1.2}$$

Equation (1.1.2) expresses the principle of the conservation of mass. In particular, in Eq. (1.1.2), one could invoke Gauss's divergence theorem and express the surface integral as

$$\int_{\mathrm{CS}} \rho(\mathbf{u} \cdot \mathbf{n}) \mathrm{d}s = \int_{\mathrm{CV}} \nabla \cdot (\rho \mathbf{u}) \mathrm{d}\mathbf{x}.$$

Consequently, the principle of conservation of mass for any choice of the CV may expressed as

$$\int_{\mathrm{CV}} \frac{\partial \rho}{\partial t} \mathrm{d}\mathbf{x} + \int_{\mathrm{CV}} \nabla \cdot (\rho \mathbf{u}) \mathrm{d}\mathbf{x} = 0. \tag{1.1.3}$$

In differential form, the principle of conservation of mass is

$$\frac{\partial \rho}{\partial t} + \nabla \cdot (\rho \mathbf{u}) = 0. \tag{1.1.4}$$

Considering the time rate of change of the total momentum within the CV and employing Newton's second law of motion, the Eq. (1.1.1) for the general transport dynamics of a flow variable may be expressed as

$$\frac{\mathrm{d}\Phi_f}{\mathrm{d}t} = \int_{\mathrm{CV}} \frac{\partial (\rho \mathbf{u})}{\partial t} \mathrm{d}\mathbf{x} + \int_{\mathrm{CS}} \rho \mathbf{u}(\mathbf{u} \cdot \mathbf{n}) \mathrm{d}s = \mathbf{F}. \tag{1.1.5}$$

In the above, $\mathbf{F}$ is the total external force acting within the CV which is made up of two components: a body force component $\boldsymbol{\chi}$ distributed through the CV and a distributed stress $\boldsymbol{\sigma}$ which gives rise to a surface force acting over the CV. Hence, $\mathbf{F}$ may be expressed as

$$\mathbf{F} = \int_{\mathrm{CV}} \boldsymbol{\chi} \mathrm{d}\mathbf{x} + \int_{\mathrm{CS}} (\mathbf{n} \cdot \boldsymbol{\sigma}) \mathrm{d}s. \tag{1.1.6}$$

Again invoking the Gauss divergence theorem,

$$\int_{\mathrm{CV}} \left(\frac{\partial (\rho \mathbf{u})}{\partial t} + \nabla \cdot (\rho \mathbf{u} \mathbf{u}) \right) \mathrm{d}\mathbf{x} = \int_{\mathrm{CV}} (\boldsymbol{\chi} + \nabla \cdot \boldsymbol{\sigma}) \mathrm{d}\mathbf{x}.$$

Employing the mass balance Eq. (1.1.3),

$$\int_{CV} \rho\left(\frac{\partial \mathbf{u}}{\partial t} + \mathbf{u}\cdot\nabla\mathbf{u}\right)\mathrm{d}\mathbf{x} = \int_{CV} (\boldsymbol{\chi} + \nabla\cdot\boldsymbol{\sigma})\mathrm{d}\mathbf{x} \tag{1.1.7}$$

In differential form, the principle of conservation of momentum is

$$\rho\left(\frac{\partial \mathbf{u}}{\partial t} + \mathbf{u}\cdot\nabla\mathbf{u}\right) = \boldsymbol{\chi} + \nabla\cdot\boldsymbol{\sigma}. \tag{1.1.8}$$

Equation (1.1.8) is vector equation for each component $\mathbf{u}$. Since the vector gradient ∇ is also a vector, the quantity $\nabla\mathbf{u}$ is a second-order tensor and the quantity $\mathbf{u}\cdot\nabla\mathbf{u}$ is a vector.

Considering the time rate of change of the total energy within the CV, the Eq. (1.1.1) for the general transport dynamics of a flow variable may be expressed as

$$\frac{\mathrm{d}\Phi_e}{\mathrm{d}t} = \int_{CV} \frac{\partial(\rho e_t)}{\partial t}\mathrm{d}\mathbf{x} + \int_{CS} \rho e_t(\mathbf{u}\cdot\mathbf{n})\mathrm{d}s \equiv \int_{CV} \left(\frac{\partial(\rho e_t)}{\partial t} + \nabla\cdot(\rho e_t\mathbf{u})\right)\mathrm{d}\mathbf{x}. \tag{1.1.9}$$

In Eq. (1.1.9), e_t is the specific total energy. The rate of generation of the total energy ($= \mathrm{d}\Phi_e/\mathrm{d}t$) consists of three groups of contributions: the first due to energy generation throughout the CV, the second due to heat energy transferred out of the CV across the boundary, and the third due to mechanical work done on the control surface by the stress field. The rate of generation of the total energy is expressed as

$$\frac{\mathrm{d}\Phi_e}{\mathrm{d}t} = \int_{CV} Q\mathrm{d}\mathbf{x} - \int_{CS} \mathbf{q}\cdot\mathbf{n}\mathrm{d}s - \left(-\int_{CS} \mathbf{u}\cdot(\boldsymbol{\sigma}\cdot\mathbf{n})\mathrm{d}s\right).$$

Invoking the Gauss divergence theorem,

$$\frac{\mathrm{d}\Phi_e}{\mathrm{d}t} = \int_{CV} (Q - \nabla\cdot\mathbf{q} + \nabla\cdot(\mathbf{u}\cdot\boldsymbol{\sigma}))\mathrm{d}\mathbf{x}.$$

It follows that

$$\frac{\mathrm{d}\Phi_e}{\mathrm{d}t} = \int_{CV} \left(\frac{\partial(\rho e_t)}{\partial t} + \nabla\cdot(\rho e_t\mathbf{u})\right)\mathrm{d}\mathbf{x} = \int_{CV} (Q - \nabla\cdot\mathbf{q} + \nabla\cdot(\mathbf{u}\cdot\boldsymbol{\sigma}))\mathrm{d}\mathbf{x}. \tag{1.1.10}$$

In differential form, the principle of conservation of energy is

$$\frac{\partial(\rho e_t)}{\partial t} + \nabla\cdot(\rho e_t\mathbf{u}) = Q - \nabla\cdot\mathbf{q} + \nabla\cdot(\mathbf{u}\cdot\boldsymbol{\sigma}). \tag{1.1.11}$$

Using the principle of mass conservation

$$\rho\left(\frac{\partial e_t}{\partial t} + \mathbf{u}\cdot\nabla e_t\right) = Q - \nabla\cdot\mathbf{q} + \nabla\cdot(\mathbf{u}\cdot\boldsymbol{\sigma}). \tag{1.1.12}$$

It is often convenient to express the energy conservation equation in terms of the specific enthalpy h rather the specific total energy, e_t. This results in the thermal energy equation. The enthalpy is the sum of the internal energy, e, of the system and the product of the pressure of the fluid in the system and its volume. Consequently, when the outside pressure on a system is held constant, a change in enthalpy entails a change in the system's internal energy and a change in the system's volume as the system exchanges energy with the outside. Thus, the enthalpy is a partial measure of the internal energy of the system and is the amount of energy in a system capable of doing mechanical work. The specific total energy, e_t, is related to the specific enthalpy h by the relation

$$h = e + pv, \quad e_t = h + gz + (\mathbf{u} \cdot \mathbf{u})/2, \tag{1.1.13}$$

where the second term represents the gravitational potential energy and the third represents the kinetic energy. In terms of the specific enthalpy, h, the thermal energy equation is

$$\rho\left(\frac{\partial h}{\partial t} + \mathbf{u} \cdot \nabla h\right) = Q - \nabla \cdot \mathbf{q} + \nabla \cdot (\mathbf{u} \cdot \boldsymbol{\sigma}) - \mathbf{u} \cdot (\nabla \cdot \boldsymbol{\sigma}). \tag{1.1.14}$$

When heat addition is such that it results in mechanical work that can be completely recovered, the process is said to reversible and isentropic. For such a reversible and isentropic process, considering an infinitesimal change

$$dq \equiv de + pdv \equiv dh - vdp = 0. \tag{1.1.15}$$

1.1.2 Stress Rate of Strain Relations and Viscosity

It is convenient to eliminate the stress tensor $\boldsymbol{\sigma}$ from the momentum conservation equation. The most general relationship between stress tensor and the rate of strain tensor representing the fluid material's response to applied force field for a homogeneous isotropic fluid can be shown to be of the form:

$$\sigma_{ij} = (-p_{th} + \lambda\varepsilon_{kk})\delta_{ij} + 2\mu\varepsilon_{ij}. \tag{1.1.16}$$

In Eq. (1.1.16), $\boldsymbol{\varepsilon}$ is the rate of strain tensor defined as

$$\boldsymbol{\varepsilon} = \frac{1}{2}\left[(\nabla\mathbf{u}) + (\nabla\mathbf{u})^T\right]. \tag{1.1.17}$$

Moreover, $\sum_{k=1}^{3} \varepsilon_{kk} = \nabla \cdot \mathbf{u}$ and is simply denoted as ε_{kk} as the repeated index is assumed to imply the summation over it. Furthermore,

$$\frac{\sigma_{kk}}{3} = (-p_{th} + \lambda\varepsilon_{kk}) + \frac{2}{3}\mu\varepsilon_{kk}. \tag{1.1.18}$$

In Eq. (1.1.18), p_{th} is known as the pressure. Thus, if σ_{kk} is denoted as $-3p_s$, where p_s is known as the static pressure,

$$p_{th} - p_s = \left(\lambda + \frac{2}{3}\mu\right)\nabla \cdot \mathbf{u}. \tag{1.1.19}$$

Thus, the difference between p_{th} and p_s vanishes if either $\nabla \cdot \mathbf{u} = 0$ or when $\lambda + (2/3)\mu = 0$. The latter result is known to be true for most fluids based on the kinetic theory of gases and is known as Stokes' hypothesis. Thus, it is normal to assume that $\lambda = -(2/3)\mu$ and when this is done

$$\sigma_{ij} = -p\delta_{ij} + \mu\left(2\varepsilon_{ij} - (2/3)\nabla \cdot \mathbf{u}\delta_{ij}\right). \tag{1.1.20}$$

In Eq. (1.1.20), μ is known as the dynamic viscosity of the fluid and p is the static pressure in the fluid. With Eqs. (1.1.20) and (1.1.9), the principle of conservation of momentum reduces to

$$\rho\left(\frac{\partial \mathbf{u}}{\partial t} + \mathbf{u} \cdot \nabla \mathbf{u}\right) = \chi - \nabla p + 2\nabla \cdot \mu\left(\varepsilon_{ij} - (1/3)\nabla \cdot \mathbf{u}\delta_{ij}\right). \tag{1.1.21}$$

Equation (1.1.21) is known as the *Navier–Stokes* or *conservation of momentum* equation and may be expressed in several alternate forms. In particular, in the case of an inviscid fluid, when $\mu = 0$, the last term in Eq. (1.1.21) is absent and the Eq. (1.1.21) is known as the *Euler* equation.

1.1.3 Perfect Gases and Acoustic Waves

The equation of state of a system composed of a finite volume of gas of mass m occupying the volume V at an absolute temperature T and at a pressure p can be stated by an equation in terms of the molecular weight M_w and a universal constant R as

$$pV = \frac{m}{M_w}\mathsf{R}T = n\mathsf{R}T. \tag{1.1.22}$$

The ratio of the mass of the gas to the molecular weight is the number of moles of gas in the volume V. Since the mass per unit volume is the density, ρ, of the gas and the specific volume is $v = 1/\rho$,

$$\frac{p}{\rho} = pv = \frac{\mathsf{R}}{M_w}T = RT \text{ and } R = \mathsf{R}/M_w. \tag{1.1.23}$$

The enthalpy per unit mass or the specific enthalpy may be expressed as

$$h = e + pv = e + RT. \tag{1.1.24}$$

Hence,

$$\frac{\partial (pv)}{\partial T} = R, \quad \frac{\partial h}{\partial T} = \frac{\partial e}{\partial T} + R. \tag{1.1.25}$$

Consider a constant-volume process and for such a process, one may write

$$\mathrm{d}q \equiv \mathrm{d}e + p\mathrm{d}v = \mathrm{d}e = C_v \mathrm{d}T. \tag{1.1.26}$$

It follows that for a constant-volume process

$$\frac{\partial e}{\partial T} = \left.\frac{\partial e}{\partial T}\right|_{v=\text{constant}} = C_v. \tag{1.1.27}$$

For a constant-pressure process

$$\frac{\partial h}{\partial T} = \left.\frac{\partial h}{\partial T}\right|_{p=\text{constant}} = C_p. \tag{1.1.28}$$

The specific heat capacity under constant pressure, C_p, and the specific heat capacity under constant volume, C_v, are related to each other by the relation

$$C_p - C_v = R. \tag{1.1.29}$$

Furthermore, for a reversible or isentropic process

$$\frac{\mathrm{d}q_{\text{rev}}}{T} \equiv C_v \frac{\mathrm{d}T}{T} + R\frac{\mathrm{d}v}{v} \equiv C_p \frac{\mathrm{d}T}{T} - R\frac{\mathrm{d}p}{p} = 0. \tag{1.1.30}$$

For an irreversible process, the quantity on the left-hand side of Eq. (1.1.30) may be identified as the change in entropy ds and it follows that

$$\frac{\mathrm{d}q_{\text{irrev}}}{T} \equiv \mathrm{d}s = C_p \frac{\mathrm{d}T}{T} - R\frac{\mathrm{d}p}{p} \geq 0 \tag{1.1.31}$$

However, since

$$\frac{\mathrm{d}p}{p} + \frac{\mathrm{d}v}{v} = \frac{\mathrm{d}T}{T}, \tag{1.1.32}$$

it follows that

$$C_v \frac{\mathrm{d}p}{p} + C_p \frac{\mathrm{d}v}{v} = 0 \text{ and that } \frac{\mathrm{d}p}{\mathrm{d}v} = -\frac{C_p}{C_v}\frac{p}{v} = -\gamma\frac{p}{v}. \tag{1.1.33}$$

Consequently, integrating either of the Eqs. (1.1.33), it can be shown that

$$pv^\gamma = p_0 v_0^\gamma = \text{constant}.$$

Since $v = 1/\rho$, $\frac{d\rho}{\rho} + \frac{dv}{v} = 0$, and

$$p\left(\frac{1}{\rho}\right)^{\gamma} = p_0\left(\frac{1}{\rho_0}\right)^{\gamma} = \text{constant.} \tag{1.1.34}$$

Equation (1.1.33) may also be expressed as

$$\frac{dp}{d\rho} = \frac{C_p p}{C_v \rho} = \gamma\frac{p}{\rho} = \gamma RT. \tag{1.1.35}$$

Using the definition of the ratio of specific heat capacities $\gamma = C_p/C_v$, Eq. (1.1.29) may be expressed as,

$$C_p - C_v = (\gamma - 1)C_v = R$$

or

$$C_v = R/(\gamma - 1) \text{ and } C_p = \gamma R/(\gamma - 1). \tag{1.1.36}$$

Consider an acoustic wave propagating into a quiescent gas. Although the wave front is necessarily spherical in three dimensions and in general, at any point on the wave front the flow is essentially one dimensional as the radius of curvature of the wave front is much larger than the distance over which the flow properties change. In a reference frame moving with wave front, the wave appears stationary and the flow approaches the wave with a velocity equal to the wave speed in the moving reference frame but moves away from the wave with a different velocity. As a result of moving through the acoustic wave, the flow properties change by an infinitesimal amount. Applying the principles of conservation of mass and momentum to points on either side of the wave front

$$\rho_1 u_1 = (\rho_1 + d\rho_1)(u_1 + du_1)$$

and

$$p_1 + \rho_1 u_1^2 = p_1 + dp_1 + (\rho_1 + d\rho_1)(u_1 + du_1)^2 = p_1 + dp_1 + \rho_1 u_1(u_1 + du_1).$$

Upon neglecting higher-order terms, the first of these equations reduces to

$$\rho_1 du_1 + u_1 d\rho_1 = 0$$

while using this relation, the second of the two equations reduces to

$$dp_1 + \rho_1 u_1 du_1 = dp_1 - u_1^2 d\rho_1 = 0.$$

As the acoustic wave front must move with the speed of sound, it follows that

$$u_1^2 = a^2 = \frac{dp_1}{d\rho_1}.$$

Using Eq. (1.1.35)

$$a = \sqrt{\mathrm{d}p_1/\mathrm{d}\rho_1} = \sqrt{\gamma RT}. \tag{1.1.37}$$

The significance of this relation is that it represents the speed of a traveling wave of disturbance in the flow and is given by the square root of the derivative with respect to the density of the adiabatic expression for the pressure. A somewhat different speed of sound formula is appropriate when the heat conduction rate is so large and fast so as to maintain a constant temperature in the fluid. The isothermal speed of sound may be shown to be

$$a_i = \sqrt{p_1/\rho_1} = \sqrt{RT}. \tag{1.1.38}$$

1.1.4 Reference States

In the study of flows associated with energy generation systems, it is convenient to define a set of reference states. While there a number of reference states that must be defined in the analysis of the dynamics of such systems, two classes of reference states appear repeatedly in the analysis. The first of these is the state of the flow when the flow velocity is equal to the local speed of sound, which is sometimes referred to as the sonic state. Since the flow velocity is equal to the local speed of sound at the sonic state, a non-dimensional variable, the Mach number M, defined as the ratio of the flow velocity at a point to the local speed of sound, plays a key role and $M = 1$ at the sonic state. The importance of the sonic state lies in the fact that it separates the subsonic flow domain characterized by $M < 1$ and the supersonic flow domain characterized by $M > 1$. In fact, the region that separates the two domains is often not distinct, and therefore, it is often important to identify the region when $M \approx 1$ or the region defined by $1 - \varepsilon \leq M \leq 1 + \varepsilon$ where ε is usually a very small number of the order of 0.02 although its precise value depends on the nature of the particular application problem. Such a region is often known as the transonic flow region, a region that is characterized by a number of anomalous behaviors.

The second class of abundantly occurring reference states is the stagnation state. Consider a state of the fluid flow where all the state variables describing that state are known, i.e., the pressure, the density, the temperature, and the velocity at that point are known. Conceptually, it is now assumed that the flow decelerated slowly and without any losses whatsoever to a state where all the velocity components are equal to zero. Such a state is said to be a stagnation state. It is only necessary that the deceleration process that brings the state from where it is fully known to a stagnation state be isentropic, i.e., without any energy losses. If one denotes the pressure, the density, and the temperature at the stagnation state in a particular flow field by the subscript "0", then one may derive the relationship between the static pressure and the corresponding temperature at a particular point in the flow field and the stagnation pressure and temperature of the flow.

To derive the relationship between the static pressure and the corresponding temperature at a particular point in the flow field and the stagnation pressure, consider the specific total energy, e_t, at any two points in the flow corresponding to two flow states. When the energy is conserved as the flow transits from one flow state to the other without any loss of energy, the specific total energy is conserved. Thus,

$$e_t = h_1 + gz_1 + (\mathbf{u}_1 \cdot \mathbf{u}_1)/2 = h_2 + gz_2 + (\mathbf{u}_2 \cdot \mathbf{u}_2)/2. \tag{1.1.39}$$

Assume that the state "1" corresponds to the static state which is indicated with no subscript and "2" to the stagnation state and that there is no change in the gravitational potential between these two states. Hence,

$$e_t = h + gz + (\mathbf{u} \cdot \mathbf{u})/2 = h_0 + gz. \tag{1.1.40}$$

It follows that

$$h_0 - h = \Delta h = (\mathbf{u} \cdot \mathbf{u})/2, \tag{1.1.41}$$

and

$$(T_0 - T) = (\mathbf{u} \cdot \mathbf{u})/(2C_p) = \left(\frac{\gamma - 1}{2}\right) TM^2. \tag{1.1.42}$$

Hence, we may denote the ratio of the stagnation temperature to the static temperature corresponding to the static pressure as

$$\frac{T_0}{T} = 1 + \left(\frac{\gamma - 1}{2}\right) M^2. \tag{1.1.43}$$

Using the second of Eq. (1.1.30) and integrating, the ratio of the stagnation pressure P_0 to the static pressure is related to the ratio of the stagnation temperature to the static temperature by

$$\frac{P_0}{p} = \left(\frac{T_0}{T}\right)^{\frac{\gamma}{\gamma-1}}. \tag{1.1.44}$$

Finally, one obtains for the ratio of the stagnation pressure to the static pressure

$$\frac{P_0}{p} = \left(1 + \left(\frac{\gamma - 1}{2}\right) M^2\right)^{\frac{\gamma}{\gamma-1}}. \tag{1.1.45}$$

1.1.5 Normal Shock Relations

It is now possible to distinguish precisely between two different kinds of flows, one associated with expansion and the other with compression. Together they form the basis of all types of flows that can occur in one dimension. The differences between them are profound. A traveling sound signal in expanding flow can never propagate in front of the region due to a continuous decrease in the sound speed. While a succession of expansion waves can produce an ever-widening expansion fan, a succession of compression waves with each propagating faster than its predecessor, the waves eventually cause a "pile up" somewhere up front, forming almost a discontinuity in the flow variables known as a "shock." In addition to "shocks," there is another type of discontinuity that can occur, the contact discontinuity. A contact discontinuity moves with fluid and generally occurs at the boundary between two different fluids although it may be generated in one of them. The shock, on the other hand, moves relative to the fluid, changing the state of each fluid element as it sweeps by it. Considering a one-dimensional flow field, the shock may be thought of as an infinite plane dividing two semi-infinite uniform regions with different fluid properties in each. Moreover, the relative velocity of the shock plane is normal to it across the whole of the plane. In general, the shock plane's velocity, v_s, normal to it is generally greater than both the fluid velocity u_- behind the shock plane and the fluid velocity u_+ ahead the shock plane. Furthermore, $u_- > u_+$, which follows from the fact that all shocks are compressive. The velocity of the shock, v_s, is also supersonic as it travels faster than sound. In addition to the equation of state, there are three conditions derived from the three laws relating to mass, momentum, and energy conservation at the shock plane, known as the Rankine–Hugoniot relations. A key difference between a shock wave and an acoustic wave is that the changes in the properties across a shock wave are large and irreversible, while changes in properties across an acoustic wave are infinitesimal and isentropic. Moreover, an acoustic wave always travels at the local speed of sound, while the shock wave travels much faster. Since one is interested in the conservation of mass, momentum, and energy in the shock plane, it is best to consider a reference frame moving with the shock and apply the conservation conditions,

$$\rho_1 u_1 = \rho_2 u_2 \tag{1.1.46a}$$

$$p_1 + \rho_1 u_1^2 = p_2 + \rho_2 u_2^2 \tag{1.1.46b}$$

$$h_1 + u_1^2/2 = h_2 + u_2^2/2 \tag{1.1.46c}$$

where $u_1 = u_- - v_s$, $u_2 = u_+ - v_s$, $\rho_1 = \rho_-$, $\rho_2 = \rho_+$, $p_1 = p_-$, $p_2 = p_+$, $h_1 = h_-$, $h_2 = h_+$ and the subscript "−" refers to the semi-infinite region upstream of the shock and the subscript "+" refers to the semi-infinite region downstream of the shock. Considering the changes in the entropy across the shock plane

$$s_2 - s_1 = C_p \ln\left[\frac{T_2}{T_1}\right] - R\ln\left[\frac{p_2}{p_1}\right] = C_p \ln\left[\frac{v_2}{v_1}\right] + C_v \ln\left[\frac{p_2}{p_1}\right]. \tag{1.1.46d}$$

From the above governing equations and the equation of state $pv = RT$ which is satisfied independently in each of the semi-infinite regions, it can be shown that

$$M_2 = +\sqrt{\frac{(\gamma-1)M_1^2+2}{2\gamma M_1^2-(\gamma-1)}}, \tag{1.1.47}$$

and that

$$\left(\frac{T_2}{T_1}\right) = \frac{\left(1+\frac{(\gamma-1)}{2}M_1^2\right)\left(\frac{2\gamma}{\gamma-1}M_1^2-1\right)}{\left(\frac{(\gamma+1)^2}{2(\gamma-1)}\right)M_1^2}, \quad \frac{p_2}{p_1} = \frac{2\gamma M_1^2}{\gamma+1} - \frac{\gamma-1}{\gamma+1},$$

$$\frac{\rho_2}{\rho_1} = \frac{(\gamma+1)M_1^2}{(\gamma-1)M_1^2+2}. \tag{1.1.48}$$

1.1.6 One-Dimensional Flow: Rayleigh Flow

In this section, we consider one-dimensional flows in constant- or variable-area ducts either with heat addition or those which are adiabatic but with friction at the walls of the ducts. These types of flows are encountered in jet engine combustors which involve substantial heat addition and in flows emerging from high-pressure combustion chambers which are adiabatic but with friction at the walls of the ducts and nozzles. The governing equations relating the one-dimensional flow boundary conditions at the two extreme ends of the flow in a constant-area duct are

$$\rho_1 u_1 = \rho_2 u_2, \tag{1.1.49a}$$

$$p_1 + \rho_1 u_1^2 = p_2 + \rho_2 u_2^2, \tag{1.1.49b}$$

$$h_1 + q_{h\_\text{in}} + u_1^2/2 = h_2 + u_2^2/2, \tag{1.1.49c}$$

and

$$s_2 - s_1 = C_p \ln\left[\frac{T_2}{T_1}\right] - R\ln\left[\frac{p_2}{p_1}\right] = C_p \ln\left[\frac{v_2}{v_1}\right] + C_v \ln\left[\frac{p_2}{p_1}\right]. \tag{1.1.49d}$$

Furthermore, the temperatures at the two extreme ends can be related to the respective stagnation conditions. Hence,

$$T_{01} = T_1 + u_1^2/(2C_p), \quad T_{02} = T_2 + u_2^2/(2C_p).$$

Assuming that the specific heat coefficient under constant pressure, C_p, is a constant, the energy conservation equation may be expressed as

$$T_{02} - T_{01} = q_{h\_\text{in}}/C_p = \Delta T_0. \tag{1.1.50}$$

Flows for which the specific heat coefficient under constant pressure, C_p, is a constant are said to be **calorically perfect**. Before attempting to solve the governing Eq. (1.1.49a–d), it is worth trying to gain a physical insight into the dynamics of the flow. In order to do this, the first two Eq. (1.1.49a) and (1.1.49b) may be expressed as

$$\rho_1 u_1 = \rho_2 u_2 = G = \text{a constant} \tag{1.1.51a}$$

$$\text{or as } p_1 - p_2 = G^2(v_2 - v_1). \tag{1.1.51b}$$

The first Eq. (1.1.51a) expresses the fact that the mass flow rate is a constant, while the second Eq. (1.1.51b) can be expressed diagrammatically as straight line on a (p, v) diagram. The straight line relating the fall in the pressure to the raise in the volume on the (p, v) diagram is known as the Rayleigh line on the (p, v) diagram. Expressing the equations diagrammatically either on a (p, v) diagram and on a (T, s) diagram allows one to gain a physical insight into the dynamics of the flow. However, to express the equations diagrammatically either on a (p, v) diagram and on a (T, s) diagram, the governing equations must be related to differential changes in the state variables as they evolve from one end of the flow to the other.

From the energy Eq. (1.1.49c),

$$\frac{\gamma R}{\gamma - 1} T_1 + q_{h\_\text{in}} + G^2 v_1^2/2 = \frac{\gamma R}{\gamma - 1} T_2 + G^2 v_2^2/2. \tag{1.1.52}$$

Rearranging and eliminating G^2 from Eq. (1.1.51b)

$$\frac{\gamma R}{\gamma - 1} T_1 \left(1 - \frac{T_2}{T_1}\right) + q_{h\_\text{in}} = \frac{p_1 - p_2}{v_2 - v_1}\left(v_2^2 - v_1^2\right)/2 = \frac{(p_1 - p_2)(v_1 + v_2)}{2}.$$

Using the equation of state, $p_i v_i = RT_i$

$$\frac{\gamma}{\gamma - 1}\left(1 - \frac{p_2 v_2}{p_1 v_1}\right) + \frac{q_{h\_\text{in}}}{RT_1} = \frac{1}{2}\left(1 - \frac{p_2}{p_1}\right)\left(1 + \frac{v_2}{v_1}\right).$$

Rearranging and solving for the pressure ratio

$$\frac{p_2}{p_1} = \left\{\frac{\gamma + 1}{\gamma - 1} + \frac{2q_{h\_\text{in}}}{RT_1} - \frac{v_2}{v_1}\right\} \Big/ \left\{\left(\frac{\gamma + 1}{\gamma - 1}\right)\left(\frac{v_2}{v_1}\right) - 1\right\}. \tag{1.1.53}$$

Equation (1.1.53) is known as the Rankine–Hugoniot equation and represents a family of curves resembling a hyperbola on a (p, v) diagram, for different values of the heat input, q_{h_in}. A typical curve in this family is often referred to as an ***H*-curve**. Together with the Rayleigh line, the Rankine–Hugoniot equation determines whether or not a particular state is physically realizable or not.

In differential form, the first of the governing Eq. (1.1.49a) is

$$\mathrm{d}(\rho u) = \rho \mathrm{d}u + u \mathrm{d}\rho = 0;$$

Hence,

$$d\rho = -\frac{G}{u}\frac{du}{u}. \tag{1.1.54}$$

Since $G = \rho u$ is a constant, the second of the governing Eq. (1.1.49b) is,

$$dp = -G du. \tag{1.1.55}$$

From the equation of state $p = \rho RT$ and since $u^2 = M^2 \gamma RT$

$$dT = T\frac{dp}{p} - T\frac{d\rho}{\rho} = -u^2\frac{du}{Ru} + T\frac{du}{u} = \left(1 - \gamma M^2\right)T\frac{du}{u}.$$

From (1.1.49d), the change in entropy may be expressed as

$$ds = C_v\frac{dp}{p} - C_p\frac{d\rho}{\rho} = \left(C_p - \gamma M^2 C_v\right)\frac{du}{u} = \gamma C_v\left(1 - M^2\right)T\frac{du}{u}. \tag{1.1.56}$$

Furthermore, from the definition of the stagnation temperature (Eq. 1.1.43)

$$dT_0 = dT + \frac{u}{C_p}du = dT + \frac{M^2\gamma RT}{C_p}\frac{du}{u}. \tag{1.1.57}$$

The latter may be expressed as

$$dT_0 = dT + \frac{M^2\gamma RT}{C_p}\frac{du}{u} = \left\{\left(1 - \gamma M^2\right) + M^2(\gamma - 1)\right\}T\frac{du}{u} = \left(1 - M^2\right)T\frac{du}{u}. \tag{1.1.58}$$

Finally, since $u = M\sqrt{\gamma RT}$ and

$$\frac{dM}{M} = \frac{du}{u} - \frac{1}{2}\frac{dT}{T}, \tag{1.1.59}$$

$$dM = M\left(\frac{1 + \gamma M^2}{2}\right)\frac{du}{u}. \tag{1.1.60}$$

Thus, the six state equations may be expressed in the form.

$$\frac{du}{u} = \frac{dM}{M} + \frac{1}{2}\frac{dT}{T}, \tag{1.1.61a}$$

$$dx_i = a_i\frac{du}{u}, \tag{1.1.61b}$$

where the states M and T play a fundamental role and when they are updated first all the rest may be updated sequentially. The six coefficients a_i $i = 1, 2, \ldots 6$ are summarized in Table 1.1. From Table 1.1, one could summarize the increases or decreases in the six state variables given either an increase in du or a decrease in du.

Table 1.1 The six coefficients a_i

i	1	2	3	4	5	6
x_i	M	T	p	ρ	T_0	s
a_i	$M\left(\frac{1+\gamma M^2}{2}\right)$	$(1-\gamma M^2)T$	$-G$	$-\frac{G}{M\sqrt{\gamma RT}}$	$(1-M^2)T$	$\gamma C_v(1-M^2)T$

Equation (1.1.50) governing the heat addition may now be considered which increases the stagnation temperature T_0, with heat addition and which is accompanied by a proportional increase in entropy.

Two distinct cases must now be considered since the case when the flow is subsonic ($M < 1$) corresponds to an increase in du, while in the case when the flow is supersonic, ($M > 1$) corresponds to a decrease in du. With an increase in du, both the pressure and the density fall, while there is an increase in the Mach number. With an increase in du, the static temperature increases when $M < \sqrt{1/\gamma}$ and decreases when $M > \sqrt{1/\gamma}$. With a decrease in du, both the pressure and the density will increase, while there is a decrease in the Mach number. With a decrease in du, the static temperature decreases when $M < \sqrt{1/\gamma}$ and increases when $M > \sqrt{1/\gamma}$. The situation can be illustrated on a (T, s) diagram. If one starts with the case of low subsonic flow with heat addition, both the static temperature and the entropy increase till $M = \sqrt{1/\gamma}$, and thereafter, there is a fall in the static temperature with increases in entropy till $M = 1$. When flow is supersonic, ($M > 1$), heat addition, both the static temperature and entropy increase, but this is accompanied by a reduction in the Mach number till $M = 1$. Since the flow corresponds to the case of constant inlet mass flow rate, the locus corresponds to the Rayleigh line on the (p, v) diagram and is therefore referred to as the Rayleigh curve on a (T, s) diagram (Fig. 1.1).

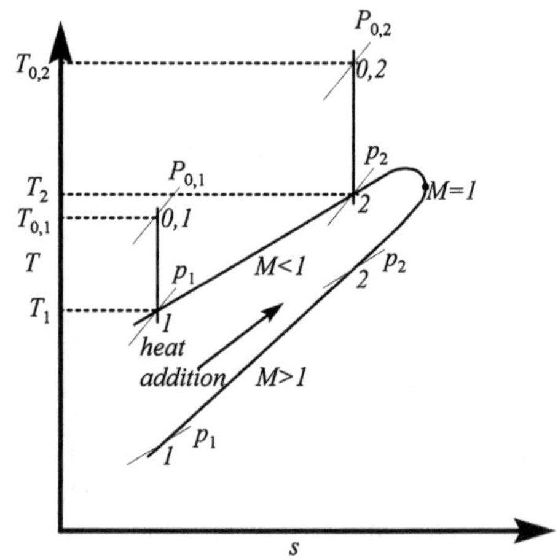

Fig. 1.1 The Rayleigh curve for heat addition on a (T, s) diagram

At sonic conditions, $M = 1$ and the change in entropy is zero, indicating that the entropy is a maximum. In practice, this represents a limiting state on a (T, s) diagram for both initially subsonic and supersonic flows. In the subsonic case, the amount of heat needed for the flow in an initial state to reach sonic conditions is finite and further addition of heat is pointless. In fact, in real applications such as jet engines, too much heat addition can result in undesirable features such as normal shocks, which can move upstream, and pressure oscillations. Furthermore, with heat addition, there is always a loss in stagnation pressure. In fact, it can be shown from Eq. (1.1.44) that

$$\frac{\mathrm{d}P_0}{P_0} = -\frac{\gamma M^2}{2}\frac{\mathrm{d}T_0}{T_0}. \tag{1.1.62}$$

Thus, loss in stagnation pressure is inherent to heat addition and does arise both in supersonic flows and in subsonic flows immaterial of flow being reversible or not. Irreversibility, however, adds to the loss in stagnation pressure, and considerations of the loss in stagnation pressure are important to the design of combustors for jet engines.

One can now consider the need to satisfy the constraints imposed by both the Rayleigh curve and the Rankine–Hugoniot equation on a (p, v) diagram. If the initial state is treated as the origin on the (p, v) diagram, and considering subsonic flow, a typical Rayleigh line represents a straight line passing through the origin and with a negative slope. It therefore lies entirely in the fourth quadrant of the diagram as shown in Fig. 1.2. On the other hand, in supersonic flow, it lies entirely in the second quadrant. Consequently, the other two quadrants may be considered to be forbidden regions as any states that lie in these quadrants are not physically realizable. The H-curve with no heat addition passes through the origin, while the H-curve with a certain fixed heat addition represents a parallel curve above the H-curve passing through the origin. The sonic condition ($M = 1$) is approached from either the subsonic side or the supersonic side and is represented by the point of tangency of the subsonic and supersonic Rayleigh lines to the H-curve. When $M = 1$ at the exit of the duct, the flow velocity reaches a peak value which is the local speed of sound. The corresponding heat added is a critical value, and flow is said to be thermally *choked*. When the heat added is more than the critical value, the intersection of the Rayleigh line with the corresponding H-curve is not possible. Thus, states beyond the point of tangency where $M = 1$ on the H-curve for the critical heat input are not accessible. On the other hand, if the heat added is less than the critical value, the Rayleigh line intersects the H-curve at two points. Only the first point is accessible through continuous heat addition; although in combustion when local heat releases are possible, the second point is also accessible under certain conditions. One way of operating such that the Rayleigh line intersects the H-curve at two points is to reduce G and consequently reduce the slope of the Rayleigh line. However, this means changing the inlet conditions which would in turn modify the Rayleigh curve on the (T, s) diagram. In real applications such as aircraft gas turbines, excess heat addition with adjustments in

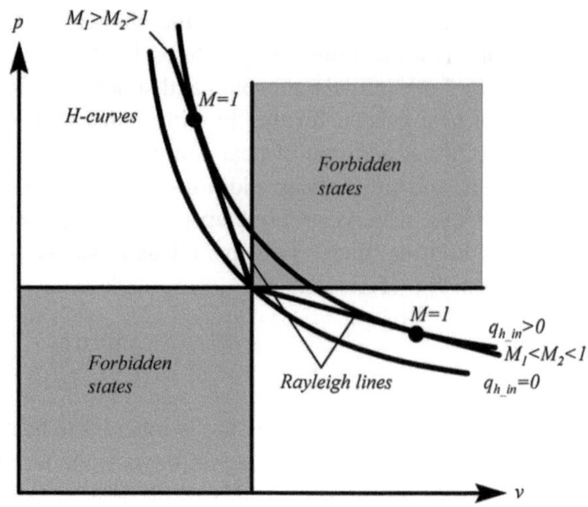

Fig. 1.2 Heat addition on a (p, v) diagram with the Rayleigh line and the H-curves

the mass flow rates must done with great care so as to avoid highly undesirable pressure oscillations. In supersonic applications involving ramjets or scramjets, there is the possibility of inducing normal shocks. Such situations must avoided, and in order to do this, the slope of the Rayleigh line may have to be altered by changing the cross-sectional area of the duct without altering any of the other inlet state conditions.

1.1.7 One-Dimensional Flow: Fanno Flow

A typical Fanno flow involves one-dimensional adiabatic flow in a duct with frictional losses at the walls of the duct. In this case, rather than modifying the energy balance equation as in a Rayleigh flow, it is the momentum balance equation that is altered as the friction force is modeled as an externally imposed force. Thus, the flow is still considered to be inviscid and effects such as viscous dissipation and boundary layers are ignored. The governing equations relating the one-dimensional flow boundary conditions at the two extreme ends of the flow in a constant-area duct with friction at the walls of the duct are

$$\rho_1 u_1 = \rho_2 u_2, \tag{1.1.63a}$$

$$p_1 + \rho_1 u_1^2 = p_2 + \rho_2 u_2^2 + \frac{\mathsf{P}}{A} \int_0^L \tau_{\mathrm{wall}} \mathrm{d}x, \tag{1.1.63b}$$

$$h_1 + u_1^2/2 = h_2 + u_2^2/2, \tag{1.1.63c}$$

and

$$s_2 - s_1 = C_p \ln\left[\frac{T_2}{T_1}\right] - R \ln\left[\frac{p_2}{p_1}\right] = C_p \ln\left[\frac{v_2}{v_1}\right] + C_v \ln\left[\frac{p_2}{p_1}\right]. \tag{1.1.63d}$$

In the above Eq. (1.1.63b), P is the wetted perimeter, A is the cross-sectional area of the duct, L is the duct length, and τ_{wall} is the wall shear stress. One approach to simplifying the momentum balance equation is to assume that the wall shear stress is proportional to the dynamic pressure and introduce a friction factor f such that

$$\tau_{\text{wall}} = f \times \frac{1}{2}\rho u^2.$$

For an equivalent circular duct of diameter D_h,

$$\frac{\mathsf{P}}{A} = \frac{4}{D_h}.$$

The friction factor f is often known as Darcy's friction factor and can be in the range of $0 \leq f \leq 0.05$. Equation (1.1.63b) can be expressed as

$$p_1 + \rho_1 u_1^2 = p_2 + \rho_2 u_2^2 + \frac{2}{D_h}\int_0^L f\rho u^2 \mathrm{d}x, \qquad (1.1.64).)$$

To illustrate the equations diagrammatically, it is convenient to express them in differential form. Thus, from Eqs. (1.1.54) and (1.1.57), since the process is adiabatic and so $\mathrm{d}T_0 = 0$

$$\mathrm{d}\rho = -\frac{G}{u}\frac{\mathrm{d}u}{u}. \tag{1.1.54}$$

$$\mathrm{d}T = -M^2(\gamma - 1)T\frac{\mathrm{d}u}{u}. \tag{1.1.65}$$

From the equation of state $p = \rho RT$

$$\mathrm{d}p = \frac{p\mathrm{d}\rho}{\rho} + \frac{p\mathrm{d}T}{T} = -p\left(1 + (\gamma - 1)M^2\right)\frac{\mathrm{d}u}{u}. \tag{1.1.66}$$

Thus, from Eqs. (1.1.56) and (1.1.59)

$$\mathrm{d}s = C_v\frac{\mathrm{d}p}{p} - C_p\frac{\mathrm{d}\rho}{\rho} = R\left(1 - M^2\right)\frac{\mathrm{d}u}{u} \tag{1.1.67}$$

$$\mathrm{d}M = M\left(1 + \frac{\gamma - 1}{2}M^2\right)\frac{\mathrm{d}u}{u}. \tag{1.1.68}$$

Thus, the five state Eqs. 1.1.54 and 1.1.65–1.1.68 may be expressed in the form:

$$\mathrm{d}x_i = b_i \frac{\mathrm{d}u}{u}. \tag{1.1.69}$$

The five coefficients b_i, $i = 1, 2, \ldots 5$ are summarized in Table 1.2.

Since the flow is adiabatic, it follows that $\mathrm{d}s \geq 0$ and it follows that $\mathrm{d}u$ (and $\mathrm{d}M$) must be positive for subsonic flow and negative for supersonic flow. Consequently, ρ, p, and T must decrease for subsonic flow and increase in the case of supersonic flow. This can be illustrated on a (T, s) diagram. It is possible to represent on the diagram the locus of all possible states on the diagram, and the locus is known as the Fanno curve (Fig. 1.3). Moreover, as friction makes the process irreversible, the stagnation always decreases.

It is possible to show that

$$\frac{\mathrm{d}P_0}{P_0} = \frac{\mathrm{d}p}{p} - \frac{\gamma}{\gamma - 1}\frac{\mathrm{d}T}{T}. \tag{1.1.70a}$$

Hence, it follows that

$$\frac{\mathrm{d}P_0}{P_0} = \left(M^2 - 1\right)\frac{\mathrm{d}u}{u} = \frac{\left(M^2 - 1\right)}{\left(1 + \frac{\gamma-1}{2}M^2\right)}\frac{\mathrm{d}M}{M}. \tag{1.1.70b}$$

In the subsonic flow, the Mach number increases, and in the supersonic case, it decreases. Consequently, the numerator of Eq. (1.1.70b) is always negative and the

Table 1.2 The five coefficients b_i

i	1	2	3	4	5
x_i	M	T	p	ρ	s
b_i	$M\left(1 + \frac{\gamma - 1}{2}M^2\right)$	$-M^2(\gamma - 1)T$	$-p(1 + (\gamma - 1)M^2)$	$-\frac{G}{M\sqrt{\gamma RT}}$	$R(1 - M^2)T$

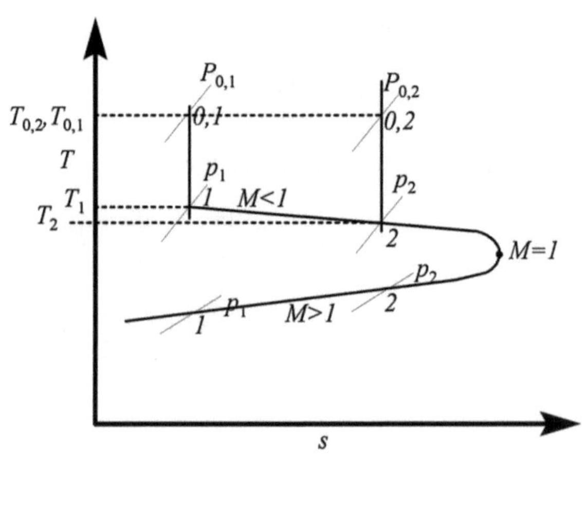

Fig. 1.3 The Fanno curve on a (T, s) diagram

stagnation pressure always decreases. The type of choking experienced in Fanno flow is known as friction choking, and if such choking is experienced in a duct, careful reduction in the mass flow rate may have to be initiated while avoiding pressure fluctuations or normal shocks, to eliminate the possibility of choking. There is critical length at which choking can occur, and for this reason, it is important to keep the length of the duct below the critical value. Increasing the area of the duct increases the critical length and is one way of avoiding friction choking.

1.1.8 Quasi-One-Dimensional Flows

In the flows considered in the preceding two sections, the area of the duct was assumed to be constant. The difficulty in considering a variable-area duct is that the flow is no longer one dimensional as there are components of the fluid velocity in the directions normal to the direction of the flow albeit are small in magnitude. However, since the magnitudes of these components are small, they are generally ignored and such flows are known as quasi-one-dimensional flows. The flows in variable-area ducts are generally close to being one-dimensional and variations in the cross-sectional areas must be considered together with the heat addition and wall friction. The general unsteady equilibrium dynamics, of quasi-one-dimensional flow with no heat addition or frictional losses, may be expressed by the continuity, conservation of momentum, and conservation of energy equations as

$$\frac{\partial}{\partial t}(\rho A) + \frac{\partial}{\partial x}(\rho A u) = 0, \tag{1.1.71a}$$

$$\frac{\partial}{\partial t}(\rho A u) + \frac{\partial}{\partial x}(\rho A u^2) = -A\left(\frac{\partial p}{\partial x} + \rho g_u\right), \tag{1.1.71b}$$

$$\frac{\partial}{\partial t}(\rho A e_t) + \frac{\partial}{\partial x}(\rho A u e_t) = 0. \tag{1.1.71c}$$

In Eq. (1.1.71c), e_t is the specific total energy (see Eq. 1.1.13), g_u is the component of the acceleration due to earth's gravity in the flow direction, while the other variables are as defined earlier. In this section, the influence-independent variations of area will be considered with the further assumption that the flow is isentropic. Hence, it follows that in these cases

$$s_2 - s_1 = \mathrm{d}s = 0. \tag{1.1.72a}$$

The remaining governing equations relating the one-dimensional flow boundary conditions at the two extreme ends of the flow in a variable-area duct are obtained from Eq. (1.1.71a–c) and are

$$\rho_1 u_1 A_1 = \rho_2 u_2 A_2, \tag{1.1.72b}$$

$$\left(p_1 + \rho_1 u_1^2\right)A_1 + \int_1^2 p\mathrm{d}A = \left(p_2 + \rho_2 u_2^2\right)A_2, \tag{1.1.72c}$$

$$h_1 + u_1^2/2 = h_2 + u_2^2/2. \tag{1.1.72d}$$

The quantity conserved in Eq. (1.1.72b) is the mass flow rate. In Eq. (1.1.72c), there is an additional term in the momentum equation due to the pressure force wall. Before attempting to solve the governing equations, it is important to consider the evaluation of the forces due to the flow on the walls of the duct. To facilitate the evaluation of the forces due to the flow on the walls of the duct, the impulse function is defined as

$$I = \int_A \mathrm{d}(pA) + \int_A \mathrm{d}\left(\rho u^2 A\right). \tag{1.1.73}$$

From Newtonian mechanics, it follows that the net change in impulse evaluated at the two ends of the flow is the net force exerted on the duct. Thus,

$$\mathsf{T} = I_2 - I_1. \tag{1.1.74}$$

Hence, from Eqs. (1.1.72c), (1.1.73), and (1.1.74)

$$\mathsf{T} = I_2 - I_1 = \int_1^2 p\mathrm{d}A. \tag{1.1.75}$$

Thus, Eq. (1.1.75) may be used to evaluate T.

Consider an aircraft flying with uniform velocity U_∞ and assume that the reference frame is fixed to the aircraft. In this situation, it follows that the flow is approaching the aircraft with the velocity U_∞, and consequently, the jet engines intake velocity is also U_∞. The subscript “1” in this application refers to the intake, while the subscript “2” refers to the exit. If we let the mass flow rate be denoted as $\dot{m}$, then it follows that

$$\dot{m} = \rho_1 u_1 A_1 = \rho_2 u_2 A_2. \tag{1.1.76}$$

Assuming that the static pressure is measured relative to the free stream and that $p_1 = p_\infty$ the thrust is

$$T = \dot{m}(u_2 - u_1) + (p_2 - p_\infty)A_2 - (p_1 - p_\infty)A_1 = \dot{m}(u_2 - u_1) + (p_2 - p_\infty)A_2. \tag{1.1.77}$$

The first term is the difference between the exit momentum and intake momentum, while the second is the thrust due to the pressure at the exit being greater than the ambient pressure.

In order to develop useful solutions to the complete set of governing equations, it is convenient to express them in differential form. Thus,

$$d\dot{m} = d(\rho u A) = 0, \tag{1.1.78a}$$

$$dp + \rho u du = 0, \tag{1.1.78b}$$

$$dh + d\left(u^2/2\right) = 0. \tag{1.1.78c}$$

From Eq. (1.1.78a),

$$\frac{d\dot{m}}{\dot{m}} = \frac{d\rho}{\rho} + \frac{du}{u} + \frac{dA}{A} = 0, \tag{1.1.79}$$

From Eq. (1.1.78b)

$$\frac{dp}{\rho} + u^2 \frac{du}{u} = 0. \tag{1.1.80}$$

Furthermore,

$$\frac{d\rho}{\rho} = \frac{dp}{\rho}\frac{d\rho}{dp} = \frac{dp}{\rho}\frac{1}{a^2} = -\frac{u^2}{a^2}\frac{du}{u} = -M^2\frac{du}{u}. \tag{1.1.81}$$

Consequently, it follows that the relationship between the changes in velocity induced by the change in area is

$$\frac{dA}{A} = -\left(1 - M^2\right)\frac{du}{u}. \tag{1.1.82}$$

Equation (1.1.82) leads to conclude that subsonic flow accelerates in converging duct or nozzle and decelerates in a diverging duct or diffuser, while in supersonic flow, it decelerates in converging duct or nozzle and accelerates in a diverging duct or diffuser.

Equation (1.1.82) also implies that when $dA = 0$, either $M = 1$, $du = 0$ or $u = \infty$. Thus, when $M = 1$, dA must be equal to zero although the converse needs not be true. This means that in an isentropic flow in a duct of varying cross-section, the sonic state can only be attained at a location where the area of the duct is stationary. This may well be the minimum throat at the end of a nozzle or the maximum area at the end of a diffuser. When the Mach number in a flow approaches unity, the flow experiences choking, and when the choking is due to area variations, it is referred to as geometric choking in contrast to thermal and friction choking which was encountered earlier.

It is essential in this type flow to be able to relate the state at any point to the sonic state. In particular, one generally considers the area A, the Mach number M, the density ρ, the velocity u, and the temperature T. All these quantities are denoted with the superscript "*" at the sonic state and $M^* = 1$. The mass flow rate satisfies

$$\dot{m} = \rho u A = \rho^* u^* A^*. \tag{1.1.83}$$

Consequently,

$$\frac{A}{A^*} = \frac{\rho^* u^*}{\rho u} = \frac{\rho_0}{\rho}\frac{\rho^*}{\rho_0}\frac{u^*}{u}. \tag{1.1.84}$$

From Eq. (1.1.45), the ratio of the density to its stagnation value is

$$\frac{\rho_0}{\rho} = \left(\frac{P_0}{p}\right)^{\frac{1}{\gamma}} = \left(1 + \left(\frac{\gamma - 1}{2}\right)M^2\right)^{\frac{1}{\gamma-1}}. \tag{1.1.85}$$

When $M = M^* = 1$, $\frac{\rho_0}{\rho^*} = \left(\frac{\gamma+1}{2}\right)^{\frac{1}{\gamma-1}}$,

$$\frac{\rho^*}{\rho} = \frac{\rho_0}{\rho}\frac{\rho^*}{\rho_0} = \left(\left(\frac{2}{\gamma + 1}\right)\left(1 + \left(\frac{\gamma - 1}{2}\right)M^2\right)\right)^{\frac{1}{\gamma-1}}. \tag{1.1.86a}$$

Similarly,

$$\frac{T^*}{T} = \frac{T^*}{T_0}\frac{T_0}{T} = \left(\frac{2}{(\gamma + 1)}\left(1 + \frac{\gamma - 1}{2}M^2\right)\right). \tag{1.1.86b}$$

Since $u = M\sqrt{\gamma RT}$

$$\frac{u^*}{u} = \frac{1}{M}\sqrt{\frac{T^*}{T}} = \frac{1}{M}\left(\frac{2}{(\gamma + 1)}\left(1 + \frac{\gamma - 1}{2}M^2\right)\right)^{\frac{1}{2}}. \tag{1.1.86c}$$

Hence,

$$\frac{A}{A^*} = \frac{1}{M}\left(\left(\frac{2}{\gamma + 1}\right)\left(1 + \left(\frac{\gamma - 1}{2}\right)M^2\right)\right)^{\frac{\gamma+1}{2(\gamma-1)}}. \tag{1.1.86d}$$

When the ratio of the area A to the throat area when the flow is choked A^* is known, the Mach number corresponding to the area A may be found. Equation (1.1.86d) yields two solutions for the Mach number, one that is valid for supersonic flow and the other that is valid for subsonic flow.

The mass flow rate for choked flow may be established from the Eq. (1.1.86a–d) and can be shown to be

$$\dot{m} = P_0 A_{\text{throat}}\sqrt{\frac{\gamma}{RT_0}}\left(\frac{2}{\gamma + 1}\right)^{\frac{\gamma+1}{2(\gamma-1)}}. \tag{1.1.87}$$

Equation (1.1.87) is an extremely important equation for the design of intakes and nozzles. Equations (1.1.86a–d) and (1.1.87) are used to establish the flow states when considering the flows through nozzles and diffusers in conjunction with Rayleigh and Fanno flows.

1.1.9 Oblique Shocks

The simplest example of shock reflection is that of a one-dimensional shock wave. In considering the reflections of normal shock waves, four equations were used: the first being the equation of state and the other three conditions derived from the three laws relating to mass, momentum, and energy conservation at the shock plane. They were used to derive a set of shock relations capable of predicting the strength of a shock. In the case of oblique shock waves, unlike the case of normal shock waves, the flow direction changes after passing through the shock wave. Since the normal component of velocity is always less than the magnitude of the velocity itself, for a given velocity and temperature before the shock, the actual Mach number of the flow interacting with the shock wave is less for an oblique shock. As a direct consequence, the stagnation pressure loss in the case of an oblique shock wave is also less than the corresponding loss for a normal shock wave. In this section, the reflection of a shock wave from a solid boundary at a finite angle of incidence other than normal to it is considered. It is assumed that plane shock wave is traveling in an ideal fluid of negligible viscosity and with negligible heat conduction which is incident at an angle α upon an infinite plane rigid wall, causing a reflected shock to arise from the wall. Oblique shock reflection may take either of two qualitatively different forms: regular reflection which is associated with small angles α and weak shocks so that the waves from the points behind the shock move more slowly relative to the wall than does the shock itself or Mach reflection which is associated with larger angles α and strong shocks which are supersonic with respect to the wall and attached to the leading edge forming a bow wave. Only the oblique shock's normal component undergoes the "shock." The tangent component does not change because it does not "move" across the shock line. Hence, the mass and momentum balance equations for the case of the normal shock wave apply for the forward velocity upstream and downstream of the shock front replaced by their normal components. Thus, the governing equations take the same form as Eq. (1.1.46a–d) and are

$$\rho_1 u_{1n} = \rho_2 u_{2n} \tag{1.1.88a}$$

$$p_1 + \rho_1 u_{1n}^2 = p_2 + \rho_2 u_{2n}^2 \tag{1.1.88b}$$

$$h_1 + u_{1n}^2/2 = h_2 + u_{2n}^2/2 \tag{1.1.88c}$$

The tangential components satisfy

$$u_{1t} = u_{2t}. \tag{1.1.88d}$$

Assuming that the fluid is calorically perfect, it is seen from the energy equation that the stagnation temperature is constant across the shock front and $T_{02} - T_{01} = \Delta T_0 = 0$. It is important to note the direction associated with the shock wave. Since the tangential components remain the same, while the normal component decreases across the shock, the flow is deflected by angle θ toward the shock front after passing by it. Thus, the normal and tangential components of velocity before and after passing by the shock front are

$$u_{1n} = u_1 \sin\alpha, \quad u_{2n} = u_2 \sin(\alpha - \theta), \tag{1.1.89a}$$

$$u_{1t} = u_1 \cos\alpha = u_{2t} = u_2 \cos(\alpha - \theta). \tag{1.1.89b}$$

From Eqs. (1.1.88a), (1.1.89a), and (1.1.89b), it follows that the ratio of the normal velocity components is

$$\frac{u_{1n}}{u_{2n}} = \frac{\tan\alpha}{\tan(\alpha - \theta)} = \frac{\rho_2}{\rho_1}.$$

But from Eq. (1.1.48)

$$\frac{\rho_2}{\rho_1} = \frac{(\gamma + 1)M_{1n}^2}{(\gamma - 1)M_{1n}^2 + 2}. \tag{1.1.90}$$

It follows that

$$\frac{u_{1n}}{u_{2n}} = \frac{\tan\alpha}{\tan(\alpha - \theta)} = \frac{\rho_2}{\rho_1} = \frac{(\gamma + 1)M_{1n}^2}{(\gamma - 1)M_{1n}^2 + 2}. \tag{1.1.91}$$

Recognizing that

$$M_{1n} = M_1 \sin\alpha, \quad M_{2n} = M_2 \sin(\alpha - \theta), \tag{1.1.92}$$

$$\frac{\tan\alpha}{\tan(\alpha - \theta)} = \frac{(\gamma + 1)M_1^2 \sin^2\alpha}{(\gamma - 1)M_1^2 \sin^2\alpha + 2}. \tag{1.1.93}$$

After some algebraic manipulations, Eq. (1.1.93) can be expressed as

$$\tan\theta = 2\cot\alpha\left(\frac{M_1^2 \sin^2\alpha - 1}{M_1^2(\gamma + \cos 2\alpha) + 2}\right). \tag{1.1.94}$$

Equation (1.1.94) determines θ given M_1 and α. From Eq. (1.1.48)

$$\left(\frac{T_2}{T_1}\right) = \frac{\left(1 + \frac{(\gamma-1)}{2} M_{1n}^2\right)\left(\frac{2\gamma}{\gamma-1} M_{1n}^2 - 1\right)}{\left(\frac{(\gamma+1)^2}{2(\gamma-1)}\right) M_{1n}^2}, \quad \frac{p_2}{p_1} = \frac{2\gamma M_{1n}^2}{\gamma + 1} - \frac{\gamma - 1}{\gamma + 1}. \tag{1.1.95}$$

Given θ and M_1, Eq. (1.1.94) can be rearranged as cubic equation for $x = \sin^2\alpha$ which is

$$x^3 + a_1 x^2 + a_2 x + a_1 = 0, \tag{1.1.96}$$

where

$$a_1 = -\frac{M_1^2 + 2}{M_1^2} - \gamma \sin^2 \theta,$$

$$a_2 = \frac{2M_1^2 + 1}{M_1^4} + \left(\frac{(\gamma + 1)^2}{4} + \frac{\gamma - 1}{M_1^2} \right) \sin^2 \theta \text{ and}$$

$$a_3 = -\frac{\cos^2 \theta}{M_1^4}.$$

However, the solution for the cubic results in three possible values for α, one of which is not physically meaningful, while the other two correspond to the weak and strong shocks. To obtain the solutions corresponding to the weak and strong shocks, the cubic equation is expressed in terms of $\tan \alpha$ as

$$\left(1 + \frac{\gamma - 1}{2} M_1^2\right) \tan \theta \tan^3 \alpha - \left(M_1^2 - 1\right) \tan^2 \alpha + \left(1 + \frac{\gamma + 1}{2}\right) \tan \theta \tan \alpha + 1 = 0. \tag{1.1.97}$$

The solutions for the weak and strong shocks can be shown to satisfy

$$\tan \alpha_{\text{weak}} = \frac{M_1^2 - 1 + 2f_1(M_1, \theta) \cos\left(\frac{4\pi + \cos^{-1}(f_2(M_1, \theta))}{3}\right)}{3\left(1 + \frac{\gamma - 1}{2} M_1^2\right) \tan \theta}, \tag{1.1.98a}$$

$$\tan \alpha_{\text{strong}} = \frac{M_1^2 - 1 + 2f_1(M_1, \theta) \cos\left(\frac{\cos^{-1}(f_2(M_1, \theta))}{3}\right)}{3\left(1 + \frac{\gamma - 1}{2} M_1^2\right) \tan \theta}, \tag{1.1.98b}$$

where

$$f_1(M_1, \theta) = \sqrt{\left(M_1^2 - 1\right)^2 - 3\left(1 + \frac{\gamma - 1}{2} M_1^2\right)\left(1 + \frac{\gamma + 1}{2} M_1^2\right) \tan^2 \theta}, \tag{1.1.99a}$$

and

$$f_2(M_1, \theta) = \frac{\left(M_1^2 - 1\right)^3 - 9\left(1 + \frac{\gamma - 1}{2} M_1^2\right)\left(1 + \frac{\gamma - 1}{2} M_1^2 + \frac{\gamma + 1}{2} M_1^4\right) \tan^2 \theta}{f_1(M_1, \theta)^3}. \tag{1.1.99b}$$

The relations given in Eqs. (1.1.90–1.1.99b) are used to solve application problems involving detached and reflected shocks associated with wings and bodies in supersonic flows.

1.1.10 Prandtl–Meyer Flow

Consider a non-viscous, non-conducting, semi-infinite polytropic fluid flowing parallel to flat surface and approaching a corner in the surface. If the surface bends into the flowing fluid, the theory of oblique shocks presented in the preceding section may be applied to it. However, if the surface bends away, a steady expanding flow occurs which is known as a Prandtl–Meyer flow.

When the flow is supersonic, $M > 1$, the velocity of the free stream past a body is greater than the speed of sound at all points in the flow. When a body with an appropriate profile such as an aerofoil moves in a flow at subsonic speeds, stream receives a signal of the body's impending arrival well before the body arrives at a point in the flow. In the case of the body moving at supersonic speeds, the flow has absolutely no prior knowledge of the body's arrival and the body cuts through a completely undisturbed and "unsuspecting" flow. The situation can be described in terms of propagation of spherical disturbance waves in (1) an incompressible flow, (2) compressible flow at subsonic speeds, (3) transonic speeds, and (4) supersonic speeds (see Fig. 1.4).

Considering Fig. 1.4a, the case when the source of the disturbance is at rest relative to the flow field, the disturbance propagates with uniform velocity in all directions and wave fronts propagate in the form of concentric circles (two-dimensional case). When the source of the disturbance is in motion with a speed

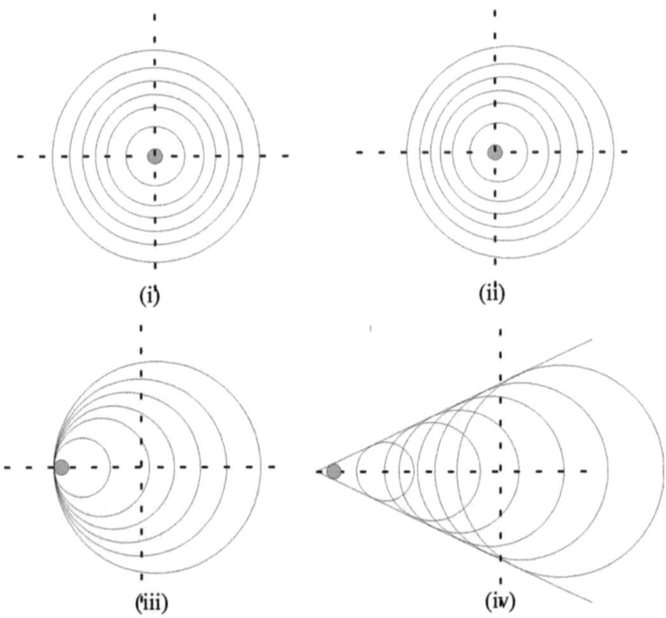

Fig. 1.4 Propagation of spherical disturbances from a point source in **a** an incompressible flow, **b** compressible flow at subsonic speeds, **c** transonic speeds, and **d** supersonic speeds

well below the speed of sound, wave fronts continue to propagate with a speed equal to the difference between the speed of sound and the speed of the disturbance, in the direction of the motion of the disturbance. In the opposite direction, the disturbance propagates at the sum of the two speeds and the wave fronts are no longer concentric circles. This situation is illustrated in Fig. 1.4b. When the disturbance moves with the speed of sound, as in Fig. 1.4c, the wave front is stationary with respect to the source, and in the course of time, it evolves into a *normal shock* boundary: a surface of discontinuity across which there is discrete change in the pressure and density. Finally, when the source moves faster than the speed of sound, it is always ahead of the wave front and the result is an *oblique shock* boundary as shown in Fig. 1.4d. The sine of the half angle of the cone, known as the *Mach angle*, is equal to the *inverse of the Mach number*. The surface of the cone forms a shock boundary. The shock boundary separates the free undisturbed flow and the region of the disturbed flow.

Supersonic expansion and isentropic compression are an extension of the Mach cone concept. The Mach line shows that a disturbance in a field of supersonic flow moves in an angle of μ (not to be confused with the dynamic viscosity which is also represented by the same symbol), which is defined as

$$\mu = \sin^{-1}(1/M) = \tan^{-1}\left(1/\sqrt{M^2-1}\right). \tag{1.1.100}$$

In the case of supersonic flow around a smooth convex corner, it can be deduced that accompanying the turning flow are a series of shock boundaries generated from the corner with each one turning an infinitesimal amount with the flow. Thus, the shock boundaries not only coalesce but action forms an expansion process as the shock fronts diverge from each other. In the special case, when the convex corner is a sharp convex corner, an expansion fan centered at the corner is generated and the flow expands as it traverses the fan. Although the expansion and compression processes are both isentropic, the case of a convex corner is not analogous to the case of concave corner which is characterized by an oblique shock.

Consider two expansion wave fronts with the angle between them equal to $\mathrm{d}v$, while the initial fan angle is v. The change in the velocity corresponding to a change in the direction $\mathrm{d}x$ is

$$\mathrm{d}u_x = (u + \mathrm{d}u)\cos v - u \approx \mathrm{d}u$$

The change in the velocity corresponding to a change in the direction $\mathrm{d}y$ is

$$\mathrm{d}u_y = (u + \mathrm{d}u)\sin \mathrm{d}v \approx u\mathrm{d}v.$$

Hence,

$$\tan \mu = \frac{1}{\sqrt{M_1^2 - 1}} = \frac{\mathrm{d}u_y}{\mathrm{d}u_x} = \frac{1}{u}\frac{\mathrm{d}u}{\mathrm{d}v}.$$

Rearranging

$$\mathrm{d}\nu = \sqrt{M_1^2 - 1}\frac{\mathrm{d}u}{u}. \tag{1.1.101}$$

The stagnation temperature is given by

$$T_{01} = T_1 + u_1^2/(2C_p). \tag{1.1.102}$$

Thus,

$$\gamma RT_{01} = \gamma RT_1 + \frac{(\gamma - 1)}{2}u_1^2 = a_1^2 + \frac{(\gamma - 1)}{2}u_1^2 = u_1^2\left(\frac{1}{M_1^2} + \frac{\gamma - 1}{2}\right). \tag{1.1.103}$$

Since T_{01} is constant, differentiating Eq. (1.1.103)

$$2u_1\mathrm{d}u_1\left(\frac{1}{M_1^2} + \frac{\gamma - 1}{2}\right) - 2u_1^2\frac{\mathrm{d}M_1}{M_1^3} = 0.$$

Hence,

$$\mathrm{d}\nu = \sqrt{M_1^2 - 1}\frac{\mathrm{d}u_1}{u_1} = \frac{1}{2M_1^2}\frac{\sqrt{M_1^2 - 1}}{\left(1 + \frac{\gamma-1}{2}M_1^2\right)}\mathrm{d}M_1^2.$$

Integrating $\mathrm{d}\nu$

$$\nu = \nu(M) = \int_0^{M^2} \mathrm{d}\nu = \frac{1}{2}\int_0^{M^2} \frac{1}{M_1^2}\frac{\sqrt{M_1^2 - 1}}{\left(1 + \frac{\gamma-1}{2}M_1^2\right)}\mathrm{d}M_1^2.$$

The deflection angle ν, which is the change in angle of the Mach line due to increase in the Mach number beyond unity, is defined by the Prandtl–Meyer function and is

$$\nu(M) = \sqrt{\frac{\gamma + 1}{\gamma - 1}}\tan^{-1}\left(\sqrt{\frac{\gamma - 1}{\gamma + 1}}\sqrt{M^2 - 1}\right) - \tan^{-1}\left(\sqrt{M^2 - 1}\right). \tag{1.1.104}$$

The deflection angle ν is a monotonic function of the Mach number M, and if a supersonic flow is deflected by an angle θ at an expansion corner,

$$\theta = \nu(M_2) - \nu(M_1). \tag{1.1.105}$$

When any two of the three quantities, θ, M_1, and M_2 are known, the third may be estimated using the Prandtl–Meyer function given by Eqs. (1.1.104) and (1.1.105).

1.2 Thermodynamics

Thermodynamics (dynamics of heat) is concerned with the transformation of any form of electromechanical energy of one kind into heat energy or the transformation of heat energy into useful electromechanical energy. Thermodynamics is the science that deals with heat and work and those properties of matter that relate to heat and work. This branch of dynamics is based on the adoption of fundamental laws (known as the *Four Laws of Thermodynamics*), which have initially been derived "empirically" from experimental observation but which have been eventually proven to be true from first principles. The basis of the study of thermodynamics lies in the observations and evidence collected over a long period of time which has been succinctly summarized in the form of the laws of thermodynamics. The study of thermodynamics allows the prediction of whether a given process of transformation of heat energy or to heat energy (thermodynamic evolution of a system) can occur at all and if it does occur whether it is spontaneous or reversible. It also allows one to obtain specific relationships between the physical thermodynamic properties of a system or the change in these thermodynamic properties (heat capacity, density, melting temperature, energy, expansion coefficient, etc.) and the change in the experimental conditions (perturbation) imposed on the system (temperature, pressure). For further details on the fundamental principles of thermodynamics, the reader is referred to Fermi (1956), while the applications of thermodynamics to energy systems are covered by Hill and Peterson (1992) and by Mattingly (2005).

The most important aspects of any thermodynamic calculation are (1) the precise definition of the thermodynamically related properties of the system to be studied and (2) the determination of the thermodynamically related properties of the surroundings for that system. The surroundings are generally defined to include only the part of the universe which exchanges energy (and matter that relates to the energy) with the system. Considering "larger" surroundings obviously enables a more accurate prediction of the behavior of the thermodynamically related properties of the system during a given process, but also leads to more extensive and involved dynamics.

To visualize what is meant by a system, its surroundings, and the associated thermodynamic processes, one should consider a number of examples such as a thermos flask, a chemical reactor, a combustion engine, a motor car, a laptop computer, the human body, the central processing unit of a laptop computer, or a living cell within the human body. In all of the above examples, one can observe that the "system boundary" is the key to its interaction with its surroundings. As one defines thermodynamic system and surroundings for any given problem, it becomes extremely important to understand the nature of the "boundaries" between the thermodynamic system and the surroundings. These boundaries can be artificial or can be very real, as in most examples mentioned above. The type of exchanges that can take place between a thermodynamic system and its surroundings and the thermodynamic system itself may be described by different qualifiers. An **open thermodynamic system** refers to a system where both matter and energy can be exchanged between the thermodynamic system and its surroundings. A **closed thermodynamic system** refers to a system where matter cannot flow either in or out of the thermodynamic system (a closed thermodynamic system is therefore characterized by a fixed mass. Yet the number of moles and the composition are not necessarily constant during a process in a closed thermodynamic system, as one may envision the process to be a chemical reaction where the total number of moles and the concentration are changing). An **isolated thermodynamic system** is one where neither mass nor energy can flow in or out of it.

The state of the system is defined a set of state variables or states. Extensive state variables are proportional to the quantity of matter being considered (e.g., the volume, total C_p). Intensive state variables are independent of quantity (concentration, viscosity, density, molar C_p). Thus, an extensive property of a composite system is equal to the sum of its values for the component subsystems. A thermodynamic system is said to be in equilibrium if the state variables defining the state of the system are not changing with time. A system in equilibrium will return to that state after being disturbed, provided it is stable.

Work is said to be done by a thermodynamic system when there is a transfer of energy from it to another mechanical system. The work done is equivalent to the energy that flows across the system boundary in response to a force moving through a distance or a pressure exerted on a system boundary changing the volume. Energy can also be transferred even when no work is done. The transfer of heat energy results from a temperature differences. Thus, the energy that flows across a system boundary in response to a temperature gradient is that part of the energy that is transferred which is not accounted for by mechanical work. This type of transfer of energy is spontaneous. The heat capacity is the amount of energy transferred across a system boundary in response to a unit temperature gradient. The increase in enthalpy of a system is equal to the heat absorbed at constant pressure, assuming the system is subjected only to pressure distribution and that it can only do work by virtue of its change in volume. The entropy is a measure of the loss of the capacity of the thermodynamic system to do work. When there is a spontaneous loss of energy due to the presence of a temperature gradient, the entropy of the system losing heat decreases, while the entropy of the system gaining heat increases by a larger amount. Thus, when the two systems are taken together, there is a net increase in entropy. Spontaneous changes in an isolated system lead to an increase in entropy. An ideal process for which the entropy is constant and there is no net increase in entropy for an isolated system is said to be a reversible process. Thus, for an isolated system, among all of the possible spontaneously attainable or allowable states that require the highest increase in entropy characterize the equilibrium state.

1.2.1 The Zeroth Law of Thermodynamics

According to the zeroth law of thermodynamics, two bodies in thermal equilibrium with a third are in thermal equilibrium with each other. The zeroth law is the basis for the concept and definition of temperature. As a consequence of the zeroth law, it is possible to state that the thermodynamic state variable, the temperature T, exists and can be used to describe the current state of any thermodynamic system and that any two or more systems that are in thermal equilibrium with each other will all have the same temperature, T.

1.2.2 The First Law of Thermodynamics

The first law of thermodynamics which is entirely based on observations states that the algebraic sum of all energy changes in an isolated system is zero. Thus, energy is conserved and although it can be converted from one form to another it cannot be created or destroyed. Thus,

$$\mathrm{d}u = \mathrm{d}q + \mathrm{d}w \tag{1.2.1}$$

where u is the internal energy, q is the heat transferred to a system from the surroundings, and w is the work done on a system by the surroundings. Work is the transfer of energy from one mechanical system to another. Work done on a system is positive, and the work done by the system is negative. An example of work by the system is the expansion of a piston in a cylinder against the atmosphere, resulting in an increase in the volume enclosed by the piston and cylinder.

$$W = \int_{1}^{2} \mathrm{d}w = -\int_{v_1}^{v_2} P\mathrm{d}v. \tag{1.2.2}$$

For a system doing work by the expansion

$$\mathrm{d}u = \mathrm{d}q - P\mathrm{d}v. \tag{1.2.3}$$

In Eq. (1.2.3), $P\mathrm{d}v$ is the reversible expansion work done on the system by the surroundings.

1.2.3 The Second and Third Laws of Thermodynamics

The second law of thermodynamics may be stated in one of several ways. It deals with thermodynamically spontaneous natural processes. Spontaneous thermodynamic changes are those which, if carried out under the proper conditions, can be made to do work. If carried out reversibly, they yield a maximum amount of work. In natural processes, spontaneous thermodynamic changes can never be carried out reversibly. The second law is concerned with the increase in entropy (ds), which is related to that part of the total energy of the system that is not available to do useful work. Thus, one statement of the second law is that "The entropy of an isolated system cannot decrease." Another more useful form of the second law is known as Clausius' postulate which is that heating (or "transferring heat to") a warmer body using energy from a colder body without doing work is not possible. Another alternate statement of the second law is Kelvin's postulate which is that it is not possible to convert the internal energy ("heat") from a body at temperature T_1 into work, without transfer of some thermal energy to a body at a lower temperature T_2 (i.e., it is not possible to build a thermodynamic engine with a 100 % efficiency).

There is also a third law of thermodynamics (known also as Nernst's theorem), which is also stated in various ways: (1) The entropy of any system at absolute zero temperature is zero; (2) it is not possible to reach $T = 0$ (on the Kelvin scale) through a finite set of cooling steps, which is also known as the principle of the "non-attainability of absolute zero." While significant progress has been made at getting to very, very low temperatures, absolute zero has never been attained.

1.2.4 Heat Required to Do Work in Real Gases

Consider a real gas which is assumed to satisfy the polytropic gas law, $pv^n = p_0 v_0^n$. From Eq. 1.1.30, for an irreversible expansion of a real gas, the heat input required may be expressed as

$$\mathrm{d}q = \frac{C_v T}{p}\left(\frac{\mathrm{d}p}{\mathrm{d}v} + \gamma\frac{p}{v}\right)\mathrm{d}v = \frac{RT}{p(\gamma-1)}\left(\frac{\mathrm{d}p}{\mathrm{d}v} + \gamma\frac{p}{v}\right)\mathrm{d}v. \tag{1.2.4}$$

But for a real gas

$$\frac{\mathrm{d}p}{\mathrm{d}v} = -n\frac{p}{v}. \tag{1.2.5}$$

Hence,

$$\mathrm{d}q = \frac{(\gamma-n)}{\gamma-1}p\mathrm{d}v. \tag{1.2.6}$$

It follows that

$$q = \frac{(\gamma-n)}{\gamma-1}\int p\mathrm{d}v. \tag{1.2.7}$$

The heat input to the gas is proportional to the polytropic work done during expansion, and sign of this quantity depends on whether $\gamma > n$ or $\gamma < n$. When $n = 1$, the heat input is exactly equal to the work done and this corresponds to isothermal expansion. When $n = \gamma$, the heat required is zero and this corresponds to adiabatic expansion.

1.2.5 Spontaneous Processes and Thermodynamic Potentials

It has already been mentioned that for an isolated system, among all of the possible spontaneously attainable or allowable states that require the highest increase in entropy characterize the equilibrium state. However, no criteria have been stated to identify possible spontaneously attainable states. However, starting with the total entropy change of the world, ΔS_{total}, induced by a thermodynamic process, which is already known to be the sole criterion for spontaneous change, one can develop certain thermodynamic potentials that could provide such criteria. Consider the total entropy change of the world, ΔS_{total}, induced by a thermodynamic process and note that it can be expressed as the sum of two terms,

$$\Delta S_{\text{total}} = \Delta S_{\text{surr}} + \Delta S_{\text{sys}} \tag{1.2.8}$$

The first of the two terms, ΔS_{surr}, is the entropy increase due to the *surroundings*. Because most reactions are either exothermic (release heat) or endothermic (absorb heat), they are accompanied by a flow of heat Δq_p across the system boundary. The enthalpy change in the reaction ΔH is defined as the flow of heat into the system from the surroundings when the reaction is carried out at constant pressure, so the heat drawn from the surroundings can be assumed to be $-\Delta q_p$. The addition of heat to a system at a relatively lower temperature can increase the level of disorder within that system in terms of our knowledge of the position of each individual molecule more than the addition of the same amount of heat to the same system at a relatively higher temperature. Thus, assuming that the process is reversible the heat drawn from the surroundings will cause the entropy of the surroundings to change by

$$\Delta S_{\text{surr}} = (-\Delta q_p)/T = -\Delta H/T. \tag{1.2.9}$$

We can therefore rewrite Eq. 1.2.8 as

$$\Delta S_{\text{total}} = (-\Delta H/T) + \Delta S_{\text{sys}}. \tag{1.2.10}$$

Multiplying Eq. 1.2.10 by the temperature T and denoting ΔS_{sys} as ΔS

$$T\Delta S_{\text{total}} = -\Delta H + T\Delta S. \tag{1.2.11}$$

Equation 1.2.11 expresses the entropy change of the world exclusively in terms of the changes in and the thermodynamic states (H, T, S) of the *system*. If $-T\Delta S_{total}$ is denoted by ΔG, then we have

$$\Delta G = \Delta H - T\Delta S, \tag{1.2.12}$$

which defines the change in the *Gibbs free energy* for the process. In Eq. 1.2.12 ΔG is the energy available to do work for a system operating at the constant pressure p, and the constant temperature T. The two terms on the right hand side can be interpreted to have the following meaning: ΔH is the heat exchanged between system and surroundings when pressure is constant at each point along the path between the initial and final states; $T\Delta S$ is the isothermally unavailable energy, which is energy that cannot be converted to work by any system that is held at constant temperature at each point along the path between the initial and final states. The definition of the change in the *Gibbs free energy potential* was based on the assumption that the process is *reversible*, therefore the ΔG is the maximum energy available to do work. On this basis it is possible to infer that if $\Delta G < 0$, work will be done spontaneously and the corresponding state attained spontaneously attainable state, for a system operating at the constant pressure p, and the constant temperature T. Moreover when $\Delta G = 0$, the system will be in a state of equilibrium and when $\Delta G > 0$, the process cannot do work spontaneously. The *Gibbs free energy potential* is defined as,

$$G = H - TS. \tag{1.2.13}$$

Moreover it follows that,

$$\left(\frac{\partial G}{\partial T}\right)_p = -S. \tag{1.2.14}$$

Thus the *Gibbs free energy potential* may be expressed as,

$$G = H + T\left(\frac{\partial G}{\partial T}\right)_p. \tag{1.2.15}$$

The following relation may also be obtained:

$$\left(\frac{\partial (G/T)}{\partial T}\right)_p = \frac{1}{T}\left(\frac{\partial G}{\partial T}\right)_p - \frac{G}{T^2} = -\frac{H}{T^2}. \tag{1.2.16}$$

Equation 1.2.16 is known as the *Gibbs Helmholtz equation.* Furthermore,

$$\delta G = v\delta p - S\delta T. \tag{1.2.17}$$

Hence it follows that,

$$\left(\frac{\partial G}{\partial p}\right)_T = v. \tag{1.2.18}$$

From Eqs. 1.2.14 and 1.2.18, it follows that,

$$\left(\frac{\partial p}{\partial T}\right)_{\Delta G=0}=\frac{\Delta s}{\Delta v}=\frac{\Delta H}{T\Delta v}. \tag{1.2.19}$$

When ΔH and Δv are constants,

$$\Delta p=\frac{\Delta H}{\Delta v}\ln\left(\frac{T}{T_0}\right). \tag{1.2.20}$$

For an ideal gas, Δv is given by the ideal gas law and

$$\left(\frac{\partial p}{\partial T}\right)_{\Delta G=0}=\frac{\Delta H}{T(RT/p)}. \tag{1.2.21}$$

Integrating Eq. 1.2.21

$$\ln\left(\frac{p}{p_0}\right)=\frac{\Delta H}{R}\left(\frac{1}{T}-\frac{1}{T_0}\right). \tag{1.2.22}$$

Equation 1.2.22 is known as the *Clausius–Clapeyron equation* which allows one to estimate the vapor pressure at some temperature, if the vapor pressure is known at another temperature provided the enthalpy of vaporization is known.

At constant temperature T, $\delta G=v\delta p$ and it follows that

$$G=G_0+\int_{p_0}^{p} v\mathrm{d}p. \tag{1.2.23}$$

For an ideal gas, at constant temperature T,

$$G=G_0+\int_{p_0}^{p}\frac{RT}{p}\mathrm{d}p=G_0+RT\int_{p_0}^{p}\frac{\mathrm{d}p}{p}=G_0+RT\ln\left(\frac{p}{p_0}\right). \tag{1.2.24}$$

When expressed in molar quantities and with $p_0=1$ atmosphere, Eq. (1.2.24) is the standard molar Gibbs energy and is the chemical potential ($\mu=G/n$) with reference to a chemical reaction. It plays a key role in the spontaneous occurrence of chemical reactions and is applied extensively in electrochemistry. To predict how the *Gibbs free energy potential* changes with changes in the pressure p, one generally invokes an important principle known as *Le Chatelier's Principle* and enunciated by *Henri Le Chatelier* in 1888: "Every change of one of the factors of an equilibrium occasions a rearrangement of the system in such a direction that the factor in question experiences a change in a sense opposite to the original change." It is particularly useful in predicting changes in the electrochemical potential due to changes in concentration in electrolytic cells.

It is important to stress that the *Gibbs free energy potential* is defined only for a system operating at the constant pressure p and the constant temperature T. For a system operating at the constant volume v and the constant temperature T another potential function, the *Helmholtz's free energy potential* is defined. It is

$$F = U - TS. \tag{1.2.25}$$

Generally, it is the *Gibbs free energy potential* that is useful for liquids, while in the case of solids operating under both constant pressure and volume, $\Delta F = \Delta G$.

1.2.6 Reversibility, Energy, and Cycles

A process is said to be reversible if at any time during the process both the system and the environment can be returned to their initial states. Reversibility must be considered to be an ideal situation. In most real systems, effects due to energy dissipative and finite temperature differences (non-quasi-static) effects are present. However, many processes are close to being reversible and the maximum efficiency of the process when it is reversible is of interest because it defines an upper limit in performance.

One of the applications of thermodynamics is to the study of reversible, cyclic processes, which are more commonly known as idealized and real "heat engines". Examples of real heat engines are the Otto cycle for automobiles and other motors, refrigerators, heat pumps, etc. The thermodynamic (or thermal) efficiency of any of these is defined as the ratio of the net amount of work done by the system during one cycle to the amount of energy that is input to the system through heating (e.g., combustion of diesel in a reciprocating heat engine, combustion of kerosene in a gas turbine jet engine, etc.). The Kelvin–Plank statement implies that for a single reservoir of heat, a system cannot develop work. Consequently, it follows that

$$\oint \delta w_{\mathrm{closed}} > 0. \tag{1.2.26}$$

From principle of conservation of energy, Eq. (1.2.26) is equivalent to

$$\oint \delta q_{\mathrm{closed}} \leq 0. \tag{1.2.27}$$

In the case of reversible processes, in Eqs. (1.2.26) and (1.2.27), the equality applies.

Imagine a system with energy from reservoir 1 at T_1 being used to heat the cyclic system with an amount of energy $\mathrm{d}q_1$, the cyclic system doing a net amount of work $\mathrm{d}w$, and being cooled by $\mathrm{d}q_2$, which in turn heats reservoir 2 at T_2. We know from the first law that

$$\mathrm{d}q_1 - \mathrm{d}q_2 - \mathrm{d}w = \mathrm{d}u_{\mathrm{cycle}} = 0. \tag{1.2.28}$$

From the second law, since the system is cyclic and reversible, in one cycle, it follows that

$$\mathrm{d}s = 0. \tag{1.2.29}$$

The entropy of the cyclic system is changed at two points, both during the heating and during the cooling. Thus,

$$\mathrm{d}s = \frac{\mathrm{d}q_1}{T_1} - \frac{\mathrm{d}q_2}{T_2} = 0. \tag{1.2.30}$$

Using this relation in Eq. (1.2.28)

$$\mathrm{d}q_1 - \left(\frac{T_2}{T_1}\right)\mathrm{d}q_1 - \mathrm{d}w = \mathrm{d}u_{cycle} = 0. \tag{1.2.31}$$

Thus,

$$\mathrm{d}w = \left(1 - \frac{T_2}{T_1}\right)\mathrm{d}q_1. \tag{1.2.32}$$

When applying the Kelvin–Plank statement to a system in thermal contact with two thermal reservoirs, it can be easily shown that the following statements are true.

1. If the work is negative (i.e., power cycles, the system does useful work), heat will flow from the reservoir at the higher temperature to the system and from the system to the lower-temperature reservoir.
2. The opposite occurs when the work is positive (in the case of a heat pump cycles).
3. The thermal efficiency of any heat engine is defined as'

$$\eta_{th} = \frac{|\Delta w_{\text{net}}|}{\Delta q_{\text{init}}} = 1 - \frac{\mathrm{d}q_2}{\mathrm{d}q_1} = 1 - \frac{T_2}{T_1}. \tag{1.2.33}$$

4. In addition, the maximum attainable efficiency is the efficiency of a reversible cycle:

$$\eta_{th\_\text{rev}} = \max(\eta_{th}). \tag{1.2.34}$$

1.2.7 Application of the Second Law

Three important corollaries to the second law statements, known as Carnot's principles, may be deduced, and these are particularly important in applying the second law to engineering of energy generation. These are summarized as follows:

1. The thermal efficiency of any heat engine is always less than that of a reversible engine operating between the same two heat reservoirs.
2. Any two reversible engines operating between the same thermal reservoirs have the same efficiency.
3. An absolute temperature scale can be defined which is independent of the nature of the measuring substance.

The third corollary follows from the second. If the thermal efficiencies are the same regardless of the design of the heat engine, then the only variable is the type of heat reservoir, which is characterized only by its temperature. Accordingly, using three independent heat engines operating between the same two heat reservoirs, it can be shown that

$$\frac{q_2}{q_1} = \frac{f(T_2)}{f(T_1)}, \tag{1.2.35}$$

where any monotonic function can be used to define the thermodynamic temperature. The Kelvin scale is defined using $f(T) = T$ and was adopted by the *International General Conference on Weights and Measures* as the standard scale. The absolute zero then corresponds to the case in which unit efficiency can be achieved.

Consider a hypothetical process defined by a cycle of quasi-equilibrium processes. The first is adiabatic compression with the temperature increasing, followed by isothermal expansion accompanied by heat input, followed by adiabatic expansion with temperature decreasing and finally isothermal compression with heat release. Such a cycle is known as the Carnot cycle. There are four processes involved to at constant entropy and two at constant temperature. The Carnot cycle can be illustrated on a *P–v* diagram (Fig. 1.5), with isotherms representing curves at constant temperatures T_1 (1–2) and T_2 (3–4), and adiabatic expansion and contraction connecting 2–3 and 4–1, respectively. The heat input and output during the isothermal processes are denoted by dq_1 and dq_2, respectively. The area enclosed by the two bounding isotherms and the two adiabatic characteristics represents the total work output of the hypothetical "Carnot engine" based on the Carnot cycle. Equation 1.2.33 gives the maximum efficiency of such a hypothetical engine.

1.2.8 Basic Power Generation Cycles

The piston of a four-stroke petrol engine, used in most motor cars, executes four strokes within the cylinder when the crankshaft completes two revolutions for each thermodynamic cycle. The thermodynamic cycle can be modeled as an ideal Otto cycle. The *P–v* and *T–s* diagrams of an ideal Otto cycle are shown in Figs. 1.6 and 1.7. In an Otto cycle, process 1–2 is an isentropic compression process, while the

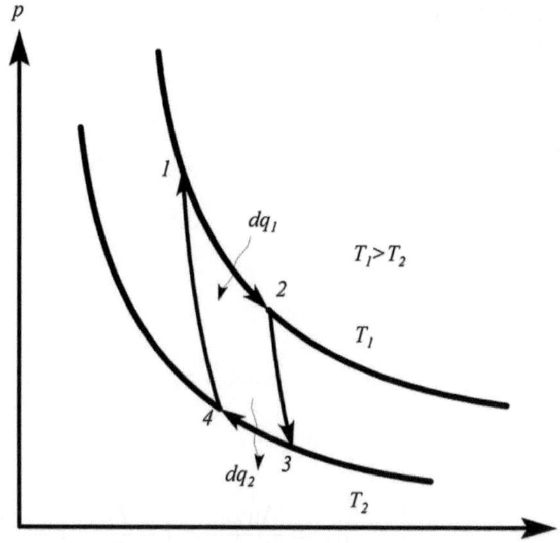

Fig. 1.5 The Carnot cycle illustrated on a *P–v* diagram

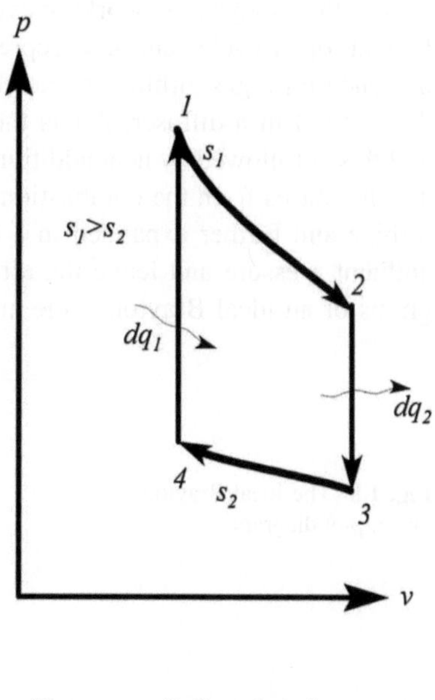

Fig. 1.6 The ideal Otto cycle: *p–v* diagram

process 3–4 is an isentropic expansion process. Processes 2–3 and 4–1 are constant-volume processes where heat is supplied and removed, respectively. Also, the *T–s* diagram shows the temperature of the air reaching its maximum value at state 3. In the four processes of the Otto cycle, only processes 1–2 and 3–4 have work interaction. It can be shown that the maximum efficiency of the Otto cycle engine is

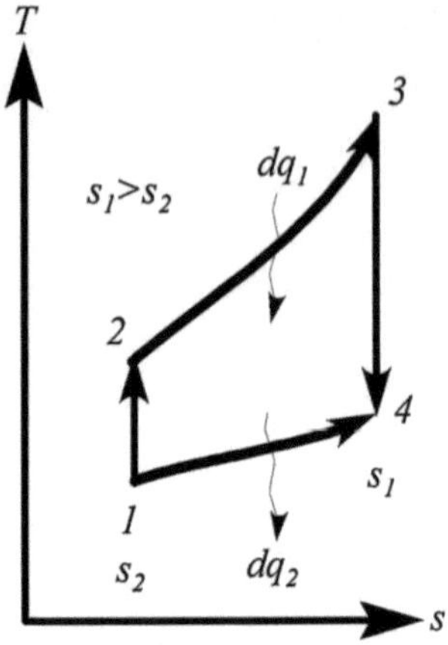

Fig. 1.7 The ideal Otto cycle: *T–s* diagram

$$\eta_{th} = \frac{|\Delta w_{\text{net}}|}{\Delta q_{\text{init}}} = 1 - \frac{T_2}{T_1} = 1 - \left(\frac{v_2}{v_1}\right)^{\gamma - 1}. \tag{1.2.36}$$

The thermodynamic working cycle for the jet engine is based on Brayton cycle. It is an open cycle (and also represented as closed for thermodynamic analysis) designed for a gas turbine engine. The pressure of the air rises slightly as it is first decelerated in a diffuser. Air is then compressed isentropically in a compressor, and this is followed by heat addition at constant pressure in a combustion chamber. The hot gases from the combustion chamber are then isentropically expanded in a turbine and further expanded in a nozzle. The gases expand in the nozzle to the ambient pressure and leave the aircraft at a high velocity. The *p–v* and *T–s* diagrams of an ideal Brayton cycle are shown in Figs. 1.8 and 1.9.

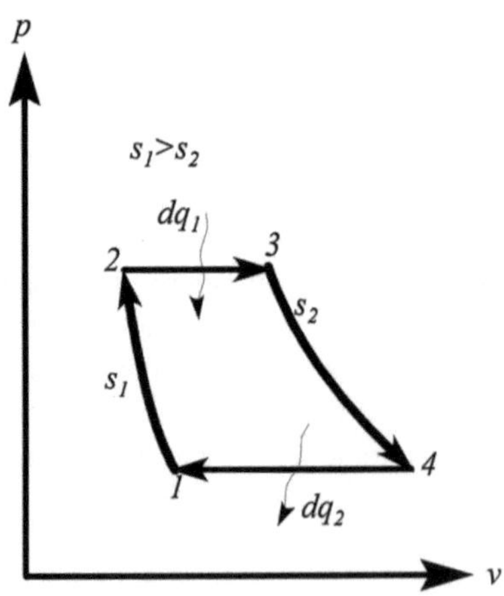

Fig. 1.8 The ideal Brayton cycle: *p–v* diagram

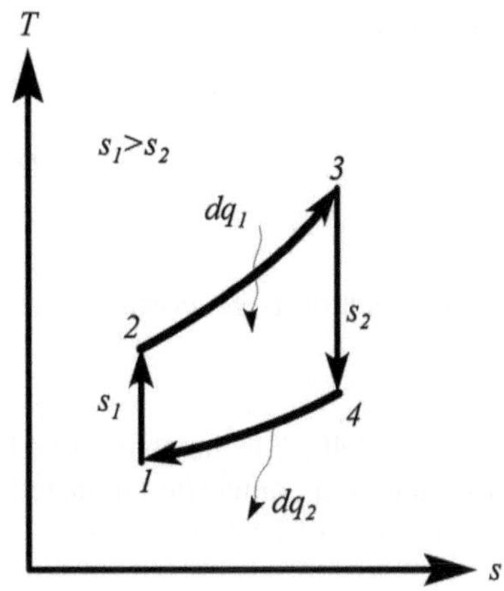

Fig. 1.9 The ideal Brayton cycle: *T–s* diagram

The fact that the line 3–4 on the *T–s* diagram is longer than the line 1–2 makes it possible extract more power with the turbine than is consumed by the compressor.

In the ideal cycle, processes of compression (1–2) and expansion (3–4) are assumed to be isentropic, while the processes of heat addition (2–3) and heat rejection (4–1) are assumed to be constant-pressure processes.

1.2.9 Heat Transfer: Conduction, Radiation, Convective Heat Transfer

The fundamental laws governing heat conduction, convection, and radiation are Fourier's law of conduction, Newton's law of cooling, and Stefan–Boltzmann's law of radiation, respectively. Fourier's law defining the flow of heat through a rectangular element of area A which is maintained at a temperature gradient $\partial T/\partial x$ is

$$\dot{Q} = -kA(\partial T/\partial x). \tag{1.2.37}$$

In Eq. 1.2.37, $\dot{Q}$ is the heat conducted in one dimension, considered positive in the direction of the surface at the higher temperature, and k is the coefficient of thermal conductivity. For a thin rectangular slab maintained at a uniform temperature gradient, the heat conducted away from the surface at the higher temperature is

$$\dot{Q} = kA(\Delta T/\Delta x). \tag{1.2.38}$$

For a composite laminated wall with N layers, where each of the layers has a different thickness Δx_i, different temperature gradients, and different thermal conductivities k_i, the heat conducted away from the surface at the higher temperature is

$$\dot{Q} = \frac{\Delta T}{\sum_{i=1}^{N} \Delta x_i / k_i A}. \qquad (1.2.39)$$

Convection is governed by the law of cooling which is

$$\dot{Q} = h\Delta T. \qquad (1.2.40)$$

In Eq. 1.2.40, h is the convective heat transfer coefficient. Considering the earlier example of a composite laminated wall and assuming that the first and last layers are capable of heat convection

$$\dot{Q} = \frac{\Delta T}{\frac{1}{h_1} + \sum_{i=1}^{N} \frac{\Delta x_i}{k_i A} + \frac{1}{h_N}}. \qquad (1.2.41)$$

In Eq. 1.2.41, h_1 and h_N are the convective heat transfer coefficients on either side of the wall.

Radial conduction across a cylindrical tube of length L with no convection possible either inside or outside the tube is governed by

$$\dot{Q} = \frac{\Delta T}{\frac{1}{2\pi kL} \ln\left(\frac{R_2}{R_1}\right)}. \qquad (1.2.42)$$

When convection within or outside the tube is possible

$$\dot{Q} = \frac{\Delta T}{\frac{1}{h_1} + \frac{1}{2\pi kL} \ln\left(\frac{R_2}{R_1}\right) + \frac{1}{h_2}}. \qquad (1.2.43)$$

In Eq. 1.2.43, h_1 and h_2 are the convective heat transfer coefficients inside and outside the tube. Generally, there exists a relationship between the heat transfer coefficient for convection and heat transfer coefficient the thermal conductivity. This relationship is expressed by specifying a non-dimensional number relating the two known as the Nusselt number which is defined as

$$Nu = hD/k. \qquad (1.2.44)$$

In Eq. 1.2.44, D is a characteristic length or diameter which is problem dependent. For problems involving forced convection, the Nusselt number value is a function of the specific values of the Reynolds number of the flow and the Prandtl number and is expressed by a functional relationship defined as

$$Nu = f(Re, Pr). \qquad (1.2.45)$$

The Prandtl number Pr is a dimensionless number, the ratio of momentum diffusivity (kinematic viscosity) to thermal diffusivity. It is defined as

$$Pr = \frac{v}{\alpha} = \frac{\text{viscous diffusion rate}}{\text{thermal diffusion rate}} = \frac{v}{k/\rho c_p}. \quad (1.2.46)$$

In Eq. (1.2.46), v is the kinematic viscosity, α is the thermal diffusivity, k is the thermal conductivity, c_{p} is the specific heat coefficient, and ρ is the density of the convection medium. For problems involving free or natural convection, the Nusselt number value is a function of the Grashof number and the Prandtl number and is expressed by a functional relationship defined as

$$Nu = f(Gr, Pr). \quad (1.2.47)$$

The Grashof number, Gr, is a dimensionless number in fluid dynamics and heat transfer which approximates the ratio of the buoyancy to viscous force acting on a fluid. It frequently arises in the study of situations involving natural convection. For vertical flat plates, it is

$$Gr_L = \frac{g\beta(T_s - T_\infty)L^3}{v^2}. \quad (1.2.48)$$

In Eq. (1.2.48), g is the acceleration due to earth's gravity, β is the volumetric thermal expansion coefficient (equal to approximately $1/T$, for ideal fluids, where T is an absolute temperature), T_s is the surface temperature, T_∞ is the bulk temperature, v is the kinematic viscosity, and L is a characteristic length. For pipes with a diameter is a characteristic length D, it is,

$$Gr_D = \frac{g\beta(T_s - T_\infty)D^3}{v^2}. \quad (1.2.49)$$

For example, for turbulent forced convection in a tube, the Nusselt number is

$$Nu = 0.023\left(Re^2\, Pr\right)^{0.4}. \quad (1.2.50)$$

For laminar free convection from a vertical plate is governed by

$$Nu = 0.59(Gr \times Pr)^{0.25}. \quad (1.2.51)$$

The Stefan–Boltzmann law governing black body radiation for the total energy radiated per unit surface area of a black body per unit time, which is directly proportional to the fourth power of the black body's thermodynamic temperature T, is

$$q_b = \sigma T^4. \quad (1.2.52)$$

In Eq. 1.2.52, the Stefan–Boltzmann constant is σ, which can be obtained as function of the Boltzmann constant, Planck's constant, and the speed of light in vacuum. Thus, for any body or enclosure,

$$\dot{Q} = \sigma \varepsilon_1 A \left(T_2^4 - T_1^4\right). \tag{1.2.53}$$

In Eq. 1.2.53, ε_1 is defined as the *emissivity* of the body. Thus, the *emissivity* is equal to unity for a black body when $T_1 = 0$.

1.3 Electrochemistry: Introduction

Electrochemistry is the study of the electrochemical interactions involving the exchange of chemical and electrical energy. For a detailed exposition on electrochemistry, the reader is referred to Hamann et al. (2007). Electrochemistry involves oxidation–reduction reactions that can be brought about by electricity or used to produce electricity. The study of electrochemistry involves the study of chemical reactions which could be a continuous source of electrons. A controlled source of electrons can be used to generate a current in a circuit. Chemical reactions can be controlled depending on the nature of energy available to them. Energy is released when chemical bonds holding a distribution of electrons around an atom are broken and vice versa. Oxidation is a chemical reaction process which can occur at an interface of a metal or electrode and an electrolyte. The metal or electrode releases electrons, and the metal or electrode is electron rich. Since metal oxidation necessarily results in supply of electrons, a second reduction process takes place in the vicinity of the metal oxidation site. Oxidation and reduction always occur together. A substance is oxidized, while another is reduced. Reduction involves a gain in electrons. For example, the reaction between potassium and chlorine involves a loss of electrons by potassium (oxidation of potassium) and a gain of electrons of chlorine (reduction of chlorine). The reactions are

$$\mathrm{K} \Leftrightarrow \mathrm{K}^+ + e^- \ (\text{oxidation}), \ \mathrm{Cl}_2 + 2e^- \Leftrightarrow 2\mathrm{Cl}^- \ (\text{reduction}). \tag{1.3.1}$$

In the above example, potassium is oxidized and chlorine is reduced. Following oxidation, positively charged ions at the electrode or metal flow into the electrolyte. For this reason, the electrode is referred to as an anode. In the electrolyte, these positively charged ions or anions migrate to another electrode, known as the cathode. Following another chemical reaction, the cathode becomes positively charged. At the cathode, there is generally a reduction reaction. The positive cathode accepts electrons from an external circuit to maintain the charge balance. Alternately, a reduction process at the cathode generates negative charged ions at the cathode–electrolyte interface (cations) which migrate to the anode. At the anode, following an oxidation reaction, there is a release of electrons which in turn flow into the cathode via an external circuit to maintain the charge balance.

In defining the properties of fuel cells, extensive use is made of the terms mole and molar mass. The atomic weight is the abundance-weighted average mass of an element (an average mass of all the isotopes that exist for that element, weighted by how abundant each isotope is in the element). Atomic weight although generally reported in atomic mass units is commonly also reported in grams per mole. A mole is a number of atoms or molecules of a substance or element equal to 6.02×10^{23} which is known as Avogadro's number. By definition, a mole is the unit of the quantity or amount of substance of matter, which contains as many elementary entities as there are atoms in 12 g of carbon-12 (i.e., carbon consisting only of the isotope 12C). Avogadro's number is the number of molecules of a substance it takes in order to have a mole of the substance, and it is equal to 6.02×10^{23}. For example, a mole of oxygen is just 6.02×10^{23} oxygen atoms. The molar mass of a molecule is what the mass would be if you had 6.02×10^{23} of those molecules. It is the total mass of all the protons, neutrons, and electrons within the atom/molecule. It is expressed in grams per mole (g/mol). The molar mass is the atomic mass and is the number just under the element's letter on the periodic table. For example, in the case of oxygen, the molar mass indicated in the periodic table is 15.9994 (this number comes directly off the periodic table underneath the O symbol for oxygen). This implies that 1 mol of oxygen = 15.9994 g. Consequently, the molar mass of oxygen is 15.9994 g/mol. The molar mass is the mass of one mole of matter and can refer to both a single element and molecules and other compounds. The molecular weight is essentially the same thing as the molar mass except that, as the name implies, it refers to molecules rather than just elements. The molar mass and molecular weight are typically given in units of grams per mole. The molecular weight of a compound is found by adding the atomic weights of all of the atoms in the compound. Water, for example, has a molecular weight of 18.0152 g/mol, as there are two hydrogen atoms that each has an atomic weight of 1.00079 g/mol and one oxygen atom with an atomic weight 15.9994 g/mol.

1.3.1 Fuel Cell Thermodynamics

To best understand the basics of fuel cell electrochemistry, the basic thermodynamic functions and their interrelationships must be defined. In what follows the temperature T and pressure p of the reactants are assumed to be constant. The best understood thermodynamic functions under these conditions are the Gibbs energy (G) and the enthalpy (H),

$$G = H - TS, \quad H = U + pV, \tag{1.3.2}$$

where U is the internal energy, V is the volume, and S is the entropy. One is particularly interested in the overall change in the Gibbs energy in a fuel cell as this can be related to the spontaneously generated electrical energy. The potential electricity output of a chemical reaction is given by

$$\Delta G = \Delta H - T\Delta S \tag{1.3.3}$$

where ΔG is the free energy change in the reaction, ΔH is the enthalpy change for the reaction, and ΔS is the entropy change for the reaction. The free energy change in the reaction ΔG is the maximum useful work associated with a chemical reaction

$$\Delta G = -nFE_{\text{rev}} \tag{1.3.4}$$

where n is the number of moles of reactants involved and F is the Faraday constant that represents the total electronic charge in coulombs in a mole. The maximum heat associated with a chemical reaction, ΔH is

$$\Delta H = -nFE_{\text{thermo}}. \tag{1.3.5}$$

In the overall reaction, this is associated with a net loss of entropy, ΔS.

When a number of reactant components are involved, the change in the Gibbs energy (ΔG) by one of its reaction components can be expressed as

$$\Delta G = \mu_k \Delta n_k, \quad \mu_k = \Delta G / \Delta n_k, \tag{1.3.6}$$

where μ_k is the chemical potential of the component k and n_k is the number of moles of the component k. The chemical potential of a gaseous component is dependent on the partial pressure of the gas (p_k), which can be written as

$$\mu_k = \mu_k^0 + RT \log(p_k) \tag{1.3.7}$$

where R is the universal gas constant and μ_k^0 the chemical potential at the standard gas pressure, usually taken to be 1 atmosphere. In a fuel cell, the overall current production occurs spontaneously without external supply of energy. In the battery mode of operation, the Gibbs energy will be negative and can be separated into two electrode parts with subscript 1 and 2 as

$$\pm \Delta G = \sum_{1,j} v_j \mu_j - \sum_{2,j} v_j \mu_j \tag{1.3.8}$$

where v_j is the stoichiometric coefficient. The maximum value of electrical work that can be done by n moles of charge carriers and consequently the standard reversible voltage E_{rev}^0 may be written as

$$E_{\text{rev}}^0 = \frac{\sum_{1,j} v_j \mu_j - \sum_{2,j} v_j \mu_j}{nF} = -\frac{\Delta G}{nF}. \tag{1.3.9}$$

Introducing the dependence of μ_k on the partial pressure of the gas (p_k),

$$E_{\text{rev}} = E_{\text{rev}}^0 + \frac{RT}{nF}\left(\sum_1 v_j \log(p_j) - \sum_2 v_j \log(p_j)\right) \tag{1.3.10}$$

where p_k is best measured in atmospheres.

1.3.2 Electrode Electrochemistry

The electrode is one of the most common sensors to measure and record potentials and currents. Metal electrodes immersed in an electrolyte give rise to metal–electrolyte junctions that are an important class of junctions and can generate a potential known as the *half-cell potential.* Consider a typical electrode–electrolyte junction where a single electrode is immersed into an electrolyte. The electrode only has one type of charge carrier, the electron, whereas the electrolyte has two types of charge carriers, the cations arising at the electrode and migrating into the electrolyte as well as anions already in the electrolyte. If we assume that the metal electrode is homogeneous of the same metal C and is made up of atoms of the same type, an electrolyte with a concentration of anions of type A^- which are contact with cat-ions in the metal of type C^+ then the following type of electronic reactions may be associated with the metal–electrolyte junction:

$$C \leftrightarrow C^{n+} + n \times e^-, \quad A^{m+} \leftrightarrow A + m \times e^- \tag{1.3.11}$$

where n, m represent the valence of the cations and anions, respectively. Thus, the material in the electrode at the interface can become oxidized to form a cation and release one or more free electrons and an anion coming to the electrode–electrolyte interface can be oxidized to a neutral atom, relinquishing one or more free electrons to the electrode. When the current equals zero, the acquisition of electrons by the cations is balanced by the release of electrons by the anions and the net electronic current is zero. When a current passes, an imbalance persists as one type of reaction dominates. The magnitude and direction of the current are determined by the magnitude and stability of the reaction, the type of electrode material, the type and concentration of the electrolyte, and the load on the circuit. The ionic availability, $a_{C^{n+}}$, per unit mole of the cation, C^{n+}, may be expressed as

$$a_{C^{n+}} = \exp(-\Delta G/RT) = \exp(-n(V_0 - V)F/RT) \tag{1.3.12}$$

where ΔG is the energy barrier associated with ionic dissociation across the interface that must be overcome, n is the valence of the relevant ion, V_0 is the standard half-cell potential for hydrogen, R is the gas constant (8.315 J/K mol), and F is Faraday's constant (96,480 C/mol). The electrical potential is the voltage at any point in space that is needed to move a unit charge of 1 C from infinity to that point. Consequently, the energy of a system with n moles of electrons moving through a potential difference ΔV changes by $q\Delta V = (-nF)\Delta V$. The electrical energy associated with the energy barrier is the product of free charge and the electrical potential (i.e., half-cell potential). The ionic availability is the availability of the ions in the solution to participate in the reaction and is proportional to the concentration, where the proportionality constant is less than or equal to unity.

Hence, rearranging the Eq. (1.3.12) for the ionic availability, standard half-cell voltages are defined relative to hydrogen which is assumed to be zero volts and can be estimated from the equation:

Table 1.3 Table of standard half-cell (redox) potentials

Material	Reaction	V
Al	$Al^{3+} + 3 \times e^-$	−1.705
Au	$Au^{3+} + 3 \times e^-$	1.498
Zn	$Zn^{2+} + 2 \times e^-$	−0.7628
H_2	$2H^+ + 2 \times e^-$	0
$Ag + Cl^-$	$AgCl + e^-$	0.223
Ag	$Ag^{1+} + e^-$	0.7996
Cu	$Cu^{2+} + 2 \times e^-$	0.3419
Cu	$Cu^{1+} + e^-$	0.522
Na	$Na^{1+} + e^-$	−2.714
Fe	$Fe^{2+} + 2 \times e^-$	−0.447
K	$K^{1+} + e^-$	−2.925
Mg	$Mg^{2+} + 2 \times e^-$	−2.363
Pt	$Pt^{2+} + 2 \times e^-$	1.2

$$V = V_0 + (RT/nF)\ln(a_{C^{n+}}). \tag{1.3.13}$$

Equation (1.3.13) was first obtained by Nernst over 100 years ago and is therefore known as the *Nernst equation*. Typical standard half-cell voltages are listed in Table 1.3 and typical reduction–oxidation reaction (redox) potentials in Table 1.4. Equation (1.1.13) can be used to analyze the electrode–electrolyte interactions. Consider a general reduction–oxidation reaction, at a biopotential electrode–electrolyte interface described by the equation,

$$\alpha A + \beta B \Leftrightarrow \gamma C + \delta D + n \times e^- \tag{1.3.14}$$

where n is the number of electrons relinquished in the reaction. The net reaction rate may be expressed as

$$k_r = (a_{A^{\alpha+}} a_{B^{\beta+}})/(a_{C^{\gamma+}} a_{D^{\delta+}}) \tag{1.3.15}$$

where $a_{X^{\omega+}}$ are the ionic availabilities of species "X". Thus, the potential across the electrode–electrolyte interface may be expressed as

$$V = V_0 + \frac{RT}{nF}\ln\left(\frac{a_{A^{\alpha+}} a_{B^{\beta+}}}{a_{C^{\gamma+}} a_{D^{\delta+}}}\right). \tag{1.3.16}$$

Table 1.4 Table of standard reduction/oxidation potentials

Redox reaction	V
$O_2 + 4H^+ + 4e^- \leftrightarrow 2H_2O$	+1.229
$O_2 + 2H_2O + 4e^- \leftrightarrow 4OH^-$	+0.401
$2H_2O + 2e^- \leftrightarrow H_2 + 2OH^-$	−0.83

The standard half-cell potential, V, is the potential for 1M concentration solution at 25 °C to that of a hydrogen electrode in 1M solution. The hydrogen electrode potential V_0 is based on the reaction

$$H_2 \Leftrightarrow 2H \Leftrightarrow 2H^+ + 2e^-, \tag{1.3.17}$$

where the source of hydrogen is H_2 gas bubbled over a platinum electrode at a pressure 1 atm.

Although the hydrogen electrode is assumed to be the reference electrode for most measurements, it is not always suitable or even ideal in practice. For biological applications, a more usual reference electrode is the calomel electrode in a saturated KCl solution with a potential of +242 mV against the hydrogen electrode. The calomel electrode is a reference electrode based on the reaction between the element mercury and mercury chloride (Hg_2Cl_2 or calomel). The electrode is based on the ionization of mercury on the electrode surface,

$$2Hg \Leftrightarrow Hg_2^{2+} + 2e^-. \tag{1.3.18}$$

In the electrolyte, one has

$$Hg_2^{2+} + 2Cl^- \Leftrightarrow Hg_2Cl_2. \tag{1.3.19}$$

When a fluid contains more than one molecular species and when there exist differences in the composition characterized by varying concentrations, the molecules are set in spontaneous motion in directions which tend to make the compositions "uniform." When two compartments c and s which differ in chemical concentration characterized by the ion concentration are brought together, there is a subsequent exchange of the charge-carrying ions. Generally, in such, there are two physically dominant forces which try to balance each other; in a semiconductor *p–n junction*, it is the force of electronic diffusion and those due to the presence of an electrostatic force field. In the current situation, it is the diffusion of ionic charge carriers from one compartment to the other. The net result of this type of diffusion is the generation of a potential difference E in volts, across the two compartments which can be described by the *Nernst equation* given by

$$E = V_c - V_s = -\frac{RT}{zF}\ln\frac{[C]_{i,c}}{[C]_{i,s}} \tag{1.3.20}$$

where $R = 8.31\,\mathrm{J\,mol^{-1}K^{-1}}$ is the universal gas constant, T is the absolute temperature in K, $F = 96484.7$ C/mol is the Faraday constant representing the total charge on a mole of electrons, $[C]_{i,c}$ is the ionic concentration in moles per unit volume of the ion, $[C]_i$ in compartment c, and $[C]_{i,s}$ is the ionic concentration in moles per unit volume for the same ion, $[C]_i$ in compartment s. The two compartments are in contact with each other through a semi-permeable membrane, and the equilibrium condition that concentrations in the two compartments are moving toward is known as the *Donnan* equilibrium.

1.3.3 Relationship to the Change in the Gibbs Potential

In Sect. 1.2.5, the change in the Gibbs potential (Eq. 1.2.19) was shown to be of the form:

$$\Delta G = \Delta G^0 + RT \ln\left(\frac{p}{p_0}\right). \tag{1.3.21}$$

Consider two fluid reactants A and B in a state of equilibrium.

$$\Delta G(A) = \Delta G(B). \tag{1.3.22}$$

Hence,

$$\Delta G_A^0 + RT \ln\left(\frac{p_A}{p_0}\right) = \Delta G_B^0 + RT \ln\left(\frac{p_B}{p_0}\right), \tag{1.3.23}$$

$$\Delta G_A^0 - \Delta G_B^0 = RT \ln\left(\frac{p_B}{p_0}\right) - RT \ln\left(\frac{p_A}{p_0}\right) = RT \ln\left(\frac{p_B}{p_A}\right), \tag{1.3.24}$$

$$\frac{p_B}{p_A} = \exp\left(-\frac{\Delta G_B^0 - \Delta G_A^0}{RT}\right). \tag{1.3.25}$$

Thus, the difference in the potentials when the partial pressure of the reactant is equal to the standard pressure determines the ratio of the partial pressures of the two gases. However, the ratio of the partial pressures may be expressed as the ratio of the concentrations or the ratio of the mole fractions or the ration of ionic availabilities in the case of solutions. Hence, in the case of the general redox reaction, the equation for the change in the Gibbs potential is

$$\Delta G = \Delta G^0 + RT \ln\left(\frac{a_{C^{\gamma+}} a_{D^{\delta+}}}{a_{A^{\alpha+}} a_{B^{\beta+}}}\right) = \Delta G^0 + RT \ln Q \tag{1.3.26}$$

The potential across the electrode–electrolyte interface may be written as

$$-nFV = -nFV_0 + RT \ln\left(\frac{a_{C^{\gamma+}} a_{D^{\delta+}}}{a_{A^{\alpha+}} a_{B^{\beta+}}}\right) = -nFV_0 + RT \ln Q. \tag{1.3.27}$$

Thus, one may relate the change in the Gibbs potential

$$\Delta G^0 = -nFV_0 \tag{1.3.28}$$

and

$$\Delta G = -nFV. \tag{1.3.29}$$

1.3.4 Fuel Cell Efficiency

Fuel cells are not limited by the Carnot cycle efficiency. The fuel cell thermodynamic efficiency is given by the ratio of the change in the Gibbs function to the change in the enthalpy in the overall cell reaction. The Gibbs function change measures the electrical work, and the enthalpy change is a measure of the heating value of the fuel. The difference between the heat of formation of the reactants and heat of formation of the products is known as the heat of reaction. The equation for the reaction of hydrogen and oxygen is exothermic (heat producing), and the enthalpy change, d*H*, is given by the heat of formation of water minus the heats of formation of hydrogen and oxygen which can be assumed to be zero as they are assumed to be made available. Thus, the enthalpy change, d*H*, is the heat of formation of water itself.

The maximum ideal efficiency of a fuel cell under standard conditions = (dG^0/dH^0). For the hydrogen–oxygen reaction at 25 °C, $dH^0 = -68{,}317$ cal/g mol of H_2 and $dG^0 = -56{,}690$ cal/g mol of H_2. The maximum efficiency of the ideal fuel cell at 25 °C is therefore equal to (56,690/68,317) = 83 %.

It may be observed that in the case of a fuel cell, the maximum ideal efficiency decreases with increasing temperature, while the limiting maximum Carnot efficiency of a heat engine increases with temperature. The heat engine tends to be more efficient at high temperatures above 700 °C, while the fuel cell is more efficient at lower temperatures.

Another measure of the fuel cell efficiency is known as the "voltage efficiency" and is the ratio of the actual generated voltage under operating conditions to the theoretical cell voltage.

Voltage Efficiency = (Actual Voltage)/(Theoretical Voltage) = $(V_A/1.23)$

1.3.5 Electrode Overpotentials

When an electrode is inserted into an electrolyte and a circuit is connected to the electrode, a current will flow, but the observed potential is different from the standard half-cell potential. This difference is known as the overpotential. An important effect that plays a role in the generation of the observed half-cell potential is the current flow through the circuit. This is not only dependent on the ohmic resistance of the electrolyte but also on the extent of polarization of the electrode. An excess charge results from a flow of current which in turn changes the ionic concentration near the electrode–electrolyte interface. For metal ions to be oxidized, they must overcome an energy barrier. If the current flow is one directional, then either the oxidation or the reduction dominates, and the height of the barrier or the activation potential changes. Both the concentration of the electrolyte and activation potential contribute to the polarization at the electrode and to the overpotential which adds to the contribution of the ohmic overpotential.

The total overpotential impedes the current flow across the interface. Metal oxide electrodes are often used in several electrochemical cell applications.

In an activation-controlled system, the rate of reaction is controlled solely by the rate of the electrochemical charge transfer process. For such activation-controlled systems, the Butler Volmer equation determines the relation between the electrical current density on an electrode and the electrode potential, considering that both a cathodic reaction and an anodic reaction can occur at the same electrode. The electrode current density is a function of the rate constants for the forward and reverse reactions k_f and k_r, respectively, the concentrations of oxidized and reduced species c_f and c_r, respectively, the symmetry factor for the forward electrode reaction β, the electrode potential V, the electrode temperature T, the number of moles of charge carriers n per mole of the product species, and the Faraday and universal gas constants. Physically it captures the fact that the reactants interact with the electrode and that the potential energy of those interactions will depend on the voltage applied to the electrodes. It also accounts for the fact that in an anodic reaction, the electron moves to the electrode which is induced by increasing the voltage and that increasing the forward reaction rate decreases backward rate of reaction. The Butler–Volmer equation is given by

$$i = F\left\{k_f c_f e^{n\beta VF/RT} - k_r c_r e^{-n(1-\beta)VF/RT}\right\} \tag{1.3.30}$$

The Butler–Volmer equation (1.3.30) may be used to determine the exchange current density, the equilibrium current density developed at an electrode. Thus,

$$0 = F\left\{k_f c_f e^{n\beta V_R F/RT} - k_r c_r e^{-n(1-\beta)V_R F/RT}\right\} \tag{1.3.31}$$

where V_R is the reversible electrode potential required for the electrode equilibrium current density to develop and the equilibrium current density is given by

$$i_0 = Fk_f c_f e^{n\beta V_R F/RT} = Fk_r c_r e^{-n(1-\beta)V_R F/RT}. \tag{1.3.32}$$

Thus, the Butler–Volmer equation (1.3.30) may be expressed in terms of the difference between the actual and reversible electrode potential and the electrode exchange current density as

$$i = i_0\left\{e^{n\beta(V-V_R)F/RT} - e^{-n(1-\beta)(V-V_R)F/RT}\right\}. \tag{1.3.33}$$

Generally, anodic reactions require a positive value for the overpotential to increase the rate of reaction, while cathodic reactions require a negative value for the overpotential to increase the rate of reaction. However, the net effect is a reduction in the overall voltage to increase the reaction rates. As the overpotential, either positive or negative, becomes larger, one of the terms of the Butler–Volmer equation (1.3.33) becomes negligible. Thus, when the overpotential is large and positive,

$$i = i_0 e^{n\beta(V-V_R)F/RT}. \tag{1.3.34}$$

It follows that the Eq. (1.3.34) can be rewritten as the so-called Tafel equation,

$$V - V_R = a + b \log i = b \log(i/i_0) \tag{1.3.35}$$

which is a relationship between the activation polarization and the current density and in which a and b are constants in volts.

There is yet another effect that must be accounted for while making real measurements of the electrode potentials. Generally, one is interested in measuring the build up of the potential that is entirely due to the accumulation of one particular ion on the surface of the electrode. However, in practice, other ions will interfere with this process and this interference of the ion selectivity may be described by a modification of the Nernst equation, known as the Nicolsky–Eisenman equation. The Nicolsky–Eisenman modification of the Nernst equation is given by

$$V = V_0 + (RT/nF) \ln\left(a_{C^{n+}} + \sum_{j=1}^{J} k_{C,D_j}^{\text{pot}} a_{D_j}^{(n/m_j)} \right) \tag{1.3.36}$$

where k_{C,D_j}^{pot} is the potentiometric selectivity coefficient of the ion $D_j^{m_j+}$ with respect to the principal ion C^{n+}, a_{D_j} is the availability of the jth interfering ion, and m_j is the corresponding charge number of the jth interfering ion. For the special case when $m_j = n$, the simplified Nicolsky–Eisenman equation is

$$V = V_0 + (RT/nF) \ln\left(a_{C^{n+}} + \sum_{j=1}^{J} k_{C,D_j}^{\text{pot}} a_{D_j} \right). \tag{1.3.37}$$

The selectivity coefficients depend on the site availabilities of the ion-selective electrode.

1.3.6 Examples of Half-Cell Electrodes

An example of generic gas electrode is shown in the Fig. 1.10. Consider the case of the hydrogen electrode. In this case, the gas is hydrogen (H_2) and the electrolyte is an aqueous solution of hydrochloric acid (HCl). At the electrode surface

$$2H^+ + 2e^- \Leftrightarrow H_2. \tag{1.3.38}$$

In the electrolyte, one has

$$HCl \Leftrightarrow H^+ + Cl^-. \tag{1.3.39}$$

From the Nernst equation

$$V = V_0 + \frac{RT}{F} \ln\left(\frac{p_{H^+}}{p_{H_2}^{1/2}} \right). \tag{1.3.40}$$

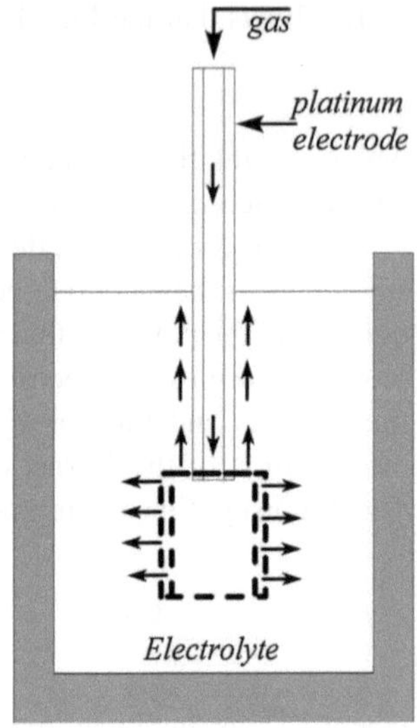

Fig. 1.10 Typical generic gas electrode

In the case of the chlorine electrode, the gas is chlorine (Cl_2) and the electrolyte is potassium chloride solution (KCl). At the electrode surface,

$$Cl_2 + 2e^- \Leftrightarrow 2Cl^-. \tag{1.3.41}$$

In the electrolyte, one has

$$KCl \Leftrightarrow K^+ + Cl^-. \tag{1.3.42}$$

From the Nernst equation

$$V = V_0 + \frac{RT}{F}\ln\left(\frac{p_{Cl_2}^{1/2}}{p_{Cl^-}}\right). \tag{1.3.43}$$

The Nernst equation predicts that the cell voltage becomes more positive if the partial pressure of chlorine gas is increased or if the concentration of Cl^- is decreased which is quite the opposite of the behavior of the hydrogen electrode.

In the case of the silver electrode, there is no gaseous input, but the electrode is coated with a porous layer of silver chloride (AgCl) so that the electrolyte is an aqueous solution containing negatively charged chlorine ions (Cl^-). At the silver/silver chloride boundary

$$AgCl + e^- \Leftrightarrow Ag + Cl^-. \tag{1.3.44}$$

From the Nernst equation

$$V = V_0 + \frac{RT}{F}\ln\left(\frac{1}{[C]_{Cl^-}}\right) = V_0 - \frac{RT}{F}\ln[C]_{Cl^-}. \tag{1.3.45}$$

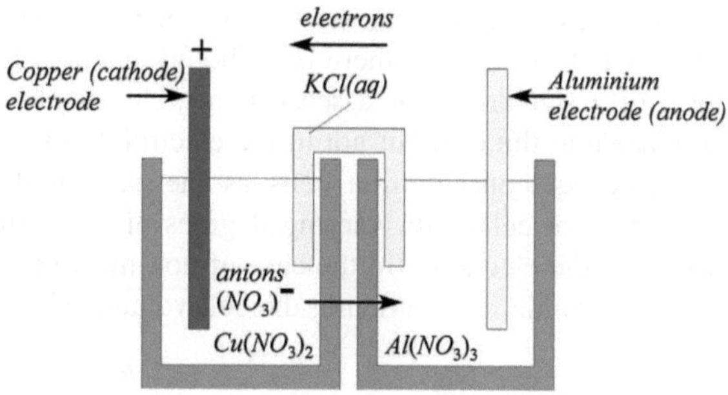

Fig. 1.11 Example of a salt bridge and its role

1.3.7 The Salt Bridge

The salt bridge is a tube containing a compatible salt such as KCl which places the two half cells in electrical contact while preventing the two different aqueous solutions in the two half cells from mixing. A typical example is shown in Fig. 1.11.

In the example illustrated in Fig. 1.11, a galvanic cell is formed with two electrodes, one copper and the other aluminum. At the cathode (electrode) surface, copper ions are reduced

$$Cu^{2+} + 2e^- \Leftrightarrow Cu. \tag{1.3.46}$$

At the anode (electrode) surface, aluminum ions are oxidized

$$Al \Leftrightarrow Al^{3+} + 3e^-. \tag{1.3.47}$$

Overall, one has

$$3Cu^{2+} + 2Al \Leftrightarrow 2Al^{2+} + 3Cu. \tag{1.3.48}$$

From the Nernst equation, assuming equimolar concentrations of the electrolyte in both half cells,

$$V_{cell} = V_0^{Cu} - V_0^{Al} = 0.3419 - (-1.705) = 2.0424v. \tag{1.3.49}$$

The role of the salt bridge is to act as a conduit for the anions only.

1.3.8 Types of Electrodes and Cells

There are five main types of electrodes: (1) metal/metal ion (e.g., Cu, Zn, Cd), (2) non-metal, non-gas and ion (Eq. iodine), (3) metal and non-metal amalgam or hybrid and ion, (4) gas (H_2) and (5) metal/metal salt (e.g., Pb/$PbSO_4$, Ag/AgCl). By connecting two electrodes, an electrolytic cell may be formed. The cells can be broadly classified into two groups depending on whether or not the two electrolytes form a liquid junction. For example, when one electrode is H_2 gas, the other is a silver electrode in an aqueous solution of HCl. The Ag electrode functions as an Ag/AgCl electrode, and there is no liquid junction. Liquid junctions can be in the form of a salt bridge or a semi-permeable membrane, i.e., a membrane that is permeable to the ions but not to the electrolyte itself. Within each category, cells could be equiconcentration cells as the case of the example in the preceding subsection or cells with varying degrees of concentration of the either the electrolyte or the electrode (in the case of non-metallic and gas electrodes). Thus, the general formula for estimating the cell voltage takes the form:

$$V_{cell} = V^0_{half-cell1} - V^0_{half-cell2} + \frac{RT}{nF}\ln\left(\frac{Q_{half-cell1}}{Q_{half-cell2}}\right). \quad (1.3.50)$$

In Eq. (1.3.50), $V^0_{half-cell}$ is the standard half-cell potential and $Q_{half-cell}$ is the concentration ratio of each half cell.

1.3.9 Examples of Batteries and Fuel Cells

There are three classes of cells:

1. Primary cells

In primary cells, chemical reaction occurs irreversibly. These type cells cannot be recharged. Typical examples are the commonly known as dry cell and the battery. This is due to several reasons, including the design of the cell and the chemical properties of the reactants. Examples include standard alkali batteries or their less common precursors, the carbon/zinc battery.

(a) The dry cell (Leclanché cell)

This is the common dry cell available in most stores in various sizes and shapes that we use to power toys, torches, and a variety of electronic goods. A zinc casing (anode) with a carbon rod in the center (cathode) is used to enclose a paste of MnO_2, NH_4Cl, $ZnCl_2$, and graphite powder. The reactions involved are tabulated in Table 1.5. In this type of cell, the by-products can slowly build up and cause the overall cell potential to drop.

(b) The basic alkaline cell

In the alkaline cell, the product of the cathode reaction is used up again in the anode reaction, and consequently, the lack of any build up of the by-products means the voltage drops (as water is consumed) more slowly than the dry cell. Since the cell voltage is also about 1.5 V, this alkaline dry cell can be used in the same devices as the normal dry cell. Notice, the only real difference between these is the fact that the reaction in this case occurs in a basic medium.

Table 1.5 Reactions in a selection of cells

Electrode or type	Reaction or name	Voltage
Cell type	*Leclanché cell*	
Anode	$Zn\ (s) \rightarrow Zn^{2+} + 2\ e^-$	
Cathode	$2MnO_2 + 2NH_4^{+} + 2e^- \rightarrow Mn_2O_3 + 2NH_3 + H_2O$	
Overall	$Zn + 2MnO_2 + 2NH_4^{+} \rightarrow Mn_2O_3 + 2NH_3 + H_2O + Zn^{2+}$	~1.5 V
Cell type	*Alkaline cell*	
Anode	$Zn\ (s) + 2OH^- \rightarrow Zn(OH)_2(s) + 2\ e^-$	
Cathode	$2MnO_2 + 2H_2O + 2e^- \rightarrow 2MnO(OH)(s) + 2OH^-$	
Overall	$Zn + 2MnO_2 + 2H_2O \rightarrow 2MnO(OH)(s) + Zn(OH)_2(s)$	~1.5 V
Cell type	*Mercury cell*	
Anode	$2Zn\ (s) + 4OH^- \rightarrow 2ZnO(s) + 2H_2O + 2\ e^-$	
Cathode	$HgO + H_2O + 2e^- \rightarrow Hg + 2OH^-$	
Overall	$Zn + HgO \rightarrow ZnO(s) + Hg$	~1.35 V
Cell type	*NiCad batteries*	
Anode	$Cd\ (s) + 2OH^- \rightarrow Cd(OH)_2(s) + 2\ e^-$	
Cathode	$NiO(OH)(s) + H_2O + e^- \rightarrow Ni(OH)_2(s) + OH^-$	
Overall	$Cd(s) + 2NiO(OH)(s) \rightarrow 2Ni(OH)_2(s) + Cd(OH)_2(s)$	~1.35 V

(c) Ruben–Mallory (mercury) cell

This type of cell is used in application where the cell potential must remain very constant over the life of the battery (commonly computers and watches, etc.). The products and reactants are all pure substances (standard state). Since the voltages of these cells are not the same as the previous two, this type of battery is not interchangeable with the first two types.

2. Secondary cells

In these cells, the chemical reaction is reversible. The battery can be recharged. Examples include nickel–cadmium (NiCad) batteries and lead/acid batteries used in cars. These cells are also known as storage cells or rechargeable cells. In theory, alkaline batteries can also be recharged, the process, however, is dangerous and can lead to explosions if not done with the proper equipment. Similarly, lead/acid car batteries can also explode if they are charged too fast albeit for a different reason.

(a) The lead/acid battery

The lead/acid battery is typical example of a cell with no liquid junction. The electrolyte used is a dilute mixture of aqueous sulfuric acid. The overall reaction is given by

$$Pb(s) + PbO_2(s) + 4H^+(aq) + 2SO_4^{2-} \rightarrow 2PbSO_4(s) + 2H_2O(l) \quad (1.3.51)$$

The cell voltage is just over 2 V. Hence, six or twelve cells in series will make a battery of cells with a combined voltage of $6 \times 2\ \text{V} \sim 12\ \text{V}$ or $12 \times 2\ \text{V} \sim 24\ \text{V}$. These are used in most motor cars and light aircraft.

(b) Nickel hydroxide batteries

These cells use nickel hydroxide as the positive electrode. The electrolyte is aqueous potassium hydroxide (KOH). During the charging process, the electrode oxidizes to nickel oxyhydroxide.

$$Ni(OH)_2 + OH^- \rightarrow NiOOH + H_2O + e^-. \quad (1.3.52)$$

At the negative electrode, platinum, water is reduced according to

$$2H_2O + 2\,e^- \rightarrow H_2 + 2\,OH^-. \quad (1.3.53)$$

When discharging, the nickel oxyhydroxide electrode is reduced as

$$NiOOH + H_2O + e^- \rightarrow Ni(OH)_2 + OH^-, \quad (1.3.54)$$

while the hydrogen is oxidized as

$$H_2 + 2OH^- \rightarrow 2H_2O + 2e^-. \quad (1.3.55)$$

Combining and balancing the reactions give

$$2Ni(OH)_2 \leftrightarrow 2NiOOH + H_2. \quad (1.3.56)$$

These cells generate hydrogen gas and can be used with fuel cells which could consume the hydrogen.

(c) Nickel/cadmium batteries

These cells use cadmium hydroxide for the negative electrode. During charging,

$$Cd(OH)_2 + 2\,e^- \rightarrow Cd + 2\,OH^-. \quad (1.3.57)$$

When discharging, the nickel oxyhydroxide electrode is reduced as

$$NiOOH + H_2O + e^- \rightarrow Ni(OH)_2 + OH^-, \quad (1.3.58)$$

while the cadmium is oxidized as

$$Cd + 2\,OH^- \rightarrow Cd(OH)_2 + 2e^-. \quad (1.3.59)$$

An interesting feature of these cells is that none of the reactants or products are in solution, so the voltage will remain very constant during the discharge period. However, the cell voltage is lower than the standard dry cell and the cells are therefore not interchangeable.

(d) Nickel/metal hydride batteries

The NiMH cell is similar to the NiCad cell. The primary difference is that the cadmium hydroxide electrode is replaced with a metal alloy. As before, during the charging process, the electrode oxidizes to nickel oxyhydroxide.

$$Ni(OH)_2 + OH^- \rightarrow NiOOH + H_2O + e^-. \quad (1.3.60)$$

At the negative electrode, the metal alloy, M, combines with a hydrogen atom to become metal hydride as

$$M + H_2O + e^- \rightarrow MH + OH^-. \quad (1.3.61)$$

The overall reaction satisfies

$$Ni(OH)_2 + M \leftrightarrow NiOOH + MH. \quad (1.3.62)$$

Typical metals used in these cells are alloys of metals, such as V, Ti, Zr, Ni, Cr, Co, and Fe.

(e) Lithium-ion Batteries

Li-ion cells use a lithium metal oxide for the positive electrode. In these metallic compounds, lithium ions can be inserted in and extracted out of the host lattice without structural modification of the electrode. Common compounds include lithium cobalt oxide, $LiCoO_2$, or lithium manganese oxide, $LiMn_2O_4$. The negative electrode of the Li-ion cell is typically made of a graphite carbon material. Liquid electrolytes such as LiPF6 or LiBF4 are commonly used. For a lithium cobalt oxide electrode, the chemical reaction of a Li-ion cell may be written as

$$LiCoO_2 \rightarrow CoO_2 + Li^+ + e^- \quad (1.3.63)$$

while the reaction at the negative electrode is

$$C_6 + Li^+ + e^- \rightarrow LiC_6. \quad (1.3.64)$$

The overall reaction of a Li-ion cell is

$$LiCoO_2 + C_6 \leftrightarrow LiC_6 + CoO_2. \quad (1.3.65)$$

3. Fuel cells

Fuel cells are devices that produce electric power by direct conversion of a fuel's chemical energy. They are essentially energy converters. Fuel cells resemble batteries in many ways, but in contrast to them, they do not store the chemical energy. Fuel has to be continuously provided to the cell to maintain the power output.

Various designs for fuel cells have been proposed, the most popular pairs are the proton exchange membrane fuel cell (PEM) (usually with a polymer electrolyte membrane or Nafion membrane), operating at temperatures up to about

100 °C and the solid oxide fuel cell (SOFC), operating at temperatures of about 800 °C or higher. Although the underlying principles are always to extract electricity without combustion, each design presents different problems and advantages and has unique characteristics that make it more appropriate for different environments. A detailed exposition of these and other fuel cells may be found in O'Hayre et al. (2006).

In a fuel cell, the reactants flow through the cell, as they are either pumped or allowed to flow by gravity. Although the reactions involved are irreversible, by continuously circulating the fuel and the oxidant through of fuel cell, the cell is made to generate a voltage continuously. Examples of fuel cells which use two reactants, hydrogen or a hydrogen-rich fuel fed to the anode and oxygen fed to the cathode and a variety of catalysts, include

1. Polymer electrolyte membrane (PEM) fuel cells (polymer membrane is a catalyst)
2. Solid oxide (SO) fuel cells (solid oxide is a catalyst)
3. Molten carbonate fuel cells (coal or natural gas is the catalyst)
4. Direct methanol fuel cells (hydrogen-rich methanol is the fuel)
5. Regenerative fuel cells (auxiliary solar power used to catalyze water)
6. Phosphoric acid fuel cells (phosphoric acid is the catalyst)
7. Alkaline fuel cells (potassium hydroxide is the catalyst).

The modeling and control of PEM and SO fuel cells are discussed in Chap. 6.

References

Babu V (2008) Fundamentals of gas dynamics. CRC Press
Fermi E (1956) Thermodynamics. Dover
Hamann CH, Hamnett A, Vielstich W (2007) Electrochemistry, 2nd edn. John Wiley
Hill P, Peterson C (1992) Mechanics and thermodynamics of propulsion, 2nd edn. Addison Wesley Publications, Reading
Liepmann HW, Roshko A (2002) Elements of gas dynamics. Dover
Mattingly JD (2005) Elements of gas turbine propulsion. McGraw-Hill Education (India) Pvt. Limited, India
O'Hayre R, Cha SW, Colella W, Prinz FB (2006) Fuel cell fundamentals. Wiley, New York
White FM (2003) Fluid mechanics, 5th edn. McGraw Hill Higher Education

Chapter 2
Principles of Energy Conversion

2.1 Preliminary Concepts of Electric Generator and Electric Motor Action

The issue in the generation of electricity associated with an energy conversion device is the efficient transfer of mechanical power to electrical power followed by the transmission of it to the user, all with minimum loss. The entire process may be viewed as integrated system involving not only the mechanical generation of power in system, such as a turbine of some sort in the first instance, but also the subsequent conversion of mechanical power to electrical power in an electric generator followed by its transmission via appropriate transmission lines to the end user. To model and study the entire process as an integrated system, it is important to understand the dynamics, modeling, and control of each of the individual subsystems.

2.1.1 Introduction

Electric motors play a key role in electricity generation and are most extensively used as servomotors. Kenjo (1991) covers the basic concepts related to electric motors, while the more advanced modeling and control concepts are covered by Wach (2011). We shall turn our attention first to electric motors in this section.

2.1.2 Principles of Electric Generator and Motor Operation

The principles of electric generator and motor operations are briefly reviewed in the rest of this subsection. To do this, we introduce some basic concepts and definitions.

1. **Magnetic Flux**: A magnetic field may be represented by continuous lines called lines of flux. The direction of flux lines at any point in the field is the

R. Vepa, *Dynamic Modeling, Simulation and Control of Energy Generation*, Lecture Notes in Energy 20,
DOI: 10.1007/978-1-4471-5400-6_2,

direction of the force acting on a North Pole placed at that given point. Lines of flux emerge from a North Pole and enter a South Pole.

2. **Electromagnetic Induction**: When the total magnetic flux threading and enveloping a coil undergoes a change, an electromotive force is induced in the coil. It is directly proportional to the time rate of change of the flux. One volt is generated in a coil of one turn when the time rate of change of flux threading and enveloping the coil is 10^8 lines per second.
3. **Magnitude of the induced EMF**: Consider a U-shaped permanent magnet with a North and a South Pole facing each other and let $\Delta\Phi$ represents the magnetic flux passing from the North to the South Pole. A coil, in a plane perpendicular to the direction of the lines of flux that initially envelopes this flux, is moved at a constant rate so that the flux enveloped by the coil reduces to zero in time Δt seconds. The average EMF generated in the coil is given by,

$$E = (\Delta\Phi/\Delta t) \times 10^{-8}\ \text{V}. \tag{2.1.1}$$

When the conductor in the coil rotates through angle $\Delta\theta$, the change in lines of magnetic flux, $\Delta\Phi$, is the product of the 1cylindrical surface area covered by the flux lines and the flux density, where the latter is assumed a constant. Thus, if the coil is of radius R, and length L, the cylindrical surface area covered by the flux lines is, $R \times L \times \Delta\theta$ and the magnetic flux is given by,

$$\Delta\Phi = R \times L \times \Delta\theta \times B \tag{2.1.2}$$

where B is the flux density. Thus, in the limit as, $\Delta t \to 0$,

$$E = BLR\frac{\mathrm{d}\theta}{\mathrm{d}t} = BLR\omega = BLv \tag{2.1.3}$$

where v is the peripheral speed of the conductor relative to the magnetic field.

4. **The Direction of the generated EMF**: The direction of the generated EMF may be found by a simple rule that the flux lines extend around a current carrying conductor in a clockwise direction, when the current flows away from a person viewing the end of the conductor and vice versa. This is the celebrated corkscrew rule. When a corkscrew is turned clockwise, in the direction of the flux, it advances away, in the direction of the current.

Thus, when a conductor travels and cuts through a bunch of flux lines, the direction of the current depends on whether it forces the flux around it to be clockwise or anticlockwise.

5. **Force on a Conductor in a Magnetic Field**: When a conductor is placed at right angles to the flux lines in a magnetic field, without any current in it, there is no electromagnetic force on this conductor. If, now, a current is forced to flow through the conductor, lines of magnetic force would encircle the conductor depending on the direction of the current flow. The conductor will have force acting on it proportional to the magnitude of the flux field and the strength

of the current in the conductor. Moreover, this results in a motion governed by Newton's second law. A conductor L meters long and carry a current of I Amperes perpendicular to a magnetic flux of B in units of 10^8 lines per square meter (1 Wb/m^2) is acted upon by a force of

$$F = I \cdot (L \times B) \text{ N}. \tag{2.1.4}$$

To determine the direction of the force, we observe that when the direction of the flux lines, due to the current flowing in the conductor and encircling it, is in the same direction as the flux lines emanating from the North Pole and ending in the South Pole, they act to repel each other. Concurrently, on the other side of the conductor, the lines of flux due to the current in the conductor and those emanating at the North Pole are in opposite direction and tend to attract each other. The direction of both these forces is the same, and they tend to reinforce each other. The above law is associated with the name of *Lorentz* and is the basis for the *motor action.*

If the conductor is carrying no current but is mechanically moved downwards with no effort, the field flux lines will tend to be pushed around the conductor in either a clockwise or counterclockwise direction and the result would be the generation of a voltage across the conductor. This can be seen when the circuit connecting the conductor is closed and current flows through it. The result would also be a force that would oppose the initial mechanical motion of the conductor. Thus, a force is required to sustain the mechanical motion through the external flux lines. This is known as the *generator action.*

2.2 Electric Motors: DC Motors

Electric motors play a key role in electricity generation and are most extensively used as servomotors. We shall turn our attention first to direct current (DC) electric motors in this section. A typical DC motor consists of a stator, which is an active field system consisting of an even number of poles, arranged alternately North and South, so the faces of the Poles lie symmetrically on a circle. The stator also supports on bearing an armature or rotor whose axis of rotation coincides with the axis of the polar circle. The armature is constructed from an assemblage of laminated disks of alloy steel with slots on the surface parallel to the axis of rotation. Conductors forming coils are housed in the slots to form a complete winding. The EMF induced in a conductor passing under a North Pole is in the opposite direction from that induced in a conductor situated under a South Pole. The whole purpose of the armature winding is to connect all the conductors so that the various induced EMF's shall add up to a combined whole. No two conductors must be connected together in a way which would cause their induced EMF‘s to oppose each other as such connection would lead to an overall reduction in the induced EMF's. The ends of the windings are connected to a current collecting device to transfer the current from a stationary supply to the rotating windings. Slip rings, usually segmented with an even number of segments, are used to connect the rotating windings via the connection made by stationary brushes rubbing on the rotating slip rings. The entire assembly is known as the commutator.

To understand the action of a DC motor, consider the magnetic field of a machine in which there is no current in the armature conductors. The lines of force emanating from the North Poles terminate on the opposite side, at a South Pole, and are distributed in an axially symmetric fashion. If the armature conductors are made to carry current now, each conductor has been associated with a magnetic field, which when superposed on the main field, causes the distribution of magnetic lines to be distorted and are no longer axially symmetric. As these lines of force have the tendency to shorten themselves by seeking a path of minimum distance, they may be regarded as being tension. The net result is that each conductor would experience a force which accumulates to generate a torque on the armature. Consequently, the armature rotates under the influence of the torque.

In the windings of the motor illustrated in Fig. 2.1, the current distribution is indicated as follows: when the current flowing in the conductors to the left of the symmetric axis AA' is in the direction away from the reader, ($\otimes$), then the current in the conductors to the right flows in the opposite direction, toward the reader ($\odot$). The brushes and commutator assembly are always responsible for distributing the current in this way. As the conductors in the right half are under the influence of the South Pole and those in the left half are under the influence of the North Pole, all conductors are acted upon by a counterclockwise force according to the left-hand rule, thus producing a net torque on the rotor shaft.

The forces on a single loop in the armature are shown in Fig. 2.2. A constant magnetic flux is generated by employing either a field-controlled electromagnet or a permanent magnet. A current I_a is assumed to be passing through the rectangular coil as shown. The length of the coil is assumed to be L. The net force on each side of the coil has a magnitude F. However, the two forces are in opposite directions. The magnitude of the force is given by the Lorentz law,

$$F = I_a \cdot (L \times B). \tag{2.2.1}$$

The resulting torque generated by a single coil is then given by,

$$T = DF = BAI_a, \tag{2.2.2}$$

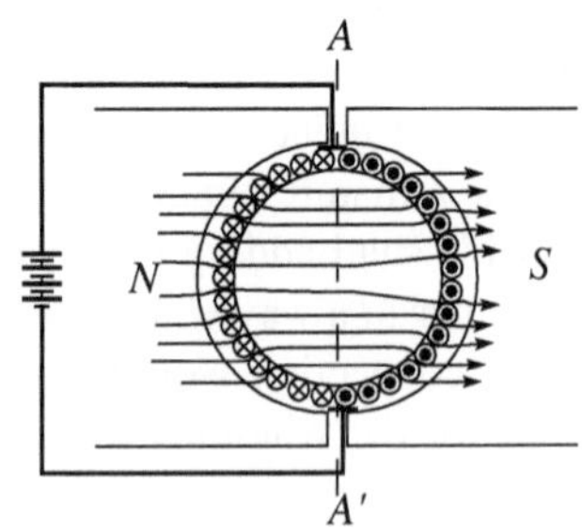

Fig. 2.1 Electric current distribution in the armature coils in a DC motor

Fig. 2.2 Torque generation in a DC motor

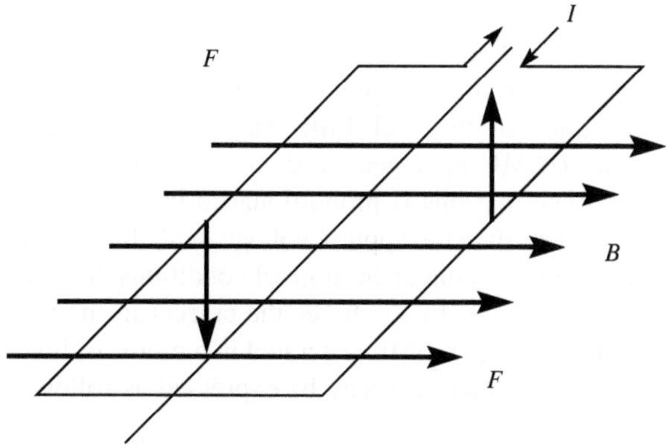

where D is the diameter of the coil and A is the area of cross-section of the coil. The magnetic flux density may be related to the magnetic intensity due to the current flow as,

$$B = \mu_0 \mu_r H, \tag{2.2.3}$$

where μ_0 is the permeability of free space and μ_r, the relative permeability. Alternately, the product of the magnetic flux density and the area is the total external flux. For a large number of such rectangular coils, N, the number of coils is directly proportional to the circumference or diameter of the armature. Hence, for a general motor armature, the torque generated is,

$$T = K_t DBAI_a = K_t D\Phi I_a. \tag{2.2.4}$$

Thus, we may define a *motor area coefficient*, a_m, as,

$$a_m = A/D^2 \tag{2.2.5}$$

and write the torque as,

$$T = K_t a_m D^3 B I_a. \tag{2.2.6}$$

The torque generated is directly proportional to the cube of the linear dimension, and hence, it must be said that size matters. Thus, small motor with a high-torque capability requires permanent magnets with high magnetic field intensity. The discovery of magnetic alloy materials such as Alnico and Ferrite, rare-earth magnetic materials like cobalt and samarium as well as the development of commercial products such as Neodymium–Iron–Boron (NdFeB) has greatly aided the development of small high-torque permanent magnet motors.

When the armature rotates the conductors in the slots cut the lines of force of the magnetic field in which they revolve, so that an EMF is induced in the armature winding as in a generator. The induced EMF acts in opposition to the current in the machine and, therefore, to the applied voltage. Thus, it is customary to refer to this voltage as the back-EMF. The existence of the back-EMF may be deduced by Lenz's law, which states that the direction of an induced EMF is such as to oppose the change causing it, which is, of course the applied voltage. The magnitude of the back-EMF may be calculated by using the formula for the induced EMF in a generator, but it is important, in the case of a motor, to appreciate that this is proportional to the product of the flux and the speed. It is always less than the applied voltage, and although the difference is small when the machine is running under normal conditions, it is the difference between these two quantities that actually drives the current in the armature winding.

Since the back-EMF generated in a motor is directly proportional to the angular speed of the motor, it may be expressed as follows:

$$E_b = K_b \omega. \tag{2.2.7}$$

Given that the total resistance of the armature circuit is R_a and that the armature is driven by a constant external voltage, V_a, the current in the armature may be expressed as follows;

$$I_a = \frac{V_a - E_b}{R_a} = \frac{V_a - K_b \omega}{R_a} \tag{2.1.12}$$

and the torque acting on the armature is given as follows:

$$T = K_t a_m D^3 B \left(\frac{V_a - K_b \omega}{R_a} \right). \tag{2.2.8}$$

Theoretically, in view of the duality of electromagnetic induction and the Lorentz law, it may shown that, K_b, is also given by

$$K_b = K_t a_m D^3 B.$$

Thus, we may express K_b in practice as,

$$K_b = k_b K_t a_m D^3 B \tag{2.2.9}$$

and the torque generated by the armature is:

$$T = K_t a_m D^3 B \left(\frac{V_a - K_b \omega}{R_a} \right) = \frac{K_t a_m D^3 B}{R_a} \left(V_a - k_b K_t a_m D^3 B \omega \right). \tag{2.2.10}$$

Thus, the torque-speed characteristics are approximately linear and are illustrated in Fig. 2.3.

2.2.1 Loading a DC Motor

The principal loads on a motor are friction, inertia and other constant or varying torque loads. Although spring loaded motors are used as electro-dynamic vibration

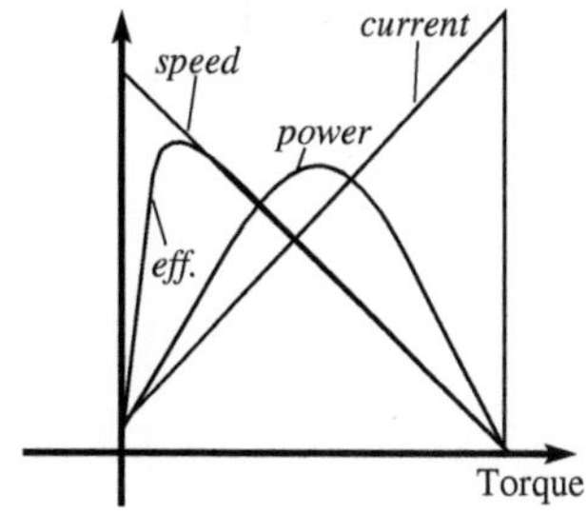

Fig. 2.3 Typical DC motor characteristics

exciters, viscous-friction and rotary inertia loads are the most common loads that one encounters in practice. Hence, considering a motor with a rotary inertia and viscous-friction load, the conditions for dynamic equilibrium are as follows:

$$T = \frac{K_t a_m D^3 B}{R_a}\left(V_a - k_b K_t a_m D^3 B\omega\right) = J\frac{d\omega}{dt} + B_f\omega. \tag{2.2.11}$$

Assuming the motor is in steady state and that the viscous friction is negligible in comparison with the back-EMF, the motor speed satisfies the relation,

$$\omega = \frac{V_a}{k_b K_t a_m D^3 B}. \tag{2.2.12}$$

Thus, when the voltage across the armature is constant, one can expect the speed of the motor to also be constant. It also follows that the speed of the motor may be controlled by appropriately controlling the voltage across the armature.

2.2.2 Types of DC Motors

DC motors may be classified in a number of different ways. They may be classified according to

1. number of poles: Bipolar, Multipolar and Homopolar
2. the method by which they drive the application device and the speed of the application: belt-driven application, Reduction-Gear driven application, directly coupled to a low-speed application or directly coupled to a high-speed application.
3. special construction features,
4. the method of field and armature excitation: Shunt and Series Motors
5. the nature of the application and
6. the method of speed control.

Broadly, however, the different types of DC motors are illustrated in Fig. 2.4.

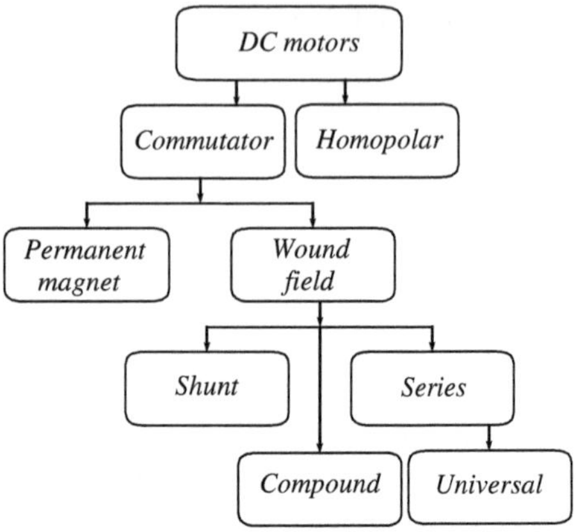

Fig. 2.4 Types of DC motors

2.3 AC Motors

Alternating-current (AC) motors may be divided broadly into two types, namely (1) synchronous motors that run at a constant speed dependent only on the supply frequency (2) asynchronous motors that run at speeds not only dependent on the supply frequency but also the load and in some cases, the position of the brushes. The latter may be broadly further sub-divided into (2a) induction motors (IM) and (2b) commutator motors. From the application point of view, the commonly used principal AC motors are the synchronous and the IMs. Commutator motors are usually also of two types: the synchronous IM, a DC exciter-driven IM, which is used extensively in fans, blowers, generators, air compressors, machinery with line shafting in industrial works such as cement mills, rolling mills, flour mills, paper mills, textile mills, and rubber works; the repulsion motor that consists of a field comprising a distributed winding, housed in slots in a smooth cored stator, an armature carrying a distributed winding connected to a commutator, a mains supply to the stator winding and a set of externally short-circuited brushes set an angle to the direction of the main flux, which is used in cranes and hoists, in printing machines, pressure pumps, and laundromats.

2.3.1 Synchronous Motors

A synchronous motor has the same relationship to an AC generator or alternator as a DC motor has to a DC generator. In other words, if an AC generator is supplied with alternating current, it is capable of rotating as a motor and doing mechanical work. However, there is one important difference between the DC motor and the synchronous motor, apart from the fact that the roles of the stator and rotor are interchanged. This is that while in the former case the applied DC voltage supplies both the field and the armature windings, and in the latter case, two independent voltage sources, one DC and one AC, are required to excite the field or rotor windings and armature or stator windings, respectively. On one hand, this feature is a disadvantage of a synchronous motor. But the advantages in using these fixed-speed motors in certain applications far outweigh this disadvantage. One method of providing the DC supply in large machines is to use an exciter driven directly by the shaft of the synchronous motor and coupled to it at one end, remote from the main motor. Thus, the exciter is connected through a commutator system to the field or rotor windings, while the main AC supply is connected to the armature or stator winding.

Now, in order to maintain the driving torque on the rotor, the latter must rotate at a speed such that its poles change positions at the same rate that the current in the stator alternates with the supply frequency. Thus, a synchronous motor rotates at a speed which is determined by the number of pole pairs in the rotor and the frequency of the supply. This speed is called the synchronous speed and the motor runs only and exclusively at this speed. Hence, the name synchronous motor for this type of motor.

It is natural to expect that the action of the motor is not self-starting and certain distinctive methods have to be adopted to raise the speed from zero to synchronism. The synchronous speed is given by,

$$\omega_s = 2\pi f_s\left(\frac{2}{P}\right) \tag{2.3.1}$$

where f_s is the supply frequency and P is the number of stator pole pairs.

2.3.2 Types of Synchronous Motors

Broadly, the different types of synchronous motors are illustrated in Fig. 2.5.

2.3.3 Operation of a Synchronous Motor

Before discussing the operation of such motors, it is necessary to consider the production of a rotating field by means of polyphase alternating currents. Let the system shown in Fig. 2.6a represents an arrangement of stator windings in which the pair of horizontal poles HH is excited by a current which passes through values, as shown by the sinusoid h, passing through the origin, in Fig. 2.6b, while the poles VV are excited by the current which passes through values represented by the sinusoid v. At the time instant $t = 0$, labeled 1 in Fig. 2.6b, current v is zero, while the current h is a negative maximum. Hence, the resultant field is shown by vector 1 in Fig. 2.6c. At instant 2, h is zero and v is a positive maximum, so that the field is represented by the vector 2, in Fig. 2.6c. If this process is followed, it

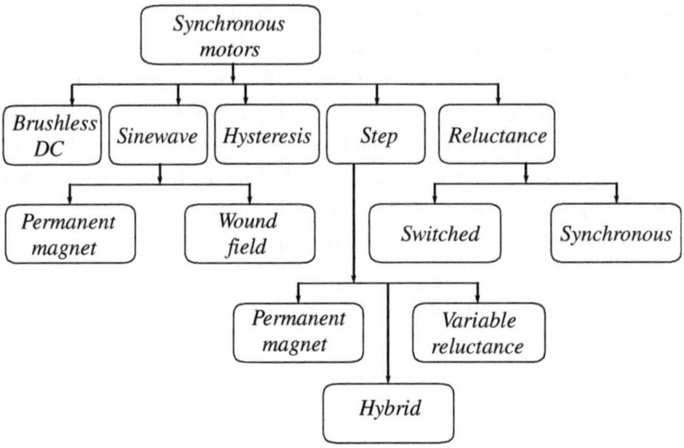

Fig. 2.5 Types of synchronous motors

will be found that the magnetic field rotates in synchronism with the alternations of the current which excites the field and as a result the rotor rotates in manner illustrated in Fig. 2.6d.

The rotating field so established is analogous to that which would be produced by a permanent magnet rotating uniformly about its midpoint, or by a revolving system of poles excited by DC as in the case of the rotor of a synchronous motor. In the case of a three-phase synchronous motor, the stator has a set of three pole pairs, with each pole located at the vertex of a regular hexagon. The stator of such a motor produces a rotating field just as in the case of the two-phase system shown in Fig. 2.6. The synchronous speed of rotation is the same for both the stator and the rotor provided the supply frequency to the number of pole pairs are the same. It is clear that if the rotor in a such a motor is brought up to the synchronous speed of the rotating field, it will remain locked to it and rotate synchronously with it.

The operation of a synchronous motor may now be followed by referring to Fig. 2.7 which represents a three-phase two-pole rotor synchronous machine. The rotor has two poles, and the stator has two poles per phase. In practice, it is usual to have salient or projecting poles on the rotor, but the armature winding is housed in slots in the concave periphery of the stator. Considering the instant at which the diagram is drawn, the rotor South Pole is attracted to the stator North Pole. It is essential that the rotor moves through one-sixth of a revolution by the time the next stator pole reaches its maximum North polarity. If the stator pole A reaches its maximum North polarity before the pole B on the left, the rotor will attempt to rotate so the rotor South Pole aligns with the stator pole, A. Assuming that the current stator pole with maximum North polarity is supplied by the second of the three phases, interchanging the connections to the first and third phase supplying the stator poles A and B, respectively, would result in reversal in the direction of rotation of the rotor. A further point about the windings of the stator coils is that

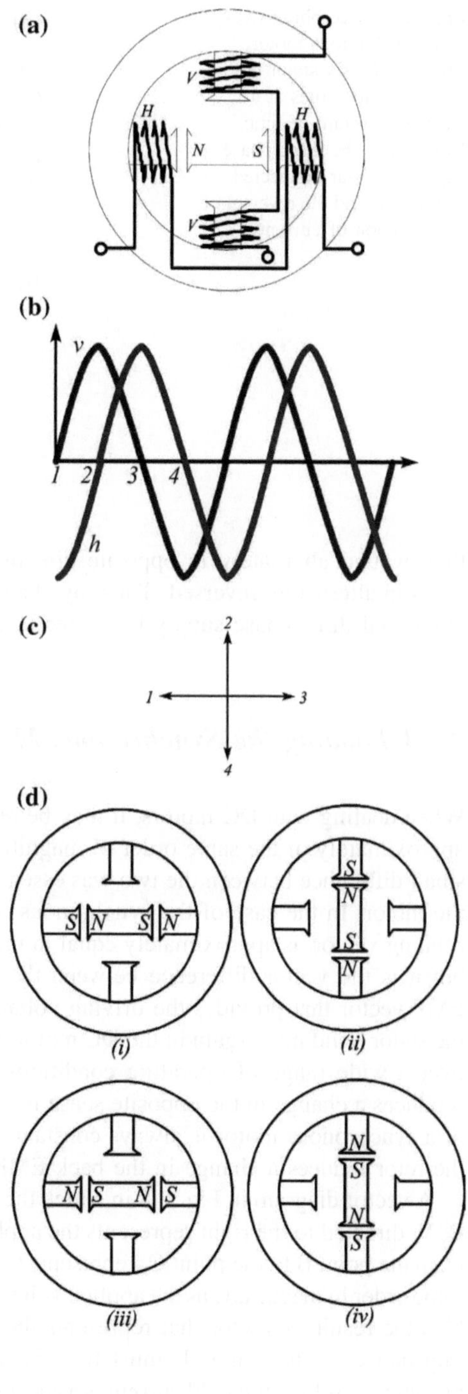

Fig. 2.6 **a** Arrangement of field windings for producing a rotating field by polyphase alternating currents. **b** Curves showing phase relationships of currents in field windings. **c** Vector diagram of a rotating field, indicating the directions of the resultant field at successive instants. **d** Sequence of the rotor positions as the currents in the stator winding complete a cycle

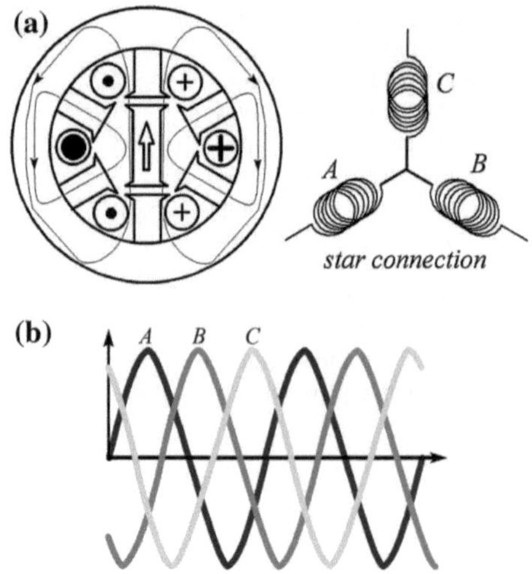

Fig. 2.7 Three-phase two-pole synchronous motor, showing the six stator poles and two rotor poles at a particular instant of time. Notice also the three-phase supply, the star-connected stator coils and the reversal of the direction of current

they wound alternately in opposite directions to ensure that the polarity of the poles in alternately reversed. Thus, by changing over any two of the stator's star-connected three-phase supply leads, the direction of rotation may be reversed.

2.3.4 Loading the Synchronous Motor

When dealing with DC motors, it may be recalled that the back-EMF was always approximately of the same order of magnitude as the applied voltage and that the small difference between the two was essentially the voltage driving the current in the motor. In the case of the synchronous motor, the back-EMF, which is now a rotating vector, is approximately equal in magnitude to the applied voltage vector, but it is the vector difference between the applied voltage vector and the back-EMF vector that provides the driving voltage, which in turn drives the current in the stator windings. Again in the DC motor, the back-EMF is practically a constant over a wide range of operating conditions so that a change in the field current produces a change in the opposite sense in the speed. On the other hand, the speed of a synchronous motor is always constant so that a change in the field current in the rotor induces a change in the back-EMF in the stator.

A vector diagram in Fig. 2.8 in which the reference vector from point A to point B, V, directed to the right represents the applied voltage and the directed vector, E, from the point B to the point C, represents the back-EMF. It is approximately of the same order in magnitude as the applied voltage vector. The directed vector from A to C is the resultant vector that represents the voltage driving a current through the impedance of the stator. It must be said that the magnitude of the vector E is unchanged as long the field current is a constant. Moreover, the vertical height of the triangle ABC is approximately proportional to the input power to the motor. In a

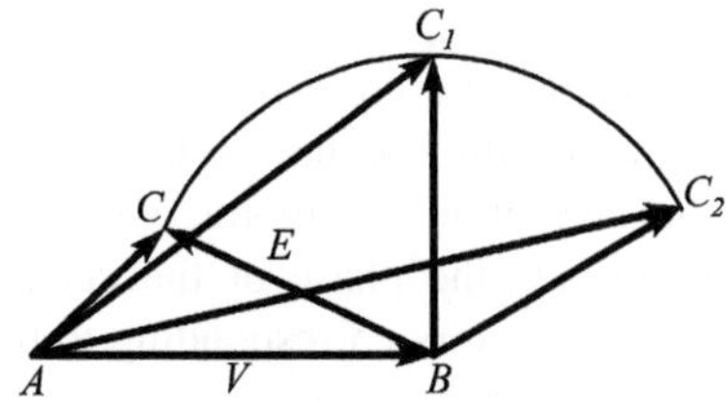

Fig. 2.8 Vector diagram showing the effect of a load on a synchronous motor with a constant field current

no-load situation, the input power represents the motor losses. The addition of mechanical load tends to increase this height without altering the magnitude of the back-EMF vector. Thus, the tip of vector E swings round in a clockwise direction, to a new position C_1, about the point B. The increased height of the triangle ABC_1 reflects the increased power required to sustain the additional mechanical load.

However, when the load on the synchronous motor is excessive, the point C continues to swing around but without any increase in the height of the triangle ABC. A further increase in the load begins to reduce the height of the triangle requiring a decrease in the input power. Such a situation is, however, untenable so when this condition is approached the rotor will fall out of synchronism and the machine comes to rest. The torque that causes the rotor to fall out of synchronism is called the pull-out torque. If instead of maintaining a constant field current, the external load on the motor is held a constant and the excitation is varied by increasing the field current, the point C moves along a horizontal line as shown in Fig. 2.9a.

Initially, the triangle ABC is an acute angled triangle and the voltage driving the current in the stator is represented by the side AC. As the point C moves along a horizontal line, the distance AC becomes a minimum before increasing again. When the distance AC is a minimum, the angle CAB is a right angle.

If the graph is plotted showing the variation of stator current as the excitation is varied, a family of 'V' shaped curves are obtained as illustrated in Fig. 2.10; the

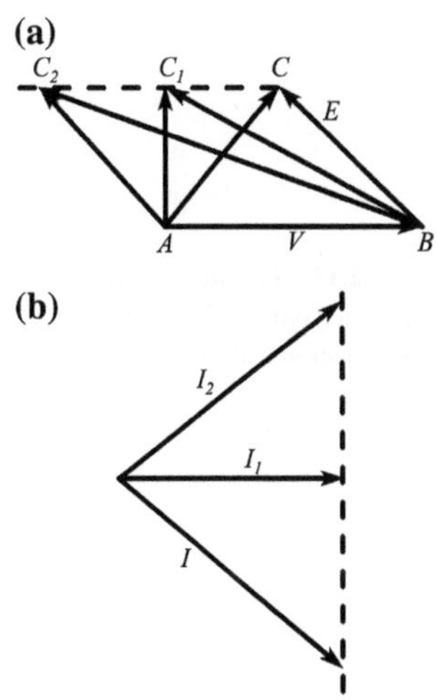

Fig. 2.9 Vector diagram showing the effect of an increasing field current on a synchronous motor with a constant load. **a** Vector diagram of voltages. **b** Current vectors corresponding to the three cases illustrated in Fig. 2.9a

greater the load on the machine, the higher is the current in the stator, and consequently, the corresponding curve is higher in the family of curves. Referring again to Fig. 2.9, the lengths of the vector AC is proportional to the magnitude of the current in the motor, while the direction, the angle, 90°—*the angle CAB*, represents the phase of the current vector relative to the applied voltage. The current vectors corresponding to the three cases illustrated in Fig. 2.9a are shown in Fig. 2.9b.

Thus, initially, when the field excitation is low, the current vector lags the applied voltage, and when the excitation is relatively large, the current vector leads the applied voltage. When the angle CAB is equal to 90°, the current vector is in phase with the applied voltage. This is an important property as the synchronous motor, when used as a servoactuator in control applications, provides for lead compensation as well. In terms of an equivalent circuit, it is equivalent to connecting a capacitor in series with a inductive circuit with a low resistance.

A typical torque–speed characteristic of a synchronous motor is illustrated in Fig. 2.11. An important feature is the non-linearity when the rotor lock into synchronism with the magnetic field rotating at the synchronous speed.

2.3.5 Induction Motors

We are now in a position to discuss the construction principles and operation of IMs. They may be designed to operate as single-phase, two-phase, or as three-phase

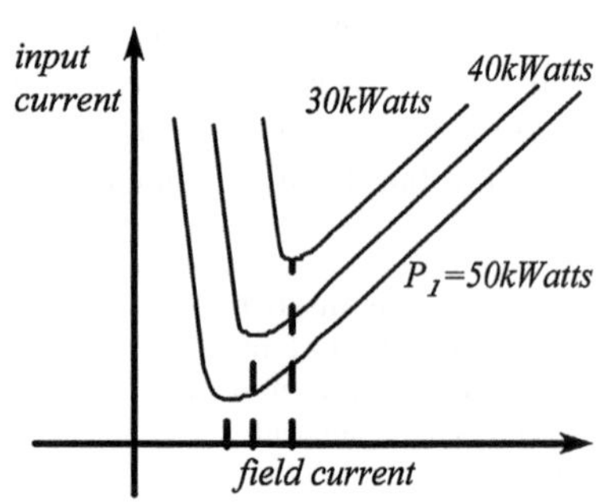

Fig. 2.10 Characteristic curves showing the variation of stator current with excitation for different values of load

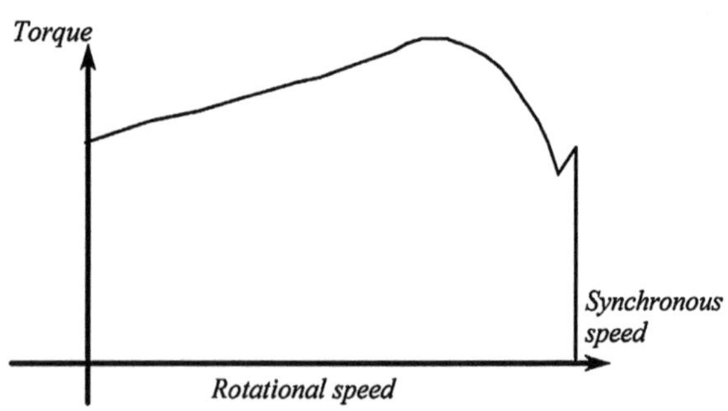

Fig. 2.11 Torque–speed characteristics of a typical synchronous motor

machines; the latter machine will be considered as a typical example. While single-phase machines may be *capacitor* or *shaded pole* types, polyphase machines are *wound rotor* or *squirrel-cage* types.

The stator of an IM is constructed as an assemblage of slotted stampings that house the stator windings. The ends of the windings are connected to a terminal box attached externally to the stator casing. The rotor is supported by a shaft that is fitted to a pair of bearings on either end. The bearings are supported by bearing housings attached to end shields that are bolted on both sides of the stator casing.

There are two types of rotor windings used in the construction of IMs. These are known as the squirrel-cage type and the phase wound or wound-type windings. The wound-type rotor is provided with a distributed winding, usually consisting of three windings one corresponding to each of three phases, the ends of which are typically connected to three insulated slip rings mounted on the rotor shaft. The object of this type of winding is to be able to connect to three-phase star-connected resistance at the time starting. The starter resistance is eventually disconnected, and the slip rings are short-circuited during the normal operation of the IM. The squirrel-cage type of rotor winding is usually confined to small IMs due to the large current this type of motor requires during starting. An associated feature of this type of motor is the low starting torque provided by it. The main advantage of this type rotor winding is that motor is simple construct, mechanically robust and efficient in operation.

The action of the squirrel-cage IM, with a three-phase stator winding connected to a three-phase supply, also relies on the rotating field set up by the supply voltage. The rotating field induces an EMF in the conductors housed in the rotor. However, since the rotor winding is a closed circuit in normal operation, a relatively large current flows in it, the direction of which by Lenz's law, is such as to oppose the change causing it. As the change causing the current is the rotation of the stator field, rotor attempts to run in the same direction as the field thereby reducing the relative angular velocity of the stator field with respect to the rotor. It is essential that in order to maintain the rotation, a torque must be produced to overcome the load and cause the rotation, and this torque is as a result of the currents flowing in the rotor conductors that are situated in and at right angles to the magnetic field. Now, if the rotor succeeds in completely catching up with the rotating field as in the case of an synchronous motor, there would be no relative motion, no induced EMF, no current in the rotor windings and therefore no torque to maintain the rotation. The rotor of an induction never really runs as fast as the rotating field; there always a small difference between the rotor speed and the synchronous speed of the rotating field. This difference is known as the slip speed and is usually expressed as a percentage of the synchronous speed of the rotating field. Thus, if for example, the synchronous speed of the rotating field 3,000 r. p. m. and the actual rotor speed is 2,940 r. p. m., the difference is 60 r. p. m. which is 2 % of the synchronous speed. The induced EMF in the conductors housed in the rotors is directly proportional to the slip. It follows that the torque required to maintain the rotation is also proportional to slip. Under no-load conditions, when the rotor only needs to overcome the friction torque to maintain a constant speed of rotation, the

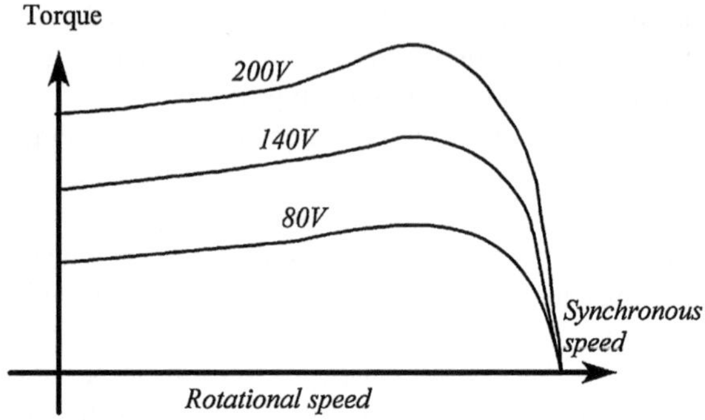

Fig. 2.12 Torque-speed characteristics of a typical induction motor

slip is relatively small. When the load on the motor is increased, the slip also increases proportionately. The result is a reduced speed of operation. Thus, the torque–speed characteristics of an IM is, therefore, similar to that of DC shunt motor. If the torque is increased beyond a certain value, called the pull-out torque, the motor ceases to run. This pull-out torque is approximated about 2–3 times the rated full-load torque. The torque-speed characteristics of a typical IM are illustrated in Fig. 2.12.

The IM depends for its operation, as we have seen, on the production of a rotating field by the multi-phase stator winding which is fed by a multi-phase supply. A simple single-phase motor is not capable of producing a rotating phase and is not self-starting, although it will continue to run if it is started and brought up to speed by an auxiliary device. This usually takes the form of an auxiliary stator winding spaced at 90° (electrical degrees), i.e., half a pole pitch from the main winding and connected in series with an impedance, to the main supply. This impedance is chosen to produce a phase difference as close as possible to 90° between the currents in the main and auxiliary windings, so the motor starts virtually as a two-phase machine. To reverse the direction of rotation of such a motor, it is necessary to simply reverse the connections to the auxiliary windings. A centrifugally operated switch is usually incorporated to cutout the auxiliary winding once the rotor attains a fair speed and is capable of running as a single-phase motor. The impedance is usually a high-capacitance condenser so as to reduce the lag induced by the stator impedance by almost 90°. Such a technique not only improves the power factor but also does de-stabilize the system when used as a servomotor in control applications. Also, since a condenser involves almost negligible power losses, it may be left permanently wired in the circuit connecting to the stator windings of the motor.

2.3.6 AC Servomotors

An AC servomotor is essentially a two-phase IM of either the squirrel-cage or drag cup type, except for certain special design features. A two-phase IM consists of two stator windings capable of generating magnetic fields that are physically oriented 90° apart in space and are excited by alternating current generating voltages that are out of phase by 90°. Thus, the resulting magnetic field is equivalent to magnetic field of constant magnitude, rotating at the corresponding synchronous speed. While the direction of rotation of the magnetic field depends upon the phase relationship between the two driving voltages, the voltages induces a current in the short-circuited rotor. The current, in turn, interacts with the magnetic field to produce a torque on the rotor, in the direction of the rotation of the field.

The AC servomotor differs from a conventional IM in two ways. The rotor of a servomotor is of low inertia and is built with a high resistance so that the ratio of the rotor reactance to the rotor resistance is as small as possible. Generally, in the design of IMs, this ratio is made as large as possible to operate with maximum torque. This operating point is usually around 5 % slip. However, if the servomotor is operated in this region and is part of a feedback controlled system, the closed loop system is generally not very stable. Hence, in the case of servomotors, the ratio of the rotor reactance to the rotor resistance is made as small as possible. The torque-speed characteristics of a typical AC servomotor and a typical high-torque IM are compared in Fig. 2.13.

The other important difference between an AC servomotor and an IM stems from the fact that in a control application, there is an inherent need to be able to control the speed or position of the rotor. Thus, in an AC servomotor, one of the two phases is employed as a *reference phase,* while the second is amplitude modulated by the control signal and is termed the *control phase*. In addition, the two phases are mounted at right angles to each other and the signal in the control phase is out of phase with respect to the signal in the reference phase by 90° (lag or lead).

For all practical purposes, the speed-torque characteristic of servomotor may be approximated by the linear relation,

$$T = K_V(V - K_b\omega). \tag{2.3.2}$$

When $\omega = 0$, (the motor rotor is held stationary), the corresponding torque is the motor stall torque; i.e.,

$$T_{\text{stall}} = K_V V. \tag{2.3.3}$$

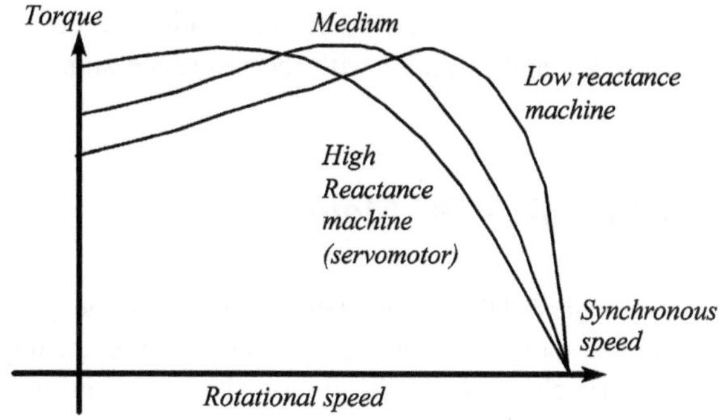

Fig. 2.13 Torque-speed characteristics of a typical induction motor and a typical AC servomotor

On the other hand, the no-load speed is defined by,

$$0 = K_V(V - K_b\omega_{\text{no-load}}) \tag{2.3.4}$$

and

$$\omega_{\text{no-load}} = \frac{V}{K_b}. \tag{2.3.5}$$

Hence, the torque may be expressed as follows:

$$T = T_{\text{stall}}\left(1 - \frac{\omega}{\omega_{\text{no-load}}}\right). \tag{2.3.6}$$

In practice, however, the torque-speed slope in the low-speed region is only fraction of the same slope at rated voltage, the relation is expressed as follows:

$$T = T_{\text{stall}}\left(1 - \frac{\alpha\omega}{\omega_{\text{no-load}}}\right) \tag{2.3.7}$$

where α is the ratio of the torque-speed slope in the low-speed region to the torque-speed slope at rated voltage.

2.3.7 AC Tachometer

The principle of operation of an AC tachometer, which is essentially a two-phase AC drag cup type generator. The rotor is generally a light weight highly conducting, short-circuited secondary conductor. It is usually a thin aluminum cup that rotates in an air gap between the stator poles. Like the AC servomotor, one of the two phases of the AC tachometer is employed as a *reference phase,* while the second is termed as the *quadrature phase.* In addition, the two phases are mounted at right angles to each other.

The reference coil is excited by a sinusoidal voltage that generates a reference flux with a phase angle that leads to the reference phase by about 90°. The reference flux induces a current proportional to itself in the rotor. In turn, the rotor current induces an emf in the quadrature coil that lags the current in phase by about 90°. Thus, the emf induced in the quadrature coil is directly proportional to the rotor speed and in phase with the reference signal.

2.4 Brushless DC Motors

The brushless DC motor operates on the same physical principles as a conventional DC motor. The only difference is the reversal of the roles of the rotating and stationary elements of the motor. The conventional DC motor has a stationary magnetic field generated either by a permanent magnet or by a field wound electromagnet, and a rotating armature. A brushless DC motor has a rotating permanent magnet assembly, while the armature windings are stationary. The term "brushless" is employed to indicate that the rotor is electronically commutated by sensing the rotor position. Commutation of the electronic current in the stationary armature windings by sequentially switching the current on an off in the windings, by solid-state devices, so as to set up a rotating field. The switching sequence depends on the rotor's position which is sensed electronically. The brushless DC motor is therefore said to be an electronically commutated DC motor.

Although a variety of devices such as inductive and capacitive transducers, magneto resistors and optical encoders are potentially available for sensing the angular position of the rotor, Hall effect sensors have emerged as the primary systems for this task. Apart from the fact that these sensors develop a polarized voltage depending on the control current and the magnetic field passing through it, they are highly sensitive and reliable microscale semiconductor devices that are eminently suitable for measuring the angular position of the rotating magnetic field of the permanent magnet rotor, relative to a stationary body.

A great advantage of brushless DC motors is the fact that they dispense with the need for brushes and commutator contacts or slip rings. Thus, the problems associated with sliding contacts and arcing are completely eliminated. The current in the armature circuit may be expressed in the usual manner as in a DC motor with an additional term to include the inductive impedance of the armature. Thus,

$$L_a \frac{\mathrm{d}}{\mathrm{d}t} I_a + R_a I_a = V_a - E_b \equiv V_a - K_b \omega. \tag{2.4.1}$$

The torque acting on the rotor satisfies the relation,

$$T = K_T I_a - T_{\mathrm{load}} \tag{2.4.2}$$

where T_{load} is the load torque acting on the rotor. Assuming the load torque is entirely due to the rotating inertial load on the rotor and the friction torque, the load torque, T_{load} is given by,

$$T_{\mathrm{load}} = J \frac{\mathrm{d}\omega}{\mathrm{d}t} + B_f \omega. \tag{2.4.3}$$

Thus,

$$I_a = \frac{T}{K_T} + \frac{J}{K_T} \frac{\mathrm{d}\omega}{\mathrm{d}t} + \frac{B_f}{K_T} \omega \tag{2.4.4}$$

and

$$\frac{L_a}{R_a} \frac{\mathrm{d}}{\mathrm{d}t} I_a + I_a = \frac{V_a - E_b}{R_a} = \frac{V_a - K_b \omega}{R_a}. \tag{2.4.5}$$

Hence, it follows that the speed–torque relationship is governed by,

$$\frac{L_a}{R_a}\frac{\mathrm{d}}{\mathrm{d}t}\left(\frac{J}{K_T}\frac{\mathrm{d}\omega}{\mathrm{d}t}+\frac{B_f}{K_T}\omega\right)+\frac{J}{K_T}\frac{\mathrm{d}\omega}{\mathrm{d}t}+\left(\frac{B_f}{K_T}+\frac{K_b}{R_a}\right)\omega=\frac{V_a}{R_a}-\frac{T}{K_T}-\frac{L_a}{R_a}\frac{\mathrm{d}}{\mathrm{d}t}\frac{T}{K_T}. \tag{2.4.6}$$

Assuming steady operating conditions, a negligible friction torque and assuming that the rotor is in torsional equilibrium ($T = 0$),

$$\omega = V_a/K_b \tag{2.4.7}$$

and the motor can be seen to operate at a constant speed.

Generally, brushless DC motors are current controlled and so the dynamics of the motor is usually expressed in terms of the armature current. The primary control input to the motor stator is the armature current which is measured and feedback by pulse-width modulation (PWM) of the input current. The resultant DC voltage is typically switched by a six-switch inverter to a three-phase motor stator winding. The inverter switching sequence results in the commutation of the current and is determined by sensing the rotor position. The sensing of the rotor position is done by a distribution of Hall effect sensors. The commutation is achieved by a set of switching transistors that are operated sequentially, to switch on and switch off the current in each phase, in such a way so as to keep the current flow in the appropriate direction necessary to maintain the rotor motion. The current feedback in each phase is a trapezoidal pulse with the linear portion of the pulse overlapping in two of the three phases. This permits a smooth transition from the one phase to the next. In terms of the electrical angular domains, each switching duration is for 120° and the six ON regions are, 180–300°, 0–120°, 300–60°, 120–240°, 60–180°, and 240–360°. A single cycle of the trapezoidal periodic pulse spans 360°, starts at zero, is flat with maximum and minimum magnitude over the 60–120° and 240–300° regions, respectively, ends at zero and is linearly varying everywhere else. Although the principal component of the torque is sinusoidal, there is a higher harmonic torque ripple. Ignoring the torque ripple, one can show that the rotor is generally maintained at constant speed. A block diagram of a typical drive system for a three-phase brushless DC motor is illustrated in Fig. 2.14.

2.5 Stepper Motors

A stepper motor is an electro-mechanical drive system that actuates a train of angular stepped movements in response to a continuous sequence of input pulses. Each individual stepped movement corresponds to a single pulse input. Stepper motors are employed as actuators in incremental motion control systems such as drives associated with printers, fax machines, photo-copiers, etc. The three most widely used types of stepper motor are as follows: (1) variable reluctance motors; (2) permanent magnet motors; and (3) hybrid stepper motors.

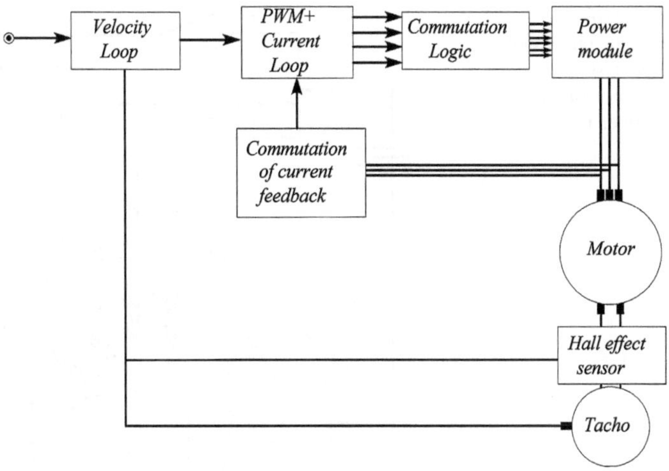

Fig. 2.14 Block diagram of a typical drive system for a three-phase brushless DC motor

The operation of the variable reluctance motor is based on the principle that a magnetic circuit in the motor always drives the rotor so as to maintain a minimum reluctance flux path. The stator windings are excited by switching a DC source in sequence. The rotor, which merely provides a path for the magnetic flux, is then force to align itself so the position it seek is a minimum reluctance flux path. A typical example of a four-phase eight-pole VR stepper stator is shown in Fig. 2.15a and the operation of three-phase VR motor is illustrated in Fig. 2.15b.

In the case of a permanent magnet stepper motor, the rotor is replaced by a permanent magnet. Thus, the rotor locks into a stationary position even when it is not excited. The operation is similar in principle to the VR type stepper motor.

2.6 The Design and Development of High Performance Electric Actuators

The design of high-performance electric drives and actuators leads to a number of challenging non-linear control synthesis problems. The requirements of high torque to inertia ratio, high efficiency, robustness, disturbance rejection and matched fast enough dynamic response have to met by any practical and viable control system. Electric drives are essentially non-linear systems. Although linear models may be developed and applied to non-linear control laws, it is suspected that they lead to very rigorous and that there is possibly an unnecessary wastage of control effort. In order to develop truly optimal high-performance actuators and drives, it is essential that the non-linear problem is dealt with rather than a linearized model.

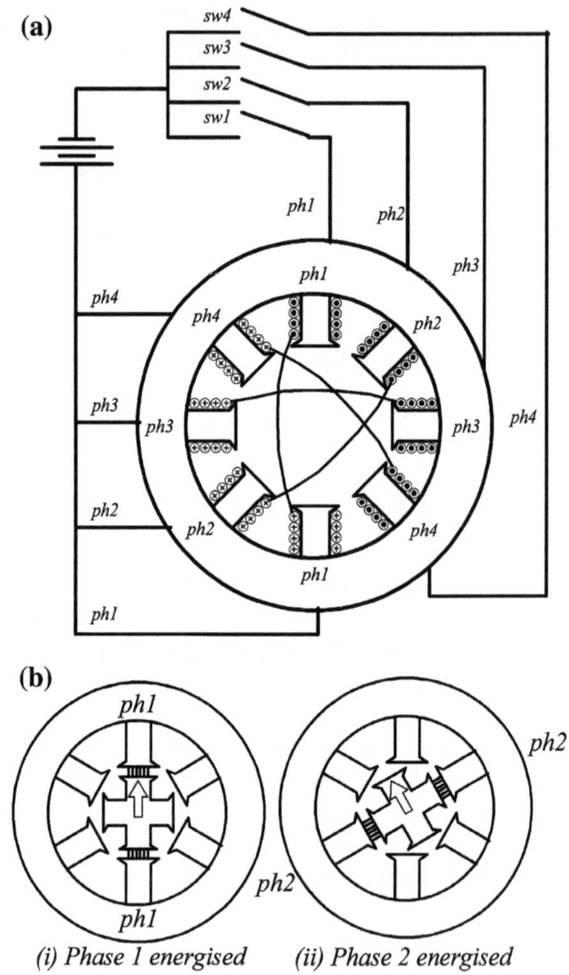

Fig. 2.15 **a** A typical example of a four-phase eight pole VR stepper stator. **b** Rotation in steps of a three-phase VR stepper motor

In non-linear control, it is customary to regard most of the synthesis techniques based on linearization and linearized models as validation tools.

Moreover, in the development of high-integrity actuators for aerospace applications, it is essential to design triplex and quadruplex actuators with torque or displacement summing. These are usually triplicated or quadruplicated redundant meeting specified reliability and safety requirements which may take the form of a two-fail/operate and fail/safe specification. This requires that the actuator continues to operate safely with two like failures and only fail to safe operating mode with a third failure. Thus, they must possess a degree of failure absorption along with a degree of graceful degradation in performance. The spatial integration of such actuators is also an important issue, and it is essential to eliminate undesirable non-linear characteristics.

2.7 Driving a DC Motor: Speed Control of a DC Motor

The principal control task associated with any motor is the control of the its rotational speed while also driving the motor. Considering a DC motor, probably simplest of control schemes involves the use of a tachogenerator or tachometer to generate a small voltage directly proportional to the speed and a feedback circuit built around the motor-tachometer plant. A tachogenerator's construction is almost identical to a DC motor but is designed to generate an output voltage proportional to the rotor speed. It may be considered to be a DC motor, operating in reverse. The speed control scheme is best explained with the block diagram illustrated in Fig. 2.16.

A simplified circuit diagram illustrates the practical construction and operation of the controller circuit.

In the circuit illustrated in Fig. 2.17, the comparator generates a input proportional to the difference between the input(+) and tacho feedback voltages. The diode ensures it is never negative. The gain of the transistor amplifier is $1 + R_B/R_A$. The speed of the motor increases till the input to the base of the transistor $Tr1$ is zero. The amplifier maintains the motor speed at this equilibrium condition. Thus, the speed control of the motor is achieved. The simple circuit shown above is not capable of bi-directional control of the rotor speed. Thus, with the simple type of circuit shown above, reverse operation is impossible. Reverse operation requires a bi-direction circuit.

An example of a bi-directional servo amplifier is illustrated in Fig. 2.18 which in addition to the resistors and capacitors uses both *pnp* and *npn* type bi-polar transistors. The amplifier illustrated in this circuit is a typical bi-directional voltage amplifier that must be capable of supplying both positive and negative currents. From a practical standpoint, a current amplifier is generally preferred. With this modification, the current to the motor is directly proportional to the input voltage.

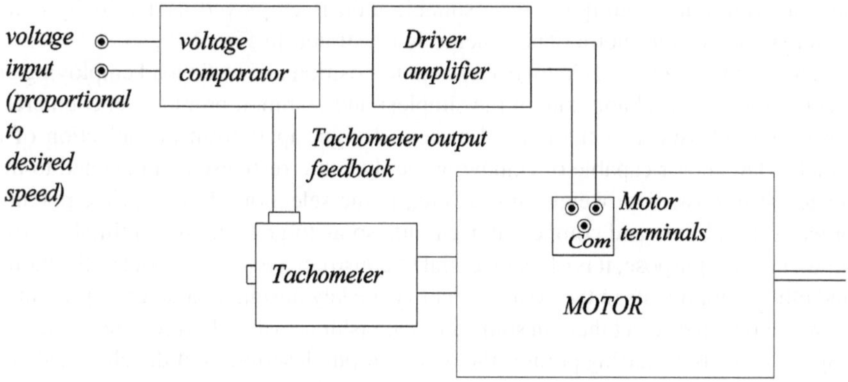

Fig. 2.16 Block diagram of the speed control of a DC motor

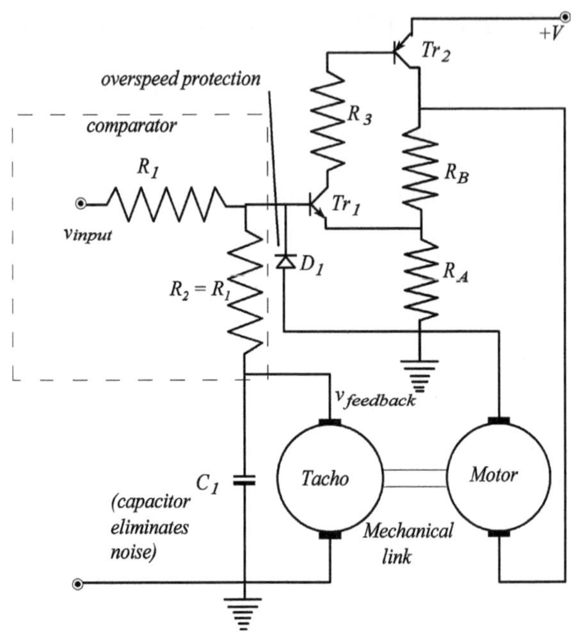

Fig. 2.17 Speed control circuit using a tachogenerator, gain of the transistor-based voltage amplifier is determined by resistors R_A and R_B

2.7.1 Controlling a DC Motor: Position Control Servo

A classical method of positioning an inertia load involves yet another feedback loop enclosing a speed controlled DC motor. The angular position of the rotor shaft is measured using an analog potentiometer and as in the case of the velocity measurement is feedback to comparator. The comparator determines the difference between the commanded or desired position signal, generated by an identical potentiometer as the one used for measuring the angular position of the motor shaft and this difference multiplied by a suitable gain is employed as a velocity command signal for the motor. The scheme is illustrated in Fig. 2.19.

The design of a feedback control system to position an inertia load employing an electric motor is probably one of the simplest and common problems encountered. An important aspect of the control system design, apart from the selection of a suitable DC motor capable of achieving the dynamic response and matched to the objective in cost, size, weight, torque etc., is the selection of appropriate potentiometers, amplifiers and compensator circuits so as to generate the optimal control input. For this purpose, it is often essential to construct a realistic model of the motor and other components in the system, identify the key design parameters represented in the model, and select them in some optimal fashion so as to best realize the design objective. Thus, a starting point is the input–output description of the electrical and mechanical entities that represent the electro-mechanical energy conversion processes within the motor. Considering the case of transient armature currents, the mathematical model of the motor may be set up from the earlier equations, involving

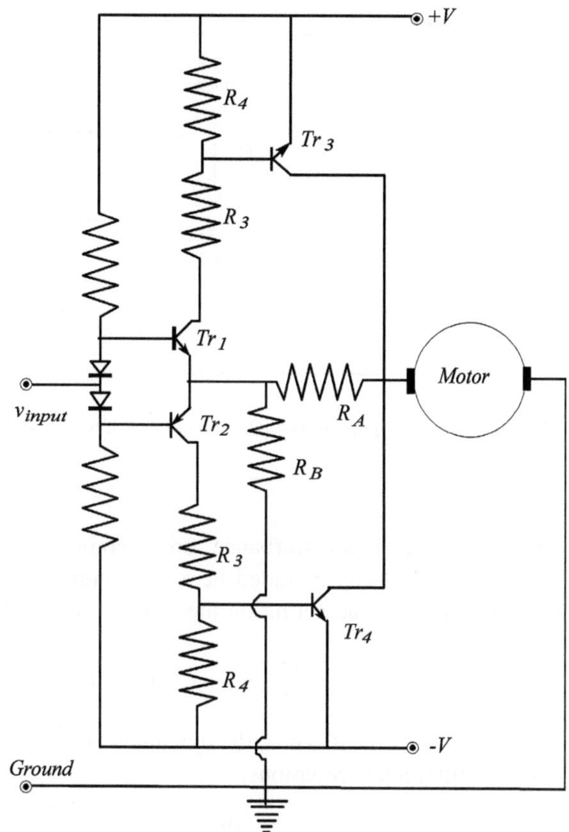

Fig. 2.18 An example of a bi-directional voltage-controlled voltage servo amplifier using *npn*- and *pnp*-type bipolar transistors

the applied voltage, armature current, motor speed, and output torque. Given that the total resistance of the armature circuit is R_a, the total inductance of the armature circuit is L_a and that the armature is driven by a constant external voltage, V_a, the current in the armature may be expressed as,

$$L_a \frac{\mathrm{d}I_a}{\mathrm{d}t} + R_a I_a = V_a - K_b \omega \quad (2.7.1)$$

and the torque acting on the armature is:

$$T = K_t a_m D^3 B \times I_a - T_{\text{load}} \quad (2.7.2)$$

where T_{load} is the load torque acting on the rotor. Assuming the load torque is entirely due to the rotating inertial load on the rotor and the friction torque, the load torque, T_{load} is given by,

$$T_{\text{load}} = J \frac{\mathrm{d}\omega}{\mathrm{d}t} + B_f \omega - T_{\text{disturbance}} \quad (2.7.3)$$

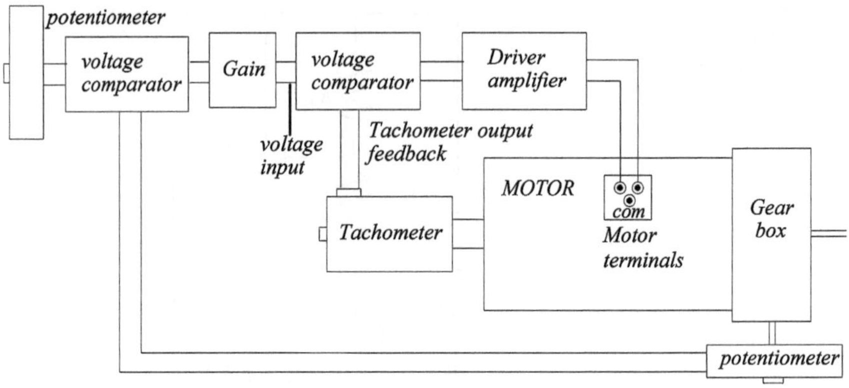

Fig. 2.19 Block diagram of the position control of a DC motor employing a position measuring potentiometer

where $T_{\text{disturbance}}$ is a disturbance torque component of the load which is usually ignored but has been included here for completeness. Under equilibrium conditions, the total torque acting on the rotor is zero and it follows that,

$$J\frac{\mathrm{d}\omega}{\mathrm{d}t} + B_f\omega = K_T \times I_a + T_{\text{disturbance}} \tag{2.7.4}$$

where $K_T = K_t a_m D^3 B$. Thus, the dynamics of the motor is described by the pair of coupled differential equations,

$$L_a\frac{\mathrm{d}I_a}{\mathrm{d}t} + R_a I_a + K_b\omega = V_a \tag{2.7.5}$$

$$J\frac{\mathrm{d}\omega}{\mathrm{d}t} + B_f\omega - K_T I_a = T_{\text{disturbance}}. \tag{2.7.6}$$

Considering the velocity and position feedback loops, the applied voltage, V_a, may be expressed as,

$$V_a = -K_p(\theta - \theta_d + K_d\omega) \tag{2.7.7}$$

where $\omega = d\theta/dt$.

The above mathematical model of the motor and the position control servo incorporating the motor may be expressed in the form of a *block diagram*. The use of such block diagram representations is commonplace in control engineering, as it provides a simple tool to assess certain properties of the closed loop control system. A block diagram representation of the mathematical model of the position control of an inertia load using a DC motor is shown in Fig. 2.20. The motor equations may be expressed as,

$$I_a = \frac{1}{L_a D + R_a} e = \frac{1}{L_a D + R_a}(V_a - K_B \omega), \ \omega = \frac{1}{JD + B_f}(K_T I_a + T_D). \quad (2.7.8)$$

and consequently the motor block diagram representation takes the form shown in Fig. 2.20a. In Fig. 2.20a, "D" is the operator, d/dt, $G_a(D)$ is the armature *transfer function* and

$$G_a(D) = \frac{1}{L_a \frac{d}{dt} + R_a} \equiv \frac{1}{L_a D + R_a}. \quad (2.7.9)$$

$G_m(D)$ is the rotor mechanical *transfer function* with,

$$G_m(D) = \frac{1}{J \frac{d}{dt} + B_f} \equiv \frac{1}{JD + B_f}. \quad (2.7.10)$$

Eliminating the current I_a from the two equations for ω and I_a,

$$\omega = \frac{1}{JD + B_f}(K_T I_a + T_D) = \frac{1}{JD + B_f}\left(\frac{K_T}{L_a D + R_a}(V_a - K_B \omega) + T_D\right) \quad (2.7.11)$$

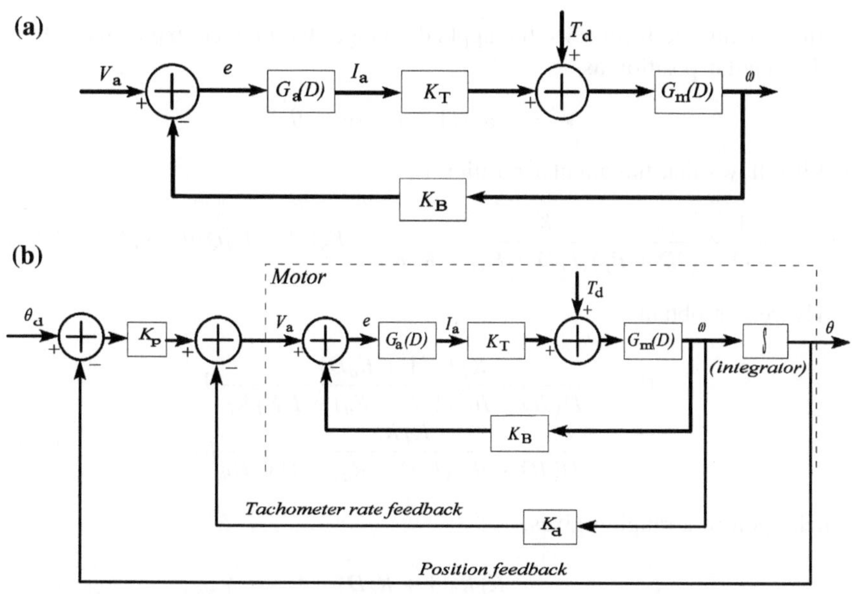

Fig. 2.20 Mathematical representation of the block diagram of the position control of an inertia load using a DC motor. **a** Block diagram of motor. **b** Block diagram of motor and position control feedbacks

and re-arranging the terms,

$$\left(1+\frac{1}{JD+B_f}\frac{K_TK_B}{L_aD+R_a}\right)\omega=\frac{1}{JD+B_f}\left(\frac{K_TV_a}{L_aD+R_a}+T_D\right). \tag{2.7.12}$$

Thus,

$$\omega=\frac{K_TV_a+T_D(L_aD+R_a)}{(JD+B_f)(L_aD+R_a)+K_TK_B}. \tag{2.7.13}$$

Assuming that the disturbance torque, T_D, is negligible,

$$\frac{\omega}{V_a}=\frac{K_T}{(JD+B_f)(L_aD+R_a)+K_TK_B}. \tag{2.7.14}$$

The above expression represents the *transfer function* relating the output angular velocity to the input armature voltage. The transfer function relating the angular position to the armature voltage is then given by,

$$\frac{\theta}{V_a}=\frac{1}{D}\times\frac{K_T}{(JD+B_f)(L_aD+R_a)+K_TK_B}. \tag{2.7.15}$$

Introducing the feedbacks, the applied voltage, V_a, may be expressed in terms of the angular position as,

$$V_a=-K_p((1+K_dD)\theta-\theta_d) \tag{2.7.16}$$

and it follows that the angular position is,

$$\theta=-\frac{1}{D}\times\frac{K_T}{(JD+B_f)(L_aD+R_a)+K_TK_B}\times K_p((1+K_dD)\theta-\theta_d). \tag{2.7.17}$$

Hence, we obtain

$$\begin{aligned}\theta=&-\frac{K_TK_p(1+K_dD)}{D(JD+B_f)(L_aD+R_a)+DK_TK_B}\theta\\&+\frac{K_TK_p}{D(JD+B_f)(L_aD+R_a)+DK_TK_B}\theta_d\end{aligned} \tag{2.7.18}$$

which upon re-arranging gives,

$$\begin{aligned}&\left(1+\frac{K_TK_p(1+K_dD)}{D(JD+B_f)(L_aD+R_a)+DK_TK_B}\right)\theta(t)\\&=\frac{K_TK_p}{D(JD+B_f)(L_aD+R_a)+DK_TK_B}\theta_d(t).\end{aligned} \tag{2.7.19}$$

The transfer function of the closed loop system relating the angular position of the armature rotor to the commanded position,

$$\frac{\theta(t)}{\theta_d(t)} = \frac{K_T K_p}{D(JD + B_f)(L_a D + R_a) + DK_T K_B + K_T K_p(1 + K_d D)}. \tag{2.7.20}$$

The transfer function has the form of a ratio of two polynomials in the D operator. The stability of the closed loop system will be considered in a latter section.

Under steady state conditions, we may set the time derivatives of $\theta(t)$ and $\theta_d(t)$ to zero. This is equivalent to setting, $D = 0$ and in this case, we have

$$\underset{t\to\infty}{\text{Limit}}\,\theta(t) = \underset{t\to\infty}{\text{Limit}}\,\theta_d(t) \tag{2.7.21}$$

provided the response of the closed loop system is stable and tends to a limit.

In the case when the armature inductance, $L_a \approx 0$, the closed loop transfer function reduces to,

$$\frac{\theta(t)}{\theta_d(t)} = \frac{K_T K_p}{D\big(R_a JD + R_a B_f + K_T(K_B + K_p K_d)\big) + K_T K_p}. \tag{2.7.22}$$

Further, if the armature inertia, J, is also negligible, we have the simplified first-order transfer function,

$$\frac{\theta(t)}{\theta_d(t)} = \frac{K_T K_p}{D\big(R_a B_f + K_T(K_B + K_p K_d)\big) + K_T K_p} \equiv \frac{1}{\tau_m D + 1} \tag{2.7.23}$$

where,

$$\tau_m = \frac{R_a B_f + K_T(K_B + K_p K_d)}{K_T K_p}. \tag{2.7.24}$$

2.8 Driving and Controlling AC Motors

The two basic types of AC servo drives are synchronous servomotors and asynchronous or IMs. Among the former, the most important is permanent magnet (PM) type. Attention is restricted to these two types of motors. The most common type of AC servomotor is the two-phase IM considered in Sect. 1.4.1.

Probably, the simplest method of controlling an AC induction motor is to increase the reactance of the stator circuit control phase by including a resistance in series with it. While such a method is suitable for certain low-power applications, it is generally extremely energy inefficient. Another method is based on using a thyristor-based "chopper" to chop and limit a sine wave to create a pulse. A thyristor is essentially a three-junction/three-terminal semi-conductor (four layer *pnpn*) device where a input to a *gate terminal* controls the flow of current between an *anode* and a *cathode*. When the pulse is created by this method, its width is completely controllable and the resulting method of control is completely

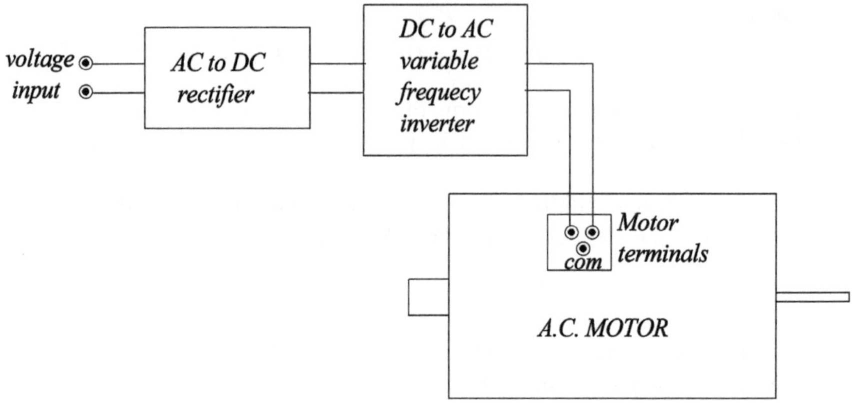

Fig. 2.21 Basic principle of inverter-based variable speed drive with controllable frequencies.

equivalent to pulse-width modulation. However, the most versatile method is the variable voltage-variable frequency approach which employs an inverter to generate the control voltage. The basic principle of inverter based variable speed drive with controllable frequencies for speed regulation of IMs is illustrated in Fig. 2.21.

As far as single phase AC motors are concerned, a simple method of control is based on the conduction angle modulation technique. The technique is illustrated in Fig. 2.22. The waveforms of the applied voltage are illustrated in Fig. 2.22a, b. The whole of the 180° half-cycle is not applied to the motor; it is only applied over part of the 180° half-cycle. The part of the cycle over which it is applied is known as the *conduction angle,* β and may vary from 0° to 180°. The larger the conduction angle, the faster the motor runs. The implementation, illustrated in Fig. 2.22c, is based on a three-layer (*pnp*) *trigger diode*, a device designed to generate the trigger current necessary to switch a *triac*, a multilayered semiconductor device like a thyristor but with ability to conduct current bi-directionally.

The basic principle in controlling PM synchronous servomotors is based on field orientation. In a frame of reference rotating with the rotor shaft, the magnetic flux generated by the PM rotor is fixed and the flux position in the rotor fixed co-ordinates can be determined by either by a shaft position sensor or directly by measuring the motor voltage and current signals. The general structure of a PM synchronous motor (PMSM) drive system, including the PWM-based three-phase control inputs for speed control, flux control and flux angle control, is illustrated in Fig. 2.23a. The method of control can, in general, be also applied to a generator where the three-phase supply is from a PM synchronous generator and the output is to a grid. The actual synthesis of the control laws are discussed in the next chapter.

Induction motor drives are quite similar to PMSM drives. A wide variety of estimators and feedback controllers could be designed for speed and position control of such motors. They generally consist of a stator current controller which includes two feedback loops to control the stator current precisely. These loops

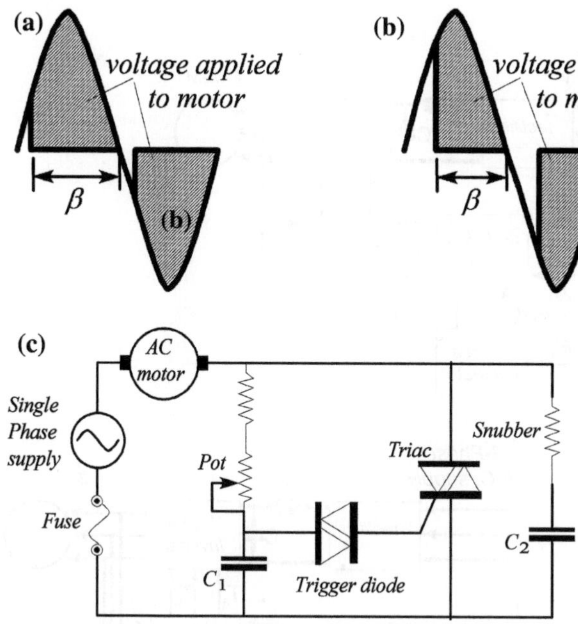

Fig. 2.22 Principle of speed adjustment of a single phase induction motor by the technique of conduction angle modulation. **a** Higher mean voltage. **b** Lower mean voltage. **c** Idealized circuit diagram

involve transformations to and from the stationary frame to the rotor fixed rotating frame, which are discussed in the next chapter. A typical IM drive is also illustrated in Fig. 2.23b. The outer loop is generally absent in a PMSM controller as precise control of the stator currents is unnecessary in the case of a synchronous motor.

2.9 Stability of Electric Servo-Actuators

An issue of paramount importance is the design of globally stable multimode controllers (torque summed mode, displacement summed mode, one-failure mode, two-failure mode or degraded mode} in order to guarantee the safe performance of such actuators. There are essentially two classes of methods of control law synthesis for non-linear plants. The first class of methods are based on stability analysis while the second on optimization. Each class of methods is associated with its own set of advantages and disadvantages.

The analysis of the asymptotic systems was pioneered by A. M. Lyapunov in his theses that was published almost 100 years ago. Lyapunov's method is not only a basic tool in the analysis of system stability but also a very valuable tool for the

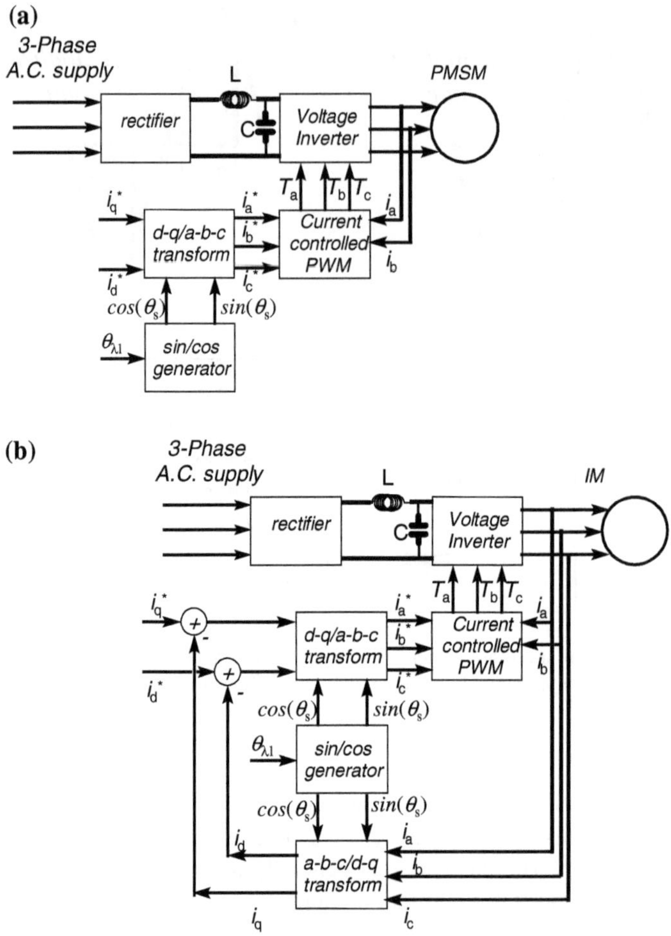

Fig. 2.23 **a** PM synchronous motor drive system showing the power and control inputs. **b** Induction motor drive system, including feedback control of stator currents, showing the power and control inputs

synthesis of controllers for non-linear systems. Lyapunov divided the problem of stability analysis of non-linear systems into two groups. The first group of problems involved non-linear systems which could be either solved exactly or reduced by some means to a linear system.

These not only included exact methods but also, in some cases, approximate techniques where the stability of the linearized system yields useful properties about the stability of equilibrium of the non-linear system.

In the case of a linear system, it is possible to analyze the stability of the solution without having to derive elaborate general solutions to the problem. Methods of evaluating stability such as the Routh Hurwitz method and Routh's tabulation allow the control system designer to establish bounds on important system gains to guarantee stability.

Consider the definition of *stability*. If the system is in a *state of equilibrium*, any *disturbance* of finite magnitude applied to it will cause a *free motion* following the disturbance. If the free motion ultimately disappears, the motion is said to be *stable*. On the other hand, if the free motion gradually degenerates to a finite motion that eventually becomes larger limitlessly, the motion is said to be *unstable*. If the free motion neither disappears nor becomes large, the system is *neutrally stable*.

To assess the stability of a system, one must consider the input–output transfer function. Given a *transfer function*, in the form,

$$\frac{y(t)}{u(t)} = G(D) = \frac{Y(D)}{U(D)}, \tag{2.9.1}$$

in the case of free motion we can assume that $u(t) = 0$.

Rationalizing the numerator and the denominator, the equations for the free response, in the absence of any inputs, takes the form:

$$P(D)y(t) = 0 \tag{2.9.2}$$

where $P(D)$ is generally a polynomial in the 'D' operator.

To solve the equations one assumes that,

$$y(t) = Ae^{pt}. \tag{2.9.3}$$

Differentiating both sides, repeatedly, one obtains, $Dy(t) = pAe^{pt}$, $D^2y(t) = p^2Ae^{pt}$, $D^3y(t) = p^3Ae^{pt}$, $D^4y(t) = p^4Ae^{pt}$, etc. Hence, it follows that,

$$P(D)y(t) = P(p)Ae^{pt} = P(p)y(t). \tag{2.9.4}$$

But,

$$P(D)y(t) = P(p)y(t) = 0. \tag{2.9.5}$$

Since the response is not assumed to be trivial and hence is non-zero, $y(t) \neq 0$ and $P(p) = 0$. The equation $P(p) = 0$ is known as the characteristic equation. The roots $p_1, p_2, p_3, \ldots, p_n$ of the characteristic equation will in general be complex quantities of the form, $p = q + jr$, where $j = \sqrt{-1}$. The response $y(t)$ is given by:

$$y(t) = \sum A_k e^{p_k t} \tag{2.9.6}$$

or by

$$y(t) = A_1 e^{p_1 t} + A_2 e^{p_2 t} + A_3 e^{p_3 t} + \cdots + A_n e^{p_n t}. \tag{2.9.7}$$

For the response to be stable, the real parts of the values of p_k must be negative for all k. Each of the cases of quadratic, cubic, and quartic equations are independently considered.

Quadratic equation, $p^2 + a_1 p + a_0 = 0$: The two roots have negative real parts, if a_1, a_0 are both greater than zero.

Cubic equation, $p^3 + a_2 p^2 + a_1 p + a_0 = 0$: A cubic has in general one real root and two complex conjugate roots, $p_1 = q_1$, $p_2 = q_2 + jr_2$, and $p_3 = q_2 - jr_2$.

The characteristic equation may be written as: $(p - q_1)(p - q_2 - jr_2)(p - q_2 + jr_2) = 0$ and $a_2 = -(2q_2 + q_1)$, $a_1 = q_2^2 + 2q_2 q_1 + r_2^2$, $a_0 = -q_1\,(q_2^2 + r_2^2)$. It follows that when the system is stable, a_2, a_1, a_0 are all greater than 0. Further, $a_1 a_2 - a_0 = -2q_2(2q_2^2 + 2q_2 q_1 + r_2^2) > 0$.

Thus:

$$a_2 > 0,\ a_1 > 0,\ a_0 > 0,\ a_2 a_1 - a_0 > 0. \tag{2.9.8}$$

These 4 conditions guarantee stability.

Quartic equation, $p^4 + a_3 p^3 + a_2 p^2 + a_1 p + a_0 = 0$: The conditions of stability are:

$$a_3 > 0,\ a_2 > 0,\ a_1 > 0,\ a_0 > 0,\ a_3 a_2 a_1 > \left(a_1^2 - a_0 a_3^2\right) \tag{2.9.9}$$

Conditions for stability in the general case may be stated in compact form. Given an algebraic characteristic equation of the form:

$$\boldsymbol{F}(s) = a_0 s^n + a_1 s^{n-1} + \cdots + a_{n-1} s + a_n = 0 \tag{2.9.10}$$

where the coefficients a_r are real, construct the following n determinants:

$$D_1 = a_1,\ D_2 = \begin{vmatrix} a_1 & a_0 \\ a_3 & a_2 \end{vmatrix},\quad D_3 = \begin{vmatrix} a_1 & a_0 & 0 \\ a_3 & a_2 & a_1 \\ a_5 & a_4 & a_3 \end{vmatrix} \tag{2.9.11a}$$

$$D_n = \begin{vmatrix} a_1 & a_0 & 0 & 0 & & & \cdots & & & \\ a_3 & a_2 & a_1 & a_0 & 0 & & \cdots & 0 & 0 & 0 \\ a_5 & a_4 & a_3 & a_2 & a_1 & a_0 & \cdots & 0 & 0 & 0 \\ \vdots & \vdots & \vdots & \vdots & \vdots & \vdots & \vdots & & & \\ a_{2n-1} & a_{2n-2} & a_{2n-3} & & & & \cdots & & & a_n \end{vmatrix} \tag{2.9.11b}$$

A necessary/sufficient condition for the roots of the characteristic equation $\mathbf{F}(s) = 0$ to have negative real parts is:

$$D_1 > 0, D_2 > 0, D_3 > 0, \ldots, \ldots, D_n > 0. \tag{2.9.12}$$

These conditions are verified by using Routh's tabular method.

Table 2.1 Routh's tabular array

p^n	a_0	a_2	a_4	a_6	..	..
p^{n-1}	a_1	a_3	a_5	a_7	..	..
p^{n-2}	b_1	b_2	b_3	b_4	..	..
p^{n-3}	c_1	c_2	c_3	..	..	
p^{n-4}	d_1	d_2	d_3	..	..	
...			..	..		
p^0			..			

2.9.1 Routh's Tabular Method

A tabular array of the form in Table 2.1 is produced.

The first two rows are formed by writing down alternate coefficients of the characteristic equation.

Each entry in the following rows is calculated from four of the previous entries according to the following pattern:

$$b_1 = \frac{a_2 a_1 - a_0 a_3}{a_1},\ b_2 = \frac{a_4 a_1 - a_0 a_5}{a_1},\ b_3 = \frac{a_6 a_1 - a_0 a_7}{a_1} \tag{2.9.13}$$

$$c_1 = \frac{a_3 b_1 - a_1 b_2}{b_1},\ c_2 = \frac{a_5 b_1 - a_1 b_3}{b_1},\ d_1 = \frac{b_2 c_1 - b_1 c_2}{c_1}. \tag{2.9.14}$$

Coefficients are calculated until only zero's are obtained, the rows shortening until the p^0 row contains only one value.

The conditions for stability are as follows:

1. Every change of sign in the first column of this table signifies the presence of a root with a positive real part.
2. For stability, therefore, all values in the first column of this array must be positive.
3. There are special cases and these require special treatment.

2.10 Electric Generators

The principle of the action of a generator was discussed in the earlier Sect. 2.1.2 and is the basis for the design of a practical electrical generator. Generators require a prime mover that imparts motion to the conductor. A simple electrical DC motor can function as generator when operated in the reverse; i.e., the motor shaft is driven by a prime mover and the motor armature output is the generator output. Like the construction of the motor armature, the generator armature is made up of conductors are housed in slots cut in a cylindrical structure (and sometimes also made of magnetic material). A prime mover provides rotary motion in the generator. The armature is rotated about a central axis of a shaft and is supported at both ends by means of bearings attached to the shaft. The armature is rotated inside a magnetic field structure by providing a small gap between two members which maintain the magnetic field between them. This gap is known as the air gap the size of which depends on the size of the generator. If the air gap is maintained constant throughout the spread of the pole arc, we have a fairly constant flux density under it in a plane perpendicular to the plane of the conductor's motion, i.e., in a radial direction with respect to the field and armature structure. Since the EMF is also proportional to the magnetic flux density B, the flux density in the air gap of AC generators is arranged to be distributed as closely to a sine wave as possible by suitable shaping by chamfering the pole shoe. Since the relative motion between the conductors and the magnetic flux lines is responsible for the production of EMF, it is immaterial whether the conductors are rotated or the magnetic flux producing poles are rotated. In most of the AC generators or alternators, it is the field that is rotated rather than the conductors. In an alternator, the external connection to the load can be taken directly from the conductors since there is no need for any rectification as in a DC generator. In a DC generator, the rectification of the generated EMF is achieved through a mechanical rectifier which is made up of the commutator and brush arrangement. Moreover, the load current supplied by the alternator can be easily supplied from stationary coils without any difficulty as there will be no sparking and wear and tear of the brushes and slip rings, whereas the low values of DC excitation current to the field coils can be easily sent through the slip rings and brush arrangement.

2.10.1 Synchronous AC Generators

The usual arrangement in an elementary permant magnet synchronous generator is as shown in Fig. 2.6a. The conductors are housed in slots cut in the field generating structure. In general, these are polyphase machines, although only two pairs of coils, indicated in its cross-section by each coil pair placed in diametrically opposite poles on the inner periphery of the stator (i.e., the armature, as it is a stationary member here) is shown in Fig. 2.6a. The conductors forming these coils are parallel to the shaft of the machine and are connected in series by end connections. The coils are actually formed by taking a continuous, insulated copper wire of suitable cross-section round a diamond shaped bobbin.

Typically, on a synchronous AC generator, the armature winding is on the stator and the field system is on the rotor. The field current, if present as in electromagnet generators, is supplied from the exciter via two slip rings, while the armature current is directly supplied to the load. This type is employed universally since very high power can be delivered. Unless otherwise stated, the subsequent discussion refers specifically to this type of rotating field type synchronous machines. According to the shape of the field, synchronous machines may be classified as a salient-pole machine or a cylindrical-rotor (non-salient pole) machine. The cylindrical-rotor construction is used in generators that operate at high speeds, such as gas-turbine generators (usually two-pole machines). This type of machine usually has a small diameter-to-length ratio, in order to avoid excessive mechanical stress on the rotor due to the large centrifugal forces. The salient-pole construction is used in low-speed generators (such as wind-turbine generators) and is similar in construction to synchronous motors. This type of machine usually has a large number of poles for low-speed operation and a large diameter-to-length ratio. The field coils are wound on the bodies of the projecting poles.

2.10.2 Dynamic Modeling of the Synchronous AC Generators

To dynamically model synchronous electrical AC generators, one needs to consider the voltage–current relationship in the stator or armature winding. When a synchronous generator is excited by a rotating permanent magnet or a field-excited electromagnet and is driven at a constant speed, a voltage is generated in the armature phase winding. If a resistive load is connected across the armature phase, an armature phase current will flow in the armature phase circuit.

Since the frequency of generated EMF is related to the rotor speed, while the speed of the armature phase reactance is related to the frequency of the current, it follows that the product of the current and the armature phase reactance is related to the rate of change of the flux linkage which rotates synchronously with the rotor field. Moreover, a change in rotor speed results in a proportional change in the frequency of EMF and current, while the power factor is determined by the nature of the total load across the phase.

The effect of the rate of change of the flux linkage on the resultant field distribution results in an armature phase reaction. Since the flux linking the armature and the rotor rotates at the same speed as the rotor, it produces a corresponding EMF in the armature winding. The total voltage across the armature phase therefore consists of the sum of a terminal voltage which is obtained from the time rate of change of the flux linkage with reference to the rotor, the voltage drop across the resistive load R_L, and an equivalent voltage drop due to the rate of change, with rotor rotation, of the flux linkage λ_a, which as modeled as an equivalent reactive load. The armature current I_a, and the EMF developed across the armature phase terminals E_g, are related to each other by the relation,

$$E_g = V_t + R_L I_a + \frac{\mathrm{d}\lambda_a}{\mathrm{d}\theta} \equiv V_t + R_L I_a + iX_a I_a \equiv V_t + Z_a I_a. \tag{2.10.1}$$

In general, the terminal voltage V_t, lags behind the generator (armature phase) EMF E_g, by an angle δ while the armature phase current I_a, lags behind the

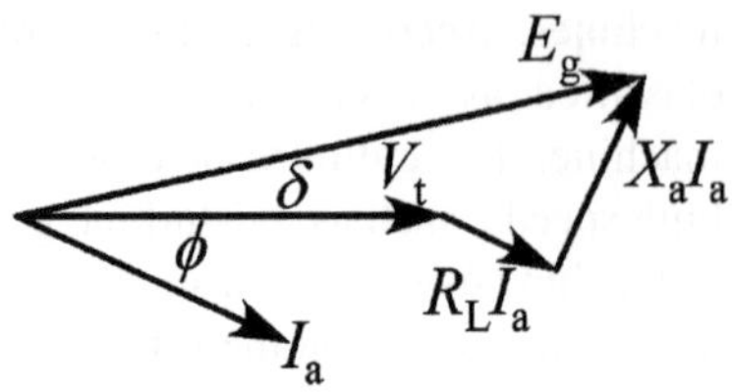

Fig. 2.24 The phase angle relationships between generator EMF, the terminal voltage and the armature current

terminal voltage V_t, by an angle φ. Thus, the phase angle relationships between the generator phase EMF E_g, the terminal phase voltage V_t and the armature phase current I_a may be illustrated by the phasor diagram in Fig. 2.24.

The active power output of the phase is,

$$P_w = E_g I_a \cos(\delta + \varphi) - V_t I_a \cos \varphi. \tag{2.10.2}$$

Assuming that the resistive load is small, i.e., $R_L \cong 0$,

$$P_w = \left| X_a I_a^2 \cos(\varphi + 90°) \right| = X_a I_a^2 \sin \varphi. \tag{2.10.3}$$

Applying the laws of mechanics to the rotational motion of a synchronous machine is governed by the equation,

$$I_r \frac{\mathrm{d}^2\theta}{\mathrm{d}t^2} = T_{\mathrm{net}}. \tag{2.10.4}$$

In Eq. 2.10.4, I_r is the rotor moment of inertia about the rotor axis of rotation, θ is the rotor angular displacement and T_{net} is the net torque acting on the rotor which is the difference between the driving mechanical torque and the electromagnetic braking torque. Generally, the electromagnetic braking torque is obtained from the product of the armature phase current and the change in the flux linkage due to the rotational displacement and the change in the stored energy in the permanent magnet due to the rotational displacement. Thus, the electromagnetic braking torque T_{el} is,

$$T_{el} = I_a \frac{\mathrm{d}\lambda_a}{\mathrm{d}\theta} - \frac{\mathrm{d}W_e}{\mathrm{d}\theta}. \tag{2.10.5}$$

The governing equations are often best expressed in terms of a set of rotating coordinates, rotating with the rotor. Thus, it is possible to relate quantities in a fixed reference frame to the corresponding quantities referred to in the rotating frame of reference. The transformation relating two orthogonal vector quantities in a fixed reference frame to the corresponding quantities referred to in the rotating frame of reference, rotating anticlockwise, is,

$$\begin{bmatrix} X \\ Y \end{bmatrix} = \begin{bmatrix} \cos \omega_0 t & -\sin \omega_0 t \\ \sin \omega_0 t & \cos \omega_0 t \end{bmatrix} \begin{bmatrix} U \\ V \end{bmatrix}. \tag{2.10.6}$$

Thus,

$$\frac{d}{dt}\begin{bmatrix} X \\ Y \end{bmatrix} = \left(\frac{d}{dt}\begin{bmatrix} \cos\omega_0 t & -\sin\omega_0 t \\ \sin\omega_0 t & \cos\omega_0 t \end{bmatrix}\right)\begin{bmatrix} U \\ V \end{bmatrix} + \begin{bmatrix} \cos\omega_0 t & -\sin\omega_0 t \\ \sin\omega_0 t & \cos\omega_0 t \end{bmatrix}\frac{d}{dt}\begin{bmatrix} U \\ V \end{bmatrix} \tag{2.10.7}$$

which may be expressed as,

$$\frac{d}{dt}\begin{bmatrix} X \\ Y \end{bmatrix} = \begin{bmatrix} \cos\omega_0 t & -\sin\omega_0 t \\ \sin\omega_0 t & \cos\omega_0 t \end{bmatrix}\left(\frac{d}{dt}\begin{bmatrix} U \\ V \end{bmatrix} - \omega_0\begin{bmatrix} V \\ -U \end{bmatrix}\right). \tag{2.10.8}$$

The last equation may be expressed in the form of two equivalent equations given by,

$$\frac{d}{dt}\begin{bmatrix} X \\ Y \end{bmatrix} = \begin{bmatrix} \cos\omega_0 t & -\sin\omega_0 t \\ \sin\omega_0 t & \cos\omega_0 t \end{bmatrix}\frac{d}{dt}\begin{bmatrix} U' \\ V' \end{bmatrix} \tag{2.10.9}$$

$$\frac{d}{dt}\begin{bmatrix} U' \\ V' \end{bmatrix} = \frac{d}{dt}\begin{bmatrix} U \\ V \end{bmatrix} - \omega_0\begin{bmatrix} V \\ -U \end{bmatrix} = \frac{d}{dt}\begin{bmatrix} U \\ V \end{bmatrix} - \omega_0\begin{bmatrix} 0 & 1 \\ -1 & 0 \end{bmatrix}\begin{bmatrix} U \\ V \end{bmatrix}. \tag{2.10.10}$$

The left-hand side of Eq. 2.10.10 represents the time derivative in the rotating frame.

For quantities that are stationary in the rotating frame,

$$\frac{d}{dt}\begin{bmatrix} U' \\ V' \end{bmatrix} = -\omega_0\begin{bmatrix} 0 & 1 \\ -1 & 0 \end{bmatrix}\begin{bmatrix} U \\ V \end{bmatrix} = -\omega_0 \boldsymbol{J}\begin{bmatrix} U \\ V \end{bmatrix}, \ \boldsymbol{J} = \begin{bmatrix} 0 & 1 \\ -1 & 0 \end{bmatrix}. \tag{2.10.11}$$

Consequently,

$$\frac{d}{d(\omega_0 t)}\begin{bmatrix} U' \\ V' \end{bmatrix} = \frac{d}{d\theta}\begin{bmatrix} U' \\ V' \end{bmatrix} = -\boldsymbol{J}\begin{bmatrix} U \\ V \end{bmatrix}. \tag{2.10.12}$$

Since the synchronous generator must naturally contain a minimum of two phases to be practical, it will be assumed that the armature has two windings orthogonal to each other. The reference winding is often referred to as the *direct* or "d" winding, while the second winding orthogonal to the reference winding is referred to as the *quadrature* or "q" winding. The directions associated with the "d" winding and the "q" winding are referred to as the "d" axis and the "q" axis, respectively. Since the rotating frame completely aligns itself with the "d" and the "q" axis initially, it also referred to as the rotating *d–q* frame.

A two-phase, permanent magnet synchronous generator, has two identical armature windings symmetrically distributed around the air gap. Armature windings are placed on the stator. Furthermore, all copper losses and the slots in the machine as well as all losses due to the wiring, saturation and slots are neglected, the spatial distribution of the stator fluxes are considered sinusoidal and the stator and rotor permeability are assumed to be infinite.

Table 2.2 Definitions of the generator's parameters

R_s	armature phase resistance
L_{ls}	Armature phase leakage inductance
L_{md}	d axis coupling inductance
L_{mq}	q axis coupling inductance
λ_d	d axis flux linkage
λ_q	q axis flux linkage
λ_{PM}	Flux linkage due to the permanent magnet

In the rotating d–q frame, under no-load conditions, given that the terminal voltage may be obtained from the time rate of change of the flux linkage, the armature winding equations are,

$$\begin{bmatrix} v_d \\ v_q \end{bmatrix} = -\frac{\mathrm{d}}{\mathrm{d}t}\begin{bmatrix} \lambda_d \\ \lambda_q \end{bmatrix} - R_s\begin{bmatrix} i_d \\ i_q \end{bmatrix} + \omega_0 \boldsymbol{J}\begin{bmatrix} \lambda_d \\ \lambda_q \end{bmatrix}. \tag{2.10.13}$$

The flux linkages are related to the currents in the windings by the relations,

$$\begin{bmatrix} \lambda_d \\ \lambda_q \end{bmatrix} = L_{1s}\begin{bmatrix} i_d \\ i_q \end{bmatrix} + \begin{bmatrix} L_{md} & 0 \\ 0 & L_{mq} \end{bmatrix}\begin{bmatrix} i_d \\ i_q \end{bmatrix} + \begin{bmatrix} 1 \\ 0 \end{bmatrix}\lambda_{PM} = \begin{bmatrix} (L_{1s} + L_{md})i_d + \lambda_{PM} \\ (L_{1s} + L_{mq})i_q \end{bmatrix}, \tag{2.10.14}$$

The definitions of the resistances and inductances are listed in Table 2.2.

The definitions of the electrical power output, the electromagnetic load torque and the exciter model are essential to complete the definition of the synchronous generator. The output active power p_a, the reactive power q_r and the electromagnetic braking or load torque T_{el} are, respectively, given by,

$$p_a = \begin{bmatrix} v_d \\ v_q \end{bmatrix}^T \begin{bmatrix} i_d \\ i_q \end{bmatrix},\ q_r = \begin{bmatrix} v_d \\ v_q \end{bmatrix}^T \boldsymbol{J}\begin{bmatrix} i_d \\ i_q \end{bmatrix}, \tag{2.10.15}$$

$$\begin{aligned} T_{el} &= \begin{bmatrix} i_d \\ i_q \end{bmatrix}^T \frac{\mathrm{d}}{\mathrm{d}\theta}\begin{bmatrix} \lambda_d \\ \lambda_q \end{bmatrix} = -\begin{bmatrix} i_d \\ i_q \end{bmatrix}^T \boldsymbol{J}\begin{bmatrix} \lambda_d \\ \lambda_q \end{bmatrix} = \begin{bmatrix} \lambda_d \\ \lambda_q \end{bmatrix}^T \boldsymbol{J}\begin{bmatrix} i_d \\ i_q \end{bmatrix} \\ &= \lambda_{PM} i_q + \left(L_{md} - L_{mq}\right) i_d i_q. \end{aligned} \tag{2.10.16}$$

It must be reiterated that the above equations are valid for permanent magnet synchronous generator with just two poles or one pole pair. The general cae will be considered in a subsequent chapter.

2.10.3 Induction AC Generators

In an induction AC generator, the permanent magnet rotor in the synchronous AC generator is replaced, in principle, by short-circuited copper coil. Assuming that the generator in in steady operation, the armature coils generate a field which in turn induce a current in the rotor coils by virtue of Lenz's law. Consequently, due to induced current flow and the stator field, force and torque is generated which tends to maintain the rotor in rotation in the same direction as the stator. At the same time, the short-circuited current in the rotor tends induce a current in the stator armature coils and the process repeats itself. The rotating field in the stator tends to drag the rotor with it although there the rotor generally lags behind the stator by small but significant angle. The mathematical modeling of the dynamics of an induction generator is not significantly different in principle from that of the synchronous generator, the only addition feature being the need to model the voltages and flux linkages in the rotor as well. The operating principle is also closely related to the principle of operation of an IM. The typical torque-speed characteristics of an IM are illustrated in Fig. 2.22. To operate in the generator mode, the IM must be driven to a speed higher than the synchronous speed. When this is the case, the speed torque curve flips about the horizontal and vertical axes as shown in Fig. 2.25. Only a part of this curve represents stable operation in the generator mode, and it is in the stable region of the curve that electric power is generated utilizing the mechanical input power from the prime mover. As in the case of an IM, the generated power is a function of the slip.

In the generator mode, if the slip is controlled in accordance with the load requirements, the induction generator will be maintained in a stable equilibrium and will deliver the necessary power. The synchronous speed is a function of the electrical frequency applied to the generator terminal. The mechanical shaft speed is determined by the prime mover. Therefore, to generate power, the electrical frequency must be adjusted as the changes in the load and the prime mover speed occur.

To draw power from the induction generator, an excitation current must be provided to the generator stator windings for induction of the rotor EMF. The resulting voltage output from the generator is regulated, multiple phase AC.

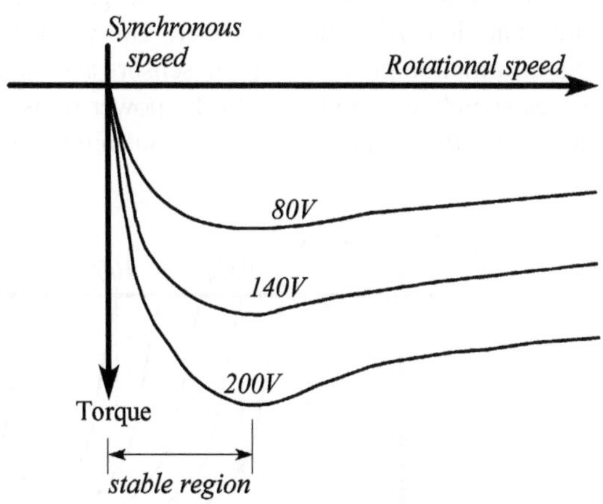

Fig. 2.25 Induction generator torque—speed characteristics

The magnitude of the excitation current will determine the magnitude of the output voltage at the bus. Thus, the excitation current must be regulated at specific levels to obtain a constant output bus voltage. The control of the output voltage is accomplished in a closed loop operation where the excitation current is adjusted to generate constant output voltage regardless of the variations of speed and load current. The controller for the induction generator seeks to adjust the electrical frequency to produce the slip corresponding to the load requirement while simultaneously adjusting the magnitude of the excitation current to provide the desirable bus voltage.

Figure 2.26 depicts the region of generator mode speed-torque characteristics for a typical induction generator. A number of torque-speed characteristic curves in the stable region of operation are shown to represent the operational characteristics. As an example, consider the situation when the prime mover is at the nominal operating speed which is denoted as 100 % of the speed. The electrical frequency must be adjusted to cater for load changes from 0 to 100 % of the load. If a vertical line is drawn along the speed of 100 %, it can be observed that the electrical frequency must be changed from 100 % at no-load to about 95 % at full load if the prime mover speed is held at 100 %.

Figure 2.27 illustrates the block diagram of a typical controller for an induction generator. The controller consists of three subsystems, the power conditioning subsystem, sensing circuits, and the controller synthesis subsystems. Power transistors using insulated gate bipolar transistors (IGBT's) or metal oxide semiconductor field effect transistor (MOSFET's), dudes and other power transistors make up the power conditioning circuits in a multiphase configuration, where the phases are matched to the generator winding. A rectifier connects the DC bus to the generator output while a power capacitor is connected across the bus. Independent power inverters are used to maintain the AC output at a constant frequency such as 60 or 50 Hz is required and to supply the inductors excitation windings. Currents and voltages are sensed both at the load and in the power section of the controller. In addition, the speed of the shaft is measured. All the parameters sensed by the sensing circuits are conditioned by the controller. The controller synthesis subsystem receives the information provided by the sensors and using a model of the generator, synthesizes switching commands to the power transistors using a control algorithm such a the PID control law. The controller output generates the

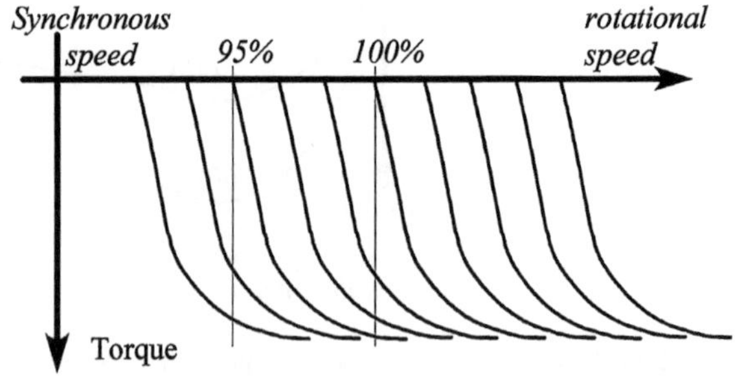

Fig. 2.26 Induction generator torque-speed characteristics in the operating range

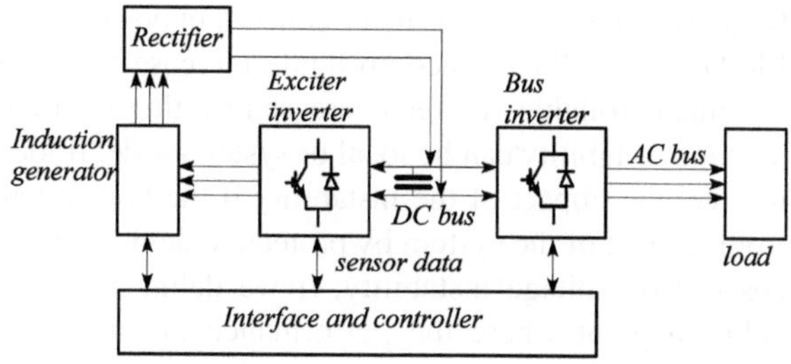

Fig. 2.27 Induction generator controller block diagram

necessary frequency and amplitude of the excitation currents that flow in the induction generator windings and the induced voltages in the squirrel-cage rotor. The controller also includes protective circuits for over-current, over-voltage, and over-temperature protection.

2.11 Power Systems

The overall task of controlling an electric power system is to maintain the balance between the electric power delivered by the generators and the power consumed by the loads, including the losses, at all time. If this balance is not maintained, it will lead to frequency deviations and large deviations in frequency will have a serious impact on the system operation. A feature of power systems is that the electric power consumption varies both in the short term and in the long term. Over long time scales, typically over an year, the peak loads of a day can be higher in the winter, so called winter peak loads, while countries with very hot summers usually have their peak loads in summer time or summer peak loads. To cope with the larger variations in the load and maintain balance, generating units must be periodically switched in or switch off. If this frequency variations are too large, the power supplied from the generators must be changed, which is done through the frequency control of the generators while they are in operation.

2.11.1 Power Systems Stability

A power system at a given operating state and subject to a given disturbance is voltage stable if the voltages approach equilibrium values which existed prior to the occurrence of the disturbances. In simple terms, the stability conditions may be stated as follows: Assuming that an equilibrium satisfying the operating constraints in the presence of variable power flow exists, the equilibrium is considered to be stable if it is so in the presence of small disturbances. Large disturbance stability is assured if the system state at the end of the disturbance lies within the region of attraction of the stable equilibrium point of the disturbance-free system. Although the above conditions apply to any type of stability, voltage stability and frequency stability are primary concern in power systems analysis (Kundur (1994), El-Hawary (2008)). Uncontrollable increase in voltage caused by self excitation and uncontrollable voltage decrease resulting in voltage collapse are also an issue. Voltage instability can be local or system-wide. If local, the rest of the system need not feel the impact of the instability if the affected area is promptly disconnected from the rest of the system by protective devices. Voltage collapse can occur in the absence of voltage instability, if we define voltage collapse as drop in voltage below a point where the performance of the connected devices will not meet design criteria. For example, consider a system serving only heating and/or lighting loads. For such loads, voltage instability cannot occur as per our definition of voltage stability. However, as more and more load is added, the voltage will keep dropping. If the loading continues, there will be a point beyond which the demanded MW will not be served, and the voltage will be too low for the devices to produce sufficient heat or light. For all practical purposes, we can say that the voltage has collapsed.

Phase angle instability can also lead to voltage collapse which might appear as voltage instability. Consider power being transmitted between two points over a transmission line. As the power transmitted approaches the phase angle stability limit (the phase angle across the line reaching 90°), the voltage at the mid-section of the line will be depressed and any load connected there will experience voltage collapse. The same situation may arise during power swings across transmission lines following a disturbance.

Power systems analysis involves voltage stability analyses considering all the relevant dynamics, and in the case of phase angle (synchronous) stability, one must consider the basics of power flow and power limits, while primarily considering the voltage. In the study of power limits in connection with phase angle stability, voltages were assumed to be either constant (generator internal voltages) or controlled so that we could be concerned only with the phase angle stability issues. Otherwise, one is concerned mostly with voltages at the load delivery point where the voltages are uncontrolled or under limited control. There are number of practical and effective corrective measures to combat voltage stability problems in system operations such as load shedding and capacitor switching which are beyond the scope of this chapter.

In the remainder of this section, the subject of power system dynamics and its control will be briefly discussed. From Eq. 2.10.4, the equilibrium of the rotating generator shaft is given by,

$$I_r \frac{\mathrm{d}^2\theta}{\mathrm{d}t^2} = T_{\mathrm{net}} = T_m - T_{el}. \tag{2.11.1}$$

In Eq. 2.11.1 T_m is the driving mechanical torque and T_{el} the electromagnetic braking torque. Under steady-state operating conditions,

$$I_r \frac{\mathrm{d}^2\theta}{\mathrm{d}t^2} = T_m - T_{el} = 0. \tag{2.11.2}$$

The angular position and velocity are measured with respect to a synchronously rotating reference frame, rotating with a constant angular velocity ω_0 rather than a stationary reference axis, and an angle δ is defined such that,

$$\delta = \theta - \omega_0 t. \tag{2.11.3}$$

Then it follows that,

$$\frac{\mathrm{d}}{\mathrm{d}t}\delta = \frac{\mathrm{d}}{\mathrm{d}t}\theta - \omega_0, \ \frac{\mathrm{d}^2}{\mathrm{d}t^2}\delta = \frac{\mathrm{d}^2}{\mathrm{d}t^2}\theta. \tag{2.11.4}$$

The governing equation of motion may now be expressed as,

$$I_r\omega_0\frac{\mathrm{d}^2\delta}{\mathrm{d}t^2} = M_{a0}\frac{\mathrm{d}^2\delta}{\mathrm{d}t^2} = \omega_0(T_m - T_{el}). \tag{2.11.5}$$

In Eq. 2.11.5, M_{a0} is the angular momentum when the rotor is rotating at the synchronous speed. If H_0 is the corresponding stored kinetic energy in the rotor,

$$H_0 = M_{a0}\left(\frac{\omega_0}{2}\right), \tag{2.11.6}$$

$$H_0\left(\frac{2}{\omega_0}\right)\frac{\mathrm{d}^2\delta}{\mathrm{d}t^2} = \omega_0(T_m - T_{el}). \tag{2.11.7}$$

The power at any speed ω is given by,

$$P = \omega(T_m - T_{el}). \tag{2.11.8}$$

Consequently, the governing equation of motion may now be expressed as,

$$H_0\left(\frac{2}{\omega_0}\right)\frac{\mathrm{d}^2\delta}{\mathrm{d}t^2} = P\frac{\omega_0}{\omega} = P\frac{\omega_0}{\omega_0 + (\omega - \omega_0)} = P\frac{1}{1 + \frac{(\omega - \omega_0)}{\omega_0}}. \tag{2.11.9}$$

Thus, Eq. 2.11.9 may be approximated as,

$$H_0\left(\frac{2}{\omega_0}\right)\frac{\mathrm{d}^2\delta}{\mathrm{d}t^2} = P\frac{1}{1 + \frac{(\omega - \omega_0)}{\omega_0}} \approx P\left(1 - \frac{(\omega - \omega_0)}{\omega_0}\right) = P\left(1 - \frac{\Delta\omega}{\omega_0}\right). \tag{2.11.10}$$

The power can be expressed as the difference between the driving mechanical power and the electromagnetic braking power and is given by,

$$P = P_m - P_{el}. \tag{2.11.11}$$

The equation for the deviation of the rotor displacement from its synchronously rotating position may be expressed in terms of the rated power P_{0R}, the driving mechanical power and the electromagnetic braking power as,

$$\left(\frac{2H_0}{\omega_0 P_{0R}}\right)\frac{d^2\delta}{dt^2} = \left(\frac{P_m - P_{el}}{P_{0R}}\right)\left(1 - \frac{\Delta\omega}{\omega_0}\right). \tag{2.11.12}$$

When the speed deviation $\Delta\omega$ can be neglected, i.e., $\Delta\omega \approx 0$,

$$\left(\frac{2H_0}{\omega_0 P_{0R}}\right)\frac{d^2\delta}{dt^2} = \frac{\Delta P}{P_{0R}}. \tag{2.11.13}$$

Equation 2.11.13 is the swing equation, which is the basic relation describing how the rotor of a synchronous machine will swing when there is an unbalance between mechanical power fed into the machine and the electrical power extracted from it. The basic physical relation is simple, as a number of simplifications have been done to make the equation useful in practice. The swing equation is of fundamental importance in the study of power oscillations in power systems. Electro-mechanical oscillations is an important phenomenon that must be considered in the analysis of most power systems, particularly those containing long transmission lines as is the case in most power distribution systems. In normal steady-state operation, all synchronous machines in the system rotate with the same angular velocity, thus maintaining synchronism. However, as a consequence of a disturbance in the mechanical driving torque or in the load, one or more generators could be accelerated or decelerated and there is risk that they can fall out of step, i.e., lose synchronism. This could have a large impact on system stability and generators losing synchronism must be disconnected otherwise they could be severely damaged.

2.11.2 Transmission Lines

Once power is generated by an electrical generator, it must be transferred to grid capable of suppling the power to one or more end users. It is essential that in doing so all losses are reduced to minimum and the voltage and frequency of the supply are maintained at there desired steady state values. Most common methods of interfacing a generator to a grid is by the use of transmission lines and transformers.

The general model of a transmission line is given by two first-order partial differential equation, for the voltage and current at any point in the line in terms of its resistive, inductive and capacitive properties Miano and Maffucci (2001). Thus, across a power line, there is a significant change in both the active and reactive power components. By restricting the interest to the conditions at the ends of the lines, the lumped-circuit line models (π-models) may be obtained, which are a family of algebraic models. The complete coupled partial differential equations could be used when studying the steady state conditions at the network nodes and the solution would reveal both the initial switching transients along the lines, and the steady-state solution after the transients have decayed. However, most engineering applications the use of the lumped-circuit line π-model would provide much of the information relative to the changes of the active and reactive power components which are of primary interest in the study of the dynamics and control of power systems.

A transmission line has four parameters; a series resistance and an inductance, and a shunt conductance and a capacitance. These are uniformly distributed along the line. In the derivation of transmission line equations, the following assumptions are usually made:

1. The transmission line is symmetrical, and the phases are balanced. In practice, this is not entirely true, but the unbalance and departure from symmetry are small.
2. The transmission line can be analyzed on a per phase basis.
3. The transmission line parameters, per unit length, are constant, i.e., they are independent of position, frequency, current, and voltage. This assumption, although approximate, is permissible for most power system studies.

The voltage and current on a transmission line depend, in general, upon both time and position. Consequently, a general mathematical description of the line involves two first-order partial differential equations relating the evolution of the voltage and the current along the line. The instantaneous voltage and current conditions at a small section of length $\mathrm{d}x$ at a distance x from the receiving end of the line are considered. The voltage and current, are denoted by v and i respectively, while the series resistance and inductance and the shunt conductance and capacitance per unit length of the line are denoted by r, l, g and c, respectively. The series voltage drop in the section $\mathrm{d}x$, is the sum of the resistive and inductive drops and is,

$$\frac{\mathrm{d}v}{\mathrm{d}x}\mathrm{d}x = (r\mathrm{d}x)i + (l\mathrm{d}x)\frac{\mathrm{d}i}{\mathrm{d}t}. \tag{2.11.14}$$

The change in the shunt current is,

$$\frac{\mathrm{d}i}{\mathrm{d}x}\mathrm{d}x = (g\mathrm{d}x)v + (c\mathrm{d}x)\frac{\mathrm{d}v}{\mathrm{d}t}. \tag{2.11.15}$$

Thus the instantaneous voltage and current, v and i, respectively, satisfy the pair of first-order partial differential equations,

$$\frac{\mathrm{d}v}{\mathrm{d}x} = ri + l\frac{\mathrm{d}i}{\mathrm{d}t}, \ \frac{\mathrm{d}i}{\mathrm{d}x} = gv + c\frac{\mathrm{d}v}{\mathrm{d}t}. \tag{2.11.16}$$

Assuming the instantaneous voltage and current, v and i, respectively, can be expressed as,

$$v(x,t) = V(x)e^{j\omega t}, \ i(x,t) = I(x)e^{j\omega t}, \ j = \sqrt{-1} \tag{2.11.17}$$

$$\frac{dV(x)}{dx} = (r + j\omega l)I(x) = zI(x) \qquad (2.11.18)$$

$$\frac{dI(x)}{dx} = (g + j\omega c)V(x) = yV(x). \qquad (2.11.19)$$

Consequently,

$$\frac{d^2V(x)}{dx^2} = (r + j\omega l)(g + j\omega c)V(x) = zyV(x) = \gamma^2 V(x) \qquad (2.11.20)$$

$$\frac{d^2I(x)}{dx^2} = (g + j\omega c)(r + j\omega l)I(x) = yzI(x) = \gamma^2 I(x). \qquad (2.11.21)$$

The general solution may be written either in terms of hyperbolic functions as,

$$V = V_s \cosh \gamma x - I_s Z_c \sinh \gamma x, \ \gamma = \sqrt{(r + j\omega l)(g + j\omega c)}, \qquad (2.11.22)$$

$$I = I_s \cosh \gamma x - (V_s/Z_c) \sinh \gamma x, \ Z_c = \sqrt{(r + j\omega l)/(g + j\omega c)}, \qquad (2.11.23)$$

The general solution may also be written either in terms of exponentials as,

$$V = (V_s + I_s Z_c)(\exp(\gamma x)/2) + (V_s - I_s Z_c)(\exp(-\gamma x)/2), \qquad (2.11.24)$$

$$I = (V_s + I_s Z_c)(\exp(\gamma x)/2Z_c) - (V_s - I_s Z_c)(\exp(-\gamma x)/2Z_c). \qquad (2.11.25)$$

Thus, the amplitude of the forward and reverse traveling voltage and current waves are, respectively, given by,

$$V^+ = (V_s + I_s Z_c)/2, \ V^- = (V_s - I_s Z_c)/2, \qquad (2.11.26)$$

and

$$I^+ = (V_s + I_s Z_c)/2Z_c, \ I^- = -(V_s - I_s Z_c)/2Z_c, \qquad (2.11.27)$$

The characteristic impedance is the ratio of the amplitude of a voltage and current wave at any one position on the transmission line. Thus, the characteristic impedance is given by,

$$Z_c = V^+/I^+ \text{ or } Z_c = -V^-/I^-. \qquad (2.11.28)$$

If at $x = 0$, $V = IZ_c$, it follows that $V_s - I_s Z_c = 0$ and only the forward traveling wave is present.

If at $x = 0$, $V = IZ_L$, $x = -L$, $V = IZ_{in}$, it follows that,

$$\frac{Z_{\text{in}}}{Z_c} = \frac{Z_L}{Z_c}\frac{1 + (Z_c/Z_L)\tanh \gamma L}{1 + (Z_L/Z_c)\tanh \gamma L}. \qquad (2.11.29)$$

For the transmission line above, we define a voltage reflection coefficient (at the load) as the ratio of reflected voltage to incident voltage, which can in general be complex. Thus if at $x = 0$, $V = IZ_L$,

$$(V_s + I_s Z_c)(1 - Z_L/Z_c) + (V_s - I_s Z_c)(1 + Z_L/Z_c) = 0. \tag{2.11.30}$$

Hence,

$$\Gamma = V^-/V^+ = (V_s - I_s Z_c)/(V_s + I_s Z_c) = -(1 - Z_L/Z_c)/(1 + Z_L/Z_c). \tag{2.11.31}$$

Thus, both the input impedance Z_{in} and the voltage reflection coefficient Γ for a given load impedance Z_L are both functions of the line length L. When $x = -L$,

$$V_R = V_s \cosh \gamma L + I_s Z_c \sinh \gamma L, \tag{2.11.32}$$

$$I_R = I_s \cosh \gamma L + (V_s/Z_c) \sinh \gamma L, \tag{2.11.33}$$

Inverting the relations results in,

$$V_s = V_R \cosh \gamma L - I_R Z_c \sinh \gamma L, \tag{2.11.34}$$

$$I_s = I_R \cosh \gamma L + (V_R/Z_c) \sinh \gamma L. \tag{2.11.35}$$

Equation *** in matrix form is,

$$\begin{bmatrix} V_R \\ I_R \end{bmatrix} = \begin{bmatrix} \cosh \gamma L & Z_c \sinh \gamma L \\ \sinh \gamma L / Z_c & \cosh \gamma L \end{bmatrix} \begin{bmatrix} V_s \\ I_s \end{bmatrix}. \tag{2.11.36}$$

For the π equivalent circuit, it can be shown that,

$$\begin{bmatrix} V_R \\ I_R \end{bmatrix} = \left[\begin{bmatrix} 1 & 0 \\ Y_\pi & 1 \end{bmatrix} + Z_\pi \begin{bmatrix} Y_\pi/2 & 1 \\ (Y_\pi/2)^2 & Y_\pi/2 \end{bmatrix} \right] \begin{bmatrix} V_s \\ I_s \end{bmatrix} \tag{2.11.37}$$

Hence,

$$Z_\pi = Z_c \sinh \gamma L, \quad \frac{Y_\pi}{2} = \frac{1}{Z_c} \tanh\left(\frac{\gamma L}{2}\right). \tag{2.11.38}$$

The impedance and conductance parameters of the π equivalent circuit of a transmission line shown in Fig. 2.28, Z_π and $Y_\pi/2$ are used to estimate the active and reactive power loss components in the transmission line. The evaluation of the active and reactive power losses in the transmission line is facilitated by rewriting transmission line lumped model as,

$$\begin{bmatrix} I_R \\ -I_s \end{bmatrix} = \frac{1}{Z_c \sinh \gamma L} \begin{bmatrix} \cosh \gamma L & -1 \\ -1 & \cosh \gamma L \end{bmatrix} \begin{bmatrix} V_R \\ V_s \end{bmatrix}. \tag{2.11.39}$$

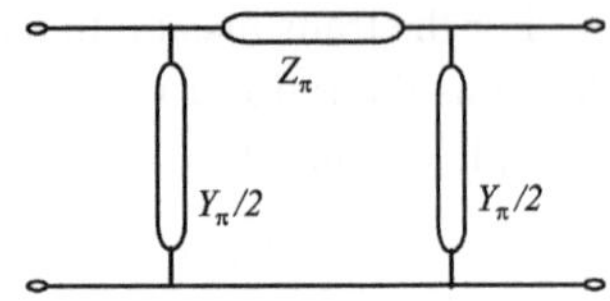

Fig. 2.28 The π equivalent circuit of a transmission line

2.11.3 Transformers

Transformers are essential to power systems, as they are used to convert electrical energy at the voltage, and it is available to some other voltage suitable for transmission or application, or vice versa, although they are typically used to control the load voltage side. Transformers are electromagnetic elements that are modeled by simplified circuit equivalent which is known as the ideal transformer. This is a two-port circuit element, shown in Fig. 2.29.

The ideal transformer as a network element constrains its terminal voltages and currents in the following way:

$$v/N_1 = v_2/N_2,\ N_1 i_1 = -N_2 i_2 \tag{2.11.40}$$

Consequently, the ideal transformer, does not consume, produce nor store energy. Thus, the sum of the power flows into a transformer is identically equal to zero:

$$p_1 + p_2 = v_1 i_1 + v_2 i_2 = 0. \tag{2.11.41}$$

However, because of the way in which an ideal transformer constrains its terminal voltages and currents, it transformers load impedances and source impedances in a special way. Thus, given that the transformer output is loaded by an impedance Z_L,

$$v_2 = -Z_L i_2,\ i_2 = -(N_1/N_2)i_1,\ v_1 = (N_1/N_2)v_2 = (N_1/N_2)^2 Z_L i_1. \tag{2.11.42}$$

Thus, the corresponding input impedance of the transformer is,

$$Z_{\text{in}} = v_1 \backslash i_1 = (N_1/N_2)^2 Z_L. \tag{2.11.43}$$

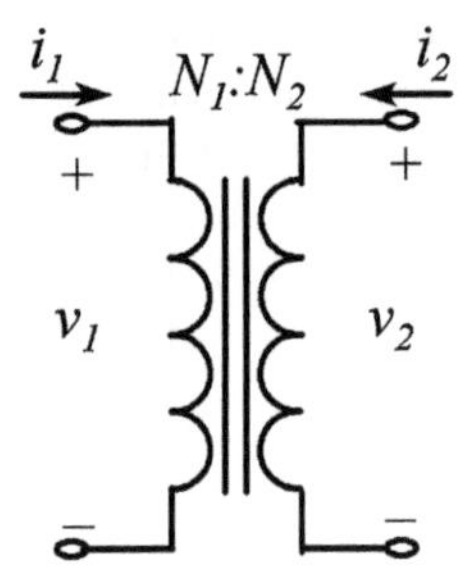

Fig. 2.29 The ideal transformer as a circuit element

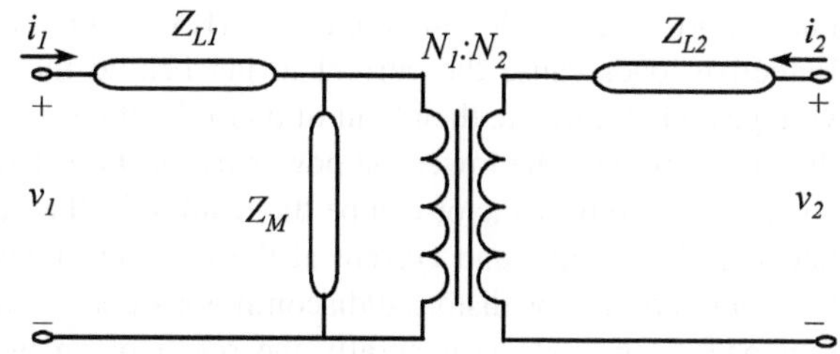

Fig. 2.30 The ideal transformer as a circuit element

The basic characteristics of a real transformer are as follows: (1) Flux leakage around the transformer input and output windings are represented by leakage inductances L_1 and L_2 which are represented by their impedances Z_{L1} and Z_{L2}; (2) The core is assumed to be made of magnetic material and is represented by a magnetization inductance ($L_{ma} \gg L_i$, $i = 1, 2$), but saturates; (3) Losses in the windings (copper wires) and core (hysteresis and induced currents) are represented with lumped resistances; (4) Steps up or down the voltage/current depending on the turn ratio as in the case of an ideal transformer. The equivalent circuit is shown in Fig. 2.30.

The dynamic equations relating the voltages and currents are,

$$\mathbf{v} = \mathbf{Ri} + \frac{\mathrm{d}\boldsymbol{\lambda}}{\mathrm{d}t} = \mathbf{Ri} + \mathbf{L}\frac{\mathrm{d}\mathbf{i}}{\mathrm{d}t}, \tag{2.11.44}$$

$$\begin{bmatrix} v_1 \\ v_2 \end{bmatrix} = \begin{bmatrix} r_1 & 0 \\ 0 & r_2 \end{bmatrix} \begin{bmatrix} i_1 \\ i_2 \end{bmatrix} + \begin{bmatrix} L_{L1} + L_{ma} & L_{ma}/m \\ L_{ma}/m & L_{L2} + L_{ma}/m^2 \end{bmatrix} \frac{\mathrm{d}}{\mathrm{d}t} \begin{bmatrix} i_1 \\ i_2 \end{bmatrix}, \quad m = N_1/N_2. \tag{2.11.45}$$

These equations may be expressed in input–output form in the Laplace transform domain or in the frequency domain.

2.11.4 Power Factor Improvement

Power factor is the ratio between actual power delivered to the apparent power where the apparent power the product of the voltage and current. For a purely resistive load, the power factor is unity. Active and reactive powers are designated by P and Q, respectively. The average power consumed by a circuit is called active power, and the power that supplies the stored energy in reactive elements is called reactive power. Active power is synonymous with "real power" or simply "power." Active power is the rate of producing, transferring, or consuming electrical energy. It is measured in watts and often expressed in kilowatts (KW) or megawatts (MW). Almost always the terms "active" or "real" power are used in place of the term "power" alone to differentiate it from reactive power. The apparent power is the product of the voltage (in volts) and the current (in amperes). It comprises both active and reactive power and is measured in "volt–amperes" but reported as "kilovolt–amperes" (KVA) or "megavolt-amperes" (MVA). Inductive loads cause the current to lag behind the voltage. The wave forms of voltage and current are then "out of phase" with each other. The more out of phase they become then the lower the power factor. Power factor is usually expressed as cos φ. A low-power factor can be the result of either a significant phase difference between the voltage and current at the load terminals or it can be due to a high harmonic content or distorted/discontinuous current waveform. A large load current phase lag angle is generally the result of an inductive load such as an IM power transformer or the Induction generator itself. The power factor decreases further with the installation of non-resistive loads such as AC motors and transformers. The power factor can be improved by using capacitors which work by supplying reactive power opposite in-phase to the reactive power stored in the load. It is usual to operate with a power factor of 0.95–0.98. The power factor can be improved by using fixed Capacitors, electronically switched capacitor banks, synchronous condensers, static synchronous compensators (STATCOMs) and modulated power filter capacitor compensators. A controlled power electronic circuit or static var compensator (SVC) is used to provide the reactive power required to control dynamic voltage swings under various system conditions and thereby improve the power system transmission and distribution performance. By installing suitably sized switched capacitors into the circuit, switched by SVCs, the power factor is improved and the value becomes nearer to 1 thus minimizing wasted energy and improving the efficiency of a plant. In addition, an AVC can dampen the fluctuations in the active power component in the circuit. The power factor can alo be increased by installing synchronous motors with loads or synchronous generators with no prime movers. These units are known as synchronous condensers and put in place primarily for improving the power factor. The STATCOM is a power electronics regulated device that can either inject or absorb reactive power. It is therefore capable of actively regulating the reactive power and maintaining near unity power factor performance. The STATCOM and the SVC belong to family of power factor compensation techniques known as flexible AC transmission systems (FACTS) family.

References

El-Hawary ME (2008) Introduction to electrical power systems, IEEE press series on power engineering, vol 50. Wiley, New York

Kenjo T (1991) An introduction electric motors and their controls. Oxford Science Publishers, Oxford University Press, Oxford

Kundur P (1994) Power system stability and control. International Edition, McGraw-Hill Professional, USA, ISBN 9780070359581

Miano G, Maffucci A (2001) Transmission lines and lumped circuits: fundamentals and applications, Academic Press series in electromagnetism. Academic Press, New York

Wach P (2011) Dynamics and control of electrical drives. Springer, Berlin

Chapter 3
Modelling of Synchronous and Induction Machines

3.1 Generic Principles of Modelling: Park Transformation and Applications

This chapter presents the synchronous and induction machine models which are currently used for control system synthesis and design. In the following section, the basic principles of modelling of both synchronous and induction machines and the park transformation are introduced. This is followed by the equivalent circuit modelling of synchronous and induction machines. Examples relating to the modelling of permanent magnet and wound-field synchronous generator and a doubly fed induction generator are considered. For an exhaustive treatment of the modelling of synchronous and induction machines, the reader is referred to Sarma (1994) and to Krause et al. (1995).

A typical AC generator is a three-phase machine. The voltages in the three phases may be expressed as,

$$v_a = V_a \cos \omega t = V_a \mathrm{Re}\left(e^{j\omega t}\right), \tag{3.1.1}$$

$$v_b = V_b \cos(\omega t - 2\pi/3) = V_b \mathrm{Re}\left(e^{j(\omega t - 2\pi/3)}\right), \tag{3.1.2}$$

Consider connecting these three voltage sources to three identical loads, each with complex impedance Z. The currents in the three phases are given by:

$$i_a = \mathrm{Re}\left(V_a e^{j\omega t}/Z\right), \tag{3.1.3}$$

$$i_b = \mathrm{Re}(V_b e^{j(\omega t - 2\pi/3)}/Z), \tag{3.1.4}$$

$$i_c = \mathrm{Re}\left(V_c e^{j(\omega t + 2\pi/3)}/Z\right). \tag{3.1.5}$$

It is useful to recognize that for any ωt,

$$\cos \omega t + \cos(\omega t - 2\pi/3) + \cos(\omega t + 2\pi/3) = 0. \tag{3.1.6}$$

R. Vepa, *Dynamic Modeling, Simulation and Control of Energy Generation*, Lecture Notes in Energy 20,
DOI: 10.1007/978-1-4471-5400-6_3, © Springer-Verlag London 2013

Hence, the sum of the currents in the three phases is,

$$i_a + i_b + i_c = 0. \tag{3.1.7}$$

The instantaneous power in the phase "a", $p_a = v_a i_a$ is,

$$p_a = \left(V_a^2/Z\right)\cos\omega t\cos(\omega t - \psi). \tag{3.1.8}$$

The instantaneous power in the three phases, $p_x = v_x i_x$, can be expressed as,

$$p_a = \left(V_a^2/2|Z|\right)((1+\cos 2\omega t)\cos\psi + \sin 2\omega t\sin\psi), \tag{3.1.9}$$

$$p_b = \left(V_b^2/2|Z|\right)((1+\cos 2(\omega t - 2\pi/3))\cos\psi + \sin 2(\omega t - 2\pi/3)\sin\psi), \tag{3.1.10}$$

$$p_c = \left(V_c^2/2|Z|\right)((1+\cos 2(\omega t + 2\pi/3))\cos\psi + \sin 2(\omega t + 2\pi/3)\sin\psi). \tag{3.1.11}$$

Hence, the total instantaneous power is,

$$p = p_a + p_b + p_c = \left(3V_a^2/2|Z|\right)\cos\psi. \tag{3.1.12}$$

The complex (active and reactive) power in each of the three phases is,

$$P_{ph} + jQ_{ph} = \left(V_a^2/2|Z|\right)(\cos\psi + j\sin\psi). \tag{3.1.13}$$

Consider the case of the control a wound-field synchronous generator with an exciter is used to generate power. The magnetic flux linked to the windings determines the total voltage drop across the windings. To estimate this voltage drop, it is necessary to evaluate the machines inductance matrix, which relates all windings' flux linkages to all windings' currents. In the case of a salient-pole synchronous machine, the inductance matrix is dependent on the rotor position. This is partially a consequence of the shape and construction of the rotor of a salient-pole machine. Thus, there exists a preferable magnetic direction which coincides with the direction of the flux produced by the field winding. This direction defines the machine's d axis. The machine's q axis is then defined at 90 electrical degrees (in a counterclockwise direction) with respect to the machine's d axis. Then, the rotor position can be expressed by means of an angle, θ, between the magnetic axis of the armature's phase a and the rotor's q axis. The currents in the phases may be transformed to the d–q frame using the Park–Blondel transformation given by,

$$\begin{bmatrix} i_0 \\ i_d \\ i_q \end{bmatrix} = \mathbf{P}\begin{bmatrix} i_a \\ i_b \\ i_c \end{bmatrix}, \quad \mathbf{P} = \sqrt{\frac{2}{3}}\begin{bmatrix} 1/\sqrt{2} & 1/\sqrt{2} & 1/\sqrt{2} \\ \sin\theta & \sin(\theta - 2\pi/3) & \sin(\theta - 4\pi/3) \\ \cos\theta & \cos(\theta - 2\pi/3) & \cos(\theta - 4\pi/3) \end{bmatrix} \tag{3.1.14}$$

where i_a, i_b and i_c are the phase currents, i_d and i_q are the currents in the d and q axes, respectively, θ is the angle between the current in the d axis, i_d and the phase current i_a, and i_0 is the "neutral" current which is zero in a **Y**-connected machine without neutral connection.

If the generator operates in sinusoidal steady state, phase voltages can be written in the *a-b-c* reference frames as,

$$[v_a \quad v_b \quad v_c] = V_p[\cos\theta_v \quad \cos(\theta_v - 2\pi/3) \quad \cos(\theta_v - 4\pi/3)].$$

The voltages in the d and q axes, v_d and v_q, may be obtained by applying the Park–Blondel transformation,

$$[v_0 \quad v_d \quad v_q]^{\mathrm{T}} = \mathbf{P}[v_a \quad v_b \quad v_c]^{\mathrm{T}} = \sqrt{\frac{3}{2}}V_p[0 \quad \sin\delta \quad \cos\delta]^{\mathrm{T}}. \tag{3.1.15}$$

The "neutral" voltage v_0 is zero. The rotor angle $\delta = \theta - \theta_v - \pi/2$, is defined as the displacement of the rotor referenced to the maximum positive value of the fundamental component of the phase voltage, in phase a. In fact, θ, θ_v, and δ may, respectively, be expressed in terms of the rotor's electrical speed ω_{e} and the rated synchronous frequency ω_r, as,

$$\theta = \omega_{\mathrm{e}}t = \theta_v + \delta + \pi/2,\ \theta_v = \omega_r t,\ \mathrm{d}\delta/\mathrm{d}t = \omega_{\mathrm{e}} - \omega_r. \tag{3.1.16}$$

The rotor windings produce a flux linkage that mainly lies in the direction of the d axis. That flux induces an EMF, which is lagging by 90°, hence in the direction of the negative q axis. For generator operation, the phasor for the induced EMF leads by an angle δ before the phasor for the terminal voltage. At $t = 0$, the negative q axis thus leads by an angle delta before the phasor for the voltage along the a axis. For $t > 0$, the d and q axes have moved by an angle $\omega_r t$ with the angular speed of the rotor ω_r. The rotor's d axis will hence be in position. The rotor traverses through one electrical cycle as it goes past a pair of magnetic poles. Thus, the rotor's electrical speed is an integral multiple of the rotor's mechanical speed. The rotor's electrical speed ω_{e} and electrical load torque T_{el}, are, respectively, related to the rotor's mechanical speed ω_m and mechanical load torque T_{mel}, and the number of poles P, via the relations,

$$\omega_{\mathrm{e}} = P\omega_m/2,\ T_{\mathrm{el}} = 2T_{\mathrm{mel}}/P. \tag{3.1.17}$$

The magnetic field generated by a winding is the sum of the fields generated by the individual coils in the winding. By Ampere's law, the magnetic field generated by N coils carrying a current i_a is proportional to $Ni_a/2$. The component of the field in a direction at an angle θ to the direction of the field is $Ni_a\cos\theta/2$. Thus, the magnetic flux generated in a direction normal to the coil can be expressed as $\lambda = L_{eqa}i_a$.

The flux linkages of phase windings a, b, and c can be expressed in terms of the self- and mutual inductances as the following,

$$\begin{bmatrix}\lambda_a\\\lambda_b\\\lambda_c\end{bmatrix}=\begin{bmatrix}\lambda_{aa}\\\lambda_{ba}\\\lambda_{ca}\end{bmatrix}+\begin{bmatrix}\lambda_{ab}\\\lambda_{bb}\\\lambda_{cb}\end{bmatrix}+\begin{bmatrix}\lambda_{ac}\\\lambda_{bc}\\\lambda_{cc}\end{bmatrix}+\begin{bmatrix}\lambda_{af}\\\lambda_{bf}\\\lambda_{cf}\end{bmatrix}$$
$$=\begin{bmatrix}L_{aa}&L_{ab}&L_{ac}\\L_{ba}&L_{bb}&L_{bc}\\L_{ca}&L_{cb}&L_{cc}\end{bmatrix}\begin{bmatrix}i_a\\i_b\\i_c\end{bmatrix}+\begin{bmatrix}L_{af}\\L_{bf}\\L_{cf}\end{bmatrix}i_f.$$

In Eq. 3.1.18, the stator self-inductances are

$$\begin{bmatrix}L_{aa}\\L_{bb}\\L_{cc}\end{bmatrix}=\begin{bmatrix}L_{aa0}\\L_{bb0}\\L_{cc0}\end{bmatrix}+L_{al}\begin{bmatrix}1\\1\\1\end{bmatrix}+\begin{bmatrix}\cos 2\theta_{\lambda a}\\\cos 2(\theta_{\lambda a}-2\pi/3)\\\cos 2(\theta_{\lambda a}+2\pi/3)\end{bmatrix}L_m \tag{3.1.19}$$

The stator mutual inductances are

$$\begin{bmatrix}L_{ab}\\L_{ac}\\L_{bc}\end{bmatrix}=\begin{bmatrix}L_{ba}\\L_{ca}\\L_{cb}\end{bmatrix}=-\frac{1}{2}\begin{bmatrix}L_{aa0}\\L_{bb0}\\L_{cc0}\end{bmatrix}+\begin{bmatrix}\cos 2(\theta_{\lambda a}-\pi/3)\\\cos 2(\theta_{\lambda a}-2\pi/3)\\\cos 2\theta_{\lambda a}\end{bmatrix}L_m, \tag{3.1.20}$$

and

$$\begin{bmatrix}L_{af}\\L_{bf}\\L_{cf}\end{bmatrix}=\begin{bmatrix}\cos(\theta_{\lambda a})\\\cos(\theta_{\lambda a}-2\pi/3)\\\cos(\theta_{\lambda a}+2\pi/3)\end{bmatrix}L_{\mathrm{afm}}. \tag{3.1.21}$$

The total flux linkage in phase a, λ_a, may be expressed as,

$$\begin{aligned}\lambda_a &= (L_{aa0}+L_{al}+L_m\cos 2\theta_{\lambda a})i_a - L_{aa0}(i_b+i_c)/2\\&\quad + i_bL_m\cos 2(\theta_{\lambda a}-\pi/3)+i_cL_m\cos 2(\theta_{\lambda a}-2\pi/3)+L_{\mathrm{afm}}i_f\cos\theta_{\lambda a}\end{aligned} \tag{3.1.22}$$

Further since,

$$i_a+i_b+i_c=0$$

$$\begin{aligned}\lambda_a &= (L_{aa0}+L_{al}+L_{aa0}/2+L_m\cos 2\theta_{\lambda a})i_a\\&\quad + i_bL_m\cos 2(\theta_{\lambda a}-\pi/3)+i_cL_m\cos 2(\theta_{\lambda a}-2\pi/3)+L_{\mathrm{afm}}i_f\cos\theta_{\lambda a}.\end{aligned} \tag{3.1.23}$$

Hence,

$$\begin{aligned}\lambda_a &= (3L_{aa0}/2+L_{al}+L_m\cos 2\theta_{\lambda a})i_a\\&\quad + i_bL_m\cos 2(\theta_{\lambda a}-\pi/3)+i_cL_m\cos 2(\theta_{\lambda a}-2\pi/3)+L_{\mathrm{afm}}i_f\cos\theta_{\lambda a}.\end{aligned} \tag{3.1.24}$$

If the synchronous inductance, L_s, is defined as,

$$L_s = 3L_{aa0}/2 + L_{\mathrm{al}}, \tag{3.1.25}$$

the total flux linkage in phase a, λ_a, may be expressed as,

$$\begin{aligned}\lambda_a &= (L_s + L_m \cos 2\theta_{\lambda a})i_a \\ &\quad + i_b L_m \cos 2(\theta_{\lambda a} - \pi/3) + i_c L_m \cos 2(\theta_{\lambda a} - 2\pi/3) + L_{\mathrm{afm}} i_f \cos \theta_{\lambda a}.\end{aligned} \tag{3.1.26}$$

Thus, the flux linkages of phase windings a, b, and c can be expressed in terms of the self- and mutual inductances as the following,

$$\begin{bmatrix} \lambda_a \\ \lambda_b \\ \lambda_c \end{bmatrix} = L_s \begin{bmatrix} i_a \\ i_b \\ i_c \end{bmatrix} + \mathbf{L}_m \begin{bmatrix} i_a \\ i_b \\ i_b \end{bmatrix} + \begin{bmatrix} \cos(\theta_{\lambda a}) \\ \cos(\theta_{\lambda a} - 2\pi/3) \\ \cos(\theta_{\lambda a} + 2\pi/3) \end{bmatrix} L_{\mathrm{afm}} i_f \tag{3.1.27}$$

where

$$\mathbf{L}_m = \begin{bmatrix} \cos 2(\theta_{\lambda a}) & \cos 2(\theta_{\lambda a} - \pi/3) & \cos 2(\theta_{\lambda a} - 2\pi/3) \\ \cos 2(\theta_{\lambda a} - 2\pi/3) & \cos 2(\theta_{\lambda a}) & \cos 2(\theta_{\lambda a} - \pi/3) \\ \cos 2(\theta_{\lambda a} - \pi/3) & \cos 2(\theta_{\lambda a} - 2\pi/3) & \cos 2(\theta_{\lambda a}) \end{bmatrix}. \tag{3.1.28}$$

The flux linkages in the three phases λ_a, λ_b, and λ_c may be transformed to and from the d–q axes flux linkage components, λ_0, λ_d, and λ_q, and can be written as,

$$\begin{bmatrix} \lambda_0 \\ \lambda_d \\ \lambda_q \end{bmatrix} = \mathbf{P} \begin{bmatrix} \lambda_a \\ \lambda_b \\ \lambda_c \end{bmatrix} \text{ and } \begin{bmatrix} \lambda_a \\ \lambda_b \\ \lambda_c \end{bmatrix} = \mathbf{P}^{-1} \begin{bmatrix} \lambda_0 \\ \lambda_d \\ \lambda_q \end{bmatrix} \text{ where } \mathbf{P}^{-1} = \mathbf{P}^T. \tag{3.1.29}$$

In particular, when $L_m = 0$,

$$\begin{bmatrix} \lambda_0 \\ \lambda_d \\ \lambda_q \end{bmatrix} = L_s \begin{bmatrix} i_0 \\ i_d \\ i_q \end{bmatrix} + \sqrt{\frac{3}{2}} \begin{bmatrix} 0 \\ \sin(\theta - \theta_{\lambda a}) \\ \cos(\theta - \theta_{\lambda a}) \end{bmatrix} L_{\mathrm{afm}} i_f \tag{3.1.30}$$

Further, if $\theta - \theta_{\lambda a} = \pi/2$, and $L_m \neq 0$,

$$\begin{bmatrix} \lambda_0 \\ \lambda_d \\ \lambda_q \end{bmatrix} = L_s \begin{bmatrix} i_0 \\ i_d \\ i_q \end{bmatrix} + \frac{3}{2} L_m \begin{bmatrix} 0 & 0 & 0 \\ 0 & 1 & 0 \\ 0 & 0 & -1 \end{bmatrix} \begin{bmatrix} i_0 \\ i_d \\ i_q \end{bmatrix} + \sqrt{\frac{3}{2}} \begin{bmatrix} 0 \\ 1 \\ 0 \end{bmatrix} L_{\mathrm{afm}} i_f. \tag{3.1.31}$$

In this case, the flux linkage due to the field is in the direction of the d axis.

The currents in a three-phase synchronous motor, defined in the conventional magnetic axes of the three machine phases, are separated by $\gamma_p = 2\pi/3$. They may be transformed to a new two-axis frame. The real or α axis of the new two-axis coordinate system is arbitrarily chosen to coincide with the a axis. Obviously, the quadrature or β axis lies in quadrature with the α axis. The pair can be represented in two dimensions and on a complex plane. The currents in the two axes are related to those in the three phases by,

$$\begin{bmatrix} i_\alpha \\ i_\beta \end{bmatrix} = \frac{2}{3}\begin{bmatrix} 1 & \cos(2\pi/3) & \cos(4\pi/3) \\ 0 & \sin(2\pi/3) & \sin(4\pi/3) \end{bmatrix}\begin{bmatrix} i_a \\ i_b \\ i_b \end{bmatrix}, \begin{bmatrix} i_a \\ i_b \\ i_b \end{bmatrix} = \begin{bmatrix} 1 & \cos(2\pi/3) & \cos(4\pi/3) \\ 0 & \sin(2\pi/3) & \sin(4\pi/3) \end{bmatrix}^T \begin{bmatrix} i_\alpha \\ i_\beta \end{bmatrix}. \tag{3.1.32}$$

The current space vector may be shown as a vector in the complex plane. The phase currents may be obtained by projecting the current space vector onto the respective phase axis. The real and imaginary components of the current space vector in the original stationary two-axis reference frame, the stationary ($\alpha\beta$) frame, are transformed to a new rotating reference frame, the dq frame, where both the current space vector and the direct and quadrature axes are rotating at a speed ω_e.

$$\begin{bmatrix} i_\alpha \\ i_\beta \end{bmatrix} = \begin{bmatrix} \cos\theta & -\sin\theta \\ \sin\theta & \cos\theta \end{bmatrix}\begin{bmatrix} i_d \\ i_q \end{bmatrix}, \begin{bmatrix} i_d \\ i_q \end{bmatrix} = \begin{bmatrix} \cos\theta & \sin\theta \\ -\sin\theta & \cos\theta \end{bmatrix}\begin{bmatrix} i_\alpha \\ i_\beta \end{bmatrix}, \frac{\mathrm{d}}{\mathrm{d}t}\theta = \omega_\mathrm{e} \tag{3.1.33}$$

3.2 Exciting a Synchronous Generator

The DC field excitation of a wound-field synchronous generator is an important factor in determining the performance of the generator. The stability of the generator, particularly when there are sudden changes in the load and the speed of response of the generator, depends to large extent on the nature of the excitation provided. Some machines generally use more than one exciter. The simplest method of providing the excitation is by a DC supply via a set of commutator split rings and brushes. The excitation current may be obtained by using a DC generator or a synchronous generator. In the latter case, it must be rectified before it is supplied to the exciter. Thus, up to 3 % of the power generated by the generator is used up in providing the excitation. To avoid wear and tear of the mechanical parts in commutator with split rings and brushes, a brushless system is desirable. The brushless system uses a second exciter to supply a stationary permanent magnet attached to a stator and rotating armature with a three-phase winding. The three-phase output is rectified by a rotating three-phase bridge rectifier, and the output is supplied to the field of the main generator. Such a system avoids the use of brushes although there is a need to provide the second smaller exciter.

3.3 Equivalent Circuit Modelling

Figure 3.1 shows the d and q axes equivalent circuits of permanent magnet synchronous generator in the d–q coordinates, which rotate synchronously with an electrical angular velocity ω. The advantage of equivalent circuit modelling is the ease with which one can include the effects of losses and hysteresis. The equivalent circuit model shown in Fig. 3.1 includes the effects of the copper loss and the core loss, where the core loss is expressed by the equivalent core loss resistance R_c. If R_c is assumed to be a function of frequency, the hysteresis loss in addition to the eddy-current loss can be considered.

3.4 The Electromechanical Model of a Permanent Magnet Synchronous Generator with Field Orientation

A three-phase permanent magnet synchronous generator consists of field generated by a permanent magnet rotating within a stator field of a set of armature coils. The armature windings are subjected to large transient currents and must have a large mechanical strength. For this reason, the armature windings are placed on the stator, while permanent magnets are mounted on the rotor. The rotating permanent magnet produces a rotating magnetic field in the air gap between the rotor and the

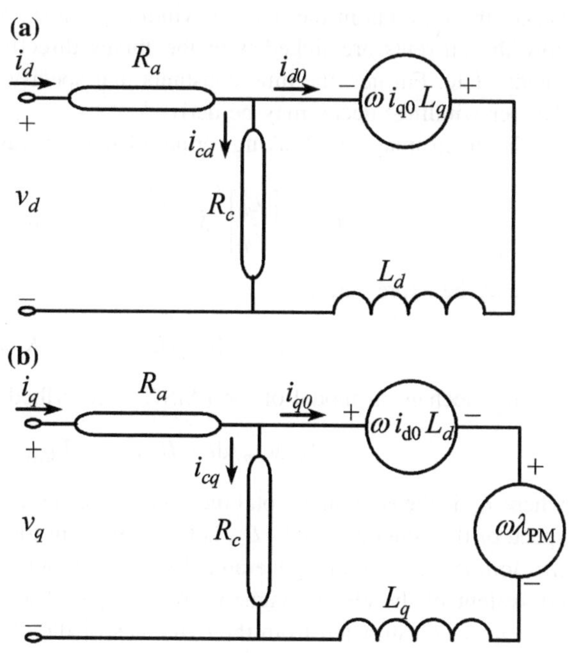

Fig. 3.1 Equivalent circuit modelling of a synchronous generator

stator. The dynamics of the voltages and currents in the three phases of the armature may be expressed directly in the d and the q axes. The dynamics in the d and the q axes are similar to that of a transformer equivalent circuit: In each of them, the windings, each characterized by some resistance and leakage inductance, are coupled through a mutual coupling inductance. The dynamical equations describing the relationship between the currents, voltages, and electrical rotational speed may be conveniently stated in the d–q frame as,

$$\begin{bmatrix} v_d \\ v_q \end{bmatrix} = -R_s \begin{bmatrix} i_d \\ i_q \end{bmatrix} + \omega_e \mathbf{J}^T \begin{bmatrix} \lambda_d \\ \lambda_q \end{bmatrix} - \frac{\mathrm{d}}{\mathrm{d}t} \begin{bmatrix} \lambda_d \\ \lambda_q \end{bmatrix}, \ \mathbf{J}^T = \begin{bmatrix} 0 & -1 \\ 1 & 0 \end{bmatrix} \tag{3.4.1}$$

$$\begin{aligned} \begin{bmatrix} \lambda_d \\ \lambda_q \end{bmatrix} &= L_{1s} \begin{bmatrix} i_d \\ i_q \end{bmatrix} + \frac{3}{2} \begin{bmatrix} L_0 - L_2 & 0 \\ 0 & L_0 + L_2 \end{bmatrix} \begin{bmatrix} i_d \\ i_q \end{bmatrix} + \begin{bmatrix} K_t \\ 0 \end{bmatrix} \\ &\equiv \begin{bmatrix} L_d & 0 \\ 0 & L_q \end{bmatrix} \begin{bmatrix} i_d \\ i_q \end{bmatrix} + \begin{bmatrix} K_t \\ 0 \end{bmatrix}, K_t = \sqrt{\frac{3}{2}} \lambda_m, \end{aligned} \tag{3.4.2}$$

v_d, v_q are the control input voltages in the direct (d) and quadrature (q) axes, i_d, i_q are the generator currents in the direct (d) and quadrature (q) axes, λ_d, λ_q d–q components of stator flux, λ_m, permanent magnet flux, K_t is the torque constant, R_s is the armature winding resistance, L_0 is the component of the self-inductance due to space fundamental air-gap flux, L_2 is the component of the self-inductance due to rotor position dependent flux, and L_{ls} is the leakage inductance. The complete description of the synchronous machine given above can be simplified and made more physical if new parameters L_d and L_q that can be expressed in terms of L_{ls}, L_0 L_2 and are introduced. In steady state and a sufficiently long time after a disturbance, the currents in the damper windings vanish, and the inductances describe how the currents are linked with the fluxes directly after a disturbance and in steady state. Further, the time constants that specify how fast the currents in the damper windings decay may be derived.

The electromagnetic braking torque of the generator is

$$T_{\mathrm{el}} = \begin{bmatrix} \lambda_d \\ \lambda_q \end{bmatrix}^T \mathbf{J} \begin{bmatrix} i_d \\ i_q \end{bmatrix}, T_{\mathrm{el}} = \left(\lambda_d i_q - \lambda_q i_d\right), \tag{3.4.3}$$

and may be expressed as,

$$T_{\mathrm{el}} = \left(i_d i_q \left(L_d - L_q\right) + K_t i_q\right). \tag{3.4.4}$$

The mechanical model of the PMSG is described by,

$$J_{\mathrm{eq}} \, \mathrm{d}\omega_m / \mathrm{d}t + B_{\mathrm{eq}} \omega_m = T_{\mathrm{drive}} - T_{\mathrm{mel}} \tag{3.4.5}$$

where J_{eq} is the equivalent total inertia of the generator shaft, ω_m is the mechanical speed of the generator shaft, B_{eq} is the equivalent total friction coefficient, T_{drive} is the torque driving the generator from the wind, and T_{mel} is the mechanical equivalent of the electromagnetic load torque. The electromagnetic load torque, T_{mel}, may be estimated from the dynamics of the electric generator.

The rotor's electrical speed ω_e and electrical load torque T_{el} are, respectively, related to the rotor's mechanical speed ω_m and mechanical load torque T_{mel}, and the number of poles P, via the relations,

$$\omega_e = P\omega_m/2,\ T_{el} = 2T_{mel}/P. \tag{3.4.6}$$

The equation governing the electrical rotor speed ω_e is

$$J_{eq}\frac{d\omega_e}{dt} + B_{eq}\omega_e = \frac{P}{2}\left(T_{drive} - \frac{P}{2}T_{el}\right) \tag{3.4.7}$$

The field orientation concept leads to a simplification of the dynamical equations of the permanent magnet synchronous generator. The permanent magnet flux linkages are aligned with the d axis, while the stator current vector is kept along the q axis direction. Hence, it follows that,

$$i_d \approx 0, \lambda_d \approx K_t, \lambda_q = \left(L_{1s} + \frac{3}{2}(L_0 + L_2)\right)i_q = L_q i_q, \tag{3.4.8}$$

$$v_d = -\omega_e L_q i_q, v_q = -R_s i_q + \omega_e \lambda_d - \frac{d\lambda_q}{dt}, \tag{3.4.9}$$

$$T_{el} = K_t i_q. \tag{3.4.10}$$

Thus, the governing equations are reduced to,

$$v_q = -R_s i_q - L_q\frac{di_q}{dt} + K_t\omega_e, \tag{3.4.11}$$

$$J_{eq}\frac{d\omega_e}{dt} + B_{eq}\omega_e = \frac{P}{2}\left(T_{drive} - \frac{P}{2}K_t i_q\right). \tag{3.4.12}$$

The governing equations are similar to those of a DC motor. The dynamic inverse model is given by,

$$\frac{P^2}{4}K_t i_q = \frac{PT_{drive}}{2} - \left(J_{eq}\frac{d}{dt} + B_{eq}\right)\omega_e, v_q = -\left(L_q\frac{d}{dt} + R_s\right)i_q + K_t\omega_e, v_d = -\omega_e L_q i_q. \tag{3.4.13}$$

Including the process input which is assumed to be given by the drive torque, the state space model is:

$$\frac{di_q}{dt} = -\frac{(R_s i_q - K_t\omega_e)}{L_q} - \frac{v_q}{L_q}, \tag{3.4.14}$$

$$\frac{d\alpha_e}{dt} = -\frac{B_{eq}}{J_{eq}}\alpha_e + \frac{P^2}{4}K_t\frac{(R_s i_q - K_t\omega_e)}{J_{eq}L_q} + \frac{P^2K_t}{4J_{eq}}\frac{v_q}{L_q} + \frac{P}{2}\frac{J_d}{J_{eq}}, \tag{3.4.15}$$

$$\frac{\mathrm{d}\omega_{\mathrm{e}}}{\mathrm{d}t} = \alpha_e, \tag{3.4.16}$$

$$\frac{\mathrm{d}\theta_e}{\mathrm{d}t} = \omega_{\mathrm{e}}, \tag{3.4.17}$$

where J_{d} is the driving torque jerk. Assuming that measurements of the quadrature axis current and electrical rotor speed are made, the state space equations are expressed as,

$$\frac{\mathrm{d}}{\mathrm{d}t}\begin{bmatrix} i_q \\ \alpha_e \\ \omega_{\mathrm{e}} \\ \theta_e \end{bmatrix} = \begin{bmatrix} -R_s/L_q & 0 & K_t/L_q & 0 \\ \frac{P^2K_tR_s}{4J_{\mathrm{eq}}L_q} & -\frac{B_{\mathrm{eq}}}{J_{\mathrm{eq}}} & -\frac{P^2K_t^2}{4J_{\mathrm{eq}}L_q} & 0 \\ 0 & 1 & 0 & 0 \\ 0 & 0 & 1 & 0 \end{bmatrix} \begin{bmatrix} i_q \\ \alpha_e \\ \omega_{\mathrm{e}} \\ \theta_e \end{bmatrix} + \begin{bmatrix} -1 \\ \frac{P^2K_t}{4J_{\mathrm{eq}}} \\ 0 \\ 0 \end{bmatrix} u_q + \begin{bmatrix} 0 \\ 1 \\ 0 \\ 0 \end{bmatrix} w_d \tag{3.4.18}$$

$$\mathbf{y} = \begin{bmatrix} 0 & 0 & 1 & 0 \\ 1 & 0 & 0 & 0 \end{bmatrix} \begin{bmatrix} i_q \\ \alpha_e \\ \omega_{\mathrm{e}} \\ \theta_e \end{bmatrix} + \begin{bmatrix} 1 & 0 \\ 0 & 1 \end{bmatrix} \begin{bmatrix} v_\omega \\ v_i \end{bmatrix} \tag{3.4.19}$$

where the control input is defined by, $u_q = v_q/L_q$, the process disturbance input by, $w_d = PJ_{\mathrm{d}}/2J_{\mathrm{eq}}$, $\mathbf{y}$ is the vector of measurements, and v_ω and v_i are the measurement disturbances. If the control input is defined by a particular control, the voltage inputs in the d and q axes are

$$\begin{bmatrix} v_d \\ v_q \end{bmatrix} = L_q \begin{bmatrix} -\omega_{\mathrm{e}} i_q \\ u_q \end{bmatrix}. \tag{3.4.20}$$

Under steady-state conditions and in the absence of noise,

$$L_q u_{\mathrm{qs}} = -R_s i_{\mathrm{qs}} + K_t \omega_{\mathrm{es}}, \tag{3.4.21}$$

$$B_{\mathrm{eq}}\omega_{\mathrm{es}} = \frac{P}{2}\left(T_{\mathrm{drive}} - \frac{P}{2}K_t i_{\mathrm{qs}}\right). \tag{3.4.22}$$

$$\begin{bmatrix} -P^2K_t/4 & 0 \\ R_s/L_q & 1 \end{bmatrix} \begin{bmatrix} i_{\mathrm{qs}} \\ u_{\mathrm{qs}} \end{bmatrix} = \begin{bmatrix} -P/2 & B_{\mathrm{eq}} \\ 0 & K_t/L_q \end{bmatrix} \begin{bmatrix} T_{\mathrm{drive}} \\ \omega_{\mathrm{es}} \end{bmatrix}, \; \alpha_e = 0. \tag{3.4.23}$$

Hence,

$$\begin{bmatrix} i_{\mathrm{qs}} \\ u_{\mathrm{qs}} \end{bmatrix} = \begin{bmatrix} -4/P^2K_t & 0 \\ 4R_s/P^2L_qK_t & 1 \end{bmatrix} \begin{bmatrix} -P/2 & B_{\mathrm{eq}} \\ 0 & K_t/L_q \end{bmatrix} \begin{bmatrix} T_{\mathrm{drive}} \\ \omega_{\mathrm{es}} \end{bmatrix}. \tag{3.4.24}$$

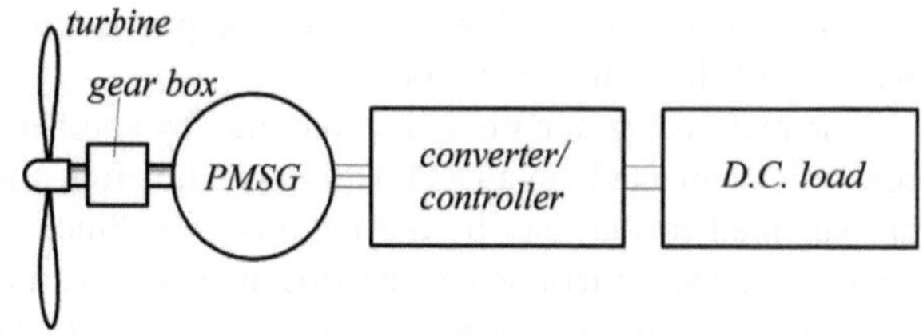

Fig. 3.2 Block diagram of a typical permanent magnet synchronous generator

or

$$\begin{bmatrix} T_{\text{drive}} \\ \omega_{\text{es}} \end{bmatrix} = \begin{bmatrix} -2/P & 2B_{\text{eq}}L_q/PK_t \\ 0 & L_q/K_t \end{bmatrix} \begin{bmatrix} -P^2K_t/4 & 0 \\ R_s/L_q & 1 \end{bmatrix} \begin{bmatrix} i_{\text{qs}} \\ u_{\text{qs}} \end{bmatrix}. \tag{3.4.25}$$

The state space equations may be expressed as,

$$\frac{\mathrm{d}}{\mathrm{d}t}\begin{bmatrix} i_q \\ \omega_{\text{e}} \end{bmatrix} = \begin{bmatrix} -R_s/L_q & K_t/L_q \\ -\frac{P^2K_t}{4J_{\text{eq}}} & -\frac{B_{\text{eq}}}{J_{\text{eq}}} \end{bmatrix} \begin{bmatrix} i_q - i_{\text{qs}} \\ \omega_{\text{e}} - \omega_{\text{es}} \end{bmatrix} + \begin{bmatrix} -1 \\ 0 \end{bmatrix} \left(u_q - u_{\text{qs}}\right) + \begin{bmatrix} 0 \\ 1 \end{bmatrix} \frac{PT_{\text{drive}}}{2J_{\text{eq}}}, \tag{3.4.26}$$

$$\mathbf{y} = \begin{bmatrix} 0 & 1 \\ 1 & 0 \end{bmatrix} \begin{bmatrix} i_q \\ \omega_{\text{e}} \end{bmatrix} + \begin{bmatrix} 1 & 0 \\ 0 & 1 \end{bmatrix} \begin{bmatrix} v_\omega \\ v_i \end{bmatrix}. \tag{3.4.27}$$

A block diagram of the complete system including the controller is shown in Fig. 3.2.

3.5 Application to a Wound-Field Synchronous Generator

A typical example of a wound-field synchronous generator is shown in the block diagram in Fig. 3.3. It is based on a 100–300 kW generator with an inverter output, in which the prime mover is a gasoline engine driving the main synchronous generator. The field excitation for the main generator is provided by a separate, smaller synchronous generator, known as an exciter. Typically, the exciter is constructed with field winding on the stator and armature winding on the rotor which makes it possible to rectify the exciter's armature AC output by a

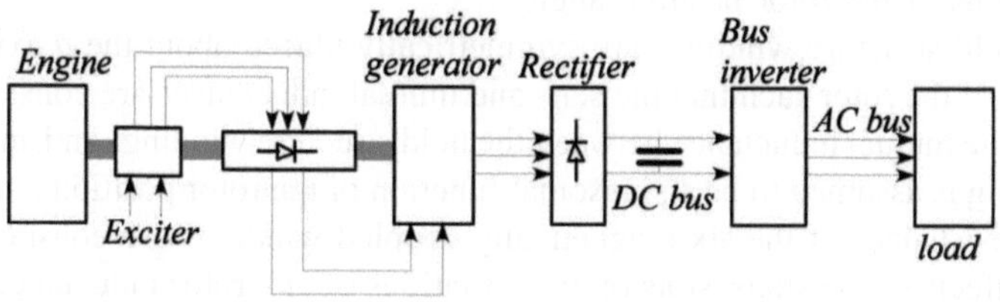

Fig. 3.3 Block diagram of a typical wound-field synchronous generator

rotating diode bridge. The rectifier's output is connected directly to the field winding of the main generator.

The main generator's output is rectified by another power diode circuit, and the output is connected via a DC link to a bus inverter. Balanced three-phase voltages are supplied to the load by the bus inverter. Since the inverter determines load frequency, the generator can be operated at a variable speed. A speed control system is used to operate the generator, so it tracks the maximum power transfer conditions based on the load versus speed characteristics that are optimal for the prime mover. Variable speed operation requires the generator to be designed with an optimum synchronous inductance. With most current wound-field synchronous generators, both the exciter and main generators have power electronic controllers to facilitate variable speed operation.

A three-phase, wound-field synchronous generator has three identical armature windings, symmetrically distributed around the air-gap and one field winding. One or more damper windings can also be present, and for convenience, it is assumed that two damper windings are present, one in each of the machine's axes. Armature windings are placed on the stator, and field and damper windings on the rotor. Rotor windings refer to the field and damper windings placed at the opposite side of the air gap with respect to the three-phase armature windings. Furthermore, all copper losses and the slots in the machine as well as all losses due to the wiring, saturation, and slots are neglected, the spatial distribution of the stator fluxes is considered sinusoidal and the stator, and rotor permeability are assumed to be infinite.

The two damper windings on the rotor are the d damper winding which has its magnetic axis in the same direction as the d axis and the q damper winding which has its magnetic axis in the same direction as the q axis. The damper windings are generally short-circuited and are generally constructed in a manner very similar to the squirrel-cage windings in an induction motor. They are used to damp out the oscillations in the rotor, to reduce over voltages under short-circuit conditions and aid in the synchronizing the machine. Although all six windings are magnetically coupled, it can be assumed in the rotating d–q axes the d axis armature winding is inductively coupled to the field and the d axis damper winding while the q axis armature winding is inductively coupled to the q axis damper winding.

The derivation of the differential equations of the machines electrical dynamics is based on the following assumptions:

1. The armature windings are distributed around the periphery of the inner face of the stator in such a manner that the self- and mutual inductances are sinusoidal functions of the rotor position angle.
2. The field structure windings are symmetrically placed about the d axis and the q axis of the rotor such that the self- and mutual inductances are constant. However, the mutual inductance between the field structure windings and an armature winding is assumed to be a sinusoidal function of the rotor position angle.
3. The resistances of the six magnetically coupled windings are constant.
4. The effects of the stator slots on the variations of any rotor inductance with the rotor position angle are neglected.

5. Saturation effects are not included in the linear model.
6. Hysteresis effects and eddy currents are neglected.

With the above assumptions, the voltages in the armature windings in the d–q frame may be expressed as,

$$\begin{bmatrix} v_d \\ v_q \end{bmatrix} = -R_s \begin{bmatrix} i_d \\ i_q \end{bmatrix} + \omega_e \mathbf{J} \begin{bmatrix} \lambda_d \\ \lambda_q \end{bmatrix} - \frac{\mathrm{d}}{\mathrm{d}t} \begin{bmatrix} \lambda_d \\ \lambda_q \end{bmatrix}, \ \mathbf{J} = \begin{bmatrix} 0 & 1 \\ -1 & 0 \end{bmatrix}. \tag{3.5.1}$$

The field equation relating the field voltage v_{fd}, the field current i_{fd}, and the field flux linkages λ_{fd} is,

$$v_{fd} = R_{fd} i_{fd} + d\lambda_{fd}/\mathrm{d}t. \tag{3.5.2}$$

The damper windings in the d–q frame, which generate no voltages, generate the damper-winding currents i_{kd} and i_{kq} due to the presence of the corresponding the flux linkages λ_{kd} and λ_{kq}. The equations relating i_{kd} and i_{kq} to the flux linkages λ_{kd} and λ_{kq} are

$$\begin{bmatrix} 0 \\ 0 \end{bmatrix} = \begin{bmatrix} R_{\mathrm{kd}} & 0 \\ 0 & R_{\mathrm{kq}} \end{bmatrix} \begin{bmatrix} i_{\mathrm{kd}} \\ i_{\mathrm{kq}} \end{bmatrix} + \frac{\mathrm{d}}{\mathrm{d}t} \begin{bmatrix} \lambda_{\mathrm{kd}} \\ \lambda_{\mathrm{kq}} \end{bmatrix}. \tag{3.5.3}$$

The flux linkages are related to the currents in the armature, field, and damper windings, by the relations,

$$\begin{bmatrix} \lambda_d \\ \lambda_q \end{bmatrix} = L_{1s} \begin{bmatrix} i_d \\ i_q \end{bmatrix} + \begin{bmatrix} L_{\mathrm{md}} & 0 \\ 0 & L_{\mathrm{mq}} \end{bmatrix} \left[\begin{bmatrix} i_d \\ i_q \end{bmatrix} - \begin{bmatrix} i_{\mathrm{kd}} \\ i_{\mathrm{kq}} \end{bmatrix} \right] - \begin{bmatrix} L_{\mathrm{md}} \\ 0 \end{bmatrix} i_{\mathrm{fd}}, \tag{3.5.4}$$

$$\lambda_{\mathrm{fd}} = \begin{bmatrix} -L_{\mathrm{md}} & (L_{1\mathrm{fd}} + L_{\mathrm{md}}) & L_{\mathrm{md}} \end{bmatrix} \begin{bmatrix} i_d & i_{\mathrm{fd}} & i_{\mathrm{kd}} \end{bmatrix}^T, \tag{3.5.5}$$

and

$$\begin{bmatrix} \lambda_{\mathrm{kd}} \\ \lambda_{\mathrm{kq}} \end{bmatrix} = -\begin{bmatrix} L_{\mathrm{md}} & 0 \\ 0 & L_{\mathrm{mq}} \end{bmatrix} \begin{bmatrix} i_d \\ i_q \end{bmatrix} + \begin{bmatrix} L_{1\mathrm{kd}} + L_{\mathrm{md}} & 0 \\ 0 & L_{1\mathrm{kq}} + L_{\mathrm{mq}} \end{bmatrix} \begin{bmatrix} i_{\mathrm{kd}} \\ i_{\mathrm{kq}} \end{bmatrix} + \begin{bmatrix} L_{\mathrm{md}} \\ 0 \end{bmatrix} i_{\mathrm{fd}}. \tag{3.5.6}$$

It is customary to refer the mechanical motion dynamics also to the electrical reference speed and write the mechanical equation of motion as,

$$J_{\mathrm{eq}} \frac{\mathrm{d}\omega_{\mathrm{e}}}{\mathrm{d}t} + B_{\mathrm{eq}} \omega_{\mathrm{e}} = \frac{P}{2} \left(T_{wt} - \frac{P}{2} T_{\mathrm{el}} \right) \tag{3.5.7}$$

The definitions of the electrical power output, the electromagnetic load torque, and the exciter model are essential to complete the definition of the synchronous generator. The output active power p_a, the reactive power q_r, and the electromagnetic load torque T_{el} are, respectively, given by,

$$p_a = \begin{bmatrix} v_d \\ v_q \end{bmatrix}^T \begin{bmatrix} i_d \\ i_q \end{bmatrix}, q_r = \begin{bmatrix} v_d \\ v_q \end{bmatrix}^T \mathbf{J} \begin{bmatrix} i_d \\ i_q \end{bmatrix}, T_{\mathrm{el}} = \begin{bmatrix} \lambda_d \\ \lambda_q \end{bmatrix}^T \mathbf{J} \begin{bmatrix} i_d \\ i_q \end{bmatrix}. \tag{3.5.8}$$

3.5.1 Modelling a Typical Exciter

The exciter is usually a synchronous machine without damper windings and that makes its modelling relatively easier. The exciter input-amplifier's output voltage, v_a, is modelled in terms of the amplifier gain k_a and amplifier time constant τ_a as

$$\tau_a \mathrm{d}v_a/\mathrm{d}t + v_a = k_a\left(v_{t-ref} - v_1 - v_3\right), \tag{3.5.9}$$

where the regulator output feedback v_1 and the generator stabilizing feedback v_3, satisfy,

$$\tau_r\, \mathrm{d}v_1/\mathrm{d}t + v_1 = k_r v_t, \tag{3.5.10}$$

$$\tau_f\, \mathrm{d}v_3/\mathrm{d}t + v_3 = k_f \mathrm{d}v_{fd}/\mathrm{d}t. \tag{3.5.11}$$

The exciter output voltage v_{fd} satisfies

$$\tau_e \mathrm{d}v_{fd}/\mathrm{d}t + k_e v_{fd} = v_a. \tag{3.5.12}$$

Since the exciter feedback is via the terminal voltage v_t and the reference the terminal voltage v_{t-ref}, the terminal voltage dynamics is also essential. In particular, one needs to include the voltage drop across the terminal. Thus, in terms of the ideal terminal *d–q* components of the voltages, the transmission line resistance R_e, and transmission line inductance L_e, the *d–q* components of the terminal voltages are

$$v_d = \sqrt{3/2}V_p \sin\delta + R_e i_d + L_e\, di_q/\mathrm{d}t, \tag{3.5.13}$$

$$v_q = \sqrt{3/2}V_p \cos\delta + R_e i_q - L_e\, di_d/\mathrm{d}t, \tag{3.5.14}$$

and

$$v_t = \sqrt{v_d^2 + v_q^2}. \tag{3.5.15}$$

Under steady-state operations, $v_d = 0$, $v_q = \sqrt{3/2}V_p$ and v_{t-ref} is assumed to be nominally given by, $v_{t-ref} = \sqrt{3/2}V_p$.

3.5.2 Model Parameter Estimation

A very important step in establishing a valid dynamic model of the synchronous generator is to establish and estimate a valid set of parameters for the synchronous generator. The purpose of the tests is to be able to estimate both the armature per phase resistance R_a and the armature per phase reactance X_a. This involves performing a series of tests and the parameters of the model are then estimated from the results of the tests. The first and simplest test is a resistance measurement test.

In this test, the resistances of the field circuit and the armature phases are measured. This test is conducted to measure winding resistance of armature with the rotor at rest and the field winding open. Alternate to the measurement of the phase resistance, the resistance between two lines may also be measured at a time. From the average of the three resistance readings, the value of the line resistance between line to line is estimated. The phase resistance is the estimated from the line resistance. If the generator is star or Y-connected, the per-phase resistance is $R_a = R_L/2$, while for a delta-connected generator, it is $R_a = 2R_L/3$. If the option to connect them either way is available, then this should be done and measurements made for both the star- and the delta-connected cases.

The second test that is performed is an open-circuit or no-load test. In this test, the generator is driven at the rated or synchronous speed. No load is connected across its stator or armature terminals. The field current is first set at the lowest possible value and increased in small steps to the maximum rated value. The open-circuit terminal voltage is measured at each setting. A plot of the open-circuit terminal voltage versus the field current gives the open-circuit characteristic of the generator. Since the current flowing in the armature is zero, this characteristic may be used to find the internally generated voltage of the generator for any given field current. This characteristic is initially linear but droops down after a certain critical value of the field current. Thereafter, the characteristic is no longer linear. The linear characteristic which is obtained by drawing a tangent to the curve at the origin is known as the air-gap line as it is represents the region where the applied magnetic field overcomes the air gap. Thereafter, saturation effects begin to come into play with the resulting loss of linearity.

The short-circuit test provides the maximum rated current capability of a synchronous generator. It is performed by driving the generator at the rated or synchronous speed. Here again, the field current is set to zero initially, and the armature is short-circuited. The armature current is measured, while the field current is increased from zero in small steps to the maximum rated value. A plot of the armature current versus the field current gives the short-circuit characteristic of the generator. Generally, the short-circuit characteristic is a straight line. From the short-circuit armature current and the open-circuit terminal voltage at the same setting of the field current, the synchronous machine impedance may be found; i.e.,

$$Z_m = \sqrt{R_a^2 + X_a^2} = E_{\mathrm{oc}}/I_{\mathrm{sc}}. \tag{3.5.16}$$

Thus,

$$X_a = \sqrt{Z_m^2 - R_a^2}. \tag{3.5.17}$$

A typical set of estimated parameters of a synchronous generator and its exciter are listed in Tables 3.1 and 3.2.

Table 3.1 Definitions and typical values assumed for the generator's parameters

P	Number of poles	4
R_s	Armature phase resistance	0.15 Ω
L_{ls}	Armature phase leakage inductance	0.024 mH
L_{md}	*d* axis coupling inductance	9.6 mH
R_{fd}	Field-winding resistance (reflected to the stator)	0.03 Ω
L_{lfd}	Field-winding leakage inductance (reflected to the stator)	0.72 mH
R_{kd}	*d* axis damper-winding resistance (reflected to the stator)	0.08 Ω
L_{lkd}	*d* axis damper-winding leakage inductance (reflected to the stator)	0.032 mH
L_{mq}	*q* axis coupling inductance	4.8 mH
R_{kq}	*q* axis damper-winding resistance (reflected to the stator)	0.08 Ω
L_{lkq}	*q* axis damper-winding leakage inductance (reflected to the stator)	0.072 mH
R_e	Transmission line resistance	0.02 Ω
L_e	Transmission line inductance	0.26 mH

Table 3.2 Definitions and typical values of the exciter parameters

k_a	Exciter input-amplifier voltage gain	100
τ_a	Exciter input-amplifier time constant	0.05
k_r	Exciter regulator output voltage gain	1
τ_r	Regulator output voltage time constant	0.3
k_f	Generator stabilizing inner-loop feedback gain	0.08
τ_f	Generator stabilizing feedback time constant	0.707
k_e	Exciter internal characteristic	1
τ_e	Exciter internal voltage time constant	0.5

3.6 Performance Characteristics of Synchronous Generators

There are three important performance features of a synchronous generator. The first is the power factor of the power delivered by the generator, and the second is its ability to be connected in parallel with other similar or dissimilar synchronous machines and run in synchronism with them. Finally, the minimum and maximum mechanical torque, essential to drive the generator, determines at what conditions the generator must be operated and the extent of overload protection that must be provided.

A synchronous generator is expected to provide an electrical network with a steady voltage under steady-state conditions. It is therefore important that the terminal voltage is kept constant by regulating the field current. Synchronous generators are run at constant speed, which is the synchronous speed that is dependent by the network frequency and the number of poles in the machine. An increase in the mechanical power input increases the torque, and consequently, there is an increase in the electrical power output. By introducing a reactive load, it is possible to establish the performance characteristics of the synchronous generator. Thus, one could estimate the power delivered by the synchronous generator for a range of input torques and a range of load power factors. In general, it is not advisable to operate the machine beyond a power factor of 60°. It is best to operate the generator so that it delivers the maximum power within the power factor constraint.

Often two or more generators are connected in parallel with supply a common load. However, before a synchronous generator is hook up in parallel with other synchronous generators, it must be "synchronized." Synchronization involves ensuring that the generator frequency is the same as the frequency of the system of generators, it is being connected in parallel with. It also means that the generator voltage is the same as the system voltage, and that the generator voltage is in-phase with the system voltage. In particular, the phase sequence of the generator must be the same as that of the system. A device known as a synchroscope is used to synchronize a generator with a system. Synchronism is achieved by adjusting the speed of the prime mover, the frequency of the output voltage, the output voltage magnitude and phase. The synchroscope facilitates the measurement of the phase difference between two voltages. Once synchronism is achieved the machine is connected to the system. The requirements are particularly stringent if the generator is being connected to an infinite bus.

3.7 Dynamic Modelling of Induction Generators

The dynamic model of the AC induction generator is obtained with a *d–q* frame derived by the application of the Park transformation to the voltages, currents, and magnetic fluxes in the three phases of the generator. The use of *d–q* frame allows one to convert the rotating stator and rotor quantities into fixed quantities in the corresponding synchronously rotating reference frame. The voltage balance equations of a three-phase induction generator can be written in a synchronously rotating-reference frame by independently considering the stator and the rotor and the mutual coupling between the two as:

$$-\begin{bmatrix} v_{ds} \\ v_{qs} \end{bmatrix} = -R_s \begin{bmatrix} i_{ds} \\ i_{qs} \end{bmatrix} + \omega_s \mathbf{J}^T \begin{bmatrix} \lambda_{ds} \\ \lambda_{qs} \end{bmatrix} - \frac{\mathrm{d}}{\mathrm{d}t} \begin{bmatrix} \lambda_{ds} \\ \lambda_{qs} \end{bmatrix}, \tag{3.7.1}$$

$$-\begin{bmatrix} v_{dr} \\ v_{qr} \end{bmatrix} = -R_r \begin{bmatrix} i_{ds} \\ i_{qs} \end{bmatrix} + s\omega_s \mathbf{J}^T \begin{bmatrix} \lambda_{dr} \\ \lambda_{qr} \end{bmatrix} - \frac{\mathrm{d}}{\mathrm{d}t} \begin{bmatrix} \lambda_{dr} \\ \lambda_{qr} \end{bmatrix}. \tag{3.7.2}$$

In Eqs. 3.7.1 and 3.7.2, the subscripts "ds" refers to the d axis of the stator, the subscripts "qs" refers to the q axis of the stator, the subscripts "dr" refers to the d axis of the rotor, the subscripts "qr" refers to the q axis of the rotor, R_s is the stator's ohmic resistance, R_r is the rotor's ohmic resistance, ω_s is the rotational speed of the synchronous reference frame, $s\omega_s = \omega_s - \omega_r$ is the slip frequency, and ω_r is the rotor's electrical speed. The equations relating the magnetic flux in the d–q frame of the stator and the rotor to the corresponding currents may be expressed as,

$$\begin{bmatrix} \lambda_{ds} \\ \lambda_{qs} \end{bmatrix} = L_{1s}\begin{bmatrix} i_{ds} \\ i_{qs} \end{bmatrix} + L_m\left(\begin{bmatrix} i_{ds} \\ i_{qs} \end{bmatrix} + \begin{bmatrix} i_{dr} \\ i_{qr} \end{bmatrix}\right) \equiv L_s\begin{bmatrix} i_{ds} \\ i_{qs} \end{bmatrix} + L_m\begin{bmatrix} i_{dr} \\ i_{qr} \end{bmatrix}, \quad L_s = L_{1s} + L_m, \tag{3.7.3}$$

$$\begin{bmatrix} \lambda_{dr} \\ \lambda_{qr} \end{bmatrix} = L_{1r}\begin{bmatrix} i_{dr} \\ i_{qr} \end{bmatrix} + L_m\left(\begin{bmatrix} i_{dr} \\ i_{qr} \end{bmatrix} + \begin{bmatrix} i_{ds} \\ i_{qs} \end{bmatrix}\right) \equiv L_r\begin{bmatrix} i_{dr} \\ i_{qr} \end{bmatrix} + L_m\begin{bmatrix} i_{ds} \\ i_{qs} \end{bmatrix}, \quad L_r = L_{1r} + L_m. \tag{3.7.4}$$

In Eqs. 3.7.3 and 3.7.4, L_{1s}, L_{1r}, and L_m are the stator leakage, rotor leakage, and mutual inductances, respectively. The electrical braking torque is given by,

$$T_{el} = \left(\lambda_{ds} i_{qs} - \lambda_{qs} i_{ds}\right) = -\left(\lambda_d r i_{qr} - \lambda_{qr} i_d r\right) = L_m\left(i_{qs} i_{dr} - i_{ds} i_{qr}\right). \tag{3.7.5}$$

Neglecting the power losses associated with the stator and rotor resistances, the active and reactive stator and rotor powers are, respectively, given by,

$$P_s = \frac{3}{2}\left(v_{ds} i_{ds} + v_{qs} i_{qs}\right), \quad Q_s = \frac{3}{2}\left(v_{qs} i_{ds} - v_{ds} i_{qs}\right) \tag{3.7.6}$$

and

$$P_r = \frac{3}{2}\left(v_d r i_d r + v_{qr} i_{qr}\right), \quad Q_r = \frac{3}{2}\left(v_{qr} i_d r - v_{dr} i_{qr}\right). \tag{3.7.7}$$

The total power exchanged between the generator and the grid is,

$$P_g = P_s + P_r \text{ and } Q_g = Q_s + Q_r. \tag{3.7.8}$$

The active power delivered to the grid by the generator is the stator active power.

Introducing the complex voltages, flux, and currents, in the stator and rotor circuits, $v_s = v_{ds} + jv_{qs}$, $j = \sqrt{-1}$, $v_r = v_{dr} + jv_{qr}$, $\lambda_s = \lambda_{ds} + j\lambda_{qs}$, $\lambda_r = \lambda_{dr} + j\lambda_{qr}$, and $i_s = i_{ds} + ji_{qs}$, $i_r = i_{dr} + ji_{qr}$, the dynamic equations may be expressed as,

$$v_s = R_s i_s + \omega_s j\lambda_s + \frac{d}{dt}\lambda_s, \tag{3.7.9}$$

$$v_r = R_r i_r + s\omega_s j\lambda_r + \frac{d}{dt}\lambda_r, \tag{3.7.10}$$

$$\lambda_s = L_s i_s + L_m i_r, \tag{3.7.11}$$

$$\lambda_r = L_m i_s + L_r i_r. \tag{3.7.12}$$

3.7.1 Equivalent Circuit Modelling

The dynamic equations expressed in terms of the complex voltages, complex flux, and complex currents in the stator and the rotor form the basis for the equivalent circuit model. Typically, the equivalent circuit model is four-terminal network as shown in Fig. 3.4.

The advantage of equivalent circuit modelling is the ability to easily include the effects of the copper loss and the core loss, where the losses may be expressed by additional equivalent resistances.

3.7.2 Parameter Estimation of Induction Generators Model

The resistance of one stator winding is estimated based on its length, cross-sectional area, and conductivity. The coil length is partially in the slots and partially in the end region. Thus, the resistance may be estimated and is the basis for estimating the stator resistance. The squirrel-cage rotor resistance is determined by adding the contributions from the bars and the end ring.

The magnetizing inductance is calculated based on the geometry of the machine, as well as the winding arrangement. The primary parameters in determining the power factor of an induction generator are the mutual and leakage inductances. The gap mutual inductance, L_m, for an induction generator may be estimated from,

$$L_m = 4\mu_0 rlN_sN_rk_ak_s/\pi gp^2. \tag{3.7.13}$$

In Eq. 3.7.13, which is derived from the equation for the magnetic flux density in the air gap, μ_0 is the permeability of the free space, g is the effective air gap width, p is the number of pole pairs, r is the rotor radius, l is the rotor length, k_a and k_s are the winding and skew factors which are functions of the geometry of the stator wings and the skew angle of the rotor bars, and N_s and N_r are the total number of turns in the stator and rotor, respectively. The winding factor of the rotor winding is a pure geometric factor, which is derived from the mesh width of two adjacent rotor bars. The mutual inductance of the machine must generally be chosen to be large so as to reduce the magnetizing current and required reactance during excitation and to maximize the power factor. Additionally, a soft saturation characteristic that decays with increasing current is desirable for increasing the operating range of the generator. The inverse dependence on the number of poles

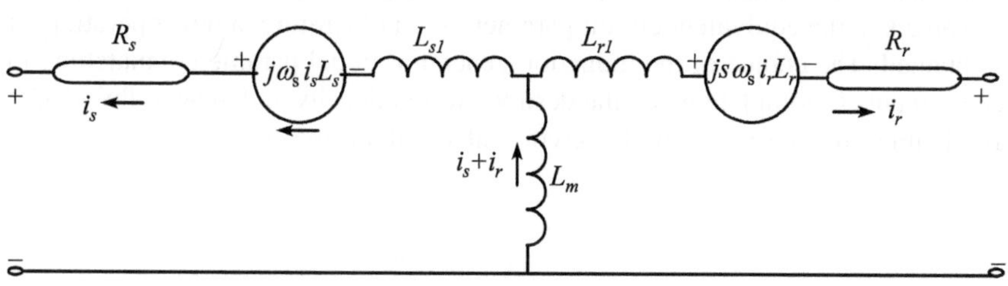

Fig. 3.4 Four terminal equivalent circuit of induction generator

demonstrates the reduction in magnetizing reactance, and the increased need for magnetizing current which are also observed in experiments as the number of poles increase. The leakage inductances, L_{1s} and L_{1r}, under normal steady operation must be as small as is possible. Otherwise, the work load of the controller such as an SVC generally goes up in cancelling the inductive reactance by increasing the capacitive reactance. Three types of leakage inductance must generally be considered in the design of a generator: slot leakage, zigzag leakage, and end-winding leakage. Slot leakage occurs due to flux which follows a path across a slot rather than the intended path, which is across the air gap via the teeth and through the stator and rotor core backs. The inductance associated with this leakage path is dependent on the permeance of the slot, which is function of the slot shape, and the number of turns being linked by the slot leakage flux. The slot leakage is generally calculated independently for the stator and rotor. Like the rotor resistance, the rotor slot leakage must be referred to the stator for use in the equivalent circuit. Zigzag leakage is due to variations in MMF between rotor and stator teeth. It is proportional to the magnetizing inductance and inversely proportional to the square of the number of slots per pole. The end-winding inductance is due to the magnetic path surrounding each end winding and is estimated empirically. The values of stator and rotor leakage inductance used in the equivalent circuit are then found as the sums of the individual leakage inductance components.

Losses in an induction generator are made up of a hysteresis loss, an eddy-current loss, and an excess loss component. These are generally estimated using empirical formulae as the measured losses are generally almost double the theoretically predicted losses. To model the losses, a conductor and representing core losses and with a specific conductance equal to the total core conductance are connected directly to the stator terminals in the equivalent circuit. The hysteresis losses account for the bulk of the losses and constitute about 70–75 % of the total losses, while the remaining losses are split evenly between the eddy current and excess losses.

When the equivalent circuit parameters have been estimated, it is possible to generate the characteristic relating the output power to the slip. Several such characteristics must be obtained to the range of values of slip and power output which would include the rated output power. For these values of slip, the stator current densities and peak air-gap flux densities must be considered to meet the application and geometry restrictions. The machine geometry must then be adjusted until an acceptable design is obtained. Adjustments in the geometry result in changes in the equivalent circuit parameters, and therefore, a new operating slip is obtained. The choice of the induction generator model parameters matching the requirements is complete when the desired current density and air-gap flux density are obtained for the value of slip giving rated output power.

3.7.3 Characteristics of Induction Generators and Experimental Determination of Parameters

The fundamental performance characteristics of induction generators are plots of variations of the generator current with a control parameter such as the conduction angle of the IGBT-type controller. The current decreases by decreasing the conduction angle while increasing the rotor speed increases the current. Another important characteristic is the harmonic factor in the generator current and its variation with a control parameter such as the conduction angle. Finally, the active and reactive power, power factor and efficiency plots are essential operating the induction generator at its maximal efficiency.

To experimentally determine the parameters of an induction generator, it is connected to perform an open-circuit test. In this test, the measurements are to be done when the machine is rotating exactly at the synchronous speed (which cannot be achieved simply by running at no-load). The induction machine is initially started as a motor with the help of an autotransformer, and once it reaches approximately the rated speed, the short-circuit switch across the ammeter and wattmeter is opened. In order to bring it to exactly the synchronous speed, a coupled DC motor is generally used to load the generator and run at the synchronous speed. The slip of the induction machine can then be measured either measuring the current frequency in the rotor circuit or by using a stroboscope to measure the rotor frequency. The active and reactive power components are also measured to obtain the open-circuit characteristics. Thus, plots of the torque, power output, power factor, and efficiency against slip are obtained. A short-circuit test may be performed with the rotor stalled or blocked. The input voltage, input current, and power are recorded. The parameters of the model may then be estimated from the observations.

3.8 Doubly Fed Induction Generator (DFIG): A Case Study

In modelling the stator of DFIG, the generator convention that positive direction of electromagnetic torque is in the direction opposing to the direction of rotation is used, while in modelling the rotor of DFIG, the motor convention is used. The dynamic equivalent circuit of generator in synchronous rotating reference frame, the *d–q* frame, is used to set up the model equations. Babu and Mohanty (2010) have discussed the modelling of a DFIG in some detail.

The four dynamical equations of the DFIG with constant coefficients in the *d–q* frame are,

$$v_{\mathrm{ds}} = R_s i_{\mathrm{ds}} + \frac{\mathrm{d}\phi_{\mathrm{ds}}}{\mathrm{d}t} - \omega_s \phi_{\mathrm{qs}}, v_{\mathrm{qs}} = R_s i_{\mathrm{qs}} + \frac{\mathrm{d}\phi_{\mathrm{qs}}}{\mathrm{d}t} + \omega_s \phi_{\mathrm{ds}}, \tag{3.8.1}$$

$$v_{dr} = R_r i_{dr} + \frac{d\phi_{dr}}{dt} - \omega_r \phi_{qr}, \quad v_{qr} = R_r i_{qr} + \frac{d\phi_{qr}}{dt} + \omega_r \phi_{dr}. \tag{3.8.2}$$

The stator fluxes are related to the stator and rotor currents in the d–q frame as,

$$\phi_{ds} = L_s i_{ds} + L_m i_{dr}, \phi_{qs} = L_s i_{qs} + L_m i_{qr}. \tag{3.8.3}$$

The rotor fluxes are related to the stator and rotor currents in the d–q frame as,

$$\phi_{dr} = L_r i_{dr} + L_m i_{ds}, \phi_{qr} = L_r i_{qr} + L_m i_{qs}. \tag{3.8.4}$$

In the above equations, R_s, R_r, L_s, and L_r are, respectively, the resistances and self-inductances of the stator and rotor windings, and L_m is the mutual inductance between a stator and a rotor phase when they are fully aligned with each other.

Furthermore, the active and reactive components of the power at stator terminals are given by,

$$P_s = v_{ds} i_{ds} + v_{qs} i_{qs}, \quad Q_s = v_{qs} i_{ds} - v_{ds} i_{qs}. \tag{3.8.5}$$

The active and reactive components of the power at rotor terminals are given by,

$$P_r = v_{dr} i_{dr} + v_{qr} i_{qr}, \quad Q_r = v_{qr} i_{dr} - v_{dr} i_{qr}. \tag{3.8.6}$$

The active and reactive powers exchanged by the generator and the grid are, respectively, the sum of the active and reactive components of the power at the stator and rotor. The electromagnetic reaction torque may be expressed as,

$$T_{el} = \left(\phi_{ds} i_{qs} - \phi_{qs} i_{ds}\right). \tag{3.8.7}$$

Assuming that the stator flux is stationary in the d–q frame and neglecting the stator's resistive voltage drop,

$$v_{ds} = 0. \tag{3.8.8}$$

The q-component of the stator voltage may be expressed as,

$$v_s = v_{qs}. \tag{3.8.9}$$

To achieve the desired stator conditions, voltage-source converters with a constant DC-link voltage are connected on the grid side to the AC grid through a transformer. Then, Eq. 3.8.1 is modified and expressed as,

$$-e_{ds} = R_{eqs} i_{ds} + \frac{d\phi_{ds}}{dt} - \omega_s \phi_{qs}, \tag{3.8.10}$$

$$v_s - e_{qs} = R_{eqs} i_{qs} + \frac{d\phi_{qs}}{dt} + \omega_s \phi_{ds}, \tag{3.8.11}$$

where e_{ds} and e_{qs} are the d–q components of the voltages generated by the voltage-source converters, and R_{eqs} is the total equivalent stator resistance. The inputs e_{ds}

and e_{qs} are expressed in terms two auxiliary control inputs, $\bar{e}_{ds}$ and $\bar{e}_{eqs}$, defined by the grid-side controller as

$$e_{ds} = -R_{eqs} i_{ds} + \bar{e}_{ds} + \frac{R_{eqs}}{L_{eqds}} \phi_{ds} + \omega_s \phi_{qs}, \tag{3.8.12}$$

$$e_{qs} = v_s - R_{eqs} i_{qs} + \bar{e}_{qs} + \frac{R_{eqs}}{L_{eqqs}} \phi_{qs} - \omega_s \phi_d s. \tag{3.8.13}$$

In Eqs. 3.8.12 and 3.8.13, L_{eqds} and L_{eqqs} are equivalent d axis and q axis stator inductances. The Eqs. 3.8.10 and 3.8.11 are decoupled and reduce to,

$$-\bar{e}_{ds} = \frac{R_{eqs}}{L_{eqds}} \phi_{ds} + \frac{d\phi_{ds}}{dt}, \quad -\bar{e}_{qs} = \frac{R_{eqs}}{L_{eqqs}} \phi_{qs} + \frac{d\phi_{qs}}{dt}. \tag{3.8.14}$$

The control inputs, $\bar{e}_{ds}$ and $\bar{e}_{qs}$, are defined in terms to optimum linear PI controllers in the inner most loop of the grid-side controller, thus facilitating the arbitrary choice of reference values for ϕ_{ds} and ϕ_{qs}.

Assuming a grid-side controller is in place and choosing a stator-flux-oriented reference frame, the d axis is aligned with the stator flux linkage vector, ϕ_s, thus $\phi_{ds} = \phi_s$ and $\phi_{qs} = 0$. It follows from Eqs. 3.8.3 and 3.8.1 that,

$$i_{qs} = -L_m i_{qr}/L_s, \quad v_{ds} = R_s i_{ds} + d\phi_s/dt. \tag{3.8.15}$$

The q-component of the stator voltage may be expressed as,

$$v_{qs} = R_s i_{qs} + \omega_s \phi_s, \tag{3.8.16}$$

$$\phi_s = L_s i_{ds} + L_m i_{dr} = \left(v_{qs} - R_s i_{qs}\right)/\omega_s = L_m i_{ms}, \tag{3.8.17}$$

where i_{ms} is defined as,

$$i_{ms} \equiv \left(v_{qs} - R_s i_{qs}\right)/\omega_s L_m. \tag{3.8.18}$$

Hence, it follows that,

$$i_{ds} = -i_{dr} L_m/L_s + \left(v_{qs} - R_s i_{qs}\right)/L_s \omega_s, \tag{3.8.19}$$

which may be expressed in terms of i_{ms} as,

$$i_{ds} = L_m (i_{ms} - i_{dr})/L_s. \tag{3.8.20}$$

The electromagnetic reaction torque given by Eq. 3.8.7 and the reactive power at the stator terminal given by the second of Eq. 3.8.5 may also be expressed in terms of i_{ms} as,

$$T_{el} = \phi_s i_{qs} = -L_m^2 i_{ms} i_{qr}/L_s, \tag{3.8.21}$$

$$Q_s = v_{qs} i_{ds} - v_{ds} i_{qs} = \omega_s \phi_s i_{ds} = \omega_s L_m^2 i_{ms} (i_{ms} - i_{dr})/L_s. \tag{3.8.22}$$

The rotor flux components given by the Eq. 3.8.4 are

$$\phi_{\mathrm{dr}} = L_r\left(\frac{L_sL_r - L_m^2}{L_sL_r}\right)i_{\mathrm{dr}} + \frac{L_m^2}{L_s}i_{\mathrm{ms}}, \tag{3.8.23}$$

$$\phi_{\mathrm{qr}} = L_r\left(\frac{L_sL_r - L_m^2}{L_sL_r}\right)i_{\mathrm{qr}}. \tag{3.8.24}$$

Defining the mutual inductance coupling coefficient σ as,

$$\sigma = \left(L_sL_r - L_m^2\right)/L_sL_r, \tag{3.8.25}$$

the rotor voltage equations given by the Eq. 3.8.2 are expressed as,

$$v_{\mathrm{dr}} = R_ri_{\mathrm{dr}} + \sigma L_r\frac{\mathrm{d}i_{\mathrm{dr}}}{\mathrm{d}t} - \sigma s\omega_sL_ri_{\mathrm{qr}}, \tag{3.8.26}$$

$$\left(R_r + s\frac{L_m^2}{L_s^2}R_s\right)i_{\mathrm{qr}} + \sigma L_r\frac{\mathrm{d}i_{\mathrm{qr}}}{\mathrm{d}t} + \sigma s\omega_sL_ri_{\mathrm{dr}}. \tag{3.8.27}$$

The total reactive power given by Eq. 3.8.22 is,

$$Q = v_{\mathrm{qr}}i_{\mathrm{dr}} - v_{\mathrm{dr}}i_{\mathrm{qr}} + \omega_sL_m^2i_{\mathrm{ms}}(i_{\mathrm{ms}} - i_{\mathrm{dr}})/L_s, \tag{3.8.28}$$

where the current i_{ms} defined by Eq. 3.8.18 is,

$$i_{\mathrm{ms}} \equiv \frac{\left(v_s - R_si_{\mathrm{qs}}\right)}{\omega_sL_m} = \frac{v_s}{\omega_sL_m} + \frac{R_s}{\omega_sL_s}i_{\mathrm{qr}}. \tag{3.8.29}$$

Hence, the total reactive power given by Eq. 3.8.28 may be expressed as,

$$Q = v_{\mathrm{qr}}i_{\mathrm{dr}} - v_{\mathrm{dr}}i_{\mathrm{qr}} + \omega_s\frac{L_m^2}{L_s} \times \left(\frac{v_s}{\omega_sL_m} + \frac{R_s}{\omega_sL_m}i_{\mathrm{qr}}\right)\left(\frac{v_s}{\omega_sL_m} + \frac{R_s}{\omega_sL_m}i_{\mathrm{qr}} - i_{\mathrm{dr}}\right). \tag{3.8.30}$$

The definitions of the resistances and inductances and their typical values assumed in this paper are listed in Table 3.3.

Table 3.3 Definitions and typical values assumed for the generator's parameters

P	Number of poles	6
R_s	Stator resistance	0.95 Ω
L_s	Stator inductance	94 mH
L_m	Magnetizing inductance	82 mH
R_r	Rotor resistance	1.8 Ω
L_r	Rotor inductance	88 mH
V_a	Stator phase voltage	380 V
	Grid frequency	50 Hz
	Nominal mechanical rotor speed	100 rads/s
	Rated maximum power	100 kW

3.8.1 The Steady-State Electromechanical Model

In steady state, assuming that,

$$v_{dr} = v_{dr}^0,\ v_{qr} = v_{qr}^0,\ i_{dr} = i_{dr}^0,\ i_{qr} = i_{qr}^0,\ s = s^0, \tag{3.8.31}$$

where the superscript "0" refers to the steady-state condition, Eqs. 3.8.26 and 3.8.27 may be expressed in matrix form as,

$$\begin{bmatrix} v_{dr}^0 \\ v_{qr}^0 - s^0 \frac{L_m}{L_s} v_s \end{bmatrix} = \begin{bmatrix} R_r & -\sigma s^0 \omega_s L_r \\ -\sigma s^0 \omega_s L_r & R_{rs} \end{bmatrix} \begin{bmatrix} i_{dr}^0 \\ i_{qr}^0 \end{bmatrix}. \tag{3.8.32}$$

In the above equations, R_{rs} is defined as,

$$R_{rs} = R_r + s^0 \frac{L_m^2}{L_s^2} R_s \tag{3.8.33}$$

The electromagnetic torque given by Eq. 3.8.21 in steady state is

$$T_{el}^0 = -\frac{L_m^2}{L_s}\left(\frac{v_s}{\omega_s L_m} + \frac{R_s}{\omega_s L_s} i_{qr}^0\right) i_{qr}^0. \tag{3.8.34}$$

It is customary to refer the mechanical motion dynamics also to the electrical reference speed and write the mechanical equation of motion given by Eq. 3.4.5 as,

$$J_{eq}\frac{d\omega_e}{dt} + B_{eq}\omega_e = \frac{P}{2}\left(T_{wt} - \frac{P}{2}T_{el}\right).$$

In steady state,

$$T_{wt} = T_{wt}^0,\ \omega_e = \omega_e^0 \text{ and } d\omega_e/dt = 0. \tag{3.8.35}$$

Since $R_s Lt; 1$, for given steady-state electrical rotor speed, ω_e^0, the solution for i_{qr}^0 may be obtained iteratively. The control inputs are selected, so the total reactive power given Eq. 3.8.28 is equal to zero in steady state. Expanding and rearranging the terms in Eq. 3.8.28, one obtains i_{dr}^0. Thus, steady-state control inputs v_{dr}^0, v_{qr}^0 may be found, and the solutions for i_{qr}^0 and i_{dr}^0 may iteratively updated.

3.8.2 The Nonlinear Perturbation Dynamics

Subtracting from Eqs. (3.8.32) from (3.8.26) and (3.8.27), respectively

$$\Delta v_{dr} = R_r \Delta i_{dr} + \sigma L_r \frac{d\Delta i_{dr}}{dt} - \sigma s^0 \omega_s L_r \Delta i_{qr} - \sigma \Delta s \omega_s L_r i_{qr} \tag{3.8.36}$$

$$\Delta v_{qr} = \left(R_r + s^0\frac{L_m^2}{L_s^2}R_s\right)\Delta i_{qr} + \sigma L_r\frac{\mathrm{d}\Delta i_{qr}}{\mathrm{d}t} \\ + \sigma s^0\omega_s L_r\Delta i_{dr} + \Delta s\frac{L_m^2}{L_s^2}R_s i_{qr} + \sigma\Delta s\omega_s L_r i_{dr}. \tag{3.8.37}$$

The perturbation states, inputs, and variables are defined as:

$$\Delta v_{dr} = v_{dr} - v_{dr}^0,\ \Delta v_{qr} = v_{qr} - v_{qr}^0,\ \Delta i_{dr} = i_{dr} - i_d r^0,\ \Delta i_{qr} = i_{qr} - i_{qr}^0, \tag{3.8.38}$$

$$\Delta s = s - s^0 = -\frac{\Delta\omega_e}{\omega_s},\ \Delta\omega_e = \omega_e - \omega_e^0. \tag{3.8.39}$$

Given that $T_{wt} = T_{wt}^0$ in steady state and that $\Delta T_{wt} = T_{wt} - T_{wt}^0$, $\Delta\omega_e$ satisfies the equation,

$$J_{eq}\frac{\mathrm{d}\Delta\omega_e}{\mathrm{d}t} + B_{eq}\Delta\omega_e = \frac{P}{2}\left(\Delta T_{wt} - \frac{P}{2}T_{el} + \frac{P}{2}T_{el}^0\right). \tag{3.8.40}$$

The electromagnetic torque is

$$T_{el} = T_{el}^0 - \frac{L_m^2}{L_s}\left(\frac{v_s}{\omega_s L_m} + \frac{R_s}{\omega_s L_s}\left(2i_{qr}^0 + \Delta i_{qr}\right)\right)\Delta i_{qr}. \tag{3.8.41}$$

The wind turbine perturbation torque $\Delta T_{wt} = \Delta T_{wt}(U, \omega_e)$, is function of two variables, the wind speed U, and the rotor speed ω_e. Given that the wind speed, $U = U_0 + \Delta u_f$ is the sum of a mean component, U_0 and a fluctuating component, Δu_f, the wind turbine perturbation torque can considered to made of two components,

$$\Delta T_{wt}(U, \omega_e) = (T_{wt}(U, \omega_e) - T_{wt}(U_0, \omega_e)) \\ + \left(T_{wt}(U_0, \omega_e) - T_{wt}^0\left(U_0, \omega_e^0\right)\right) = \Delta T_{wt}|_{\omega_e} + \Delta T_{wt}|_{U=U_0}, \tag{3.8.42}$$

where the first component is evaluated at the current rotor speed, ω_e. Hence, without making any assumptions that the perturbations are small, the equation for mechanical motion may be expressed as,

$$J_{eq}\frac{\mathrm{d}\Delta\omega_e}{\mathrm{d}t} = \left(\frac{P}{2}\frac{\Delta T_{wt}}{\Delta\omega_e}\bigg|_{U=U_0} - B_{eq}\right)\Delta\omega_e \\ + \frac{P^2}{4}\frac{L_m^2}{L_s}\left(\frac{v_s}{\omega_s L_m} + \frac{R_s}{\omega_s L_s}\left(2i_{qr}^0 + \Delta i_{qr}\right)\right)\Delta i_{qr} + \Delta T_{wt}|_{\omega_e}, \tag{3.8.43}$$

while the electrical machine perturbation equations are

$$\sigma L_r\frac{\mathrm{d}\Delta i_{dr}}{\mathrm{d}t} = -\sigma L_r i_{qr}\Delta\omega_e - R_r\Delta i_{dr} + \sigma s^0\omega_s L_r\Delta i_{qr} + \Delta v_{dr}, \tag{3.8.44}$$

Table 3.4 Definitions and typical values of the wind turbine parameters

R	Wind turbine blade disc radius	6 m
	Number of blades	3
U_0	Nominal wind speed	10 m/s
P_n	Wind power at nominal wind speed	~10 kW
	Gearbox ratio	10
J_{eq}	Rotor inertia	40 kgm^2
B_{eq}	Viscous friction coefficient	0.07 Nms/rad
	Cut-in wind speed (m/s)	3.5 m/s

$$\sigma L_r \frac{\mathrm{d}\Delta i_{qr}}{\mathrm{d}t} = + \left(\sigma L_r i_{dr} + \frac{L_m^2}{\omega_s L_s^2} R_s i_{qr} \right) \Delta\omega_e - \sigma s^0 \omega_s L_r \Delta i_{dr} - \left(R_r + s^0 \frac{L_m^2}{L_s^2} R_s \right) \Delta i_{qr} + \Delta v_{qr} \tag{3.8.45}$$

The complete nonlinear equations for the perturbation states used for the design of the nonlinear rotor-side controller may be expressed in state space form as,

$$\frac{\mathrm{d}}{\mathrm{d}t} \begin{bmatrix} \Delta\omega_e \\ \Delta i_{dr} \\ \Delta i_{qr} \end{bmatrix} = \mathbf{A} \begin{bmatrix} \Delta\omega_e \\ \Delta i_{dr} \\ \Delta i_{qr} \end{bmatrix} + \frac{1}{\sigma L_r} \begin{bmatrix} 0 & 0 \\ 1 & 0 \\ 0 & 1 \end{bmatrix} \begin{bmatrix} \Delta v_{dr} \\ \Delta v_{qr} \end{bmatrix} + \begin{bmatrix} 1 \\ 0 \\ 0 \end{bmatrix} \alpha_u, \tag{3.8.46}$$

where

$$\alpha_u = \left(\Delta T_{wt} / \Delta u_f \big|_{\omega_e} / J_{eq} \right) \Delta u_f, \tag{3.8.47}$$

is the disturbing angular acceleration on the rotor due to wind speed fluctuating component, and **A** is a matrix of functional coefficients. The complete characteristics of the wind turbine are summarized in Table 3.4.

References

Babu BC, Mohanty KB (2010) Double-fed induction generator for variable speed wind energy conversion systems: modeling and simulation. Int J Comput Electr Eng 2(1):1793–8163

Krause PC, Wasynczuk O, Sudhoff SD (1995) Analysis of electric machinery (IEEE Press series on power engineering). Wiley-IEEE Press, Hoboken

Sarma MS (1994) Electric machines, 2nd edn. West Publishing Company, Eagan

Chapter 4
Wind Power Generation and Control

4.1 Introduction

Wind power has evolved rapidly in the last two decades with regard to both power rating and size of wind turbines that are currently being installed. In the past few years, several developments have taken place. There has not only been a continuous increase in the rated power of wind turbines being installed, but also in the development of reliable wind turbines rated around 2–10 MW which are grid safety code compliant and suitable for different installation environments—onshore and offshore, as it can be seen in the commercial products currently in the market.

Broadly, wind turbines (Manwell et al. 2010) can be classified into two groups: horizontal axis wind turbines, which are more efficient in capturing wind power particularly when it is concentrated in a particular direction and when wind speeds are relatively high, and vertical axis wind turbines, which are suitable for capturing wind power at low wind speeds and when the wind is omni-directional. Vertical axis wind turbines are mainly the Darrieus and Savonius types which either depend on the rotor blades generating lift as in the first case or on the impulse or drag of the rotor blades as in the second. Most Darrieus machines have their blades in an eggbeater configuration although several new designs including helical rotors are being introduced.

4.2 Wind Turbine Components

Almost all wind turbines that produce electricity for supply to a grid are horizontal axis wind turbines which consist of rotor blades that rotate around a horizontal hub. The hub is connected to a gearbox and a generator (when direct-drive generators are present, the gearbox is not necessary), which are located inside the nacelle. The nacelle houses some of the electrical components and is mounted on top of the tower. The electric current is then distributed by a transformer to the

R. Vepa, *Dynamic Modeling, Simulation and Control of Energy Generation*, Lecture Notes in Energy 20,
DOI: 10.1007/978-1-4471-5400-6_4, © Springer-Verlag London 2013

grid. Many different design concepts are in use. The most common ones are two- or three-bladed, stall or pitch regulated, horizontal axis machines working at variable or near fixed rotational speed. All forms of wind turbines are designed to extract power from a moving air stream. The blades have an aerofoil cross section and extract wind by a lift force caused by a pressure difference between blade sides. The aerofoil section shapes across the span of the blade are generally widely varying. For maximum efficiency, the blades are pretwisted and tapered.

4.3 Wind Turbine Aerodynamics: Momentum Theory

Consider the stream tube around a wind turbine shown in Fig. 4.1. Four stations along the flow are indicated in the diagram: station 1 is far upstream of the turbine, station 2 just before the blades, station 3 just after the blades, and station 4 far downstream of the blades. Between the stations 2 and 3, energy is extracted from the wind by the turbine disc, and there is consequently a step change in pressure.

From the continuity of mass flow and the incompressibility assumption, $V_2 = V_3$. It then follows that $p_1 = p_4$. We can also assume that between stations 1 and 2 and between stations 3 and 4, the flow is frictionless so we can apply Bernoulli's equation for the conservation of energy. Thus, the pressure differential across the disc is related to the difference in the kinetic energy between stations 4 and 1. Hence,

$$p_2 - p_3 = \frac{1}{2}\rho\left(V_1^2 - V_4^2\right) \tag{4.3.1}$$

Hence, the axial force is given by,

$$\mathrm{d}F_x = (p_2 - p_3)\mathrm{d}A = \frac{1}{2}\rho\left(V_1^2 - V_4^2\right)\mathrm{d}A. \tag{4.3.2}$$

The axial induction factor is defined as,

$$a = \frac{(V_1 - V_2)}{V_1}. \tag{4.3.3}$$

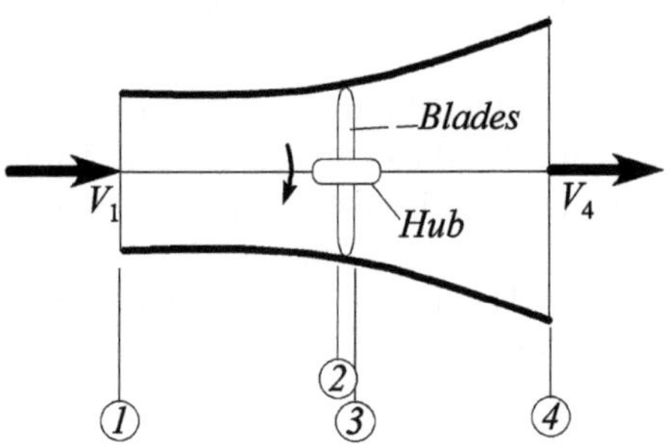

Fig. 4.1 A typical stream tube enclosing the turbine disc

Then,

$$V_2 = (1 - a)V_1. \tag{4.3.4}$$

Hence,

$$V_1 - V_4 = V_1 - (1 - 2a)V_1 = 2aV_1. \tag{4.3.5}$$

Hence, considering an annular element of the stream tube,

$$\mathrm{d}F_x = (p_2 - p_3)\mathrm{d}A = \frac{1}{2}\rho(V_1 - V_4)(V_1 + V_4)2\pi r\mathrm{d}r. \tag{4.3.6}$$

4.3.1 Actuator Disc Theory

Perhaps the simplest approach to the aerodynamic modelling of the wind turbine is the actuator disc theory which ignores the flow rotational effects. In this case, one may integrate the increment of the axial forces $\mathrm{d}F_x$, across the cross-sectional area of the turbine disc, and obtain

$$F_x = \frac{1}{2}\rho(V_1 - V_4)(V_1 + V_4)\pi R^2. \tag{4.3.7}$$

For steady-state flow, the mass flow rate across the wind turbine disc can be obtained using the equation,

$$\dot{m} = \rho A V_2. \tag{4.3.8}$$

Applying the conservation of linear momentum equation on both sides of the actuator disc and using the fact that the thrust is the product of the pressure drop and the area gives,

$$F_x = \rho A V_2(V_1 - V_4) = A\Delta p = \frac{1}{2}\rho(V_1 - V_4)(V_1 + V_4)A. \tag{4.3.9}$$

Hence,

$$V_2 = \frac{1}{2}(V_1 + V_4) \tag{4.3.10}$$

and

$$V_4 = (2V_2 - V_1) = V_1(1 - 2a). \tag{4.3.11}$$

Hence, in the case of an annular element of the stream tube,

$$\begin{aligned} \mathrm{d}F_x &= (p_2 - p_3)\mathrm{d}A = \frac{1}{2}\rho(V_1 - V_4)(V_1 + V_4)2\pi r\mathrm{d}r \\ &= \frac{1}{2}\rho V_1^2\left(1 - (1 - 2a)^2\right)2\pi r\mathrm{d}r. \end{aligned} \tag{4.3.12}$$

The expression simplifies to,

$$dF_x = \frac{1}{2}\rho V_1^2 4a\,(1 - a)2\pi r dr. \tag{4.3.13}$$

Thus, for the case of the actuator disc model,

$$F_x = \frac{1}{2}\rho V_1^2 4a(1 - a)\pi R^2. \tag{4.3.14}$$

The power absorbed by the wind turbine is

$$P = F_x V_2 = \frac{1}{2}\rho V_1^3 4a(1 - a)^2 \pi R^2. \tag{4.3.15}$$

The power in the wind is

$$P_{\text{wind}} = \frac{1}{2}\rho V_1^3 \pi R^2. \tag{4.3.16}$$

The power coefficient is

$$C_P = \frac{P}{P_{\text{wind}}} = 4a(1 - a)^2. \tag{4.3.17}$$

The axial force or thrust coefficient is

$$C_T = 4a(1 - a). \tag{4.3.18}$$

4.3.2 The Betz Limit

The maximum C_P may be determined by taking the derivative of $a(1 - a)^2$ with respect to a and setting it equal to zero. This yields $a = 1/3$, and the corresponding value of $C_P = C_{P,\max}$ is given by, $C_{P,\max} = 16/27 = 0.5926$. This result is known as the Betz limit, and it indicates that the ideal rotor must be designed such that the wind speed at the rotor is $1 - a = 2/3$ of the free stream wind speed, so it operates for maximum power extraction. This implies that the flow through the disc corresponds to an upstream cross-sectional area of $1 - a = 2/3$ the disc area that expands to twice the area far downstream. It must be emphasized that the Betz limit is not a universal limit as it is based on the approximate actuator disc theory. Thus, in principle, it is possible to extract more power than that predicted by the Betz limit, but it is important to recognize that there is indeed a real physical limit to the maximum power extracted from the wind.

Although the maximum thrust is obtained for $a = 1/2$ corresponding to a $C_{T,\max} = 1$, the thrust coefficient for maximum power transfer, based on the actuator disc model, is $C_T = 8/9$. There are a number of losses which detract the wind turbine from achieving maximum power-transfer efficiency. These include the wake vortex rotation behind the disc which consumes energy, tip losses associated with blades and aerodynamic drag which is another form of energy dissipation. One way of including the effects of wake rotation is to consider the rotation of the flow which is addressed in the next section.

4.3.3 Effects of Flow Rotation

At inlet to the blade, the flow is not rotating, and at exit from the blade row, the flow rotates at rotational speed ω. Over the blade row, wake rotation has been induced. The average rotational flow over the blade due to wake rotation is, therefore, assumed to be $\omega/2$. The blade is physically rotating with a speed of Ω. The average relative angular velocity that the blade experiences is, therefore, $\Omega + \omega/2$.

The increment in the torque due to the rotating annular flow is given in terms of the induced wake angular velocity, ω by,

$$\mathrm{d}T = \rho(\mathrm{d}A)(\Omega + \omega/2)\omega r^2 = \rho(\Omega + \omega/2)\omega r^2 \times 2\pi r\mathrm{d}r \tag{4.3.19}$$

The angular velocity induction factor, in terms of the induced wake angular velocity, ω, and the blade angular velocity, Ω, is defined as,

$$a' = \frac{\omega}{2\Omega}. \tag{4.3.20}$$

Thus, the axial force is

$$\mathrm{d}F_x = \rho(1 + a')2a'\Omega^2 r^2 \times 2\pi r\mathrm{d}r = 4\pi\rho(1 + a')a'\Omega^2 r^3\mathrm{d}r. \tag{4.3.21}$$

The torque is given by,

$$\mathrm{d}T = \rho V_1(1 - a)2a'\Omega r^2 \times 2\pi r\mathrm{d}r = 4\pi\rho V_1(1 - a)a'\Omega r^3\mathrm{d}r \tag{4.3.22}$$

At the tip of the turbine blade, losses are developed in a manner similar to those found in wind tip vortices on helicopter rotor blades. These can be accounted for by means of a correction factor first suggested by Prandtl. Prandtl's correction factor Q_{tip} varies from 0 to 1 and results in the reduction in the forces along the blade. Prandtl's tip loss correction factor Q_{tip} is defined as,

$$Q_{\mathrm{tip}} = \frac{2}{\pi}\cos^{-1}\{\exp(-(N(1-\bar{r})/2\bar{r}\sin\phi))\}. \tag{4.3.23}$$

When the tip loss correction is introduced, the increment in the axial force and torque is

$$\mathrm{d}F_x = 4\pi Q_{\mathrm{tip}}\rho(1 + a')a'\Omega^2 r^3\mathrm{d}r \tag{4.3.24}$$

and

$$\mathrm{d}T = 4\pi Q_{\mathrm{tip}}\rho V_1(1 - a)a'\Omega r^3\mathrm{d}r. \tag{4.3.25}$$

Thus, the momentum theory therefore yields expressions for the axial thrust and the torque due to an annular element of the fluid.

Equating the torque increment obtained for the case of the rotating stream tube with that obtained earlier from linear momentum analysis for the actuator disc, one may obtain a relation between the two induction factors. Thus,

$$\mathrm{d}F_x = 4\pi Q_{\mathrm{tip}}\rho(1 + a')a'\Omega^2 r^3\mathrm{d}r = 4\pi Q_{\mathrm{tip}}\rho V_1^2 a(1 - a)r\mathrm{d}r. \tag{4.3.26}$$

It follows that the inflow at a radius r is related to a and a' by,

$$\lambda_r = \frac{V_1}{\Omega r} = \sqrt{\frac{a'(1 + a')}{a(1 - a)}}. \tag{4.3.27}$$

Hence,

$$a' = \frac{1}{2}\left(\sqrt{1 + 4\lambda_r^2 a(1 - a)} - 1\right). \tag{4.3.28}$$

The power generated in each annular ring is given by,

$$\mathrm{d}P = \Omega\mathrm{d}T = 4\pi Q_{\mathrm{tip}}\rho\lambda_r(1 - a)a'\Omega^3 R^5\bar{r}^4\mathrm{d}\bar{r} \tag{4.3.29}$$

Thus, the power coefficient C_P is

$$C_P = \frac{8}{\lambda_f^3}\int_{\bar{r}_h}^{1} Q_{\mathrm{tip}}\lambda_r(1 - a)a'\bar{r}^4\mathrm{d}\bar{r}. \tag{4.3.30}$$

But,

$$\lambda_r = \frac{\lambda_f}{\bar{r}},\quad \mathrm{d}\lambda_r = -\lambda_f\frac{\mathrm{d}\bar{r}}{\bar{r}^2}. \tag{4.3.31}$$

Hence,

$$C_p = 8\lambda_f^2\int_{\lambda_f}^{\lambda_f/\bar{r}_h} Q_{\mathrm{tip}}(1 - a)a'\frac{\mathrm{d}\lambda_r}{\lambda_r^5} \tag{4.3.32}$$

For maximum power to be generated in each annular ring,

$$\lambda_r^2 = \frac{(1 - 3a_{\max})}{(1 - a_{\max})(4a_{\max} - 1)^2}. \tag{4.3.33}$$

Hence,

$$a_{\max}^3 - (3/2)a_{\max}^2 + (9/16)a_{\max} - (1/16) + (1 - 3a_{\max})/16\lambda_r^2 = 0 \quad (4.3.34)$$

and

$$a'_{\max} = \frac{(1 - 3a_{\max})}{(4a_{\max} - 1)}. \quad (4.3.35)$$

Solving for $a_{\max}(\lambda_r)$ numerically, for a given value of λ_r, the maximum power coefficient is

$$C_{p,\max} = 8\lambda_f^2 \int_{\lambda_f}^{\lambda_f/\bar{r}_h} Q_{\text{tip}}(1 - a_{\max}(\lambda_r)) \frac{(1 - 3a_{\max}(\lambda_r))\,\mathrm{d}\lambda_r}{(4a_{\max}(\lambda_r) - 1)\,\lambda_r^5}. \quad (4.3.36)$$

4.4 Blade Element Momentum Theory

The blade element momentum (BEM) theory combines the two methods of modelling the operation of a wind turbine. The first method of modelling uses the momentum balance on a rotating annular stream tube passing through turbine disc as shown in Fig. 4.2. The second method of modelling, known as the blade element theory, uses the expressions for the aerofoil lift and drag forces along various spanwise sections of the blade. The first and the second methods are used to obtain the induced flow velocities across the turbine, while the second is used to obtain the thrust and torque on the blades.

To apply the blade element theory to a wind turbine blade, the blade is divided into a large number of sections or elements along its length. One makes two assumptions:

1. There is no aerodynamic interference between different blade elements.
2. The forces on the blade elements are solely determined by the two-dimensional section lift and drag forces.

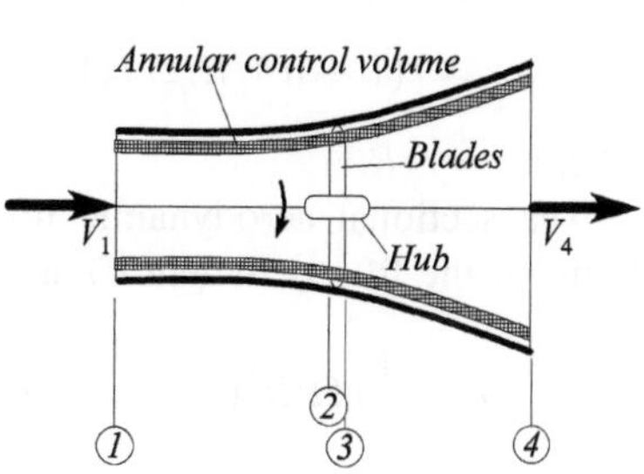

Fig. 4.2 A typical rotating annular control volume within a stream tube

Fig. 4.3 Typical wind turbine blade divided into elements

Considering a blade divided into N elements along its length as shown in Fig. 4.3. Each of the blade elements will experience a slightly different flow as they have different rotational speeds, different chord lengths, and different inflow angles. Blade element theory involves dividing the blade into a large number (usually about twenty or more) of elements and calculating the flow about each of the elements. The overall performance characteristics are determined by numerically integrating the forces and moments along the blade span.

The inflow angle far away from the turbine disc is

$$\lambda_f = V_1/\Omega R = V/\Omega R \tag{4.4.1}$$

while the inflow at the disc is given by,

$$\lambda_f + \lambda_0 = \frac{V(1-a)}{\Omega R(1+a')}. \tag{4.4.2}$$

Thus,

$$\lambda_0 = \frac{V(1-a)}{\Omega R(1+a')} - \frac{V}{\Omega R} = -\frac{V(a+a')}{\Omega R(1+a')} = -\lambda_f\frac{a+a'}{1+a'}. \tag{4.4.3}$$

To obtain expressions for a and a', one obtains expressions for $\mathrm{d}F_x$ and $\mathrm{d}T$ using blade element theory. Comparing the equations for $\mathrm{d}F_x$ and $\mathrm{d}T$ obtained by using the momentum theory and by using the blade element theory, expressions for a and a' may be obtained.

The aerodynamic forces and moments are developed in terms of the non-dimensional velocity components normal to the disc u_P and in-plane components in the tangential and radial direction, u_T and u_R, respectively. These are

$$u_T = r(1+a')/R \equiv \bar{r}(1+a'),\ u_R = 0 \text{ and } u_P = \lambda_f(1-a), \tag{4.4.4}$$

where $\bar{r} = r/R$.

The sectional aerodynamic lift and drag forces are, respectively, defined in terms of the lift coefficient C_l and the drag coefficient C_d as,

$$L = \frac{1}{2}\rho\Omega^2R^2\left(u_T^2 + u_P^2\right)cC_l,\ D = \frac{1}{2}\rho\Omega^2R^2\left(u_T^2 + u_P^2\right)cC_d. \tag{4.4.5}$$

The sectional angle of attack is $\alpha = \phi - \theta$ where θ is blade section pitch angle, and ϕ is the inflow angle defined as,

$$\phi = \tan^{-1}(u_P/u_T) = \tan^{-1}\left(\lambda_f(1-a)/\bar{r}(1+a')\right). \qquad (4.4.6)$$

The differential force normal to the rotor disc acting on a section of the rotor blade is

$$\mathrm{d}F_x = \mathrm{d}L\cos\phi + \mathrm{d}D\sin\phi = \frac{1}{2}\frac{\rho\Omega^2R^2}{\sin^2\phi}\lambda_f^2(1-a)^2c(C_l\cos\phi + C_d\sin\phi)\mathrm{d}r. \qquad (4.4.7)$$

The differential force acting tangential to the blade disc due to a section of the rotor blade is

$$\mathrm{d}F_t = -\mathrm{d}L\sin\phi + \mathrm{d}D\cos\phi = \frac{1}{2}\frac{\rho\Omega^2R^2}{\sin^2\phi}\lambda_f^2(1-a)^2c(-C_l\sin\phi + C_d\cos\phi)\mathrm{d}r. \qquad (4.4.8)$$

The torque acting on the disc is

$$\mathrm{d}T = R\mathrm{d}F_t = \frac{1}{2}\frac{\rho\Omega^2R^2}{\sin^2\phi}\lambda_f^2(1-a)^2c(-C_l\sin\phi + C_d\cos\phi)r\mathrm{d}r. \qquad (4.4.9)$$

For N blades, introducing the local solidity ratio, $\sigma' = Nc/\pi r$,

$$\mathrm{d}F_x = \frac{1}{2}\pi\sigma'\frac{\rho\Omega^2R^2}{\sin^2\phi}\lambda_f^2(1-a)^2(C_l\cos\phi + C_d\sin\phi)r\mathrm{d}r, \qquad (4.4.10)$$

$$\mathrm{d}T = \frac{1}{2}\pi\sigma'\frac{\rho\Omega^2R^2}{\sin^2\phi}\lambda_f^2(1-a)^2(-C_l\sin\phi + C_d\cos\phi)r^2\mathrm{d}r. \qquad (4.4.11)$$

Comparing these expressions with the corresponding expressions for momentum theory, the induction factors satisfy the relations,

$$\frac{a}{(1-a)} = \frac{\sigma'}{8Q_{\text{tip}}\left(\sin^2\phi\right)}(C_l\cos\phi + C_d\sin\phi), \qquad (4.4.12)$$

$$\frac{a'}{(1-a)} = \frac{\sigma'}{8Q_{\text{tip}}\left(\sin^2\phi\right)}\lambda_f(-C_l\sin\phi + C_d\cos\phi). \qquad (4.4.13)$$

Solutions for a and a', respectively, may be obtained as,

$$a = \left(1 + \left\{\frac{\sigma'}{8Q_{\text{tip}}\left(\sin^2\phi\right)}(C_l\cos\phi + C_d\sin\phi)\right\}^{-1}\right)^{-1}, \qquad (4.4.14)$$

$$a' = \frac{\sigma'(1-a)}{8Q_{\text{tip}}\left(\sin^2\phi\right)}\lambda_f(-C_l\sin\phi + C_d\cos\phi). \tag{4.4.15}$$

Hence, the factor $1 - a$ may be expressed as,

$$1 - a = \left\{1 + \frac{\sigma' C_l}{8Q_{\text{tip}}\sin\phi\tan\phi}\left(1 + \frac{C_d\tan\phi}{C_l}\right)\right\}^{-1}. \tag{4.4.16}$$

4.4.1 The Power Coefficient: The BEM Theory Expression

The power generated by an annular element is given by,

$$\mathrm{d}P = \Omega\mathrm{d}T. \tag{4.4.17}$$

Hence, the total power generated by the wind turbine disc is

$$P = \frac{1}{2}\pi\rho\Omega^3R^5\lambda_f^2\int_{\bar{r}_h}^{1}\frac{\sigma'(1-a)^2C_d}{\sin\phi\tan\phi}\left(1 - \frac{C_l\tan\phi}{C_d}\right)\bar{r}^2\mathrm{d}\bar{r}. \tag{4.4.18}$$

The power of the wind is given by,

$$P_{\text{wind}} = \frac{1}{2}\pi\rho\Omega^3R^5\lambda_f^3. \tag{4.4.19}$$

Thus, the power coefficient, based on the BEM theory, may be expressed as,

$$C_P = \frac{P}{P_{\text{wind}}} = \frac{1}{\lambda_f}\int_{\bar{r}_h}^{1}\frac{\sigma'(1-a)^2C_d}{\sin\phi\tan\phi}\left(1 - \frac{C_l\tan\phi}{C_d}\right)\bar{r}^2\mathrm{d}\bar{r}. \tag{4.4.20}$$

4.5 Aerodynamic Design of the Blade

The typical behavior of an aerofoil can be classified into three flow regimes: the attached flow regime, the high lift/stall development regime, and the flat plate/fully stalled regime (Burton et al. 2001, p. 101). In the attached flow regime, considering the flow on the upper surface of aerofoil, the lift due to the upper surface pressure distribution increases with the angle of attack. In the high lift/stall development regime, the lift coefficient reaches a peak as the aerofoil approaches a certain critical angle of attack. Stall occurs when the angle of attack exceeds the critical value (depending on the Reynolds number) and separation of the boundary layer on the upper surface takes place. The aerodynamic performances of different aerofoils are different because of different geometries of the aerofoils. Thus, an optimum choice of the aerofoil profile can be made for a particular wind turbine by enunciating the specific requirements of the desired energy conversion efficiency. For some modern wind turbines, the blade tips are designed using a thin aerofoil for high lift to drag ratio, and the root region is designed using a thick version of the same aerofoil for structural strength and stiffness. The crucial parameters for choosing an aerofoil are maximum lift to drag ratio and low pitching moment. Generally, in order to achieve the maximum lift and efficiency for some long blades, not only the chord length, thickness, and twist angle change along the span of the blade, but also the shape of aerofoil varies along it. The most popular aerofoil for wind tunnel applications in the past was NACA4412 since the lower surface of this aerofoil is flat, and consequently, the blade is easy to manufacture with glass fiber, although its aerodynamic performance is not always optimal for wind turbines. The current approach is to design a custom aerofoil shape, and many practical aerofoils have been designed for different wind turbines such as the aerofoil series mentioned latter in this section.

When the flow velocity over the upper surface of aerofoil increases, the flow which is generally laminar at low velocities corresponding to low Reynolds numbers can gradually become turbulent. This transition to turbulent flow is associated with a small range of Reynolds numbers. Further increase in the Reynolds number generally results in flow separation which is associated with loss of lift and substantial increases in drag and must therefore be avoided at all costs. The phenomenon of transition from laminar to turbulent flow is associated with the transition lag effect which results in a hysteresis loop in the lift, drag and pitch moment coefficient variations with respect to the flow Reynolds numbers. In an effort to avoid encountering this effect, the flow domain is restricted either to laminar flow or to turbulent flow domains. Natural laminar flow aerofoils have also been designed to operate at low Reynolds numbers that ensure the flow is entirely laminar over the entire aerofoil. Typical examples of the low Reynolds number turbulent and laminar flow aerofoils for wind turbine applications are the Wortmann (1974) and Selig (Giguére and Selig 1998) aerofoil series, respectively.

Low Reynolds number aerofoils are sometimes classified by their associated laminar separation bubbles. When laminar separation bubbles appear, usually on the top of the aerofoil due to the inability of the flow to make a transition to turbulent flow in the attached boundary layer, the flow separates before transition. The net result is a separation bubble that dramatically increases the drag. Separation bubbles can be avoided by ensuring that transition does occur fairly early in the flow over the aerofoil's upper surface. Drela and Giles (1987) and Drela (1989) have discussed a comprehensive method of validating the performance of an aerofoil section profile based on a hybrid inviscid and viscous analytic approach which has emerged as a standard for the design of aerofoil sections. Aerofoil design has also been discussed by Eppler (1990) and Miley (1974).

The aerodynamic design of the wind turbine blade, after the choice of the blade section profile is made, is an iterative process involving an initial sizing of the wind turbine based on the desired power output, which is related to the C_P, the expected or design power coefficient (0.35–0.38 for a modern three-bladed wind turbine), η the expected electrical and mechanical efficiency (0.85 would be a suitable value), R the tip radius, and V the expected wind velocity. Thus,

$$P = C_p\eta\frac{1}{2}\pi\rho R^5\Omega^3\lambda_f^3. \tag{4.5.1}$$

The preliminary design variable is the tip speed ratio $\lambda_{\mathrm{tsr}} = 1/\lambda_f$, which is defined as the inverse of the inflow ratio. Typically, the tip speed ratio, λ_{tsr}, is chosen in the range 6–9. A three-bladed rotor is generally considered most efficient as the interference effects are considerably minimized. Consequently, the number of blades is generally chosen to be 3. Thus, the appropriate rotor diameter is selected. The next step in the design is to choose an aerofoil shape from one of a range of available aerofoil families or from a set of custom designs. These are the NACA four, five, and six "digit" aerofoils (such as, NACA00xx, NACA44xx, NACA63-4xx, NACA64$_2$-4xx, or NACA FFA-W3-xxx, typical examples being NACA0012, NACA4412, NACA63-418, NACA64$_2$-415, or NACA FFA-W3-241), the Danish aerofoil family Risø-B1-xx (Risø-B1-18), the German MHxxx families (MH43), the American NREL aerofoil families Sxxx (S814), the Delft University family DU 9x-Wy-2zz (DU 91 W2-250), the Swedish aerofoil family FFA-W3-2zz (FFA-W3-211), and others that are suitable for wind tunnel blades. A typical family of aerofoils for wind turbine blades the Risφ-B1-xx family is illustrated in Fig. 4.4. The aerofoil profile may be varied across the span. The slimmer profile is generally used closer to the wing tip, while they get thicker as one approaches the hub. The blade is therefore assumed to be divided into a finite number of sections or elements.

The next step in the design process is the calculation of the section aerodynamic properties, in particular, the section lift, induced drag, and profile drag coefficients, for a range of angles of attack, for each of the selected aerofoil profiles. The aerodynamic lift coefficient may be found by the vortex lattice method. The aerofoil surface is divided into piecewise straight line segments or panels known as "boundary elements" and vortex sheets of strength Γ_0 are placed on each panel. The vortex sheets (miniature vortices of strength $\Gamma_0 \mathrm{d}s$, where $\mathrm{d}s$ is the length of a panel) are used, as vortices give rise to circulation, and hence lift. Discrete unsteady vortex is located at the ¼ chord point of each panel, while the downwash is matched at the ¾ chord point of each panel. Hence, on each panel, there is

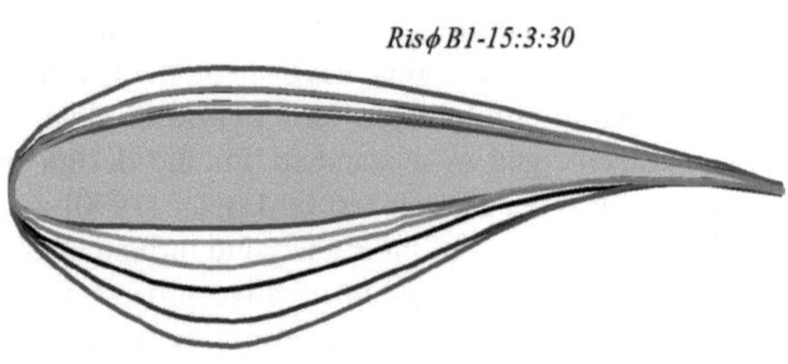

Fig. 4.4 The Risφ-B1-xx family of aerofoil profiles

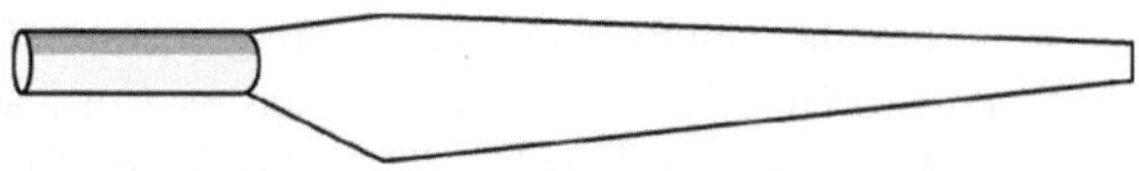

Fig. 4.5 Typical linear blade chord variation of a wind turbine blade

vortex sheet of strength d $\Gamma = \Gamma_0 \mathrm{d}s$, where d$s$ is the panel length. Each panel is defined by its two end points (panel joints) and by the control point, located at the panel center, where the boundary condition is applied. Thus, the velocity and pressure distributions on the aerofoil surface are found. The pressure distribution is integrated to obtain the lift coefficient.

The angle of attack for which the section lift is 80 % is selected as the blade section pretwist. The pretwist in the vicinity of the hub is reduced to produce a practical design of the blade. A chord distribution along the blade length is chosen to produce an almost uniform lift. This is given by,

$$c = \frac{8\pi R\lambda_f}{3N}\sin\phi. \tag{4.5.2}$$

The local solidity ratio is then given by,

$$\sigma' = \frac{8\lambda_f}{3}\sin\phi. \tag{4.5.3}$$

In practice, one may choose to approximate it by a linear variation along the span as shown Fig. 4.5. To estimate the revised rotor power coefficient, an initial estimate of the induction factors and the inflow angle is used. The induction factors are assumed to be $a = a' = 0.2$. Thus, the initial estimate of the inflow angle is assumed to be,

$$\phi = \tan^{-1}\left(\lambda_f(1 - a)/\bar{r}(1 + a')\right) \approx (0.8/1.2)\tan^{-1}\left(\lambda_f/\bar{r}\right). \tag{4.5.4}$$

From the revised power coefficient estimate, it is essential to iteratively adjust the twist and chord distribution till an acceptable design is obtained and the inductions factors are determined. The blade design is complete when the entire geometry of the blade is determined and validated.

4.6 Blade Structural Dynamics

Rotating pretwisted beams exhibit interesting coupling phenomena between extensional, torsional, and flexural vibrations and have been the subject of extensive research due to their use as wind turbine rotor blades. Due to the complex geometry introduced by the pretwist, a number of approximate theories were developed for the analysis of non-rotating and rotating pretwisted blades. An early approach developed for the vibration analysis of non-rotating pretwisted blades due to Chu (1951) where the fibers were modelled as helices and the stress field was determined along and normal to these directions. This model was shown to represent the bending-torsion coupling incorrectly, and it acted as a spur to further research by several authors such as Curti and Risitano (1979), Hodges (1980), Rosen (1983), Krenk (1983), Tsuiji (1985), Kosmatka (1992), Xu et al. (1997), and Liu et al. (2009). The method adopted by most researchers is based on the semi-inverse method. The semi-inverse method was first used by Saint–Venant for solving the elastic deformation of prismatic rods under two resultant end loads: a force and a moment [see for example, Sokolnikoff (1956)]. The governing equations for the displacements of the structure are simplified by assuming the functional form of the displacement field $[u \quad v \quad w]$ in three orthogonal directions based on appropriate physically meaningful postulates. Accordingly, the correctness and complexity of the resulting equations depend crucially on the choice and compatibility of the assumed displacement field. The semi-inverse method can be applied both to the uncoupled flexural and to the coupled axial torsional vibrations of beams. In this section, we shall first consider the uncoupled flexural and then the coupled axial torsional vibrations of beams, before considering the fully coupled flexural, axial torsional equations of motion.

The most commonly used flexural vibration theory is based on the celebrated Euler–Bernoulli postulates:

1. Cross sections remain plain and orthogonal to the neutral axis, so the average slope across the section is given by $-\partial w/\partial x \equiv -w'$;
2. The longitudinal fibers do no compress each other; consequently, the normal stress in the transverse direction is zero, and the bending moment is

$$M_y = -EI_{yy}w''. \tag{4.6.1}$$

where

$$I_{yy} = \iint_A y^2 \mathrm{d}y\mathrm{d}z \tag{4.6.2}$$

is the second moment of the area of cross section;

3. The rotational inertia is neglected; hence, the sectional shear force is

$$Q = M_y' = -EI_{yy}w'''; \tag{4.6.3}$$

4. Only the inertia forces due to transverse translation are considered, and the effects of rotary inertia are ignored. The equilibrium equations are

$$Q' = M_y'' = -EI_{yy}w'''' = -p_w = \rho h\partial^2 w/\partial t^2 \tag{4.6.4}$$

where p_w is the transverse load pressure.

A generalization of the Euler–Bernoulli theory is the Ambartsumyan–Vlasov theory which postulates that the transverse displacements of all points in the cross section are equal; thus, $\partial w/\partial z \equiv 0$. The theory assumes the two-dimensional displacement field to be,

$$u(x, y, t) = -y\frac{\partial w(x, t)}{\partial x} + \varphi(x, t)g(y)\ , \ g(0) = 0 \tag{4.6.5}$$

$$w(x, y, t) = w(x, t). \tag{4.6.6}$$

In the Vlasov theory, $g(y)$ is assumed to be quadratic function, $g(y) = y^2$. For a uniform beam, the differential equation for transverse vibrations is

$$\frac{\partial^4 w}{\partial x^4} - \left(\frac{1}{c_s^2 + ac_t^2}\right)\frac{\partial^4 w}{\partial x^2 \partial t^2} + \frac{1 - v^2}{D^2}\frac{\partial^2 w}{\partial t^2} + \frac{1}{ac_s^2 c_t^2}\frac{\partial^4 w}{\partial t^4} - p_w = 0, \tag{4.6.7}$$

where

$$c_s^2 = \frac{E}{\rho(1 - v^2)},\ c_t^2 = \frac{G}{\rho},\ D^2 = \frac{EI_{yy}}{\rho h},\ a = \frac{I_{yy}I_1}{hI_0}, \tag{4.6.8}$$

$$I_0 = \int_{-h/2}^{h/2} yg(y)\mathrm{d}y,\ I_1 = \int_{-h/2}^{h/2} \frac{\partial g(y)}{\partial y}\mathrm{d}y. \tag{4.6.9}$$

Another generalization of the Euler–Bernoulli theory in the Timoshenko theory which assumes an arbitrary shear coefficient q and the differential equation for transverse vibrations is

$$\frac{\partial^4 w}{\partial x^4} - \left(\frac{1}{c_s^2 + qc_t^2}\right)\frac{\partial^4 w}{\partial x^2 \partial t^2} + \frac{1}{D^2}\frac{\partial^2 w}{\partial t^2} + \frac{1}{qc_s^2 c_t^2}\frac{\partial^4 w}{\partial t^4} - p_w = 0. \tag{4.6.10}$$

Turning our attention to coupled axial torsional vibrations, two common assumptions are made for the coupled extension–torsion vibrations of a bar: First, that the deformations in the planes parallel to the cross sections are negligible; and second, that the axial deformation consists of a bulk displacement and cross-sectional warping. A typical displacement field for the coupled extension–torsion vibrations of a bar is

$$u(x, y, z, t) = \bar{u}(x, t) + \phi'(x, t)\psi(x, y, z) \tag{4.6.11}$$

$$v(x, y, z, t) = y\{\cos\phi(x, t) - 1\} - z\sin\phi(x, t) \tag{4.6.12}$$

$$w(x, y, z, t) = y\sin\phi(x, t) + z\{\cos\phi(x, t) - 1\} \tag{4.6.13}$$

where u, v, w are, respectively, the displacement components in the x, y, z directions, and ϕ represents the angle of elastic rotation additional to the pretwist angle. The function $\psi(x, y, z)$ is a cross-sectional warping function that is chosen to satisfy,

$$\left(\frac{\partial^2}{\partial y^2}+\frac{\partial^2}{\partial z^2}\right)\psi(x,y,z)=0. \tag{4.6.14}$$

The function $\psi(x, y, z)$ also satisfies certain boundary conditions in a direction normal to the boundary of the cross section.

The other important ingredient in the derivation of the equations of motion is the strain displacement relations which are obtained from Green's strain–displacement relations,

$$\varepsilon_{ij}=\frac{1}{2}\left[\frac{\partial u_j}{\partial x_i}+\frac{\partial u_i}{\partial x_j}+\frac{\partial u_k}{\partial x_i}\frac{\partial u_k}{\partial x_j}\right]. \tag{4.6.15}$$

The strain displacement relations are

$$\varepsilon_{xx}=\frac{\partial u}{\partial x}+\frac{1}{2}\left[\left(\frac{\partial u}{\partial x}\right)^2+r^2\theta^2\right], \tag{4.6.16}$$

$$\varepsilon_{xy}=\frac{1}{2}\theta\left[\frac{\partial\psi}{\partial y}\left(1+\frac{\partial u}{\partial x}\right)-z\right],\ \varepsilon_{xz}=\frac{1}{2}\theta\left[\frac{\partial\psi}{\partial z}\left(1+\frac{\partial u}{\partial x}\right)+y\right], \tag{4.6.17}$$

$$\varepsilon_{yy}=\frac{1}{2}\theta^2\left[\frac{\partial\psi}{\partial y}\right]^2,\ \varepsilon_{yz}=\frac{1}{2}\theta^2\left[\frac{\partial\psi}{\partial y}\frac{\partial\psi}{\partial z}\right]\text{ and }\varepsilon_{zz}=\frac{1}{2}\theta^2\left[\frac{\partial\psi}{\partial z}\right]^2. \tag{4.6.18}$$

where

$$\frac{\partial u}{\partial x}=\bar{\varepsilon}_{xx}+\theta\frac{\partial\psi}{\partial x}+\psi\frac{\partial\theta}{\partial x},\ \bar{\varepsilon}_{xx}=\frac{\partial\bar{u}}{\partial x},\ \theta=\frac{\partial\phi}{\partial x}\text{ and }r^2=x^2+y^2. \tag{4.6.19}$$

To derive the equations of motion, one uses Hamilton's variational principle,

$$\delta\int_{t_1}^{t_2}(T_{\mathrm{ke}}-U+W_{\mathrm{ext}})\mathrm{d}t=0 \tag{4.6.20}$$

where T_{ke} is the kinetic energy of the rod, U is the internal energy stored in the rod, and W_{ext} is the work done by the conservative external forces.

To express the kinetic energy in terms of the displacement field, the velocity vector of each point in the beam is used and is given by,

$$\dot{\mathbf{R}}=\dot{\mathbf{u}}=\dot{u}\hat{e}_x+\dot{v}\hat{e}_y+\dot{w}\hat{e}_z \tag{4.6.21}$$

$$\dot{\mathbf{R}}=\dot{\mathbf{u}}=\left(\dot{\bar{u}}+\dot{\theta}\psi\right)\hat{e}_x-\dot{\phi}(y\sin\phi+z\cos\phi)\hat{e}_y+\dot{\phi}(y\cos\phi-z\sin\phi)\hat{e}_z. \tag{4.6.22}$$

Expressing the Cartesian unit vectors $\left(\hat{e}_x\,\hat{e}_y\,\hat{e}_z\right)$ in cylindrical coordinates, the cylindrical unit vectors $\left(\hat{e}_x\,\hat{e}_r\,\hat{e}_\phi\right)$ are related to the Cartesian unit vectors by,

$$\hat{e}_r = \frac{1}{r}\left(y\hat{e}_y + z\hat{e}_z\right),\ \hat{e}_\phi = \frac{1}{r}\left(y\hat{e}_z - z\hat{e}_y\right). \tag{4.6.23}$$

The velocity vector of each point in the beam in cylindrical coordinates is given by,

$$\dot{\mathbf{R}} = \dot{\mathbf{u}} = \left(\dot{\bar{u}} + \dot{\theta}\psi\right)\hat{e}_x + r\dot{\phi}\left(-\sin\phi\hat{e}_r + \cos\phi\hat{e}_\phi\right). \tag{4.6.24}$$

Liu et al. (2009) have shown the equations of free vibration take the form,

$$p_T + EA\frac{\partial^2\bar{u}}{\partial x^2} + ES_1\frac{\partial^2\phi}{\partial x^2} = \rho A\frac{\partial^2\bar{u}}{\partial t^2} \tag{4.6.25}$$

$$q_T + ES_1\frac{\partial^2\bar{u}}{\partial x^2} + (EK_1 + GJ_e)\frac{\partial^2\phi}{\partial x^2} = \rho I_p\frac{\partial^2\phi}{\partial t^2} - \rho K_0\frac{\partial^4\phi}{\partial x^2\partial t^2} + EK_0\frac{\partial^4\phi}{\partial x^4} \tag{4.6.26}$$

where the ρ is the mass density per unit length of the rod, A is the area of cross section, E is the Young's modulus, G is the shear modulus of rigidity, p_T is the distributed tensile force along the axis of the beam, q_T is the distributed torsional moment along the axis of the beam,

$$S_i = \iint_A \frac{\partial^i\psi}{\partial x^i}\mathrm{d}y\mathrm{d}z,\ K_i = \iint_A \left(\frac{\partial^i\psi}{\partial x^i}\right)^2 \mathrm{d}y\mathrm{d}z,\ D_{ij} = \iint_A \frac{\partial^i\psi}{\partial x^i}\frac{\partial^j\psi}{\partial x^j}\mathrm{d}y\mathrm{d}z, \tag{4.6.27}$$

$$J_e = \iint_A \left\{\left(\frac{\partial\psi}{\partial y} - z\right)^2 + \left(\frac{\partial\psi}{\partial z} + y\right)^2\right\}\mathrm{d}y\mathrm{d}z \text{ and } I_p = \iint_A r^2\mathrm{d}y\mathrm{d}z. \tag{4.6.28}$$

The boundary conditions at the two ends, $x = 0$ and $x = L$, are

$$\frac{\partial^2\phi}{\partial x^2} = 0,\ F_T = EA\frac{\partial\bar{u}}{\partial x} + ES_1\frac{\partial\phi}{\partial x} \tag{4.6.29}$$

is the total force acting at the end in question, and

$$F_M = ES_1\frac{\partial\bar{u}}{\partial x} + (EK_1 + GJ_e)\frac{\partial\phi}{\partial x} + \rho K_0\frac{\partial^3\phi}{\partial x\partial t^2} - EK_0\frac{\partial^3\phi}{\partial x^3}, \tag{4.6.30}$$

is the total torsional moment acting on the end in question.

When the longitudinal inertia is ignored, the tension along the axis of the bar satisfies

$$\frac{\partial T}{\partial x} = EA\frac{\partial^2\bar{u}}{\partial x^2} = -ES_1\frac{\partial^2\phi}{\partial x^2} - p_T. \tag{4.6.31}$$

The decoupled equation for torsional equilibrium is

$$q_T + \frac{S_1}{A}\frac{\partial T}{\partial x} + (EK_1 + GJ_e)\frac{\partial^2 \phi}{\partial x^2} = \rho I_p \frac{\partial^2 \phi}{\partial t^2} - \rho K_0 \frac{\partial^4 \phi}{\partial x^2 \partial t^2} + EK_0 \frac{\partial^4 \phi}{\partial x^4}. \tag{4.6.32}$$

Considering the influence of blade rotation, which couples the transverse bending and torsion vibrations, the primary effect is the stiffening of the blades due to centrifugal forces acting on the blade cross section. The linear coupled out-of-plane bending and torsional displacement equations of the rotating blade that is not twisted and with all shear and rotary inertia effects ignored are

$$\frac{\partial^2}{\partial \bar{x}^2}\left(\frac{EI_{yy}}{\Omega^2 R^4}\frac{\partial^2 \bar{w}}{\partial \bar{x}^2}\right) - \frac{\partial}{\partial \bar{x}}\left(\frac{T}{\Omega^2 R^2}\frac{\partial \bar{w}}{\partial \bar{x}}\right) = mp_{Iz} + m\frac{\partial q_{Iy}}{\partial \bar{x}} + \frac{L_w}{\Omega^2 R^2} \tag{4.6.33}$$

$$-\frac{\partial}{\partial \bar{x}}\left(\frac{GJ_e}{\Omega^2 R^4}\frac{\partial \phi}{\partial \bar{x}}\right) - \frac{\partial}{\partial \bar{x}}\left(\frac{Tk_A^2}{\Omega^2 R^4}\frac{\partial \phi}{\partial \bar{x}}\right) = mq_{Ix} + mq_{Iz}\frac{\partial \bar{w}}{\partial \bar{x}} + \frac{M_\phi}{\Omega^2 R^2} \tag{4.6.34}$$

where $p_{Ix}\hat{e}_x + p_{Iy}\hat{e}_y + p_{Iz}\hat{e}_z$ is the external inertial force vector, and $q_{Ix}\hat{e}_x + q_{Iy}\hat{e}_y + q_{Iz}\hat{e}_z$ is the external inertial moment vector. In the above equations, L_w and M_ϕ are the total aerodynamic out-of-plane bending force and pitching moment distributions acting on the blade.

Given that,

$$p_{Iz} = -\left(\frac{\partial^2 \bar{w}}{\partial \psi^2} - \bar{x}_{cg}\frac{\partial^2}{\partial \psi^2}(\phi + \theta_r)\right), \tag{4.6.35}$$

$$q_{Ix} = \bar{x}_{cg}\frac{\partial^2 \bar{w}}{\partial \psi^2} - \frac{k_m^2}{R^2}\left(1 + \frac{\partial^2}{\partial \psi^2}\right)(\phi + \theta_r),\ q_{Iy} = -\bar{x}_{cg}\bar{x}(\phi + \theta_r) \text{ and } q_{Iz} = \bar{x}_{cg}\bar{x}, \tag{4.6.36}$$

the equations of motion may be expressed as,

$$\begin{aligned}
&\frac{\partial^2}{\partial \bar{x}^2}\left(\frac{EI_{yy}}{\Omega^2 R^4}\frac{\partial^2 \bar{w}}{\partial \bar{x}^2}\right) - \frac{\partial}{\partial \bar{x}}\left(\frac{T}{\Omega^2 R^2}\frac{\partial \bar{w}}{\partial \bar{x}}\right)\\
&\quad + m\left(\frac{\partial^2 \bar{w}}{\partial \psi^2} - \bar{x}_{cg}\frac{\partial^2}{\partial \psi^2}(\phi + \theta_r) + \frac{\partial}{\partial \bar{x}}\left(\bar{x}_{cg}\bar{x}(\phi + \theta_r)\right)\right)\\
&\quad = \frac{L_w}{\Omega^2 R},
\end{aligned} \tag{4.6.37}$$

$$
\begin{aligned}
&-\frac{\partial}{\partial\bar{x}}\left(\frac{GJ_e}{\Omega^2R^4}\frac{\partial\phi}{\partial\bar{x}}\right)-\frac{\partial}{\partial\bar{x}}\left(\frac{Tk_A^2}{\Omega^2R^4}\frac{\partial\phi}{\partial\bar{x}}\right)\\
&+m\left(\frac{k_m^2}{R^2}\left(1+\frac{\partial^2}{\partial\psi^2}\right)(\phi+\theta_r)-\bar{x}_{cg}\frac{\partial^2\bar{w}}{\partial\psi^2}-\bar{x}_{cg}\bar{x}\frac{\partial\bar{w}}{\partial\bar{x}}\right),\\
&=\frac{M_\phi}{\Omega^2R^2}
\end{aligned}
\tag{4.6.38}
$$

$$
\begin{aligned}
T&=\Omega^2R^2\int_{\bar{x}}^{1}m\bar{x}\mathrm{d}\bar{x},\ \bar{x}=x/R,\ \bar{w}=w/R,\ \psi=\Omega t,\ k_m^2\\
&=\frac{1}{m}\iint_A\rho r^2\mathrm{dydz},\ k_A^2=\frac{1}{A}\iint_A r^2\mathrm{dydz},
\end{aligned}
\tag{4.6.39}
$$

where θ_r is the root blade rigid rotation, $\bar{x}_{cg}$ normalized distance of section mass center aft of pitch axis, k_A^2 polar radius of gyration of cross section area about the elastic axis, and k_m^2 polar radius of gyration of cross section mass about the elastic axis.

To consider the effect of pretwist, it is important include the dynamics of in-plane bending motion as the in-plane and out-of-plane bending displacement are coupled as a result of pretwist and blade root pitching. To obtain the force and moment equilibrium equations, in the undeformed reference axes, the force and moment equilibrium equations in the deformed axes are first established and transformed to the undeformed axes. The force and moment equilibrium equations in the deformed axes are

$$
\bar{\mathbf{F}}_{,\mathbf{x}}+\bar{\mathbf{p}}=\mathbf{0},
\tag{4.6.40}
$$

$$
\bar{\mathbf{M}}_{,\mathbf{x}}+\bar{\mathbf{q}}+\hat{\mathbf{e}}'_{\mathbf{x}}\times\bar{\mathbf{F}}=\mathbf{0}
\tag{4.6.41}
$$

where $\bar{\mathbf{p}}$ is the distributed force vector acting on blade, $\bar{\mathbf{q}}$ is the distributed moment vector acting on blade, $\hat{\mathbf{e}}'_{\mathbf{x}}$ is a unit vector in the directions of the deformed axes, $\bar{\mathbf{F}}$ are the elastic force resultants generated in the blade, consisting of the tension along the blade axis and shear forces in directions normal to the blade axes, and $\bar{\mathbf{M}}$ are the moment resultants generated in the blade, consisting of the twisting moment along the blade axis and bending moments normal to the blade axis. The relationships between the deformed unit vectors and there slopes can expressed in terms of the curvatures and twist.

In deriving the foregoing equations, it is tacitly assumed that the sequence of displacements are the blade root pitch and pretwist, the in-plane blade displacement followed by the out-of-plane blade displacement and the blade torsion. The non-dimensional displacements along the elastic axis are assumed to be given by,

$(\bar{u}\,\bar{v}\,\bar{w})$, where $\bar{u}$ is the foreshortening (the axial elastic displacement is assumed to be zero) and is defined as,

$$\bar{u} = -\frac{1}{2}\int_0^{\bar{x}}\left[\left(\frac{\partial \bar{v}}{\partial \bar{x}_1}\right)^2 + \left(\frac{\partial \bar{w}}{\partial \bar{x}_1}\right)^2\right]\mathrm{d}\bar{x}_1. \tag{4.6.42}$$

The equations of motion in terms of the moment and force resultants with the nonlinear terms isolated and expressed collectively in separate expressions are

$$\begin{aligned}\frac{1}{\Omega^2 R^4}\frac{\partial^2}{\partial \bar{x}^2}\begin{bmatrix} M_{\bar{y}} \\ M_{\bar{z}} \end{bmatrix} - \frac{\partial}{\partial \bar{x}}\left(\frac{T}{\Omega^2 R^2}\frac{\partial}{\partial \bar{x}}\begin{bmatrix} \bar{w} \\ \bar{v} \end{bmatrix}\right) + \frac{\partial^2}{\partial \bar{x}^2}\left(\frac{X_{II}T}{\Omega^2 R^3}\begin{bmatrix} \sin\theta_G \\ \cos\theta_G \end{bmatrix}\right) + \frac{1}{\Omega^2 R^4}\begin{bmatrix} N_w \\ N_v \end{bmatrix} \\ = m\begin{bmatrix} p_{Iz} \\ p_{Iy} \end{bmatrix} + m\frac{\partial}{\partial \bar{x}}\begin{bmatrix} q_{Iy} \\ -q_{Iz} \end{bmatrix} + \frac{1}{\Omega^2 R}\begin{bmatrix} L_w \\ L_v \end{bmatrix},\end{aligned} \tag{4.6.43}$$

$$\begin{aligned}-\frac{1}{\Omega^2 R^4}\frac{\partial}{\partial \bar{x}}M_{\bar{x}} - \frac{\partial}{\partial \bar{x}}\left(\frac{Tk_A^2}{\Omega^2 R^4}\frac{\partial(\theta_G + \phi)}{\partial \bar{x}}\right) + \frac{TX_{II}}{\Omega^2 R^3}\left(\cos\theta_G\frac{\partial^2 \bar{w}}{\partial \bar{x}^2} - \sin\theta_G\frac{\partial^2 \bar{v}}{\partial \bar{x}^2}\right) + \frac{N_\phi}{\Omega^2 R^4} \\ = mq_{Ix} + \frac{M_\phi}{\Omega^2 R^2}\end{aligned} \tag{4.6.44}$$

Following the ordering scheme of Friedmann (1977) and Friedmann and Shamie (1975), the coupled equations of motion, which are expressed in a form to highlight the gyroscopic nature of the coupling due to lack of inertial symmetry about the blade axis, are given by,

$$\begin{aligned}\frac{\partial^2}{\partial \bar{x}^2}\left(\frac{E}{\Omega^2 R^4}\left\{\begin{bmatrix} I_{yy} & 0 \\ 0 & I_{zz} \end{bmatrix}\frac{\partial^2}{\partial \bar{x}^2}\begin{bmatrix} \bar{w} \\ \bar{v} \end{bmatrix} + (I_{zz} - I_{yy})\sin\theta_G\begin{bmatrix} \cos\theta_G & \sin\theta_G \\ -\sin\theta_G & \cos\theta_G \end{bmatrix}\frac{\partial^2}{\partial \bar{x}^2}\begin{bmatrix} \bar{v} \\ \bar{w} \end{bmatrix}\right\}\right) \\ - \frac{\partial}{\partial \bar{x}}\left(\frac{T}{\Omega^2 R^2}\frac{\partial}{\partial \bar{x}}\begin{bmatrix} \bar{w} \\ \bar{v} \end{bmatrix}\right) + \frac{\partial^2}{\partial \bar{x}^2}\left(\frac{X_{II}T}{\Omega^2 R^3}\begin{bmatrix} \sin\theta_G \\ \cos\theta_G \end{bmatrix}\right) + \frac{1}{\Omega^2 R^4}\begin{bmatrix} N_w \\ N_v \end{bmatrix} \\ = m\begin{bmatrix} p_{Iz} \\ p_{Iy} \end{bmatrix} + m\frac{\partial}{\partial \bar{x}}\begin{bmatrix} q_{Iy} \\ -q_{Iz} \end{bmatrix} + \frac{1}{\Omega^2 R}\begin{bmatrix} L_w \\ L_v \end{bmatrix},\end{aligned} \tag{4.6.45}$$

$$\begin{aligned}-\frac{\partial}{\partial \bar{x}}\left(\frac{GJ_e}{\Omega^2 R^4}\frac{\partial \phi}{\partial \bar{x}}\right) - \frac{\partial}{\partial \bar{x}}\left(\frac{Tk_A^2}{\Omega^2 R^4}\frac{\partial(\theta_G + \phi)}{\partial \bar{x}}\right) + \frac{TX_{II}}{\Omega^2 R^3}\left(\cos\theta_G\frac{\partial^2 \bar{w}}{\partial \bar{x}^2} - \sin\theta_G\frac{\partial^2 \bar{v}}{\partial \bar{x}^2}\right) + \frac{N_\phi}{\Omega^2 R^4} \\ = mq_{Ix} + \frac{M_\phi}{\Omega^2 R^2},\end{aligned} \tag{4.6.46}$$

where θ_G is the geometric pitch angle and is composed of the blade root pitch angle and the section pretwist, X_{II} is the offset of the point of action of the tension from the elastic axis, positive aft, $\bar{v}$, I_{zz}, and L_v are the in-plane blade displacement, in-plane bending section area moment of inertia, and in-plane aerodynamic load, respectively. N_w, N_v, and N_ϕ are the nonlinear terms which could be ignored in the case of linear stability analysis. The components of the external inertial force and moment vectors assuming large rotations and small deformations are

$$p_{Ix} = \bar{x} + 2\frac{\partial \bar{v}}{\partial \psi}, \tag{4.6.47}$$

$$p_{Iy} = \bar{v} - \bar{x}_{cg}\cos\theta_G - \left(\frac{\partial^2 \bar{v}}{\partial \psi^2} + \bar{x}_{cg}\sin\theta_G \frac{\partial^2}{\partial \psi^2}(\phi + \theta_G)\right) \tag{4.6.48}$$

$$p_{Iz} = -\left(\frac{\partial^2 \bar{w}}{\partial \psi^2} - \bar{x}_{cg}\cos\theta_G \frac{\partial^2}{\partial \psi^2}(\phi + \theta_G)\right), \tag{4.6.49}$$

$$\begin{aligned} q_{Ix} = {} & \bar{x}_{cg}\cos\theta_G \frac{\partial^2 \bar{w}}{\partial \psi^2} - \left(\frac{\partial^2 \bar{v}}{\partial \psi^2} - \bar{v}\right)\bar{x}_{cg}\sin(\theta_G + \phi) \\ & - \frac{k_m^2}{R^2}\left(\frac{\sin 2(\theta_G + \phi)}{2} + \frac{\partial^2(\phi + \theta_G)}{\partial \psi^2}\right), \end{aligned} \tag{4.6.50}$$

$$q_{Iy} = -\bar{x}_{cg}\bar{x}\sin(\phi + \theta_G) \text{ and } q_{Iz} = \bar{x}_{cg}\left(\bar{x} + 2\frac{\partial \bar{v}}{\partial \psi}\right)\cos\theta_G. \tag{4.6.51}$$

The non-dimensional tension is given by,

$$\frac{1}{m\Omega^2 R^2}\frac{\partial T}{\partial \bar{x}} = -p_{Ix}. \tag{4.6.52}$$

To derive the linear equations, all second- and higher-order terms in the translational and rotational elastic displacements are ignored in the equations of motion. Following the ordering scheme of Friedmann (1977), Friedmann and Shamie (1975) and Rosen and Friedmann (1977), the second- and third-order nonlinear terms in the equations of motion are

$$\begin{bmatrix} N_w \\ N_v \end{bmatrix} = \begin{bmatrix} N_{w1} \\ N_{v1} \end{bmatrix} + \begin{bmatrix} N_{w2} \\ N_{v2} \end{bmatrix} + \begin{bmatrix} N_{wI} \\ N_{vI} \end{bmatrix} + \begin{bmatrix} N_{we} \\ N_{ve} \end{bmatrix}, \tag{4.6.53}$$

$$\begin{aligned} \begin{bmatrix} N_{w1} \\ N_{v1} \end{bmatrix} = {} & \frac{\partial^2}{\partial \bar{x}^2}\left(E\phi\left\{(I_{zz} - I_{yy})\cos\theta_G \begin{bmatrix} 2\sin\theta_G & \cos\theta_G \\ \cos\theta_G & -2\sin\theta_G \end{bmatrix} \frac{\partial^2}{\partial \bar{x}^2}\begin{bmatrix} \bar{w} \\ \bar{v} \end{bmatrix}\right\}\right) \\ & + \frac{\partial}{\partial \bar{x}} GJ_e \frac{\partial \phi}{\partial \bar{x}}\frac{\partial^2}{\partial \bar{x}^2}\begin{bmatrix} -\bar{v} \\ \bar{w} \end{bmatrix}, \end{aligned} \tag{4.6.54}$$

$$\begin{bmatrix} N_{w2} \\ N_{v2} \end{bmatrix} = -\frac{\partial^2}{\partial \bar{x}^2}\left(EB_2\left(\frac{\partial \theta_G}{\partial \bar{x}}\frac{\partial \phi}{\partial \bar{x}} + \frac{1}{2}\left(\frac{\partial \phi}{\partial \bar{x}}\right)^2\right)\begin{bmatrix} \sin\theta_G \\ \cos\theta_G \end{bmatrix}\right), \tag{4.6.55}$$

$$\begin{bmatrix} N_{wI} \\ N_{vI} \end{bmatrix} = -\frac{\partial}{\partial \bar{x}} \begin{bmatrix} -q_{Ix}\frac{\partial \bar{v}}{\partial \bar{x}} \\ q_{Ix}\frac{\partial \bar{w}}{\partial \bar{x}} \end{bmatrix}, \tag{4.6.56}$$

$$\begin{bmatrix} N_{\text{we}} \\ N_{\text{ve}} \end{bmatrix} = \frac{\partial^2}{\partial \bar{x}^2} \left(X_{II} RT\phi \begin{bmatrix} \cos\theta_G \\ -\sin\theta_G \end{bmatrix} \right), \tag{4.6.57}$$

$$N_\phi = N_{\phi 1} + N_{\phi 2} + N_{\phi 3} + N_{\phi I} + N_{\phi e}, \tag{4.6.58}$$

$$\begin{aligned} N_{\phi 1} =& \left(E\left\{ (I_{zz} - I_{yy}) \left(\frac{\partial^2 \bar{w}}{\partial \bar{x}^2} \frac{\partial^2 \bar{v}}{\partial \bar{x}^2} \cos 2\theta_G + \frac{1}{2} \left(\left(\frac{\partial^2 \bar{w}}{\partial \bar{x}^2} \right)^2 - \left(\frac{\partial^2 \bar{v}}{\partial \bar{x}^2} \right)^2 \right) \sin 2\theta_G \right) \right\} \right) \\ &+ \left(E\left\{ (I_{zz} - I_{yy}) \left(\phi \left(\left(\frac{\partial^2 \bar{w}}{\partial \bar{x}^2} \right)^2 - \left(\frac{\partial^2 \bar{v}}{\partial \bar{x}^2} \right)^2 \right) \cos^2\theta_G \right) \right\} \right) \end{aligned} \tag{4.6.59}$$

$$N_{\phi 2} = -\frac{\partial}{\partial \bar{x}} \left(EB_1 \left\{ \left(\frac{\partial \theta_G}{\partial \bar{x}} \frac{\partial \phi}{\partial \bar{x}} + \frac{1}{2} \left(\frac{\partial \phi}{\partial \bar{x}} \right)^2 \right) \left(\frac{\partial \theta_G}{\partial \bar{x}} + \frac{\partial \phi}{\partial \bar{x}} \right) \right\} \right), \tag{4.6.60}$$

$$\begin{aligned} N_{\phi 3} =& \frac{\partial}{\partial \bar{x}} \left(EB_2 \left\{ \left(\frac{\partial \theta_G}{\partial \bar{x}} + \frac{\partial \phi}{\partial \bar{x}} \right) \left(\frac{\partial^2 \bar{v}}{\partial \bar{x}^2} \cos\theta_G + \frac{\partial^2 \bar{w}}{\partial \bar{x}^2} \sin\theta_G + \phi \frac{\partial^2 \bar{w}}{\partial \bar{x}^2} \cos\theta_G \right) \right\} \right) \\ &- \left(EB_2 \frac{\partial^2 \bar{w}}{\partial \bar{x}^2} \cos\theta_G \left\{ \frac{\partial \phi}{\partial \bar{x}} \left(\frac{\partial \theta_G}{\partial \bar{x}} + \frac{\partial \phi}{\partial \bar{x}} \right) - \frac{1}{2} \left(\frac{\partial \phi}{\partial \bar{x}} \right)^2 \right\} \right) \end{aligned} \tag{4.6.61}$$

$$N_{\phi I} = -q_{Iy} \frac{\partial \bar{v}}{\partial \bar{x}} - q_{Iz} \frac{\partial \bar{w}}{\partial \bar{x}} \tag{4.6.62}$$

and

$$N_{\phi e} = -TX_{II} R\phi \left(\sin\theta_G \frac{\partial^2 \bar{w}}{\partial \bar{x}^2} + \cos\theta_G \frac{\partial^2 \bar{v}}{\partial \bar{x}^2} \right), \tag{4.6.63}$$

where

$$EB_1 = \iint\limits_{A'} r'^2 (r'^2 - k_A^2) E \mathrm{d}A', \ EB_2 = \iint\limits_{A'} r'^2 (y' - e_A) E \mathrm{d}A'. \tag{4.6.64}$$

The prime refers to the deformed coordinates of a point in the cross section.

The corresponding nonlinear expression for the non-dimensional tension is given by,

$$\frac{1}{m\Omega^2 R^2} \frac{\partial T}{\partial \bar{x}} = -p_{Ix} + \frac{\partial}{\partial \bar{x}} \left(q_{Iy} \frac{\partial \bar{w}}{\partial \bar{x}} \right) - \frac{\partial}{\partial \bar{x}} \left(q_{Iz} \frac{\partial \bar{v}}{\partial \bar{x}} \right). \tag{4.6.65}$$

The additional terms due to the offset of the tension action point are present in the work of Rosen and Friedmann (1977) but the additional term in the torque equation which is proportional to k_A^2 and due to warping is considered small and ignored.

To solve the equations of motion one adopts Galerkin's method. Thus, the transverse bending, lateral bending, and torsional displacements are assumed to be,

$$\begin{bmatrix} \bar{w}(\bar{x},\psi) \\ \bar{v}(\bar{x},\psi) \end{bmatrix} = \sum_{i=1}^{N_m} \begin{bmatrix} w_{0i}(\psi) \\ v_{0i}(\psi) \end{bmatrix} \varphi_i(\bar{x}) \quad (4.6.66)$$

$$\phi(\bar{x},\psi) = \sum_{i=1}^{N_m} a_{0i}(\psi)\chi_i(\bar{x}) \quad (4.6.67)$$

$$\varphi_i(\bar{x}) = \cosh(\beta_i\bar{x}) - \cos(\beta_i\bar{x}) - \alpha_i[\sinh(\beta_i\bar{x}) - \sin(\beta_i\bar{x})] \quad (4.6.68)$$

$$\chi(\bar{x}) = \sqrt{2}\sin(\pi(i - 1/2)\bar{x}), \quad (4.6.69)$$

where

$$\beta_i = [1.875\, 4.694\, 7.855\, 10.996\, 14.137, \ldots, (i - 0.5) * \pi] \quad (4.6.70)$$

and

$$\alpha = -(\cos(\beta_i) + \cosh(\beta_i))/(\sin(\beta_i) + \sinh(\beta_i)). \quad (4.6.71)$$

The linear equations of motion, in the absence of aerodynamic loads, are reduced to standard eigenvalue problem for the squares of the blade natural frequencies. A typical set of blade natural frequencies are shown in Fig. 4.6.

4.7 Rotor Aeroelastic Dynamic Modelling

Wind turbine blade dynamics plays a significant role in the generation of wind power. Wind turbine blade dynamics can be expressed in either a blade-fixed frame or frame fixed to the rotor disc. It is convenient to represent the aerodynamic forces in a frame fixed to the blade, while the wind inputs and gusts are more easily represented in a frame fixed to the rotor disc plane which usually aligns itself normal to the wind direction. Thus, it is usual to use both frames of reference and transform from one to the other. The blades are assumed to be equi-spaced and to lie with the plane of the rotor disc. Each blade is assigned a reference index, and the reference blade is denoted by the index 1. Thus, the azimuth angle of the kth blade ψ_k in an N-bladed rotor within the disc plane in terms of the azimuth angle of the reference blade $\psi_1$is given by,

$$\psi_k = \psi_1 + 2(k - 1)\pi/N. \quad (4.7.1)$$

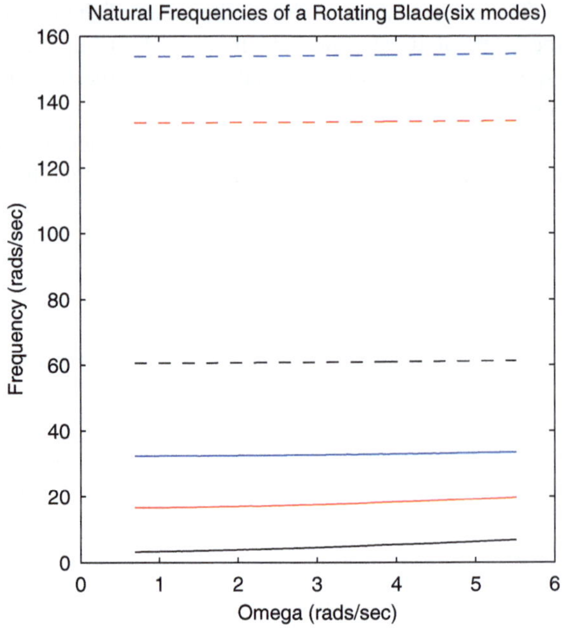

Fig. 4.6 Blade natural frequencies as function of rotor speed with the CG and tension axis offsets equal to zero

The rotor angular velocity is denoted by Ω, and consequently, the rotor period is $T = 2\pi/\Omega$, and the blade passage period is $T_b = T/N$. It is convenient to introduce the non-dimensional time variable, $\psi = \Omega t$, so the rotor period in non-dimensional time is $T = 2\pi$, and the frequencies are normalized by the rotor speed Ω.

The rotor dynamic model is developed in terms of non-dimensional quantities so that the general rotor configurations can be analyzed without the need to specify the size. The development of the model closely follows the methodology outlined by Padfield (1996) and Fox (1993). Thus, the rotor geometrical properties are normalized by the rotor radius R and its powers, while linear velocities are normalized by the blade tip speed ΩR and densities by the density of air ρ. The important rotor blade properties of interest are the aerodynamic forces and moments acting on the individual blades as well as the rotor thrust and torque which are related to the blade forces. Although the blades are not assumed to be uniform, it is assumed that there structural properties may be represented by a set of assumed modes. Each blade is assumed to fully controllable in pitch with the root of the blade offset by e from the rotor axis and with a mean aerodynamic chord length of c. The blades are assumed to identical to each other. The mass distribution and the mass moments of inertia are assumed to be known and given by the flapping inertia I_β about the blade root and the section pitching inertia I_θ. The moments of inertia are non-dimensionalized by the characteristic inertia of the blade,

$$I_b = \int_e^R mr^2 \mathrm{d}r. \tag{4.7.2}$$

All moments are non-dimensional and normalized by $I_b\Omega^2$. Sectional blade lift is assumed to be linearly proportional to the local angle of attack α with a constant lift curve slope such that $C_{l\alpha} = a_\infty$. The lock number is an important non-dimensional aerodynamic parameter defined in terms of the aerodynamic lift coefficient slope characteristic a_∞ as,

$$\gamma = \rho a_\infty c R^4 / I_b. \tag{4.7.3}$$

The distance of the blade section center of mass, the blade section aerodynamic center, and the blade section elastic axis from the blade pitch axis will be assumed to negligible. Furthermore, the blade sections are assumed to be symmetric. Thus, the flapwise bending modes and torsion modes are not coupled, the lift acts along the elastic axis, and there is no aerodynamic moment at the aerodynamic center.

The torsional motion of the blade is assumed to be composed of rigid body rotation $\theta_r(t)$ and elastic torsion $\theta_{\mathrm{et}}(t)$, so the torsional equilibrium equation is given by,

$$I_\theta \frac{\partial^2 \theta_r(t)}{\partial t^2} + I_\theta \frac{\partial^2 \theta_{et}(t)}{\partial t^2} - GJ_e \frac{\partial^2 \theta_{\mathrm{et}}(t)}{\partial r^2} = M_{\mathrm{et}} \tag{4.7.4}$$

where M_{et} is the external torsional moment consisting primarily of a centrifugal stiffening moment (sometimes known as a propeller moment) and an aerodynamic pitching moment.

In the case of flapping of the flexible blade, the equation of motion can be expressed as,

$$\begin{aligned} &\Omega^2 R^3 \int_e^R \frac{\partial^2 \beta(r)}{\partial \psi^2} m(r-e)^2 \mathrm{d}r + \Omega^2 R^3 \int_e^R \beta(r) m r (r-e) \mathrm{d}r \\ &+ \int_e^R S(\beta)(r-e)\beta(r)\mathrm{d}r \\ &= M_{e\beta} \end{aligned} \tag{4.7.5}$$

where M_{et} is the external aerodynamic flapping moment about the root which is assumed to be offset from the rotor axis be a distance e, $\beta(r)$ is the angle an element of mass on the blade at a distance r from the rotor axis makes with the root. $S(\beta)$ is defined by the equations,

$$S(\beta) = \frac{1}{\beta(r)} \frac{\partial^2}{\partial r^2} EI \frac{\partial^2((r-e)\beta)}{\partial r^2} - \frac{1}{\beta(r)} \frac{\partial}{\partial r} T \frac{\partial((r-e)\beta)}{\partial r}, \tag{4.7.6}$$

where T is the centrifugal tension in the blade. M_{et} satisfies the equation,

$$M_{e\beta} = \int_e^R L(\beta)(r - e)\mathrm{d}r. \tag{4.7.7}$$

Assuming the blade elastic flapping and torsion displacement angles can be expressed, respectively, as a summation of modes,

$$\beta(r) = \sum_{j=1}^{J} \beta_j \zeta_j(r), \ \ \theta_{\mathrm{et}}(r) = \sum_{j=1}^{J} \theta_j \xi_j(r), \tag{4.7.8}$$

the modal amplitude equations may be expressed as,

$$\begin{gathered} \frac{\partial^2 \theta_r(t)}{\partial t^2} \int_e^R I_\theta \xi_k(r)\mathrm{d}r + \sum_{j=1}^{J} \frac{\partial^2 \theta_j(t)}{\partial t^2} \int_e^R I_\theta \xi_j(r)\xi_k(r)\mathrm{d}r \\ - \sum_{j=1}^{J} \theta_j \int_e^R GJ_e \frac{\partial^2 \xi_j(r)}{\partial r^2} \xi_k(r)\mathrm{d}r = \int_e^R M_{et}\xi_k(r)\mathrm{d}r \\ k = 1, 2, 3 \ldots J, \end{gathered} \tag{4.7.9}$$

or as,

$$\sum_{j=1}^{J} I_\theta^{\mathrm{kj}} \frac{\partial^2 \theta_j(t)}{\partial t^2} + I_\theta^k \frac{\partial^2 \theta_r(t)}{\partial t^2} + \sum_{j=1}^{J} k_{\mathrm{et}}^{\mathrm{kj}} \theta_j = M_{\mathrm{et}}^k, \ \ k = 1, 2, 3 \ldots J, \tag{4.7.10}$$

and

$$\begin{gathered} \Omega^2 R^3 \sum_{j=1}^{J} \frac{\partial^2 \beta_j}{\partial \psi^2} \int_e^R \zeta_k(r)\zeta_j(r) m(r - e)^2 \mathrm{d}r + \Omega^2 R^3 \sum_{j=1}^{J} \beta_j \int_e^R \zeta_k(r)\zeta_j(r) m r(r - e)\mathrm{d}r \\ + \int_e^R S\left(\sum_{j=1}^{J} \beta_j \zeta_j(r)\right) \sum_{j=1}^{J} \beta_j \zeta_j(r)\zeta_k(r)(r - e)\mathrm{d}r = \int_e^R L\left(\sum_{j=1}^{J} \beta_j \zeta_j(r)\right)(r - e)\zeta_k(r)\mathrm{d}r, \end{gathered} \tag{4.7.11}$$

or as,

$$\Omega^2 R^3 \sum_{j=1}^{J} \frac{\partial^2 \beta_j}{\partial \psi^2} I_\beta^{\mathrm{jk}} + \Omega^2 R^3 \sum_{j=1}^{J} \beta_j I_{\beta h}^{\mathrm{jk}} + \sum_{j=1}^{J} k_\beta^{\mathrm{jk}} \beta_j = M_\beta^k. \tag{4.7.12}$$

Dividing both the torsion and the flapping equations by $I_b \Omega^2$ and assuming that the assumed modes are normal modes, results in a set of pairs of equations for flapping and torsion modal amplitudes which are

$$\sum_{j=1}^{J} \bar{I}_{\theta}^{\mathrm{kj}} \frac{\partial^2 \theta_j(t)}{\partial \psi^2} + \bar{I}_{\theta}^{k} \frac{\partial^2 \theta_r(t)}{\partial \psi^2} + \sum_{j=1}^{J} \bar{I}_{\theta}^{\mathrm{kj}} \omega_{\theta k}^2 \theta_j = \bar{M}_{\mathrm{et}}^{k}, \quad k = 1, 2, 3 \ldots J, \tag{4.7.13}$$

and

$$\sum_{j=1}^{J} \bar{I}_{\beta}^{\mathrm{jk}} \frac{\partial^2 \beta_j}{\partial \psi^2} + \sum_{j=1}^{J} \bar{I}_{\beta h}^{\mathrm{jk}} \beta_j + \sum_{j=1}^{J} \bar{I}_{\beta}^{\mathrm{jk}} \omega_{\beta k}^2 \beta_j = \bar{M}_{\beta}^{k}, \quad k = 1, 2, 3 \ldots J \tag{4.7.14}$$

where

$$\bar{I}_{\theta}^{k} = \frac{1}{I_b} \int_e^R I_\theta \xi_k(r) \mathrm{d}r, \ \bar{I}_{\theta}^{\mathrm{kj}} = \frac{1}{I_b} \int_e^R I_\theta \xi_j(r) \xi_k(r) \mathrm{d}r, \tag{4.7.15}$$

$$\bar{I}_{\beta}^{\mathrm{jk}} = \frac{1}{I_b} \int_e^R \zeta_k(r) \zeta_j(r) m(r - e)^2 \mathrm{d}r, \tag{4.7.16}$$

$$\bar{I}_{\beta h}^{\mathrm{jk}} = \frac{1}{I_b} \int_e^R \zeta_k(r) \zeta_j(r) m r(r - e) \mathrm{d}r, \ \bar{k}_{\mathrm{et}}^{\mathrm{kj}} = \frac{1}{I_b} \int_e^R G J_e \frac{\partial^2 \xi_j(r)}{\partial r^2} \xi_k(r) \mathrm{d}r, \tag{4.7.17}$$

$$\bar{k}_{\beta}^{\mathrm{jk}} = \frac{1}{I_b} \int_e^R S\left(\sum_{j=1}^{J} \beta_j \zeta_j(r) \right) \sum_{j=1}^{J} \beta_j \zeta_j(r) \zeta_k(r)(r - e) \mathrm{d}r, \tag{4.7.18}$$

$$\bar{M}_{\mathrm{et}}^{k} = \frac{1}{I_b} \int_e^R M_{\mathrm{et}} \xi_k(r) \mathrm{d}r, \tag{4.7.19}$$

and

$$\bar{M}_{\beta}^{k} = \frac{1}{I_b} \int_e^R L\left(\sum_{j=1}^{J} \beta_j \zeta_j(r) \right)(r - e) \zeta_k(r) \mathrm{d}r. \tag{4.7.20}$$

4.7.1 Multiblade Coordinates

The aerodynamically coupled flap-pitch equations of motion of a single blade have been derived in a rotating frame as function of the azimuth angle ψ, When $\psi = \psi_k$, the equations represent the motion of the kth blade. To derive the equations of motion of all the blades as a single unit, the coefficients in the equations may be expressed in terms of the so-called multiblade coordinates. This is done by expanding all trigonometric functions such as products of sine and cosine functions as the sums of relevant sine and cosine terms. Groups of terms containing factors 1, $\cos(\psi)$, $\sin(\psi)$, … may be identified and individually set equal to zero to give N equations of motion for an N-bladed rotor for the fixed frame harmonics of each blade. Thus, the fixed frame equations of motion obtained by applying multiblade coordinate transformations will represent the dynamics of the rotor disc containing N blades. For example, if the flap modal coordinates in the kth mode for an N-bladed rotor are $\beta_{k1}, \beta_{k2}, \beta_{k3}, \ldots, \beta_{kN}$, they are transformed by discrete Fourier series transformations to fixed frame coefficients, $\beta_{k0}, \beta_{k1c}, \beta_{k1s}, \ldots, \beta_{\text{kd}}$, by,

$$\beta_{k0} = \frac{1}{N}\sum_{m=1}^{N}\beta_{\text{km}},\ \beta_{\text{knc}} = \frac{2}{N}\sum_{m=1}^{N}\beta_{\text{km}}\cos(n\psi_m),\ \beta_{\text{kns}} = \frac{2}{N}\sum_{m=1}^{N}\beta_{\text{km}}\sin(n\psi_m),$$
$$\beta_{\text{kd}} = \frac{1}{N}\sum_{m=1}^{N}\beta_{\text{km}}(-1)^m, \tag{4.7.21}$$

where the last transformation exists only for even numbers of blades. For a four-bladed rotor, one has $\beta_{k0}, \beta_{k1c}, \beta_{k1s}, \beta_{\text{kd}}$, while for a three-bladed rotor, one has $\beta_{k0}, \beta_{k1c}, \beta_{k1s}$. As most wind turbine rotors are three-bladed, we shall assume that the number of blades is 3. Moreover, the above four transformations will be referred to as the collective, cosine-cyclic, sine-cyclic, and differential collective operators.

4.7.2 Equations of Motion in Multiblade Coordinates

To express the flapping and torsion blade equations of motion in terms of multiblade coordinates the collective and cyclic summation operators are applied to them. Thus, the flapping and torsion blade equations in multiblade coordinates are

$$\sum_{j=1}^{J}\bar{I}_\theta^{\text{kj}}\begin{bmatrix}1&0&0\\0&1&0\\0&0&1\end{bmatrix}\begin{bmatrix}\theta''_{j0}\\\theta''_{\text{jc}}\\\theta''_{\text{js}}\end{bmatrix} + \sum_{j=1}^{J}\bar{I}_\theta^{\text{kj}}\begin{bmatrix}0&0&0\\0&0&2\\0&-2&0\end{bmatrix}\begin{bmatrix}\theta'_{j0}\\\theta'_{\text{jc}}\\\theta'_{\text{js}}\end{bmatrix} + \bar{I}_\theta^k\begin{bmatrix}1&0&0\\0&1&0\\0&0&1\end{bmatrix}\begin{bmatrix}\theta''_{r0}\\\theta''_{\text{rc}}\\\theta''_{\text{rs}}\end{bmatrix}$$
$$+\bar{I}_\theta^k\begin{bmatrix}0&0&0\\0&0&2\\0&-2&0\end{bmatrix}\begin{bmatrix}\theta'_{r0}\\\theta'_{\text{rc}}\\\theta'_{\text{rs}}\end{bmatrix} + \bar{I}_\theta^k\begin{bmatrix}0&0&0\\0&-1&0\\0&0&-1\end{bmatrix}\begin{bmatrix}\theta_{r0}\\\theta_{\text{rc}}\\\theta_{\text{rs}}\end{bmatrix}$$
$$+\sum_{j=1}^{J}\bar{I}_\theta^{\text{kj}}\begin{bmatrix}\omega_{\theta k}^2&0&0\\0&\omega_{\theta k}^2-1&0\\0&0&\omega_{\theta k}^2-1\end{bmatrix}\begin{bmatrix}\theta_{j0}\\\theta_{\text{jc}}\\\theta_{\text{js}}\end{bmatrix} = \begin{bmatrix}\bar{M}_{et0}^k\\\bar{M}_{\text{etc}}^k\\\bar{M}_{\text{ets}}^k\end{bmatrix},\quad k = 1,2,3\ldots J, \tag{4.7.22}$$

and

$$\sum_{j=1}^{J}\bar{I}_{\beta}^{\mathrm{kj}}\begin{bmatrix}1&0&0\\0&1&0\\0&0&1\end{bmatrix}\begin{bmatrix}\beta''_{j0}\\\beta''_{\mathrm{jc}}\\\beta''_{\mathrm{js}}\end{bmatrix}+\sum_{j=1}^{J}\bar{I}_{\beta}^{\mathrm{kj}}\begin{bmatrix}0&0&0\\0&0&2\\0&-2&0\end{bmatrix}\begin{bmatrix}\beta'_{j0}\\\beta'_{\mathrm{jc}}\\\beta'_{\mathrm{js}}\end{bmatrix}+\sum_{j=1}^{J}\bar{I}_{\beta h}^{\mathrm{kj}}\begin{bmatrix}1&0&0\\0&1&0\\0&0&1\end{bmatrix}\begin{bmatrix}\beta_{j0}\\\beta_{\mathrm{jc}}\\\beta_{\mathrm{js}}\end{bmatrix}$$
$$+\sum_{j=1}^{J}\bar{I}_{\beta}^{\mathrm{kj}}\begin{bmatrix}\omega_{\beta k}^2&0&0\\0&\omega_{\beta k}^2-1&0\\0&0&\omega_{\beta k}^2-1\end{bmatrix}\begin{bmatrix}\beta_{j0}\\\beta_{\mathrm{jc}}\\\beta_{\mathrm{js}}\end{bmatrix}=\begin{bmatrix}\bar{M}_{\beta 0}^{k}\\\bar{M}_{\beta c}^{k}\\\bar{M}_{\beta s}^{k}\end{bmatrix},\quad k=1,2,3\ldots J. \tag{4.7.23}$$

4.7.3 Centrifugal Stiffening of the Torsion Modes

In the same way as the centrifugal force is responsible for stiffening the blade in flapping, it is also responsible for small but significant stiffening effect in torsion. This moment is sometimes referred to as the propeller moment and is proportional to the pitch angle for small rotations about the torsion axis. Assuming the blade to be flat, the sectional propeller pitching moment can be shown to be,

$$M_{\mathrm{pm}}=-\int_0^c \mathrm{dm}\Omega^2 x(x\sin\theta)\approx-\Omega^2 I_\theta\theta \tag{4.7.24}$$

where dm is a mass element within the section at a distance x from the pitching axis. It acts as a restoring moment and provides for a stiffening effect very similar to the centrifugal stiffening in flapping. Since the pitch angle is given by,

$$\theta(t)=\theta_r(t)+\theta_{\mathrm{et}}(t)=\theta_r(t)+\sum_{j=1}^{J}\theta_j\xi_j(r), \tag{4.7.25}$$

The non-dimensional propeller moment in the kth mode is

$$\bar{M}_p^k\approx-\bar{I}_\theta^k\theta_r-\sum_{j=1}^{J}\bar{I}_\theta^{\mathrm{kj}}\theta_j. \tag{4.7.26}$$

Hence, in multiblade coordinates,

$$\begin{bmatrix}\bar{M}_{p0}^k\\\bar{M}_{\mathrm{pc}}^k\\\bar{M}_{\mathrm{ps}}^k\end{bmatrix}=-\sum_{j=1}^{J}\bar{I}_\theta^{\mathrm{kj}}\begin{bmatrix}1&0&0\\0&1&0\\0&0&1\end{bmatrix}\begin{bmatrix}\theta_{j0}\\\theta_{\mathrm{jc}}\\\theta_{\mathrm{js}}\end{bmatrix}-\bar{I}_\theta^k\begin{bmatrix}1&0&0\\0&1&0\\0&0&1\end{bmatrix}\begin{bmatrix}\theta_{r0}\\\theta_{\mathrm{rc}}\\\theta_{\mathrm{rs}}\end{bmatrix}. \tag{4.7.27}$$

Thus, the torsion modal equations in multiblade coordinates are

$$\sum_{j=1}^{J}\bar{I}_{\theta}^{\mathrm{kj}}\begin{bmatrix}1&0&0\\0&1&0\\0&0&1\end{bmatrix}\begin{bmatrix}\theta''_{j0}\\\theta''_{\mathrm{jc}}\\\theta''_{\mathrm{js}}\end{bmatrix}+\sum_{j=1}^{J}\bar{I}_{\theta}^{\mathrm{kj}}\begin{bmatrix}0&0&0\\0&0&2\\0&-2&0\end{bmatrix}\begin{bmatrix}\theta'_{j0}\\\theta'_{\mathrm{jc}}\\\theta'_{\mathrm{js}}\end{bmatrix}+\bar{I}_{\theta}^{k}\begin{bmatrix}1&0&0\\0&1&0\\0&0&1\end{bmatrix}\begin{bmatrix}\theta''_{r0}\\\theta''_{\mathrm{rc}}\\\theta''_{\mathrm{rs}}\end{bmatrix}$$
$$+\bar{I}_{\theta}^{k}\begin{bmatrix}0&0&0\\0&0&2\\0&-2&0\end{bmatrix}\begin{bmatrix}\theta'_{r0}\\\theta'_{\mathrm{rc}}\\\theta'_{\mathrm{rs}}\end{bmatrix}+\bar{I}_{\theta}^{k}\begin{bmatrix}1&0&0\\0&0&0\\0&0&0\end{bmatrix}\begin{bmatrix}\theta_{r0}\\\theta_{\mathrm{rc}}\\\theta_{\mathrm{rs}}\end{bmatrix}$$
$$+\sum_{j=1}^{J}\bar{I}_{\theta}^{\mathrm{kj}}\begin{bmatrix}\omega_{\theta k}^{2}+1&0&0\\0&\omega_{\theta k}^{2}&0\\0&0&\omega_{\theta k}^{2}\end{bmatrix}\begin{bmatrix}\theta_{j0}\\\theta_{\mathrm{jc}}\\\theta_{\mathrm{js}}\end{bmatrix}=\begin{bmatrix}\bar{M}_{a0}^{k}\\\bar{M}_{\mathrm{ac}}^{k}\\\bar{M}_{\mathrm{as}}^{k}\end{bmatrix},\tag{4.7.28}$$

where the right-hand side represents the aerodynamic pitching moment in multiblade coordinates.

4.7.4 The Aerodynamic Moments Based on Blade Element Theory

The hub plane is assume to be normal to the rotor shaft, and velocity is assumed to always approach the hub plane in a direction normal to it. The wind velocity is normalized by the rotor blade tip speed ΩR, and the free stream inflow ratio is defined as,

$$\lambda_f = V/\Omega R.\tag{4.7.29}$$

The total inflow λ is the sum of the free stream inflow λ_f and the induced inflow λ_i. Although the inflow in general is a complex function of both the azimuth and radial position in the disc plane, it is approximated by the function,

$$\lambda = \lambda_0 + \lambda_c(r/R)\cos(\psi) + \lambda_s(r/R)\sin(\psi).\tag{4.7.30}$$

A simple finite-state model of the dynamics of the coefficients λ_0, λ_c and λ_s was developed by Pitt and Peters (1981), and this model will be adopted here. The aerodynamic forces and moments are developed in terms of the non-dimensional blade velocity components normal to the disc u_P and in-plane components in the tangential and radial direction, u_T and u_R, respectively. These are

$$u_T = r/R \equiv \bar{r},\ u_R = 0 \text{ and } u_P = \lambda + \left(\bar{r} - \bar{e}\right)\beta',\tag{4.7.31}$$

where $\bar{e} = e/R$.

The sectional aerodynamic lift and drag forces are defined as,

$$L = \frac{1}{2}\rho\Omega^2 R^2 \left(u_T^2 + u_P^2\right) c C_{l\alpha}(\phi - \theta) \tag{4.7.32}$$

and

$$D = \frac{1}{2}\rho\Omega^2 R^2 \left(u_T^2 + u_P^2\right) c C_d, \tag{4.7.33}$$

where ϕ is the inflow angle defined as

$$\phi = \tan^{-1}(u_P/u_T) \approx u_P/u_T. \tag{4.7.34}$$

The force components normal and tangential to the rotor disc acting on a section of the rotor blade are

$$F_n = L\cos\phi - D\sin\phi \approx L. \tag{4.7.35}$$

$$F_t = L\sin\phi + D\cos\phi \approx D + L\phi. \tag{4.7.36}$$

The sectional angle of attack is

$$\alpha = (\phi - \theta) = u_P/u_T - \theta = \{(\lambda - (\bar{r} - \bar{e})\beta')/\bar{r}\} - \theta_r(t) + \sum_{j=1}^{J} \theta_j \xi_j(\bar{r}) \tag{4.7.37}$$

and

$$\bar{r}\alpha = \lambda - (\bar{r} - \bar{e})\beta' - \bar{r}\theta_r(t) + \bar{r}\sum_{j=1}^{J} \theta_j \xi_j(\bar{r}). \tag{4.7.38}$$

The flapping moment in the kth mode is

$$M_\beta^k = \int_e^R L(\beta)(r - e)\zeta_k(r)\mathrm{d}r. \tag{4.7.39}$$

Thus, the non-dimensional flapping moment is

$$\bar{M}_\beta^k = \frac{M_{e\beta}^k}{I_b\Omega^2} = \frac{\gamma}{2}\int_{\bar{e}}^1 (\bar{r} - \bar{e})\bar{r}^2 \alpha \zeta_k(\bar{r})\mathrm{d}\bar{r}. \tag{4.7.40}$$

Hence,

$$\begin{aligned}\bar{M}_\beta^k = & -\frac{\gamma\theta_r}{2}\int_{\bar{e}}^1 (\bar{r} - \bar{e})\bar{r}^2 \zeta_k(\bar{r})\mathrm{d}\bar{r} - \sum_{j=1}^{J}\frac{\gamma\theta_j}{2}\int_{\bar{e}}^1 (\bar{r} - \bar{e})\bar{r}^2 \xi_j(r)\zeta_k(\bar{r})\mathrm{d}\bar{r} \\ & + \frac{\gamma}{2}\int_{\bar{e}}^1 (\bar{r} - \bar{e})\bar{r}\lambda\zeta_k(\bar{r})\mathrm{d}\bar{r} - \frac{\gamma\beta_j'}{2}\int_{\bar{e}}^1 (\bar{r} - \bar{e})^2 \bar{r}\zeta_j(\bar{r})\zeta_k(\bar{r})\mathrm{d}\bar{r}\end{aligned} \tag{4.7.41}$$

Let

$$I_{\mathrm{nm}}^{k} = \int\limits_{\bar{e}}^{1} (\bar{r} - \bar{e})^{n}\bar{r}^{m}\zeta_{k}(\bar{r})\mathrm{d}\bar{r},\ J_{\mathrm{nm}}^{\mathrm{jk}} = \int\limits_{\bar{e}}^{1} (\bar{r} - \bar{e})^{n}\bar{r}^{m}\xi_{j}(r)\zeta_{k}(\bar{r})\mathrm{d}\bar{r}.$$

and

$$I_{\mathrm{nm}}^{\mathrm{jk}} = \int\limits_{\bar{e}}^{1} (\bar{r} - \bar{e})^{n}r^{m}\zeta_{j}(r)\zeta_{k}(\bar{r})\mathrm{d}\bar{r}. \tag{4.7.42}$$

Then, it follows that

$$\bar{M}_{\beta}^{k} = -\frac{\gamma\theta_{r}}{2}I_{12}^{k} - \sum_{j=1}^{J}\frac{\gamma\theta_{j}}{2}J_{12}^{\mathrm{jk}} - \frac{\gamma\beta_{j}'}{2}I_{21}^{\mathrm{jk}} + \frac{\gamma}{2}\int_{\bar{e}}^{1} (\bar{r} - \bar{e})\bar{r}\lambda\zeta_{k}(\bar{r})\mathrm{d}\bar{r}. \tag{4.7.43}$$

Although the steady aerodynamic pitching moment is zero, an aerodynamic damping moment is included in the pitch dynamics and arises from the quasi-steady aerofoil aerodynamics (Bisplinghoff et al. 1955). The non-dimensional aerodynamic pitching moment then takes the form,

$$\bar{M}_{a}^{k} = \frac{M_{a}^{k}}{I_{b}\Omega^{2}} = -\sum_{j=1}^{J}\frac{\gamma\theta_{j}'}{16}\int_{\bar{e}}^{1} \bar{c}^{2}\bar{r}\xi_{j}(\bar{r})\xi_{k}(\bar{r})\mathrm{d}\bar{r} = -\sum_{j=1}^{J}\frac{\gamma\theta_{j}' I_{\bar{c}\bar{c}}}{16} \tag{4.7.44}$$

where

$$I_{\bar{c}\bar{c}} = \int\limits_{\bar{e}}^{1} \bar{c}^{2}\bar{r}\xi_{j}(\bar{r})\xi_{k}(\bar{r})\mathrm{d}\bar{r}. \tag{4.7.45}$$

The aerodynamic flapping and pitching moments in the multiblade coordinates are

$$\begin{bmatrix} \bar{M}_{\beta 0}^{k} \\ \bar{M}_{\beta c}^{k} \\ \bar{M}_{\beta s}^{k} \end{bmatrix} = -\frac{\gamma}{2}\sum_{j=1}^{J} J_{12}^{\mathrm{jk}} \begin{bmatrix} \theta_{j0} \\ \theta_{\mathrm{jc}} \\ \theta_{\mathrm{js}} \end{bmatrix} - \frac{\gamma}{2}I_{12}^{k} \begin{bmatrix} \theta_{r0} \\ \theta_{\mathrm{rc}} \\ \theta_{\mathrm{rs}} \end{bmatrix} + \frac{\gamma}{2} \begin{bmatrix} I_{11}^{k} & 0 & 0 \\ 0 & I_{12}^{k} & 0 \\ 0 & 0 & I_{12}^{k} \end{bmatrix} \begin{bmatrix} \lambda_{0} \\ \lambda_{c} \\ \lambda_{s} \end{bmatrix}$$
$$-\frac{\gamma}{2}\sum_{j=1}^{J} I_{21}^{\mathrm{jk}} \begin{bmatrix} \beta_{j0}' \\ \beta_{\mathrm{jc}}' \\ \beta_{\mathrm{js}}' \end{bmatrix} - \frac{\gamma}{2}\sum_{j=1}^{J} I_{21}^{\mathrm{jk}} \begin{bmatrix} 0 & 0 & 0 \\ 0 & 0 & 1 \\ 0 & -1 & 0 \end{bmatrix} \begin{bmatrix} \beta_{j0} \\ \beta_{\mathrm{jc}} \\ \beta_{\mathrm{js}} \end{bmatrix} \tag{4.7.46}$$

and

$$\begin{bmatrix} \bar{M}_{a0}^k \\ \bar{M}_{ac}^k \\ \bar{M}_{as}^k \end{bmatrix} = -\frac{\gamma}{16}\sum_{j=1}^{J} \bar{I}_{\bar{c}\bar{c}} \left\{ \begin{bmatrix} \theta'_{j0} \\ \theta'_{jc} \\ \theta'_{js} \end{bmatrix} + \begin{bmatrix} 0 & 0 & 0 \\ 0 & 0 & 1 \\ 0 & -1 & 0 \end{bmatrix} \begin{bmatrix} \theta_{j0} \\ \theta_{jc} \\ \theta_{js} \end{bmatrix} \right\}. \tag{4.7.47}$$

4.7.5 The Inflow Dynamics

The model of the inflow dynamics is based on the finite-state approximation developed by Pitts and Peters (1981). The induced inflow is modelled as,

$$\lambda_i = \lambda_{i0} + \lambda_c \bar{r}\cos(\psi) + \lambda_s \bar{r}\sin(\psi),\ \lambda_0 = \lambda_{i0} + \lambda_f \tag{4.7.48}$$

where λ_{i0}, λ_c, λ_s are the magnitudes of the uniform, fore to aft cyclic and side to side cyclic variations of the induced inflow. The unsteady variations of these magnitudes are related to unsteady variations in the thrust, pitching moment, and rolling moment coefficients of the rotor disc as a whole. These dynamic relations can be approximated by first-order finite state models, and for a horizontal axis wind turbine rotor disc, these relations reduce to,

$$\frac{16}{225\pi}\begin{bmatrix} 24 & 0 & 0 \\ 0 & 5 & 0 \\ 0 & 0 & 5 \end{bmatrix} \frac{d}{dt}\begin{bmatrix} \lambda_{i0} \\ \lambda_c \\ \lambda_s \end{bmatrix} + \frac{1}{2(2\lambda_i + \lambda_f)}\begin{bmatrix} 1 & 0 & 0 \\ 0 & 4 & 0 \\ 0 & 0 & 4 \end{bmatrix}\begin{bmatrix} \lambda_{i0} \\ \lambda_c \\ \lambda_s \end{bmatrix} = \begin{bmatrix} C_T \\ -C_M \\ C_L \end{bmatrix}. \tag{4.7.49}$$

Based on the BEM theory, it can also be shown that the induced inflow is related to the free stream inflow and the thrust coefficient by the relation,

$$\lambda_i = \sqrt{\frac{\lambda_f^2}{4} + \frac{C_T}{2}} - \frac{\lambda_f}{2}. \tag{4.7.50}$$

This relation allows one to define the initial conditions for the unsteady inflow components.

4.7.6 The Inflow Driving Hub Moments

The aerodynamic hub moments of interest are the thrust, the pitching moment, and the rolling moment. The rotor disc axes are usually oriented with the z-axis pointing into the rotor disc and normal to it, the x-axis point down and in the rotor disc plane and the y-axis pointing horizontal and in the rotor disc plane. The three axes together constitute a right-handed reference frame. The thrust force is defined as positive in the direction pointing to the z-axis, and the rolling moment is positive about the x-axis. The pitching moment is normally assumed positive "nose up" and is therefore negative about the y-axis.

To determine the thrust force, it is essential to determine the thrust due to each blade and sum over all the blades. The shear force due to the kth blade, normal to the disc plane, and in the direction of the z-axis, is

$$T_{\mathrm{Sh}\_k} = \int_{e}^{R} (F_z - m\ddot{z})_k \mathrm{d}r \approx \int_{e}^{R} \left(L(\beta) - m(r-e)\ddot{\beta}\right)_k \mathrm{d}r. \tag{4.7.51}$$

Non-dimensionalizing by $\pi\rho R^4\Omega^2$ the above relation,

$$\bar{T}_{\mathrm{Sh}} = \frac{T_{\mathrm{Sh}}}{\pi\rho R^4\Omega^2} = \frac{C_{l\alpha}c}{2\pi R}\int_{\bar{e}}^{1} \bar{r}^2\alpha \mathrm{d}\bar{r} - \frac{\beta''}{\pi\rho R^4}\int_{e}^{R} (r-e)m\mathrm{d}r. \tag{4.7.52}$$

where the subscript k has been dropped. In terms of the solidity ratio, $\sigma = Nc/\pi R$,

$$\bar{T}_{\mathrm{Sh}} = \frac{C_{l\alpha}\sigma}{2N}\int_{\bar{e}}^{1} \bar{r}^2\alpha \mathrm{d}\bar{r} - \frac{\beta''}{\pi\rho R^2}\int_{\bar{e}}^{1} (\bar{r}-\bar{e})m\mathrm{d}\bar{r} \tag{4.7.53}$$

Since

$$\bar{r}\alpha = \lambda - (\bar{r}-\bar{e})\sum_{j=1}^{J}\zeta_j(\bar{r})\beta_j' - \bar{r}\theta_r(t) - \sum_{j=1}^{J}\bar{r}\theta_j\xi_j(r), \tag{4.7.54}$$

$$\begin{aligned}\bar{T}_{\mathrm{Sh}} = &-\frac{C_{l\alpha}\sigma\theta_r}{2N}\int_{\bar{e}}^{1} \bar{r}^2\mathrm{d}\bar{r} - \frac{C_{l\alpha}\sigma\theta_j}{2N}\int_{\bar{e}}^{1} \bar{r}^2\xi_j(\bar{r})\mathrm{d}\bar{r} + \frac{C_{l\alpha}\sigma}{2N}\int_{\bar{e}}^{1} \bar{r}\lambda\mathrm{d}\bar{r} \\ &- \frac{C_{l\alpha}\sigma}{2N}\int_{\bar{e}}^{1} (\bar{r}-\bar{e})\bar{r}\zeta_j(\bar{r})\beta_j'\mathrm{d}\bar{r} - \frac{1}{\pi\rho R^2}\int_{\bar{e}}^{1} (\bar{r}-\bar{e})m\zeta_j(\bar{r})\beta_j''\mathrm{d}\bar{r}\end{aligned} \tag{4.7.55}$$

$$\bar{T}_{\mathrm{Sh}} = -\frac{C_{l\alpha}\sigma\theta_r}{2N}I_{02} - \frac{C_{l\alpha}\sigma\theta_j}{2N}J_{02}^{j} + \frac{C_{l\alpha}\sigma}{2N}\int_{\bar{e}}^{1} \bar{r}\lambda d\bar{r} - \frac{C_{l\alpha}\sigma\beta'}{2N}I_{11} - \beta'' S_\beta \tag{4.7.56}$$

where

$$I_{\mathrm{nm}} = \int_{\bar{e}}^{1} (\bar{r}-\bar{e})^n\bar{r}^m\mathrm{d}\bar{r},\ J_{\mathrm{nm}}^{j} = \int_{\bar{e}}^{1} (\bar{r}-\bar{e})^n\bar{r}^m\xi_j(\bar{r})\mathrm{d}\bar{r},\ I_{nm}^{j} = \int_{\bar{e}}^{1} (\bar{r}-\bar{e})^n\bar{r}^m\zeta_j(\bar{r})\mathrm{d}\bar{r},$$

$$S_\beta^j = \frac{1}{\pi\rho R^2}\int_{\bar{e}}^{1} \zeta_j(\bar{r})(\bar{r}-\bar{e})m\mathrm{d}\bar{r}. \tag{4.7.57}$$

The thrust coefficient in multiblade coordinates for an N-bladed rotor is

$$\begin{bmatrix} C_T \\ C_{Tc} \\ C_{Ts} \end{bmatrix} = -\frac{C_{l\alpha}\sigma}{2}\sum_{j=1}^{J} J_{02}^{j}\begin{bmatrix} \theta_{j0} \\ \theta_{jc} \\ \theta_{js} \end{bmatrix} - \frac{C_{l\alpha}\sigma}{2} I_{02}\begin{bmatrix} \theta_{r0} \\ \theta_{rc} \\ \theta_{rs} \end{bmatrix} + \frac{C_{l\alpha}\sigma}{2}\begin{bmatrix} I_{01} & 0 & 0 \\ 0 & I_{02} & 0 \\ 0 & 0 & I_{02} \end{bmatrix}\begin{bmatrix} \lambda_0 \\ \lambda_c \\ \lambda_s \end{bmatrix}$$
$$-\frac{C_{l\alpha}\sigma}{2}\sum_{j=1}^{J} I_{11}^{j}\left\{\begin{bmatrix} \beta_0' \\ \beta_c' \\ \beta_s' \end{bmatrix} + \begin{bmatrix} 0 & 0 & 0 \\ 0 & 0 & 1 \\ 0 & -1 & 0 \end{bmatrix}\begin{bmatrix} \beta_0 \\ \beta_c \\ \beta_s \end{bmatrix}\right\}$$
$$-\sum_{j=1}^{J} S_{\beta}^{j}\left\{\begin{bmatrix} 1 & 0 & 0 \\ 0 & 1 & 0 \\ 0 & 0 & 1 \end{bmatrix}\begin{bmatrix} \beta_{j0}'' \\ \beta_{jc}'' \\ \beta_{js}'' \end{bmatrix} + \begin{bmatrix} 0 & 0 & 0 \\ 0 & 0 & 2 \\ 0 & -2 & 0 \end{bmatrix}\begin{bmatrix} \beta_{j0}' \\ \beta_{jc}' \\ \beta_{js}' \end{bmatrix} + \begin{bmatrix} 0 & 0 & 0 \\ 0 & -1 & 0 \\ 0 & 0 & -1 \end{bmatrix}\begin{bmatrix} \beta_{j0} \\ \beta_{jc} \\ \beta_{js} \end{bmatrix}\right\}. \quad (4.7.58)$$

The aerodynamic moment acting on the hub at the rotor axis can resolved into two components: one due to the flapping moment at the root and the other due to the moment of the shear force at the root. Thus, the total hub moment due the kth blade may be expressed as,

$$M_{\mathrm{hub},k} = \int_e^R (F_z - m\ddot{z})_k r\,\mathrm{d}r \approx \int_e^R \left(L(\beta) - m(r-e)\ddot{\beta}\right)_k (r-e)\,\mathrm{d}r$$
$$+e\int_e^R \left(L(\beta) - m(r-e)\ddot{\beta}\right)_k \mathrm{d}r. \quad (4.7.59)$$

Thus, the first integral may be evaluated as,

$$\int_e^R \left(L(\beta) - m(r-e)\ddot{\beta}\right)_k (r-e)\,\mathrm{d}r = I_b\Omega^2\left\{\bar{M}_\beta - \sum_{j=1}^{J}\bar{I}_\beta^j\frac{\partial^2\beta_j}{\partial\psi^2}\right\}. \quad (4.7.60)$$

By employing the flapping equilibrium equation, the right-hand side may be expressed in terms of blade natural frequencies and inertia integrals as,

$$\int_e^R \left(L(\beta) - m(r-e)\ddot{\beta}\right)_k (r-e)\,\mathrm{d}r = I_b\Omega^2\left\{\sum_{j=1}^{J}\bar{I}_{\beta h}^j\beta_j + \sum_{j=1}^{J}\bar{I}_\beta^j\omega_{\beta j}^2\beta_j\right\} \quad (4.7.61)$$

where

$$\bar{I}_\beta^j = \frac{1}{I_b}\int_e^R \zeta_j(r)m(r-e)^2\mathrm{d}r,\ \bar{I}_{\beta h}^j = \frac{1}{I_b}\int_e^R \zeta_j(r)mr(r-e)\mathrm{d}r. \quad (4.7.62)$$

Thus

$$M_{\mathrm{hub},k} = \int_e^R (F_z - m\ddot{z})_k r\mathrm{d}r \approx I_b\Omega^2 \left\{ \sum_{j=1}^J \bar{I}^j_{\beta h}\beta_j + \sum_{j=1}^J \bar{I}^j_{\beta}\omega^2_{\beta j}\beta_j \right\} + eT_{\mathrm{Sh\_}k} \tag{4.7.63}$$

Non-dimensionalizing by $\pi\rho R^5\Omega^2$ the above relation,

$$C_{\mathrm{hub\_moment}} = \frac{M_{\mathrm{hub},k}}{\pi\rho R^5\Omega^2} \approx \frac{I_b}{\pi\rho R^5} \left\{ \sum_{j=1}^J \bar{I}^j_{\beta h}\beta_j + \sum_{j=1}^J \bar{I}^j_{\beta}\omega^2_{\beta j}\beta_j \right\} + \bar{e}\bar{T}_{\mathrm{Sh\_}k}. \tag{4.7.64}$$

Since,

$$\frac{I_b}{\pi\rho R^5} = \frac{C_{l\alpha}\bar{c}}{\pi\gamma} = \frac{C_{l\alpha}\sigma}{N\gamma}, \tag{4.7.65}$$

$$C_{\mathrm{hub\_moment}} \approx \frac{C_{l\alpha}\sigma}{N\gamma} \left\{ \sum_{j=1}^J \bar{I}^j_{\beta h}\beta_j + \sum_{j=1}^J \bar{I}^j_{\beta}\omega^2_{\beta j}\beta_j \right\} + \bar{e}\bar{T}_{\mathrm{Sh\_}k}. \tag{4.7.66}$$

Summing over N blades,

$$C_{\mathrm{HM}} \approx \frac{C_{l\alpha}\sigma}{\gamma} \left\{ \sum_{j=1}^J \left(\bar{I}^j_{\beta h} + \bar{I}^j_{\beta}\omega^2_{\beta j} \right)\beta_j \right\} + \bar{e}C_T. \tag{4.7.67}$$

Thus, in multiblade coordinates,

$$\begin{bmatrix} C_{\mathrm{HM0}} \\ C_{\mathrm{HMc}} \\ C_{\mathrm{HMs}} \end{bmatrix} = \bar{e}\begin{bmatrix} C_T \\ C_{\mathrm{Tc}} \\ C_{\mathrm{Ts}} \end{bmatrix} + \frac{C_{l\alpha}\sigma}{\gamma}\sum_{j=1}^J \left(\bar{I}^j_{\beta h} + \bar{I}^j_{\beta}\omega^2_{\beta j} \right) \begin{bmatrix} 1 & 0 & 0 \\ 0 & 1 & 0 \\ 0 & 0 & 1 \end{bmatrix} \begin{bmatrix} \beta_{j0} \\ \beta_{\mathrm{jc}} \\ \beta_{\mathrm{js}} \end{bmatrix}. \tag{4.7.68}$$

The components of the hub moment in the y and x directions are

$$C_M = -C_{\mathrm{HMc}} \text{ and } C_L = C_{\mathrm{HMs}}. \tag{4.7.69}$$

Finally, the rotor disc thrust coefficient is given by C_T. Hence,

$$\begin{bmatrix} C_T \\ C_M \\ C_L \end{bmatrix} = \begin{bmatrix} 1 & 0 & 0 \\ 0 & 0 & 0 \\ 0 & 0 & 0 \end{bmatrix} \begin{bmatrix} C_T \\ C_{\mathrm{Tc}} \\ C_{\mathrm{Ts}} \end{bmatrix} + \begin{bmatrix} 0 & 0 & 0 \\ 0 & -1 & 0 \\ 0 & 0 & 1 \end{bmatrix} \begin{bmatrix} C_{\mathrm{HM0}} \\ C_{\mathrm{HMc}} \\ C_{\mathrm{HMs}} \end{bmatrix}. \tag{4.7.70}$$

It follows that

$$\begin{bmatrix} C_T \\ C_M \\ C_L \end{bmatrix} = \begin{bmatrix} 1 & 0 & 0 \\ 0 & -e & 0 \\ 0 & 0 & e \end{bmatrix} \begin{bmatrix} C_T \\ C_{\mathrm{Tc}} \\ C_{\mathrm{Ts}} \end{bmatrix} + \frac{C_{l\alpha}\sigma}{\gamma}\sum_{j=1}^J \left(\bar{I}^j_{\beta h} + \bar{I}^j_{\beta}\omega^2_{\beta j} \right) \begin{bmatrix} 0 & 0 & 0 \\ 0 & -1 & 0 \\ 0 & 0 & 1 \end{bmatrix} \begin{bmatrix} \beta_{j0} \\ \beta_{\mathrm{jc}} \\ \beta_{\mathrm{js}} \end{bmatrix}. \tag{4.7.71}$$

4.7.7 The Rotor Torque Coefficient: The General Expression

The main moment produced by the rotor is the rotor torque, and this produces a dominant component about the rotor shaft axis which is converted into electrical energy. This torque can be obtained by integrating the moments of the in-plane loads about the shaft axis. For the *k*th blade, one obtains the rotor torque as,

$$N_{\text{hub},k} = \int_0^R (F_y - m\ddot{y})_k r\mathrm{d}r \approx \int_e^R (D + L(\beta)\phi)_k r\mathrm{d}r - \dot{\Omega}\int_0^R mr^2\mathrm{d}r. \tag{4.7.72}$$

Hence,

$$N_{\text{hub},k} = \frac{1}{2}\rho\Omega^2 R^5\bar{c}\int_{\bar{e}}^{1} (C_d\bar{r} + C_{l\alpha}\alpha u_P)_k\bar{r}^2\mathrm{d}\bar{r} - \dot{\Omega}R^2\int_0^1 m\bar{r}^2\mathrm{d}\bar{r}. \tag{4.7.73}$$

Non-dimensionalizing by $\pi\rho R^5\Omega^2$ the above relation,

$$\frac{N_{\text{hub},k}}{\pi\rho R^5\Omega^2} = \frac{C_{l\alpha}\sigma}{2N}\int_{\bar{e}}^{1}\left(\frac{C_d}{C_{la}}\bar{r} + \alpha u_P\right)_k \bar{r}^2\mathrm{d}\bar{r} - \frac{\dot{\Omega}}{\Omega^2}\int_0^1 \frac{m}{\pi\rho R^3}\bar{r}^2\mathrm{d}\bar{r}. \tag{4.7.74}$$

The theoretical aerodynamic rotor torque coefficient is

$$C_Q = \frac{N_{\text{aero}}}{\pi\rho R^5\Omega^2} = \frac{C_{l\alpha}\sigma}{2}\int_{\bar{e}}^{1}\left(\frac{C_d}{C_{l\alpha}}\bar{r} + \alpha u_P\right)_k \bar{r}^2\mathrm{d}\bar{r}. \tag{4.7.75}$$

The theoretical aerodynamic rotor torque coefficient simplifies to,

$$C_Q = \frac{C_{l\alpha}\sigma}{2}\frac{C_d}{C_{l\alpha}}\frac{1 - \bar{e}^4}{4} + \frac{C_{l\alpha}\sigma}{2}\int_{\bar{e}}^{1} (\bar{r}\alpha u_P)_k\bar{r}\mathrm{d}\bar{r} \tag{4.7.76}$$

where

$$\bar{r}\alpha = \lambda - (\bar{r} - \bar{e})\sum_{j=1}^{J}\zeta_j(\bar{r})\beta_j' - \bar{r}\theta_r(t) - \sum_{j=1}^{J}\bar{r}\theta_j\xi_j(r), \tag{4.7.77}$$

$$u_P = \lambda - (\bar{r} - \bar{e})\sum_{j=1}^{J}\zeta_j(\bar{r})\beta_j'. \tag{4.7.78}$$

Assuming that $u_P \approx \lambda \approx \lambda_{i0} + \lambda_f$, the theoretical aerodynamic rotor torque coefficient is

$$C_Q = \frac{C_{l\alpha}\sigma}{2}\frac{C_d}{C_{l\alpha}}\frac{1-\bar{e}^4}{4} + (\lambda_{i0} + \lambda_f)C_T. \tag{4.7.79}$$

4.7.8 The Rotor Torque Coefficient: The Case Constant Inflow and Rigid Blades with Steady Root Pitch

In the case of constant inflow, the Pitt-Peters model reduces to,

$$\frac{384}{225\pi}\frac{d\lambda_{i0}}{dt} + \frac{\lambda_{i0}}{2(\lambda_f + 2\lambda_{i0})} = C_T. \tag{4.7.80}$$

Denoting $C_{l\alpha} = a_\infty$ for brevity, the thrust equation is

$$\bar{T}_{\text{sh}} = \frac{a_\infty\sigma(\lambda_{i0} + \lambda_f)}{2N}\frac{1-\bar{e}^2}{2} - \frac{a_\infty\sigma}{2N}\frac{1-\bar{e}^3}{3}\theta_r. \tag{4.7.81}$$

The thrust coefficient for an N-bladed rotor is

$$C_T = \frac{a_\infty\sigma(\lambda_{i0} + \lambda_f)}{2}\frac{1-\bar{e}^2}{2} - \frac{a_\infty\sigma}{2}\frac{1-\bar{e}^3}{3}\theta_r \tag{4.7.82}$$

which is

$$C_T = \frac{a_\infty\sigma}{2}\frac{1-\bar{e}^2}{2}\lambda_{i0} + \frac{a_\infty\sigma}{2}\frac{1-\bar{e}^2}{2}\lambda_f - \frac{a_\infty\sigma}{2}\frac{1-\bar{e}^3}{3}\theta_r. \tag{4.7.83}$$

Hence, the aerodynamic torque coefficient and the inflow dynamics, respectively, reduce to,

$$C_Q = \frac{a_\infty\sigma}{2}\frac{C_d}{a_\infty}\frac{1-\bar{e}^4}{4} + \frac{a_\infty\sigma}{2}\left((\lambda_{i0} + \lambda_f)^2\frac{1-\bar{e}^2}{2} - \theta_r(\lambda_{i0} + \lambda_f)\frac{1-\bar{e}^3}{3}\right), \tag{4.7.84}$$

$$\begin{aligned}&\frac{384}{225\pi}\frac{d\lambda_{i0}}{dt} + \frac{1}{2}\left(\frac{1}{(\lambda_f + 2\lambda_{i0})} - a_\infty\sigma\frac{1-\bar{e}^2}{2}\right)\lambda_{i0}\\ &= \frac{1}{2}a_\infty\sigma\frac{1-\bar{e}^2}{2}\lambda_f - \frac{1}{2}a_\infty\sigma\frac{1-\bar{e}^3}{3}\theta_r.\end{aligned} \tag{4.7.85}$$

The transfer function relating λ_{i0} and λ_f is

$$\lambda_{i0} = \frac{a_\infty \sigma \frac{1-\bar{e}^2}{2}\lambda_f - a_\infty \sigma \frac{1-\bar{e}^3}{3}\theta_r}{\left\{\frac{768}{225\pi}s + \frac{1}{(\lambda_f + 2\lambda_{i0})} - a_\infty \sigma \frac{1-\bar{e}^2}{2}\right\}}. \quad (4.7.86)$$

and

$$\lambda_f + \lambda_{i0} = \frac{\left(\frac{768}{225\pi}s + \frac{1}{\lambda_f}\right)\lambda_f - a_\infty \sigma \frac{1-\bar{e}^3}{3}\theta_r}{\left\{\frac{768}{225\pi}s + \frac{1}{(\lambda_f + 2\lambda_{i0})} - a_\infty \sigma \frac{1-\bar{e}^2}{2}\right\}}. \quad (4.7.87)$$

It may be noted that the above equation is only valid for relatively small inflow. Under steady-state conditions,

$$\lambda_f + \lambda_{i0} = \frac{1 - a_\infty \sigma \frac{1-\bar{e}^3}{3}\theta_r}{\frac{1}{(\lambda_f + 2\lambda_{i0})}\left\{1 - a_\infty \sigma \frac{1-\bar{e}^2}{2}(\lambda_f + 2\lambda_{i0})\right\}} \quad (4.7.88)$$

which may be expressed as,

$$\lambda_f + \lambda_{i0} = (\lambda_f + 2\lambda_{i0})\frac{\left(1 - a_\infty \sigma \frac{1-\bar{e}^3}{3}\theta_r\right)}{\left\{1 - a_\infty \sigma \frac{1-\bar{e}^2}{2}(\lambda_f + 2\lambda_{i0})\right\}}. \quad (4.7.89)$$

Hence,

$$\lambda_{i0} + \lambda_f = \lambda_f\left(\frac{1 - a_\infty \sigma \frac{1-\bar{e}^3}{3}\theta_r}{1 - 2a_\infty \sigma \frac{1-\bar{e}^2}{3}\left(\theta_r - \frac{3}{4}(\lambda_f + 2\lambda_{i0})\right)}\right). \quad (4.7.90)$$

Equation 4.7.90 may solved iteratively for $\lambda_{i0} + \lambda_f$ and expressed as, $\lambda_{i0} + \lambda_f = \lambda_f(1 + K_{\text{tifr}})$.

The theoretical rotor torque coefficient is given by,

$$C_Q = \frac{C_{l\alpha}\sigma}{2}\frac{1-\bar{e}^2}{2}\left(\frac{C_d}{C_{l\alpha}}\frac{1+\bar{e}^2}{2} + \lambda_f^2\left(1 + K_{\text{tifr}}(\lambda_f)\right)^2 + \frac{2}{3}\frac{1-\bar{e}^3}{1-\bar{e}^2}\theta_r\lambda_f\left(1 + K_{\text{tifr}}(\lambda_f)\right)\right). \quad (4.7.91)$$

The theoretical power coefficient may be derived from the torque coefficient and

$$C_P = \frac{C_Q}{\lambda_f} = \frac{C_{l\alpha}\sigma}{2}\frac{1-\bar{e}^2}{2}\left(\frac{C_d}{C_{l\alpha}\lambda_f}\frac{1+\bar{e}^2}{2} + \lambda_f\left(1+K_{\text{tifr}}(\lambda_f)\right)^2 + \frac{2}{3}\frac{1-\bar{e}^3}{1-\bar{e}^2}\theta_r\left(1+K_{\text{tifr}}(\lambda_f)\right)\right). \tag{4.7.92}$$

The theoretical maximum power coefficient may be obtained numerically. The primary contribution to C_P for a horizontal axis wind turbine is from the second term in the Eq. 4.7.92. However, the value obtained by the use of this formula is generally small, and often a more realistic estimate of the actual power coefficient is obtained from measurements made at the wind turbine site and approximated using curve fitting procedures. One such approximation is discussed in the next section.

4.8 The Actual Power Coefficient and its Dependence on the Blade Setting Angle

Typically, the power coefficient C_P is expressed as [Rosas (2003) and Slootweg et al. (2003)]

$$C_p = 0.22\left(\frac{116}{\Gamma} - 0.4\times\theta - 5\right)e^{-12.5/\Gamma}, \tag{4.8.1}$$

$$\Gamma = \frac{1}{\left(\dfrac{1}{\lambda_{\text{tsr}}+0.08\theta} - \dfrac{0.035}{1+\theta^2}\right)},\ \frac{1}{\Gamma} = \frac{1}{\lambda_{\text{tsr}}+0.08\theta} - \frac{0.035}{(1+\theta^2)}. \tag{4.8.2}$$

A generic manner of expressing the C_P is

$$C_p = 0.22\left(\frac{c(1)\times 1450}{\Gamma} - 5(c(2)\times 12.5 + c(3)\times\theta)\right)e^{-1/(c(4)\Gamma)}, \tag{4.8.3}$$

$$\frac{1}{\Gamma} = \frac{1}{\lambda_{\text{tsr}}+c(5)\theta} - \frac{7}{16}\frac{c(6)}{\left(1+\theta^{0.16/c(7)}\right)}. \tag{4.8.4}$$

$$c(i) = 0.08,\ i = 1,\ldots,7. \tag{4.8.5}$$

In the above expressions, Γ may be considered to be the equivalent aerodynamic tip speed ratio.

To find the optimum blade pitch angle that maximizes the power coefficient C_p with respect to the blade pitch angle θ,

$$\frac{\partial}{\partial\theta}C_p = 0. \tag{4.8.6}$$

For a maximum, it can be shown that

$$\frac{\partial}{\partial\theta}\left(\frac{1}{\Gamma}\right) = -\frac{1}{\Gamma^2}\frac{\partial\Gamma}{\partial\theta} = -\left\{\frac{0.08}{(\lambda_{\mathrm{tsr}} + 0.08\theta)^2} - \frac{0.07\theta}{(1 + \theta^2)^2}\right\}. \tag{4.8.7}$$

Hence for maximum,

$$\frac{1}{2900\left(\frac{12.5}{\Gamma} - 1\right) - 312.5(0.4 \times \theta + 5)} = \frac{0.08}{(\lambda_{\mathrm{tsr}} + 0.08\theta)^2} - \frac{0.07\theta}{(1 + \theta^2)^2}. \tag{4.8.8}$$

Rearranging the above for the blade pitch setting angle,

$$\theta = \frac{116}{5}\left(\frac{12.5}{\Gamma} - 1\right) - \frac{0.008}{\frac{0.08}{(\lambda_{\mathrm{tsr}} + 0.08\theta)^2} - \frac{0.07\theta}{(1 + \theta^2)^2}} - 12.5. \tag{4.8.9}$$

The above expression is solved for the optimum blade pitch angle numerically.

The power coefficient C_P may be conveniently expressed in an alternate form. When $\theta = 0$,

$$C_{p0} = 0.22\left(\frac{116}{\Gamma_0} - 5\right)e^{-12.5/\Gamma_0}, \quad \frac{1}{\Gamma_0} = \frac{1}{\lambda_{\mathrm{tsr}}} - 0.035. \tag{4.8.10}$$

Hence,

$$\frac{C_p}{C_{p0}} = \frac{\left(\frac{116}{\Gamma} - 0.4 \times \theta - 5\right)}{\left(\frac{116}{\Gamma_0} - 5\right)} e^{-12.5\left(\frac{1}{\Gamma} - \frac{1}{\Gamma_0}\right)} \tag{4.8.11}$$

where

$$\frac{1}{\Gamma} - \frac{1}{\Gamma_0} = 0.035\left(\frac{\theta^2}{1 + \theta^2}\right) - \frac{0.08\,\theta}{\lambda_{\mathrm{tsr}}(\lambda_{\mathrm{tsr}} + 0.08\,\theta)}. \tag{4.8.12}$$

Consequently, the ratio of C_P to its value with the pitch setting angle equal to 0 is

$$\frac{C_p}{C_{p0}} = \left\{1 + \frac{\left(116\left(\frac{1}{\Gamma} - \frac{1}{\Gamma_0}\right) - 0.4 \times \theta\right)}{\left(\frac{116}{\Gamma_0} - 5\right)}\right\} e^{-12.5\left(\frac{1}{\Gamma} - \frac{1}{\Gamma_0}\right)}. \tag{4.8.13}$$

Equation 4.8.13 may be expressed as,

$$\frac{C_p}{C_{p0}} = \left\{1 - \lambda_{\text{tsr}}\,\theta\frac{(116f(\lambda_{\text{tsr}},\theta) - 0.4)}{(116 - 9.06\lambda_{\text{tsr}})}\right\}e^{12.5\theta f(\lambda_{\text{tsr}},\theta)} \tag{4.8.14}$$

$$f(\lambda_{\text{tsr}},\theta) = -\frac{1}{\theta}\left(\frac{1}{\Gamma} - \frac{1}{\Gamma_0}\right) = \frac{0.08}{\lambda_{\text{tsr}}(\lambda_{\text{tsr}} + 0.08\theta)} - 0.035\left(\frac{\theta}{1+\theta^2}\right). \tag{4.8.15}$$

With generic expression for C_P, to maximize C_P with respect t the tip speed ratio λ_{tsr},

$$\frac{\partial}{\partial\lambda_{\text{tsr}}}C_p = 0, \tag{4.8.16}$$

$$C_{p,\max} = 0.22\left(\frac{c(1)\times 1450}{\Gamma_{\text{cp max}}} - 5(c(2)\times 12.5 + c(3)\times\theta)\right)e^{-1/\left(c(4)\Gamma_{\text{cp max}}\right)}, \tag{4.8.17}$$

$$\frac{1}{\Gamma_{\text{cp max}}} = \frac{1}{\lambda_{\text{cp max}} + c(5)\theta} - \frac{7}{16}\frac{c(6)}{\left(1+\theta^{0.16/c(7)}\right)}. \tag{4.8.18}$$

Hence, for $\frac{\partial}{\partial\lambda_{\text{tsr}}}C_p = 0$,

$$\frac{1}{\Gamma_{\text{cp max}}} = \frac{5(c(2)\times 12.5 + c(3)\times\theta)}{c(1)\times 1450} + c(4). \tag{4.8.19}$$

$$\frac{1}{\lambda_{\text{cp max}} + c(5)\theta} = \frac{7}{16}\frac{c(6)}{\left(1+\theta^{0.16/c(7)}\right)} + \frac{5(c(2)\times 12.5 + c(3)\times\theta)}{c(1)\times 1450} + c(4). \tag{4.8.20}$$

The optimum tip speed ratio is

$$\lambda_{\text{cp max}} = \frac{1}{\frac{7}{16}\frac{c(6)}{\left(1+\theta^{0.16/c(7)}\right)} + \frac{5(c(2)\times 12.5 + c(3)\times\theta)}{c(1)\times 1450} + c(4)} - c(5)\theta. \tag{4.8.21}$$

When $c(i) = 0.08,\ i = 1,\ldots,7$, these expressions reduce to,

$$\frac{1}{\Gamma_{\text{cp max}}} = \frac{0.4\times\theta + 5}{116} + \frac{1}{12.5} = \frac{178.5 + 5\theta}{1450} \tag{4.8.22}$$

and

$$\lambda = \lambda_{\mathrm{cp\,max}} = \frac{1450 \times \left(1 + \theta^2\right)}{\left(1 + \theta^2\right)(178.5 + 5\theta) + 50.75} - 0.08\theta. \quad (4.8.23)$$

4.9 Maximum Power Point Tracking and Protection Against Excessive Wind Torque

All modern wind turbines are endowed with a maximum power point control and tracking systems that ensure the optimal operation of the wind turbine. To achieve maximum conversion, the turbine must necessarily operate at an optimum tip speed ratio which to a very large extent depends on the variation of the power coefficient with respect to the tip speed ratio, a relationship that can only be determined experimentally. In the case of most of the current wind turbine designs, this can be achieved by appropriately controlling the rotational speed of the electric generator, as proposed first by Connor and Leithead (1993). When the primary variables can be measured, several maximum power point tracking (MPPT) algorithms have been developed and these are discussed by Datta and Ranganathan (2003). The concept of MPPT was first introduced in the design of solar panels carrying arrays of photovoltaic cells, for spacecraft in the 1970s with objective of maximizing the power transfer from the power source. Esram et al. (2006) have discussed some of the recent problems of tracking photovoltaic arrays. MPPT algorithms generally belong to three classes of algorithms: (1) based on maximizing the power generated at the grid side; (2) based on maximizing the power transfer from the wind to the generator at the rotor side; and (3) based on maximizing the total power transfer to the grid. What is required in the latter case is to be able to set the speed of the generator at a point where the power transfer from the wind to the generator is maximized; i.e., the source impedance of the generator is matched to the output impedance of the turbine in accordance with the maximum power-transfer theorem in electrical circuit theory. Examples of MPPT algorithms for wind turbines are presented in Qiaoet al. (2008), Taraft et al. (2008), Lee et al. (2009), Örs (2009), Jou et al. (2008).

Classical versions of MPPT algorithms employ a maximum power search algorithm (i.e., hill climbing method), and typical examples of these are presented in Van den Heever et al. (1989), Wang and Chang (2004), Tan and Islam (2004) for dynamically driving the operating point to the maximum power point. The hill climbing method seeks to optimize either the estimated power transferred from the turbine to the generator or the power transferred from the wind to the turbine. Most methods have relied on maximizing the latter, and this does not guarantee that the actual power transferred from the turbine to the generator is also optimized. Another drawback of these methods is the significant estimation errors in the real-time computation of the wind power captured by the turbine or the power generated by the turbine which can often result in high-frequency power fluctuations that are highly undesirable to say the least. Several schemes have been proposed to alleviate the problems due to the errors in the estimation by using real-time estimates of the wind speed in addition to estimating the power coefficient, while other schemes use relatively robust control techniques based on fuzzy logic (see Abo-Khalil et al. (2004) for example). Another approach presented by Munteanu et al. (2009) has been to deliberately introduce a small probing signals or noise to determine the optimum point accurately.

Several strategies for achieving maximum power tracking and control have been proposed for a variety of power systems in Hohm (2003) and in Yu et al. (2003). In addition to the studies mentioned earlier, there have been a number of MPPT controllers proposed recently for wind turbines based on some form of optimal control by Bhowmik et al. (1999), Qiao et al. (2009), Koutroulis and Kalaitzakis (2006), Kawabe et al. (2007). In fact, a recent book by Munteanu et al. (2008), on the topic, has covered the optimal control-based strategies quite extensively. There have also been a few methods based on some form of optimal estimation of the wind speed (Qiao et al. 2008); Abo-Khalil and Lee 2008). A nonlinear controller-based MPPT method has also been proposed for wind turbines by Boukhezzar and Siguerdidjane (2005). Several of the optimal control strategies may be efficiently implemented for a wind turbine provided highly reliable nonlinear estimation algorithms that are used to estimate the states of the wind turbine in operation. One such approach is described and implemented by Vepa (2011).

Most of the wind turbine control systems differ in the manner in which they limit the energy converted to electricity when extreme winds are present and it is not safe to operate the MPPT controller. Thus, wind turbine control systems fall into two broad categories. Pitch-regulated wind turbines have an active control system that can vary the pitch angle of the turbine blade about its own axis to decrease the torque produced by the blades in a fixed-speed turbine and to decrease the rotational speed in variable-speed turbines. This type of control is usually employed in the presence of extreme wind speeds and when high rotational speeds and aerodynamic torques can damage the electricity generators and the power electronic converters. When wind speeds generate power above rated the power, the blades will pitch so that there is less lift and more drag due to increasing flow separation along the blade length. This will slow down the turbine's rotational speed or the torque transferred to the shaft so that the rotational speed or the torque is kept constant below a set threshold. Pitch-regulated turbines see increasing power up until the rated wind speed, beyond which it sees constant power up until a cutout speed when the pitch control is no longer able to limit the rotational speed/aerodynamic torque or where other forces like structural vibrations, turbulence, or gusts pose a threat to a rotating turbine.

The blade pitch angle is set at a desired value $\theta = \theta_d$ to protect the generator against excessive wind loads. The setting ensures that the maximum power transferred from the wind to the turbine is restricted. The blade angle set point is estimated from the maximum power coefficient set point at a certain operating wind speed range. This will ensure that when the wind turbine is tracking the maximum power point, the power transferred will be less than or equal to the maximum limit determined by the blade pitch angle. When the wind turbine is shut down, the blade pitch angle is set at a maximum permissible value, $\theta = \theta_d|_{\max}$, and reduced depending on the operating wind speeds. To define the set point for the power coefficient in terms of a predefined power setting,

$$P = \frac{1}{2}\rho V_1^3 \pi R^2 C_p,\ C_{p,\text{set\_pt}} = \frac{P_{\text{set\_pt}}}{\frac{1}{2}\rho V_1^3 \pi R^2}. \quad (4.9.1)$$

To calculate the desired collective pitch setting angle for the blades, from the maximum rated power of the wind turbine and the estimated maximum expected wind speeds over time frame, the power coefficient set point is first estimated using the formula,

$$C_{p,\text{set\_pt}} = \frac{P_{\text{rated\_max}}}{\frac{1}{2}\rho V_{\max,e}^3 \pi R^2}. \quad (4.9.2)$$

Then if

$$C_{p,\text{set\_pt}} \geq C_{p,\max}\big|_{\theta=0},\ \theta_d = 0. \quad (4.9.3)$$

If $C_{p,\text{set\_pt}} < C_{p,\max}\big|_{\theta=0}$, θ_d must satisfy,

$$C_{p,\text{set\_pt}} = C_{p,\max}\big|_{\theta=\theta_d} = 0.22\left(\frac{116}{\Gamma_{\text{cp}\max,d}} - 0.4 \times \theta_d - 5\right)e^{-12.5/\Gamma_{\text{cp}\max,d}}. \quad (4.9.4)$$

$$\frac{1}{\Gamma_{\text{cp}\max,d}} = \frac{178.5 + 5\theta_d}{1450}, \quad (4.9.5)$$

$$\lambda_{\text{cp}\max,d} = \frac{1450 \times (1 + \theta_d^2)}{(1 + \theta_d^2)(178.5 + 5\theta_d) + 50.75} - 0.08\theta_d. \quad (4.9.6)$$

Usually, this is implemented in MATLAB as a table lookup function, using a table of values of $C_{p,\max}$ for a range of values of θ. For the given value of $C_{p,\text{set\_pt}}$, the table lookup function estimates the corresponding $\theta_d = \theta$ by appropriate interpolation.

Stall-regulated wind turbines, on the other hand, have their blades designed so that when wind speeds are high, the rotational speed or the aerodynamic torque, and thus the power production, decreases with increasing wind speed above a certain stall speed (when the blades are pitched into stall at a speed that is usually not the same as the rated wind speed). The decrease in power with increasing wind speeds is due to unsteady aerodynamic stall effects in the flow over the turbine blades (regions of the blade are stalled, propagating from the hub and outwards with increasing wind speeds). The blades are designed so that they will perform worse (in terms of energy extraction) in high wind speeds to protect the wind turbine without the need for active controls. The benefit of stall-regulation over pitch-regulation is limited the capital cost of the turbine, as well as lower maintenance associated with more moving parts. Like the pitch-regulated wind turbines, stall-regulated wind turbines are also equipped with aerodynamic and mechanical brakes to bring the turbine to a halt and to reduce the lift and drag forces to a minimum in extreme wind speeds.

The difference then, between pitch-regulated and stall-regulated wind turbines, is mostly noticeable in high wind speeds. While the stall-regulated systems rely on the aerodynamic design of the blades to control the aerodynamic torque or the rotational speed of the turbine in high wind speeds, the pitch-regulated systems use an active pitch control for the blades. This allows the pitch-regulated systems to have a constant power output above the rated wind speed, while the stall-regulated systems are not able to keep a constant power output in high winds. On the other hand, stall-regulated wind turbines use very little of the generated power and are therefore more efficient in energy extraction than pitch-regulated wind turbines. The design of stall-regulated wind turbine generally requires the use of blade structural shape morphing or the use of active flow control techniques to induce stall at the most appropriate instances.

4.10 Quasi-steady Aerodynamic Loads on Flexible Rotor Blades

The blade velocity components in the blade fixed undeformed body coordinates at a point x from the blade root are

$$\begin{bmatrix} U_R \\ U_T \\ U_P \end{bmatrix} = \left(\begin{bmatrix} \dot{u} \\ \dot{v} \\ \dot{w} \end{bmatrix} + \begin{bmatrix} -v \\ x \\ v_{\text{in}} \end{bmatrix} \right). \tag{4.10.1}$$

In the above, $[u\,v\,w]$ are the elastic translational displacements, v_{in} in the total inflow velocity. If $\mathbf{T}$ is the transformation from the undeformed to the deformed coordinates, the blade velocity components in the deformed coordinates are

$$\begin{bmatrix} U_R \\ U_T \\ U_P \end{bmatrix} = \mathbf{T} \left(\begin{bmatrix} \dot{u} \\ \dot{v} \\ \dot{w} \end{bmatrix} + \begin{bmatrix} -v \\ x \\ v_{\text{in}} \end{bmatrix} \right). \tag{4.10.2}$$

The velocity components to first order in θ are

$$U_T = \Omega R\bar{x} + \dot{v}, \tag{4.10.3}$$

$$U_P = -\Omega R\bar{x} \left(\theta_G + \phi_e + \int_0^x v' w'' dx \right) + \Omega R \lambda_{\text{in}} + \dot{w} - (\theta_G + \phi_e)\dot{v} + \Omega v w'. \tag{4.10.4}$$

In the above, θ_G is the geometric pitch angle including pretwist and controlled pitch input, and ϕ_e is the elastic torsion. The local inflow angle is defined as $\phi = \tan^{-1}\left(\frac{U_P}{U_T}\right)$. Given, $\psi = \Omega t$

$$L_w = 2\pi\rho_\infty bR\left[(\theta_G + \phi_e)U_T^2 - U_P U_T + \left(\frac{3}{2} + \bar{x}_{ac}\right)U_T bR\Omega\frac{\mathrm{d}}{\mathrm{d}\psi}(\theta_G + \phi_e)\right], \tag{4.10.5}$$

$$L_v = -2\pi\rho_\infty bR\left[(\theta_G + \phi_e)U_P U_T - U_P^2 + \left(\frac{C_{d0}}{2\pi}\right)U_T^2 - \left(\frac{3}{2} + \bar{x}_{ac}\right)U_P bR\Omega\frac{\mathrm{d}}{\mathrm{d}\psi}(\theta_G + \phi_e)\right], \tag{4.10.6}$$

$$M_\phi = 2\pi\rho_\infty (bR)^2\left[\bar{x}_{ac}((\theta_G + \phi_e)U_T^2 - U_P U_T) - (1 + \bar{x}_{ac})\left(\frac{1}{2} + \bar{x}_{ac}\right)U_T bR\Omega\frac{\mathrm{d}}{\mathrm{d}\psi}(\theta_G + \phi_e)\right]. \tag{4.10.7}$$

Linearizing the above expressions,

$$\begin{aligned}
[L_w L_v M_\varphi] = {} & \frac{2\pi}{2}\rho_\infty(2b)\left(\Omega^2R^2\left[-\bar{x}\lambda_{\text{in}}\lambda_{in}^2 - \bar{x}^2\frac{C_{d0}}{2\pi}0\right] - \Omega^2R^2\left[\bar{x}^2\bar{x}\lambda_{\text{in}}0\right](\theta_G + \phi_e)\right) \\
& + \frac{2\pi}{2}\rho_\infty(2b)\Omega R\left(\Omega\bar{x}\left[(b - v)R\bar{x}\int_0^x ()w''\mathrm{d}x0000\right][w'v']\right. \\
& - \left[\bar{x}\lambda_{\text{in}} - 2\bar{x}(\theta_G + \phi_e) - \frac{3b}{2}\bar{x}\bar{x}(\theta_G + \phi_e) - 2\lambda_{\text{in}}2\bar{x}\frac{C_{d0}}{2\pi}\right. \\
& \left. + \lambda_{\text{in}}(\theta_G + \phi_e)000\frac{b^2}{2}\bar{x}\right]\left[\dot{w}\dot{v}\dot{\phi}_e\right]\Big) - \frac{2\pi}{2}\rho_\infty(2b)\left[\frac{b}{2}00000000\right]\left[\ddot{w}\ddot{v}\ddot{\phi}_e\right]
\end{aligned} \tag{4.10.8}$$

The inflow consists of the inflow due to the free stream λ_f and the induced inflow λ_i which is assumed to be uniform across the rotor disc. Thus,

$$\lambda_{\text{in}} = \lambda_f + \lambda_i. \tag{4.10.9}$$

The induced inflow is assumed to be the sum of steady component and a transient component and consequently can be expressed as,

$$\lambda_i = \lambda_{i0} + \Delta\lambda_i, \ \lambda_{\text{in}0} = \lambda_f + \lambda_{i0}. \tag{4.10.10}$$

The steady component of the induced inflow is estimated from the steady thrust coefficient as,

$$\lambda_{i0} = \sqrt{\frac{\lambda_f^2}{4} + \frac{C_{T0}}{2}} - \frac{\lambda_f}{2}. \tag{4.10.11}$$

Inserting the inflow expression and discarding all higher-order terms,

$$\begin{aligned}
[L_w L_v M_\phi] &= \frac{2\pi}{2}\rho_\infty(2b)\Bigg(\Omega^2R^2\left[-\bar{x}\lambda_{\text{in}0}\lambda_{\text{in}0}^2 - \bar{x}^2\frac{C_{d0}}{2\pi}0\right] \\
&+ \Omega^2R^2[-\bar{x}2\lambda_{\text{in}0} - \bar{x}\theta_G 0]\Delta\lambda_i - \Omega^2R^2\left[\bar{x}^2\bar{x}\lambda_{\text{in}0}0\right](\theta_G + \phi_e)\Big) \\
&- \frac{2\pi}{2}\rho_\infty(2b)\Bigg(\Omega R\Bigg[\bar{x}\lambda_{\text{in}0} - 2\bar{x}\theta_G - \frac{3b}{2}\bar{x}\bar{x}\theta_G - 2\lambda_{\text{in}0}2\bar{x}\frac{C_{d0}}{2\pi} \\
&+ \lambda_{\text{in}0}\theta_G 000\frac{b^2}{2}\bar{x}\Bigg]\left[\dot{w}\dot{v}\dot{\phi}_e\right] + \left[\frac{b}{2}00000000\right]\left[\ddot{w}\ddot{v}\ddot{\phi}_e\right]\Bigg) \\
&+ \frac{2\pi}{2}\rho_\infty(2b)\left(\Omega^2R\bar{x}[b00000][w'v']\right).
\end{aligned} \tag{4.10.12}$$

Hence,

$$\begin{aligned}
L_w &= \frac{2\pi}{2}\rho_\infty(2b)\left(\Omega^2R\bar{x}bw' - \Omega^2R^2\bar{x}\lambda_{\text{in}0} - \Omega^2R^2\bar{x}\Delta\lambda_i - \Omega^2R^2\bar{x}^2(\theta_G + \phi_e)\right) \\
&- \frac{2\pi}{2}\rho_\infty(2b)\Omega R\left\{\bar{x}\dot{w} + (\lambda_{\text{in}0} - 2\bar{x}\theta_G)\dot{v} - \frac{3b}{2}\bar{x}\dot{\phi}_e + \frac{b}{2\Omega R}\ddot{w}\right\},
\end{aligned} \tag{(4.10.13)}$$

$$\begin{aligned}
L_v &= \frac{2\pi}{2}\rho_\infty(2b)\left(\Omega^2R^2\left(\lambda_{\text{in}0}^2 - \bar{x}^2\frac{C_{d0}}{2\pi}\right) + \Omega^2R^2(2\lambda_{\text{in}0} - \bar{x}\theta_G)\Delta\lambda_i - \Omega^2R^2\bar{x}\lambda_{\text{in}0}(\theta_G + \phi_e)\right) \\
&- \frac{2\pi}{2}\rho_\infty(2b)\left(\Omega R\left\{(\bar{x}\theta_G - 2\lambda_{\text{in}0})\dot{w} + \left(2\bar{x}\frac{C_{d0}}{2\pi} + \lambda_{\text{in}0}\theta_G\right)\dot{v}\right\}\right),
\end{aligned} \tag{4.10.14}$$

The thrust due to the kth blade is

$$\text{Thrust}_k = \int_{R_h}^{R}(F_z - m\ddot{z})_k \mathrm{d}r \approx \int_{R_h}^{R}(L_w - m\ddot{w})_k \mathrm{d}r. \tag{4.10.15}$$

Non-dimensionalizing by $\pi\rho R^4\Omega^2$ the above relation, the thrust coefficient is

$$C_T = \frac{\text{Thrust}}{\pi\rho_\infty R^4\Omega^2} = \frac{1}{\pi\rho_\infty R^3\Omega^2}\int_{\bar{x}_h}^{1} L_w \mathrm{d}\bar{r} - \frac{1}{\Omega^2 R}\int_{\bar{x}_h}^{1}\frac{m}{\pi\rho_\infty R^2}\ddot{w}\mathrm{d}\bar{r}, \tag{4.10.16}$$

$$\begin{aligned}
\frac{1}{\pi\rho_\infty\Omega^2R^3}\int_{\bar{x}_h}^{1} L_w d\bar{r} &= -2\left((\lambda_{\text{in}0} + \Delta\lambda_i)\int_{\bar{x}_h}^{1}\bar{b}\bar{r}\mathrm{d}\bar{r} + (\theta_G + \phi_e)\left(\frac{1 - \bar{x}_h^3}{3}\right)\right) \\
&+ 2\int_{\bar{x}_h}^{1}\bar{b}^2 w'\bar{r}\mathrm{d}\bar{r} - \frac{2\bar{b}}{\Omega R}\int_{\bar{x}_h}^{1}\left\{\bar{r}\dot{w} + (\lambda_{\text{in}0} - 2\bar{r}\theta_G)\dot{v} - \frac{3R\bar{b}}{2}\bar{r}\dot{\phi}_e\right\}\mathrm{d}\bar{r} - \frac{1}{\Omega^2 R}\int_{\bar{x}_h}^{1}\bar{b}^2\ddot{w}\mathrm{d}\bar{r},
\end{aligned} \tag{4.10.17}$$

$$\frac{1}{\Omega^2 R}\int_{\bar{x}_h}^{1}\frac{m}{\pi\rho_\infty R^2}\ddot{w}\mathrm{d}\bar{r} = \frac{1}{\Omega^2 R}\int_{\bar{x}_h}^{1}\mu\ddot{w}\mathrm{d}\bar{r},\ \mu = \frac{m}{\pi\rho_\infty R^2}, \tag{4.10.18}$$

Hence if,

$$\bar{\bar{b}} = \frac{2}{1-\bar{x}_h^2}\int_{\bar{x}_h}^{1}\bar{b}\bar{r}\mathrm{d}\bar{r}, \tag{4.10.19}$$

$$C_T = -2\left(\left(\lambda_{\text{in0}} + \Delta\lambda_i\right)\left(\frac{1-\bar{x}_h^2}{2}\right)\bar{\bar{b}} + \left(\theta_G + \phi_e\right)\left(\frac{1-\bar{x}_h^3}{3}\right)\right) + 2\int_{\bar{x}_h}^{1}\bar{b}^2 w'\bar{r}\mathrm{d}\bar{r}$$
$$-\frac{2\bar{b}}{R}\int_{\bar{x}_h}^{1}\frac{\partial}{\partial\psi}\left\{\bar{r}w + \left(\lambda_{\text{in0}} - 2\bar{r}\theta_G\right)v - \frac{3R\bar{b}}{2}\bar{r}\phi_e\right\}\mathrm{d}\bar{r} - \frac{1}{R}\int_{\bar{x}_h}^{1}\left(\mu + \bar{b}^2\right)\frac{\partial^2}{\partial\psi^2}w\mathrm{d}\bar{r}. \tag{4.10.20}$$

In steady flow,

$$C_{T0} = -2\left(\lambda_{\text{in0}}\left(\frac{1-\bar{x}_h^2}{2}\right)\bar{\bar{b}} + \theta_G\left(\frac{1-\bar{x}_h^3}{3}\right)\right). \tag{4.10.21}$$

In the case of constant inflow with steady geometric pitch, the Pitt-Peters finite state model for the dynamic inflow reduces to,

$$\frac{384}{225\pi}\frac{\mathrm{d}\Delta\lambda_i}{\mathrm{d}t} + \frac{\Delta\lambda_i}{2\left(\lambda_f + 2\lambda_{i0}\right)} = C_T. \tag{4.10.22}$$

The quasi-steady aerodynamic loads on the blade may be expressed as,

$$\begin{aligned}\left[L_w L_v M_\phi/R\right] = {} & 2\frac{m\Omega^2 R}{\mu}\bar{b}\left(\left[-\bar{x}\lambda_{\text{in0}}\lambda_{\text{in0}}^2 - \bar{x}^2\frac{C_{d0}}{2\pi}0\right] + \left[-\bar{x}2\lambda_{\text{in0}} - \bar{x}\theta_G 0\right]\Delta\lambda_i - \left[\bar{x}^2\bar{x}\lambda_{\text{in0}}0\right]\theta_G\right) \\ & - 2\frac{m\Omega^2 R}{\mu}\bar{b}\left(\left[\bar{x}\lambda_{\text{in0}} - 2\bar{x}\theta_G - \frac{3\bar{b}}{2}\bar{x}\bar{x}\theta_G - 2\lambda_{\text{in0}}2\bar{x}\frac{C_{d0}}{2\pi} + \lambda_{\text{in0}}\theta_G 000\frac{\bar{b}^2}{2}\bar{x}\right]\right. \\ & \left.\times\frac{\partial}{\partial\psi}\left[\bar{w}\bar{v}\varphi_e\right] + \left[\frac{\bar{b}}{2}00000000\right]\frac{\partial^2}{\partial\psi^2}\left[\bar{w}\bar{v}\varphi_e\right]\right) \\ & + 2\frac{m\Omega^2 R}{\mu}\bar{b}\bar{x}\left[\bar{b}0 - \bar{x}00 - \lambda_{\text{in0}}000\right]\left[\frac{\partial}{\partial\bar{x}}\left[\bar{w}\bar{v}\right]\phi_e\right],\end{aligned} \tag{4.10.23}$$

4.11 Dynamics and Aeroelasticity of Flexible Rotor Blades

Problems of structural dynamics and aeroelasticity of wind turbine rotors are relatively well understood. Major progress in the field of wind turbine rotor dynamics was achieved during last three decades of the twentieth century thanks to improvements in performance of modern computers and application of nonlinear elastic model-based analysis to both helicopter and wind turbine rotors. An extensive summary of the current status of rotorcraft aeroelasticity is given by Friedmann (2003) and Friedmann and Hodges (2003). Friedmann and Hodges (2003) show that many different analytical methods have been developed in the field of helicopter rotor blade dynamics which are also applicable to wind turbines. A range of methods, from the simple but elegant Lagrange's equations based on an energy approach to computationally intensive but generic and powerful finite element method, have been developed for wind turbine rotors. The problem of rotor blade dynamics can be split into two major phases—formulation of blade equations of motion and their solution. Wind turbine blades, which are subjected to very high oscillatory loads, have dynamic characteristics that are generally complicated due to couplings of blade flexible deformations about two orthogonal axes with blade section rotation. The offset of center of gravity (CG) from the elastic axis plays an important role in analysis of rotor blade aeroelastic stability (Friedmann and Hodges 2003). Rotor blades are subjected to harmonic forcing caused by aerodynamic forces and centrifugal forces generated by blade rotation. Centrifugal forces are dependent upon rotor blade radius and cause the blades to stiffen. Coriolis forces are caused by combination of blade rotation and blade deformations and affect dynamic stability of rotor blades. Since rotational effects play important role in rotor blade dynamics and they are not neglected, but taken into consideration. The wind turbine blade degrees of freedom are mutually coupled, and the coupling can have a significant effect on blade dynamics and stability. Hence, equations of motion of a rotor blade are expressed in terms of a number of assumed modes which are also mutually coupled. The coupling terms cannot be ignored if one wishes to predict the blade behavior correctly. Houbolt and Brooks (1957) give derivations of combined equations of motion of bending and torsion of a rotor blade modelled as a slender beam. Aeroelastic equations of a helicopter rotor undergoing torsion and both flapwise and chordwise bending can be found in Kaza and Kvaternik (1977, 1979). Ordering schemes have been developed and are applied to equations of motion to retain the most significant terms and ignore terms which will not affect the blade behavior appreciably (Friedmann and Hodges 2003). Although wind turbine rotor blade dynamics can be described with the aid of the Newtonian approach and by the Euler's equations of motion applied to a lumped mass model of the rotor blade (Bramwell et al. 2001), it is the energy methods that represent the most convenient way of derivation of equations of motion. Energy methods are based on the principle of virtual work and Hamilton's principle. Lagrange's equations are the result of the application of these principles. Lagrange's method allows the derivation of equations of motion via partial differentiation of the Lagrangian which may be expressed in terms of the kinetic and potential energy of a dynamic system. Rosen and Freidmann (1979) have extended the dynamical equations of motion of a rotating rotor blade to large amplitudes in the nonlinear regime, and Floros (2000) has extended the nonlinear dynamic equations of motion of a rotating rotor blade to layered composite blades with closed cross sections. The integration of all the sectional properties is performed over the closed section using normal and tangential coordinates as illustrated in Fig. 4.7.

Since aspect ratios of wind turbine rotor blades are high, they can be regarded as slender beams. Modelling of rotor blades as flexible, infinitely thin beams is sufficient for most dynamic problems. Although in reality infinite number of modes could be used to describe dynamics of a wind turbine rotor blade, sufficient approximation of the behavior can be made with the aid of the primary dominant modes. The method of assumed modes belongs to the group of global methods and represents a fairly popular way of solving for the motion of rotor blades. A series of functions (mode shapes) are used for the first approximation of blade shape. Lagrange's method and the Raleigh-Ritz method are some of the most typical of methods of assumed modes. They allow estimation of modal shapes and corresponding modal frequencies. However Galerkin's method which is based on energy considerations represents most popular of these methods. It is widely used as it can be used in the case of nonslinear or non-conservative problems that both Lagrange's and Raleigh-Ritz methods cannot solve (Bramwell et al. 2001). In Galerkin's method, an approximation function is substituted into the partial differential equation of blade motion. The most significant aspect of Galerkin's method applied to wind turbine rotors is that range of aerodynamic models of the loading and inflow dynamics can be incorporated into the analysis. If n different mode shapes and frequencies are considered, it is transformed into a system of n ordinary differential equations (Bramwell et al. 2001). The blade natural frequencies and actual mode shapes are determine from these equations. Thus, the complete set of linear equations may be obtained by combining the linear structural dynamic equations of motion with the expressions for the linearized aerodynamic loads. One may obtain a typical set of root loci for different rotor speeds (Fig. 4.8) or with different inflow velocities (Fig. 4.9).

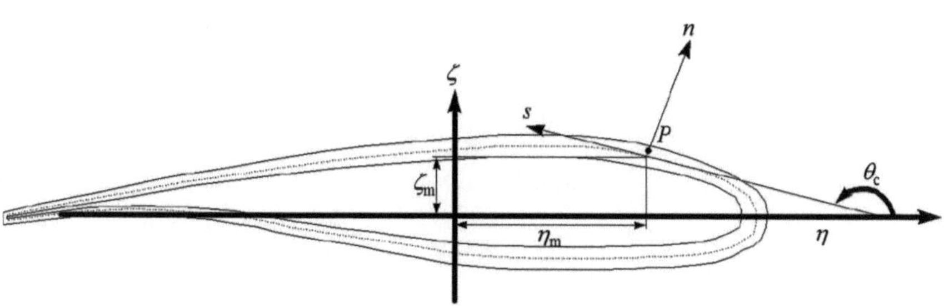

Fig. 4.7 Definition of coordinates for thin-walled sections

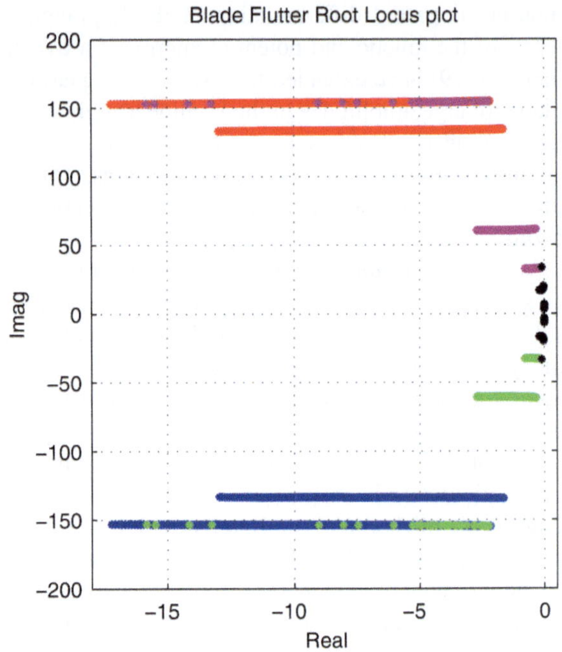

Fig. 4.8 Blade root locus for changing rotor speeds, with the CG and tension axis offsets equal to zero

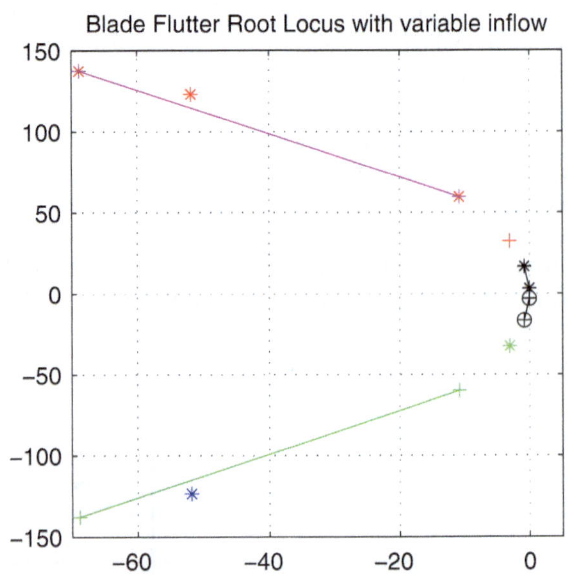

Fig. 4.9 Blade root locus for increasing inflow velocities, with the CG and tension axis offsets equal to zero

4.12 Wind Field Velocity Distributions and Spectrum

The general wind field distribution over a land mass is discussed by Burton et al. (2001). Winds are large-scale movements of air masses in the atmosphere. These movements of air are created on a global scale primarily by differential solar heating of the Earth's atmosphere. The wind speed at a given location is continuously varying. Turbulence refers to wind speed fluctuations over short time scales. Wind speed can be described as the sum of a mean wind speed, U_0, and a fluctuating component, Δu_f. The standard deviation of the turbulent component σ_u depends on the turbulence intensity I_U and the average wind speed and is given by the product, $\sigma_u = I_U \times U_0$. The turbulence intensity model is due to Rosas (2003) where I_U is modelled as,

$$I_U = I^*((a + 15/U_0)/(a + 1)), \tag{4.12.1}$$

where $I^* = 0.18$ for higher turbulence sites and $I^* = 0.16$ for lower turbulence sites with corresponding values for a of 2 and 3, respectively.

Rosas (2003) used a model for the power spectral density of turbulence which is known as the Kaimal model and is a derivative of the von Karman spectral model of the wind turbulence,

$$S(f) = \sigma_u^2 \frac{x_L}{U_0}\left(1 + \frac{3}{2}\left(\frac{f x_L}{U_0}\right)\right)^{-\frac{5}{3}} \tag{4.12.2}$$

where σ_u is the variance, x_L is the turbulence length scale (with a proposed maximum of 600 m), and U_0 is the mean wind velocity at the hub height. There is difficulty in specifying a uniformly valid approximation to the above spectrum as it is not function of f^2. Nevertheless, it is possible to approximate it and construct a shaping filter that may be used to simulate the wind. A curve fit of Kaimal's spectral approximation is

$$S(f) = \frac{\sigma_u^2 (x_L/U_0)}{\left(1 + \frac{3}{2}\left(\frac{f x_L}{U_0}\right)\right)^{\frac{5}{3}}} \approx \frac{\sigma_u^2 (x_L/U_0)}{\left(1 + k\left(\frac{\omega x_L}{2\pi U_0}\right)^2\right)} \tag{4.12.3}$$

where $k \approx 10$ to accurately model the mid-frequency range (0.01–1.0 rads/s), $k \approx 40$ to model the low-frequency range (<0.01 rads/s), and $k \approx 1$ to model the high-frequency range (>1.0 rads/s).

Given a Butterworth-type shaping filter transfer function of the form,

$$G_{vv}(s) = \sigma_u \sqrt{x_L/U_0} \times \sqrt{1/\left(1 + \sqrt{k}(s/\omega_c)\right)}, \quad \omega_c = 2\pi U_0/x_L, \tag{4.12.4}$$

the power spectral density may be approximated by,

$$\Phi_{vv}(\omega) = \sigma_u^2 \frac{x_L}{U_0} \frac{1}{\left[1 + k(\omega/\omega_c)^2\right]}. \quad (4.12.5)$$

The Kaimal spectrum given by Eq. 4.12.2 and the first-order approximation given by Eq. 4.12.5 are compared in Fig. 4.10. The parameters $k = 1$, 10, and 20 were approximately obtained by respectively matching the normalized magnitudes of 0.7, 0.5, and 0.3 of the Butterworth filter to that of the Kaimal spectrum. Although not perfect fit, it is seen that the Butterworth spectrum could be made to fit the Kaimal spectrum over a specified band of frequencies by choosing the filter constants appropriately. The Butterworth filter developed represents an approximation of the Kaimal spectrum and was chosen for the simulation. The reason for the use of this filter is that it facilitates easier simulation. The corresponding state equation of the shaping filter is

$$\dot{z}(t) = -\frac{\omega_c}{\sqrt{k}} z(t) + \omega_c \sigma_u \sqrt{\frac{x_L}{kU_0}} w(t), \ \ y = z(t), \quad (4.12.6)$$

where $w(t)$ is unit intensity white noise.

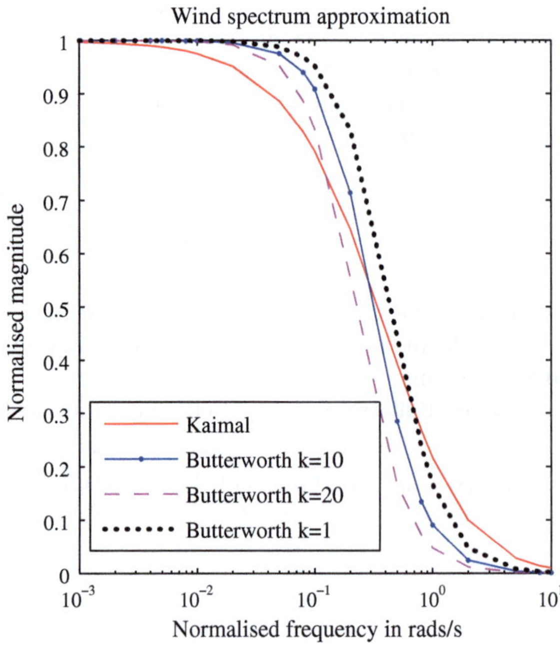

Fig. 4.10 Comparison of the Kaimal spectrum with the first-order Butterworth spectrum, $k = 1$, 10, and 20

4.13 Support Structures

4.13.1 Tower Dynamics and Aeroservoelasticity

The horizontal axis wind turbine which includes the rotor, the gear train and the generator, and other associated components must be supported at a particular operational design height at which it is expected to capture the maximum wind energy. Thus, the horizontal axis wind turbine must be supported on a tower which could be, from the ground level to the rotor hub, as high as 100 m high or even higher. The actual design height of the tower structure is determined by the rotor radius and the wind conditions at the site. Generally, the wind speed increases with altitude above the ground albeit not linearly. The higher wind speeds at higher altitudes are due to wind shear which can increase the power captured by the rotor. Thus, although increasing the tower height increases the power output, there is point beyond which the marginal increase in the power generated is offset by the costs involved in installing and maintaining a wind turbine at such a height.

The tower itself may be modelled as a non-uniform beam like structure, fixed to the ground, and excited at the other end at the rotor hub axis intersection by a non-conservative normal thrust force as well as a set of three mutually perpendicular hub moments. Generally, as the non-conservative normal thrust force and the three mutually perpendicular hub moments are functions of the blade dynamics degrees of freedom and the blade dynamics itself is a function of the hub base motion and rotation, the tower dynamics is coupled with the blade dynamics. To minimize the effects of this dynamic coupling, the natural frequencies of the tower dynamics must be chosen must chosen to be in a domain that does not overlap the natural frequencies associated with the rotor dynamics. The aerodynamic flow field around the blade is such that typically a fluid particle takes about 10–50 ms to traverse the blade chord length which corresponds to a frequency band of 20–100 Hz. A typical 20 m long rotor blade has a non-rotating frequency above 1.67 Hz. The effect of rotation and the presence of wind increase the natural frequency by factor 2–3 times, and consequently, the rotor dynamics natural frequencies are between 5 and 10 Hz. Thus, the tower structure natural frequencies are generally chosen below 1.67 Hz to minimize the dynamic coupling effects with the rotor. Coupled vibrations generally result larger amplitudes which consequently reduce the fatigue life of the machine. A turbine operating in a downwind state has also to withstand the effects of the vortices shed off the tower which are called the tower shadow. As the rotor blades pass through the tower shadow region, the shed vortices excite the blades, which tend to significantly increase the resulting noise and vibrations.

Dynamic inflow refers to the unsteady aerodynamics associated with the rotor and rotor blade flow field. The unsteady aerodynamic blade flow field results in a hysteresis effect when considering the variations of the lift coefficient with dynamic variations in the angle of attack particularly in the high angle of attack regions where the blade lift is prone to dynamic stall. Dynamic finite-state models of the dynamics of lift coefficient and the angle of attack have been proposed and validated for several rotor configurations, and these are used in the coupled aeroelastic analysis of wind turbine stability to ensure that the wind turbine is always operating under stable wind conditions. Additional stabilization is provided by using feedback controls to control the operating pitch angle of the blades so as to be operated at an angle of attack just below the onset of dynamic stall and thus provide the optimal or maximum power point operation. Because of the inclusion of feedback control of the blade pitch angle and the dynamics of blade servoactuation in the aeroelastic analysis, the analysis is referred to as aeroservoelastic analysis.

4.13.2 Offshore and Floating Support Structures

The availability of wind energy at offshore locations, particularly in the North Atlantic Ocean adjoining several northern European nations, the United Kingdom, North America as well as at several other locations on the planet offers the possibility of extracting significantly large quantities of power. A single offshore wind turbine with a rotor diameter up to 80 m can generate typically up to two megawatts and up to 5 MW for wind turbines with rotor diameters up to 120 m, while a typical wind farm can generate up to 300–500 MW of power. In general, the wind quality for power generation is better offshore than onshore as it is not only stronger and more consistent but also more easily predictable. However, there are a number of technological issues such as the design of supporting tower structures moored in the ocean or to floating platforms, the need for autonomous operation under all weather conditions, and the ability to absorb a number of failures while maintaining failsafe operation, which must be resolved prior to the installation of offshore wind power systems. Not only does one have to follow an integrated design approach, reliability, availability, maintainability, and serviceability during operation are absolutely vital in order to reduce the downtime due to a fault. There is need for complete integrated control and grid integration when the offshore wind farm is connected to (or disconnected from) the grid. The offshore environment imposes certain specific requirements on the wind turbines which are absent in the case of on shore wind turbines.

Offshore wind turbines may be classified based on the (1) structural configuration, (2) foundation type, and (3) the installation principle. Typical structural configurations include a mono-tower or a braced or a lattice tower. The foundation types can vary from a piled or gravity foundation, while the installation could be a lifted or floated installation. Mono-pile type tapered pile support structures are used in the North Sea where the water depth is about 20 m. While floating installations are generally necessary when the water depth exceeds 6–100 m, bottom mounted moored designs are viable below this depth. For large power generation, multi-unit floating offshore platforms mounted on several spar buoys which can house up to 16 independent wind turbines have proposed. As wave loadings in shallow waters are significantly lower at frequencies below 0.04 Hz and beyond 0.4 Hz, structures with there lowest natural frequency greater than 0.45–0.5 Hz are generally adopted for offshore applications. For deep water applications, structures with natural frequencies in excess of 1 Hz are normally used.

In the case of piled structures, the tower is modelled as a non-uniform beam structure with substantial portion of the beam structure embed within an elastic medium with a non-uniform translational stiffness and rotational stiffness with increasing depth and proportional translational damping and rotational damping. The damping characteristics are chosen so as to reflect the soil damping and internal friction within the soil. Generally, the stiffness and damping characteristics are different in the offshore case from the onshore case.

In the case of gravity foundations, the tower is assumed to be attached to an elastic half space with the same density, shear modulus, and Poisson's ratio as the soil. The discreet stiffness and damping parameters are largely influenced by the size and geometry of the foundation plate which is key parameter in the formulae for determining the equivalent translational stiffness, rotational stiffness, translational damping, and rotational damping characteristics.

Finally, there is the issue of hydrodynamic loadings which refers to a range of loadings including drag forces caused by vortices shed by the submerged structure, vortex shedding related oscillatory loading associated with the shed vortices that cause the drag, inertia forces due to the added mass effect of the fluid whether it is stationary or moving with a uniform current, diffraction loading due to the wave diffraction effects as the flow passes by large structures and finally, slam and slap loading due to wave breaking effects. The first two are associated with the vortex shedding, while the latter three are inertial effects associated with uniform and wave flows. Clearly, it is useful to define a non-dimensional number which can be used to determine whether the drag-related or inertia-related forces dominated on a particular structure. Such a non-dimensional number is the Keulegan–Carpenter number which is the ratio of the typical drag force to the typical inertia force. Generally, the drag force dominates when the ratio of diameter of the structural member to the wave length is less than 0.2, as beyond this ratio the degree of diffraction is significant. When the depth of the water is greater than half the wave length and the ratio of diameter of the structural member to the wave length is less than 0.2, the Keulegan–Carpenter number reduces to pi times the ratio of the wave height to the diameter of the structural member. When the Keulegan–Carpenter number is less than 1 the drag force is less than 0.05 times the inertia force. The drag force dominates when the Keulegan–Carpenter number exceeds 40. When the ratio of diameter of the structural member to the wave length is less than 0.2, the Morrison equation may be used to determine the magnitude and direction of hydrodynamic loading on a structural member. Otherwise diffraction effects dominate and potential flow methods may be used to determine the velocity potential and pressure distributions which are then integrated over the surface of wetted body to determine the forces and moments acting on it.

4.13.3 Hydrodynamic and Subsea Ice Loading

The development of approaches for consideration of the combined effects of wind and waves is the preferred approach to the analysis and design of offshore wind turbines. Time domain analysis is normally used for the design of offshore wind turbines, so the nonlinearities involved in both the loading and the structural response may be easily represented. Parallel, time domain analysis with a stochastic wind and wave description is preferred for the design of offshore wind turbines and the support structures. Such an analysis approach is well suited to reflect the dynamics of the total system although it has some specific shortcomings in representing nonlinearities of wave loading effects.

The linear wave theory is also known as the small amplitude wave theory and is most commonly used in deriving the wave loads on the floating platform (Newman 1977). The solution to the linear small amplitude wave theory can be summarized by the equations in Table 4.1 where the surface sea wave is assumed to be a simple sinusoid. The wave forces acting on a structure consist of the forces and moments due to the incoming waves known as the Froude–Krylov forces and the additional forces due to wave radiation and diffraction. The Froude–Krylov force is found by integrating the incident wave pressure over the surface of the body and resolving the resultant forces and moments into the components.

In reality, waves on the sea surface are not simple sinusoids. The surface appears to be composed of the summation of several random waves of various lengths and frequencies primarily generated by the wind. Waves are generated by the local wind, which is known as the *wind sea*. In addition, waves that were generated in other areas at other times and have propagated into the area being observed are known as the *swell*. However, swell is generally made up of waves of similar wavelength and frequency.

The actual frequency of the waves is determined from one of set of standard sea-state spectra defining a random distribution of wave amplitudes. If one assumes that the wave amplitude function $\eta(t)$ is a wide sense stationary stochastic

Table 4.1 Equations in linear wave theory

Wave property	Equation
Surface profile	$\eta = A\cos(kx - \omega t)$
Quadrature component	$\eta^{\perp} = A\sin(kx - \omega t)$
Horizontal particle velocity	$u = \frac{2\pi}{T}\frac{\cosh k(z+d)}{\sinh \mathrm{kd}}\eta$
Vertical particle velocity	$w = \frac{2\pi}{T}\frac{\sinh k(z+d)}{\sinh \mathrm{kd}}\eta^{\perp}$
Horizontal particle acceleration	$\dot{u} = \frac{4\pi^2}{T}\frac{\cosh k(z+d)}{\sinh \mathrm{kd}}\eta^{\perp}$
Vertical particle acceleration	$\dot{w} = -\frac{4\pi^2}{T}\frac{\sinh k(z+d)}{\sinh \mathrm{kd}}$
Dynamic pressure	$p = \rho g\frac{\cosh k(z+d)}{\cosh \mathrm{kd}}\eta$
Wave velocity	$c = [(g/k)\tanh \mathrm{kd}]^{1/2}$
Group velocity	$c_g = \frac{c}{2}\left\{1 + \frac{2\mathrm{kd}}{\sinh(2\mathrm{kd})}\right\}$

process, it follows that that the auto-correlation function of $\eta(t)$ is obtained by passing White noise through a linear filter. The input/output relation for a linear system may be expressed by the convolution integral. Hence, it follows that the power spectrum $S(\omega)$ of $\eta(t)$, which is the Fourier transform of the auto-correlation function, can be expressed as the product,

$$S(\omega) = H(\omega)H^*(\omega) \tag{4.13.1}$$

where $H^*(\omega)$ is the complex conjugate of $H(\omega)$, which in turn is the Fourier transform of $\eta(t)$. An approximate realization of $H(\omega)$ is known as a *sea-state filter*, and it facilitates the simulation of random seas by filtering random broadband white noise. Furthermore, a typical wave amplitude at a particular frequency ω_k may obtained from,

$$A(\omega_k) = \sqrt{2S(\omega_k)\Delta\omega} \tag{4.13.2}$$

where $\Delta\omega$ is the constant frequency interval between discrete frequencies at which the spectrum is sampled.

A typical example of the sea-state spectrum is the Bretschneider (1963) (one-sided) ocean wave spectrum given by:

$$S(\omega) = \frac{5}{16\omega}\left(\frac{\omega_m}{\omega}\right)^4 H_s^2 \exp\left(-\frac{5}{4}\left(\frac{\omega_m}{\omega}\right)^4\right), \quad \omega_m = 0.4\sqrt{\frac{g}{H_s}}, \tag{4.13.3}$$

where ω is the frequency in radians per second, ω_m is the modal (most likely) frequency of any given wave, g is the local acceleration due to gravity, and H_s is the significant wave height. Another typical example is the JONSWAP spectrum was developed to define the wave by Hasselman et al. (1973) for the Joint North Sea Wave Project. The formula is to be derived from the modified Pierson–Moskowitz spectrum formula. The Pierson–Moskowitz (1964) spectrum formula is

$$S(\omega) = \frac{8.1}{1000\omega}\left(\frac{\sqrt{g}}{\omega\sqrt{H_s}}\right)^4 H_s^2 \exp\left(-0.032\left(\frac{\sqrt{g}}{\omega\sqrt{H_s}}\right)^4\right). \tag{4.13.4}$$

The Pierson–Moskowitz spectrum may be expressed in non-dimensional form as,

$$\bar{S}(\Omega) = \frac{1}{\Omega^5}\exp\left(-\frac{\beta}{\Omega^4}\right). \tag{4.13.5}$$

It may be simulated by a discrete sea-state filter of the form,

$$y_n = -\sum_{k=1}^{l} a_k y_{n-k} + \sum_{k=0}^{l} b_k w_{n-k} \tag{4.13.6}$$

where w_k denotes a band-limited random White noise input sequence. The coefficients a_k and b_k are obtained numerically by minimizing the mean square error between the approximate and desired spectrum. A typical set of coefficients for a

Table 4.2 Digital filter coefficients for simulating the scaled, non-dimensional Pierson–Moskowitz spectrum with $\beta = 1$

Index	0	1	2	3	4	5	6	7
b_k	8.1047	−25.140	11.136	37.368	−38.057	−12.058	23.579	−0.85629
a_k	0	−92.280	2.308	132.860	−89.433	21.836	34.409	−19.221

particular value of β are listed in Table 4.2. Considering the non-dimensional Pierson-Moskowitz spectrum, it may be observed that when $\Omega = 1$, $\bar{S}(\Omega) = \exp(-\beta)$ and that when $\Omega^4 = \beta$, $\bar{S}(\Omega) = 1/e\beta^{5/4}$. Moreover $\bar{S}(\Omega)$ is maximum when $\Omega^4 = 4\beta/5$.

The JONSWAP spectrum was developed for the *limited fetch* North Sea by the offshore industry and is used extensively. The distance from a lee shore is called the *fetch*, or the distance over which the wind blows with constant velocity. The JONSWAP spectrum is defined by,

$$S(\omega) = \frac{\alpha}{\omega}\left(\frac{\sqrt{g}}{\omega\sqrt{H_s}}\right)^4 H_s^2 \exp\left(-\frac{5}{4}\left(\frac{\omega_0}{\omega}\right)^4\right)\gamma^\delta,\ \delta = \exp\left(-\frac{(\omega - \omega_0)^2}{2\tau^2\omega_0^2}\right), \tag{4.13.7}$$

where α is a parameter related to the prevailing wind velocity field with the wind velocity U_w and a fetch length x, γ is an overshooting or peakness parameter, while τ defines the shape of the spectrum. The parameters α, τ, γ, ω_0, and x_0 are defined by,

$$\alpha = 0.76(x_0)^{-0.22}, \tag{4.13.8}$$

$$\tau = 0.07 \text{ for } \omega \le \omega_0,\ \ \tau = 0.09 \text{ for } \omega > \omega_0, \tag{4.13.9}$$

$$1 \le \gamma \le 6, \tag{4.13.10}$$

$$\omega_0 = 2\pi(g/U_w)(x_0)^{-0.33} \tag{4.13.11}$$

and

$$x_0 = gx/U_w^2. \tag{4.13.12}$$

When x is not known, $\alpha = 0.0081$. A typical choice for γ is $\gamma = 3.3$. Another spectrum developed by Ochi and Hubble (1976) is the Ochi spectrum. It is a three-parameter spectrum which is an extension of the Bretschneider ocean wave spectrum allowing to make it wider for developing seas ($\lambda < 1$) or narrower for swell ($\lambda > 1$).

$$S(\omega) = \frac{1}{4\omega\Gamma(\lambda)}\left(\frac{4\lambda + 1}{4}\right)^\lambda \left(\frac{\omega_m}{\omega}\right)^{4\lambda} H_s^2 \exp\left(-\left(\frac{4\lambda + 1}{4}\right)\left(\frac{\omega_m}{\omega}\right)^4\right) \tag{4.13.13}$$

where λ is parameter that defines the relative width of the spectrum, and $\Gamma(\cdot)$ is the Gamma function.

For larger bodies relative to the wavelength, linear diffraction theory is applicable. In this case, the total velocity potential is considered to be the sum of both the incident and the scattered potential. Together they are obtained as solutions of a Laplace equation with appropriate boundary conditions on the surface of the body of the structural elements, boundary conditions at the bottom of the ocean and at the free surface at the ocean surface. The solution for the total velocity potential is reduced to a boundary integral equation for the velocity potential which is in turn solved by numerical methods. The boundary integral equation uses the free surface Green's function (Wehausen and Laitone 1960) for infinite or finite depth. By doing so, one obtains for the infinite depth case,

$$\phi(x,y,0) = \phi^i(x,y,0) + \sum_{m=-\infty}^{\infty} \int_{S_B} G(x,y,0;\xi,\eta)[k\phi(\xi,\eta,0) - w(\xi,\eta)]\mathrm{d}\xi\mathrm{d}\eta \tag{4.13.14}$$

where $G(x,y,0;\xi,\eta)$ is the free surface Green's function satisfying,

$$\nabla^2 G = 0, \ -\infty < z < 0, \tag{4.13.15}$$

$$\frac{\partial G}{\partial z} - kG = -\delta(x - \xi, y - \eta), \ z = 0 \tag{4.13.16}$$

$$G, \ |\nabla G| \to 0, \ z \to -\infty. \tag{4.13.17}$$

On $z = 0$, the Green's function is given by,

$$\begin{aligned} G(x,y,0;\xi,\eta) &= -\frac{k}{4\pi}\left(\frac{2}{Z} - \pi(\mathbf{H}_0(Z) + Y_0(Z) - 2i\pi J_0(Z))\right), \ Z \\ &= k\sqrt{(x - \xi)^2 + (y - \eta)^2}. \end{aligned} \tag{4.13.18}$$

In the above, $\mathbf{H}_0(Z)$ is the Struve function of order zero, $J_0(Z)$ is the Bessel function of the first kind and zero order, and $Y_0(Z)$ is Bessel function of the second kind and zero order (Abramowitz and Stegun 1972). From the velocity potential, the pressure distribution over the body is determined which is then integrated over the surface of the body, and the resultant forces and moments are resolved to determine the components.

When the ratio of diameter of the structural member to the wave length is less than 0.2, the Morrison equation obtains the total force from an inertia force and a drag force component. It is an empirical relation in terms of the drag and apparent mass coefficients, the fluid density, the diameter of the structural member, and the velocity and acceleration of structural member relative to the fluid, which defines the incremental force vector acting on an elemental length of the structural member. For an elemental height dz of a vertical cylinder of diameter D in a fluid flowing with a uniform velocity V, the Morrison equation is given by,

$$\mathrm{d}F = C_M \rho \frac{\pi D^2}{4} \dot{V} \mathrm{d}z + C_D \rho \frac{D}{2} |V| V \mathrm{d}z. \tag{4.13.19}$$

In the Morrison equation, C_M and C_D are the inertia and drag coefficients. The velocity V is not a constant and can represent the combination of multiple contributions to the flow. The Morrison equation may be rewritten such that multiple flow processes are explicitly included in the equation. Thus, the Morrison equation may be expressed as,

$$\begin{aligned} \mathrm{d}F = {} & \rho \frac{\pi D^2}{4} a_w \mathrm{d}z + (C_M - 1) \rho \frac{\pi D^2}{4} (a_w - \ddot{s}) \mathrm{d}z \\ & + C_D \rho \frac{D}{2} |u_w + u_c - \dot{s}| (u_w + u_c - \dot{s}) \mathrm{d}z \end{aligned} \tag{4.13.20}$$

where a_w is the wave acceleration, u_w is the wave velocity, u_c is the current velocity, and s represents the structural displacement. The force and moment acting on a finite segment of the structure of length L are given by,

$$F = \int_0^L \mathrm{d}F \text{ and } M = \int_0^L z \mathrm{d}F. \tag{4.13.21}$$

Although initially the Morrison equation was proposed only for a horizontal wave force on a vertical cylinder, it was generalized for the case of an inclined cylinder and shown to give reasonable results for the vector force acting on the member. For an inclined cylinder, the Morrison equation is expressed in component form, for the normal and tangential components of the force, as,

$$\mathrm{d}F_n = C_M \rho \frac{\pi D^2}{4} \dot{V}_n \mathrm{d}z + C_D \rho \frac{D}{2} |V_n| V_n \mathrm{d}z \tag{4.13.22}$$

$$\mathrm{d}F_t = C_f \rho \frac{\pi D}{2} |V_t| V_t \mathrm{d}z \tag{4.13.23}$$

where V_n and V_t are the normal and tangential components of the fluid relative velocity vector. Moreover, the above equations can be further generalized when both a wave and a uniform current are present.

The hydrodynamic coefficients, the drag ($\sim$0.7) and apparent mass ($\sim$2) coefficients per unit length of the structural member are functions of the Reynolds number and the Keulegan–Carpenter number and are obtained empirically or experimentally. Another important loading acting on subsea structural members is the loading due to the presence, motion, breaking and crushing of ice sheets, particularly in subsea conditions close to the poles. Under each of these conditions, the loads generated can be modelled as random external loads occurring over a finite frequency bandwidth and predefined amplitudes. Thus, the overall loading on the entire structure and its response to these loads may be obtained by standard methods.

4.13.4 Floater Dynamics and Wave Hydrodynamics

Typically, in the case floating offshore wind turbine, it is attached to a tethered spar buoy which consists of a non-uniform vertical cylinder that provides the necessary buoyancy to keep the turbine in a stable floating equilibrium. Three or more spokes are attached to the cylinder and tethers are attached to the ends of these spokes. The tethers are generally anchored at corresponding points on the ocean floor and are vertical when the floater is in equilibrium. A tethered floating wind turbine can be configured in three ways: (1) with the tethers held in tension to stabilize the platform, (2) with the tethers as catenary moorings and the spar buoy stabilized by using ballast, and (3) with spar buoy buoyancy stabilized like in barge and tethers again serving as catenary moorings. The floater in all three configurations is constrained to symmetric motions about the vertical axis, and for this reason, only the heave, surge, and pitch motions need be considered in its dynamic analysis. In pure heave, the equation of motion of a floating body is

$$(M + A_{33})\frac{\mathrm{d}^2 h}{\mathrm{d}t^2} + \rho g S_{\mathrm{wh}} h = F_h \tag{4.13.24}$$

where M is the mass of the floater, A_{33} is the added mass due to the body floating in water and radiating waves, h is the displacement in the heave degree of freedom, ρ is the density of water, g the acceleration due to gravity, S_{wh} the plane are enclosed by the waterline, and F_h is the external force exciting the heave motion due to wave forces acting on the floater, due to both the incident waves and diffraction effects. All heave damping terms are assumed to be negligible and ignored. When

$$(M + A_{33}) = \rho d_h S_{\mathrm{wh}}/c_{33}, \tag{4.13.25}$$

where d_h is the depth of the hull, and c_{33} is a constant,

$$\frac{\mathrm{d}^2 h}{\mathrm{d}t^2} + c_{33}\frac{g}{d_h}h = c_{33}\frac{F_h}{\rho d_h S_{\mathrm{wh}}}. \tag{4.13.26}$$

In the case of pure surge, the restoring forces are almost negligible and the surge equation may be simply expressed in terms of Newton's second law of motion.

In pure pitch, the equation of motion of a floating body is

$$(I_{55} + A_{55})\frac{\mathrm{d}^2 \theta}{\mathrm{d}t^2} + \rho g I_{\mathrm{xxb}}\theta = M_\theta, \tag{4.13.27}$$

where I_{55} is the pitching moment of inertia of the floater, A_{55} is the added moment of inertia in pitch due to the body floating in water and radiating waves, θ is the angular displacement in the pitch degree of freedom, I_{bxx} is the second moment of the beam of the floater, $b(x)$, defined by,

$$I_{bxx} = \int_{L} x^2 b(x)\mathrm{d}x \tag{4.13.28}$$

and M_θ is the external moment exciting the pitch motion due to wave forces acting on the floater and wind forces acting on the tower and rotor. All pitch damping terms are assumed to be negligible and ignored. When

$$(I_{55} + A_{55}) = \rho d_h I_{\mathrm{bxx}} / c_{55}, \tag{4.13.29}$$

where d_h is the depth of the hull, and c_{55} is a constant,

$$\frac{\mathrm{d}^2\theta}{\mathrm{d}t^2} + c_{55}\frac{g}{d_h}\theta = c_{55}\frac{M_\theta}{\rho I_{\mathrm{bxx}} d_h}. \tag{4.13.30}$$

The constants c_{33} and c_{55} are in the range of $0.1 \le c \le 0.16$ for most floating bodies.

For the spar buoy by itself, there is practically no excitation of the yaw mode as the cylinders are symmetric about the vertical axis. By choosing the axis along a preferred direction, sway and roll motion can be ignored as in principle they are similar to surge and pitch, provide the floater is stable in roll. In pure roll, the equation of motion of a floating body is

$$(I_{44} + A_{44})\frac{\mathrm{d}^2\phi}{\mathrm{d}t^2} + \rho g V_d GM\phi = M_\phi, \tag{4.13.31}$$

where I_{44} is the rolling moment of inertia of the floater, A_{44} is the added moment of inertia in roll due to the body floating in water and radiating waves, ϕ is the angular displacement in the roll degree of freedom, V_d is the volume of water displaced by the floater, GM is the meta-centric height, and M_ϕ is the external moment exciting the rolling motion due to wave forces acting on the floater and wind forces acting on the tower and rotor. All roll damping terms are assumed to be negligible and ignored. When

$$(I_{44} + A_{44}) = \rho V_d b_m^2 / c_{44}, \tag{4.13.32}$$

where b_m is the mean beam of the floater, and c_{44} is a constant,

$$\frac{\mathrm{d}^2\phi}{\mathrm{d}t^2} + c_{44}\frac{g\mathrm{GM}}{b_m^2}\phi = c_{44}\frac{M_\phi}{\rho V_d b_m^2}. \tag{4.13.33}$$

The constant c_{44} is in the range of $0.3 \le c_{44} \le 0.45$ for most floaters. The distance from the center buoyancy to the meta-center is given by the ratio of the second moment of the water plane area about a longitudinal axis in the water plane parallel to the roll axis to the volume of the displaced water.

The condition for static stability of floater in a sea can be summarized as follows:

The gravitational force acts downward at the CG of the structure. This point is fixed unless the mass of the floater is deliberately shifted by control forces. The net buoyancy force acts upward with its center at the center of mass of the fluid instantaneously displaced by the floater. This point unlike the CG depends on the oscillation of the floater with respect to the sea. The instantaneous center of buoyancy defines the meta-center which is the point of intersection of the instantaneous vertical passing through it and the line joining the equilibrium center of buoyancy and the center of gravity. In the case where the CG is above the instantaneous center of buoyancy, the floater is stable only when the meta-center is directly above the center of gravity. When the floater rolls slightly, a restoring moment tends to roll the floater back to equilibrium and the center of buoyancy moves back toward its equilibrium position. However, if the center of buoyancy continues to move further away from its equilibrium position, the floater is unstable. Therefore, the floater's stability increases as the width or beam increases. For structures where the beam is not wide enough or cannot be made wide enough, stability is achieved by having the CG below the center of buoyancy, and the structure has pendulum like roll stability. Buoyancy tanks can be used for achieving this stability, as is the case with submarines. The meta-centric height GM is the difference between the heights of the positions of the CG and of the meta-center. The value is positive for stable floaters.

In general, one could include all the degrees of freedom, but the use of the three degree of freedom model is adequate for most wind turbine applications. The overall system behaves like a spring restrained inverted pendulum with the stiffness in the surge direction being relatively soft in comparison with the other two directions. Consequently, the motions in both heave and pitch are relatively small. The presence of the tethers renders the coupling between the surge (and sway) motion and heave motion to be nonlinear, and therefore, it is often necessary that the tethers are long enough to minimize the destabilizing influence of the nonlinearity. The general formulation of the differential equation of motion for a floating offshore platform structural system may be written in the same format as those of any ocean going vehicles as defined by Fossen (1994) and Fossen and Smogeli (2004). The equations of motion are

$$(\mathbf{M} + \mathbf{A})\ddot{\mathbf{x}} + (\mathbf{C} + \mathbf{B})\dot{\mathbf{x}} + \mathbf{K}\mathbf{x} + \mathbf{K_R}\mathbf{x} + \mathbf{K_G}\mathbf{x} = \mathbf{F_{Wave}} + \mathbf{F_{Aero}} + \mathbf{F_{Con}} \tag{4.13.34}$$

where **M**, **C**, and **K** are the inertia matrix, damping matrix, and stiffness matrix associated with the floating offshore platform structural system, **A** is the hydrodynamic added mass matrix, **B** is the hydrodynamic added damping matrix, **x** is the vector of translational and rotational displacements in the degrees of freedom of the platform, $\mathbf{K_R}$ is the first-order hydrostatic restoring force matrix, $\mathbf{K_G}$ is the mooring stiffness matrix derived by using the Morrison equation, $\mathbf{F}_{\mathbf{Wave}}$ is the generalized wave excitation forces and moments vector, $\mathbf{F}_{\mathbf{Aero}}$ is the generalized rotor and tower aerodynamic excitation forces and moments vector, and $\mathbf{F}_{\mathbf{Con}}$ is a vector of the control forces. In defining the first-order hydrostatic restoring force matrix $\mathbf{K}_{\mathbf{R}}$, only the heave, pitching, and rolling degrees of freedom need be considered.

When the wind turbine in mounted on the spar buoy, pitching motions of the floater cause the wind turbine aerodynamic rotor thrust to generate a yawing moment, and for this reason, the yawing degree of freedom must also generally be included in the analysis of coupled dynamics of the wind turbine and the floater. Ensuring the dynamic stability of floater and wind turbine system is the primary objective of the dynamic analysis of the system. Moreover, some simulation studies seem to indicate that in the tension leg platform, where the tethers are held in tension to hold the platform in position, the rotor floater coupling is minimized.

4.13.5 Floater–Rotor Passive and Active Decoupling Control

The major technical challenge of installing an offshore wind turbine is related to minimizing the wave-induced motion of the turbine rotor and understanding the coupling between support structure and the wind turbine, while achieving exceptional static and dynamic stability. Standard control algorithms for variable blade pitch turbines aimed at MPPT in a variable wind will introduce a thrust force variation that adds negative damping for the pitch and surge motion. Further, the wave-induced motions of the floater add to the complexity of the turbine design. Modified control algorithms are thus required that ensure that the rotor is completely decoupled from the floater and vice versa. Also, the mooring system involves expensive design challenges to dampen the response amplitude of the floater to wave-induced excitation.

Normally, in the analysis of a conventional land-based wind turbine, the rotor is assumed to be a short rigid shaft that couples the rotor disc and hub motion to the tower. The rotor disc is assumed to transmit forces and moments to the tower. However, in the case of an offshore wind turbine, the rotor is assumed to be a flexible cantilever support on one end, the rotor disc and hub while being supported at the other by the tower. The rotor rotation coupled with the pitching of the floater platform results in a gyroscopic coupled torsional motion of the tower and a yawing motion of the floater platform. To minimize this coupling, the rotor dynamic motion must be completely decoupled from the pitching motion of the floater. Assuming steady rotation of the rotor disc with an angular velocity Ω, the equations of motion of the elastic rotor may be expressed as,

$$\begin{bmatrix} m_d & 0 & 0 & 0 \\ 0 & m_d & 0 & 0 \\ 0 & 0 & I_d & 0 \\ 0 & 0 & 0 & I_d \end{bmatrix} \frac{d^2}{dt^2} \begin{bmatrix} x_d \\ y_d \\ \theta_d \\ \psi_d \end{bmatrix} + \Omega \begin{bmatrix} 0 & 0 & 0 & 0 \\ 0 & 0 & 0 & 0 \\ 0 & 0 & 0 & I_p \\ 0 & 0 & -I_p & 0 \end{bmatrix} \frac{d}{dt} \begin{bmatrix} x_d \\ y_d \\ \theta_d \\ \psi_d \end{bmatrix}$$
$$+ \begin{bmatrix} k_1 & 0 & 0 & k_2 \\ 0 & k_1 & k_2 & 0 \\ 0 & -k_2 & k_3 & 0 \\ k_2 & 0 & 0 & k_3 \end{bmatrix} \begin{bmatrix} x_d \\ y_d \theta_d \\ \psi_d \end{bmatrix} = \begin{bmatrix} F_{Hx} \\ F_{\mathrm{Hy}} \\ M_{\mathrm{Hx}} \\ M_{\mathrm{Hy}} \end{bmatrix} + \begin{bmatrix} F_{Tx} \\ F_{\mathrm{Ty}} \\ M_{\mathrm{Tx}} \\ M_{\mathrm{Ty}} \end{bmatrix} \tag{4.13.35}$$

where m_d is the mass of the rotor disc, I_d is the moment of inertia of the symmetric rotor disc about an axis in the plane of the rotor disc, I_p is the polar moment of inertia of the rotor disc, k_i, $i = 1$, 2 and 3, are the rotor shaft stiffness coefficients, x_d, y_d, θ_d, and ψ_d are the rotor disc translational and rotational degrees of freedom, F_{Hx} and F_{Hy}, and M_{Hx} and M_{Hy} are the rotor disc hub forces and moments, respectively, acting on the rotor shaft, and F_{Tx} and F_{Ty}, and M_{Tx} and M_{Ty} are the tower and floater motion-induced forces and moments acting on the rotor shaft. In the case of an offshore wind turbine, the rotor disc hub forces and moments are functions of the floater displacement and rotation. Moreover, the rotor dynamics is coupled with the tower motion even if the tower is assumed to be a simple cantilever, and the rotor exerts a bending moment and twisting moment on the tower. Thus to ensure complete decoupling of the rotor disc and floater, the rotor shaft motion must be completely decoupled from the floater. This can be achieved either by active feedback control, involving feedbacks of the floater motion to the rotor shaft suspension system, or by feedback control of the floater itself to ensure vibration free support of the wind turbine. Passive damping of the floater motion can substantially alleviate the influence of the floater motion on the dynamics of the rotor shaft and tower. A number of passive and active damping concepts are currently being implemented and tested for future offshore wind turbine installations.

References

Abo-Khalil AG, Lee D-C (2008) MPPT control of wind generation systems based on estimated wind speed using SVR. IEEE Trans Ind Electron 55(3):1489–1490

Abo-Khalil AG, Lee D-C, Seok J-K (2004) Variable speed wind power generation system based on fuzzy logic control for maximum output power tracking. In: Proceedings of 2004 IEEE 35th annual power electronics specialists conference-PESC 04, vol 3. Aachen, Germany, pp 2039–2044

Abramowitz M, Stegun IA (eds) (1972) Handbook of mathematical functions with formulas, graphs, and mathematical tables. Dover Publications, New York, ISBN 978-0-486-61272-0

Bhowmik S, Spee R, Enslin JHR (1999) Performance optimization for doubly fed wind power generation systems. IEEE Trans Ind Appl 35(4):949–958

Bisplinghoff RL, Ashley H, Halfman RL (1955) Aeroelasticity. Addison-Wesley, Cambridge

Boukhezzar B, Siguerdidjane H (2005) Nonlinear control of variable speed wind turbines without wind speed measurement. In: Proceedings of 44th IEEE conference on decision and control. Seville, Spain, pp 3456–3461

Bramwell ARS, Done G, Balmford D (2001) Bramwell's helicopter dynamics, 2nd edn. Butterworth-Heinemann, Oxford. ISBN 0-7506-5075-3

Bretschneider CL (1963) A one-dimensional gravity wave spectrum. In: Proceedings of a conference on ocean wave spectra. Prentice-Hall, Englewood Cliffs, pp 41–56

Burton T, Sharpe D, Jenkins N, Bossanyi E (2001) Wind energy handbook. John Wiley & Sons Ltd, England

Chu C (1951) The effect of initial twist on the torsional rigidity of thin prismatical bars and tubular members. In: Proceedings of the 1st US national congress of applied mechanics. pp 265–269

Connor B, Leithead WE (1993) Investigation of fundamental trade-off in tracking the $Cp_{\max}$ curve of a variable speed wind turbine. In: Proceedings of the 12th British wind energy conference. pp 313–7

Curti G, Risitano A (1979) Coupled free torsional and axial vibration of pre-twisted bars. Meccanica 157–162

Datta R, Ranganathan VT (2003) A method of tracking the peak power points for a variable speed wind energy conversion system. IEEE Trans Energy Convers 18(1):163–166

Drela M (1989) XFOIL: an analysis and design system for low Reynolds number airfoils. In: Mueller TJ (ed) Low Reynolds number aerodynamics. Lecture notes in engineering, vol 54. Springer, New York, pp 1–12

Drela M, Giles MB (1987) Viscous-inviscid analysis of transonic and low Reynolds number airfoils. AIAA J 25(10):1347–1355

Eppler R (1990) Airfoil design and data. Springer, New York

Esram T, Kimball JW, Krein PT, Chapman PL, Midya P (2006) Dynamic maximum power point tracking of photovoltaic arrays using ripple correlation control. IEEE Trans Power Electron 21(5):1282–1291

Floros MW (2000) Elastically tailored composite rotor blades for stall alleviation and vibration reduction. Ph.D. dissertation, The Pennsylvania State University, Department of aerospace engineering

Fossen TI (1994) Guidance and control of ocean vehicles. Wiley, New York

Fossen TI, Smogeli ØN (2004) Nonlinear time-domain strip theory formulation for low-speed manoeuvring and station-keeping. Model Ident Control 25(4):201–221

Fox ME (1993) Blade mounted actuation for helicopter rotor control. M.Sc. thesis, Massachusetts Institute of Technology, pp 83–125

Friedmann PP (1977) Recent developments in rotary-wing aeroelasticity. AIAA J Aircr 14–11:1027–1041

Friedmann PP (2003) Rotary wing aeroelasticity: a historical perspective. In: AIAA paper no. 2003–1817, Proceedings 44th AIAA/ASME/ASCE/AHS/ASC structures, structural dynamics and materials conference. Norfolk, VA. (Special 1 hour invited paper-in the dynamics specialist part of the conference)

Friedmann PP, Hodges DH (2003) Rotary wing aeroelasticity—a historical perspective. In: Proceedings of 10th international workshop on aeroelasticity of rotorcraft systems, Atlanta, Georgia

Friedmann P, Shamie J (1975) Aeroelastic stability of trimmed helicopter blades in forward flight. Paper presented at the 1st European rotorcraft and powered lift aircraft forum, University of Southampton, England

Giguére P, Selig MS (1998) New airfoils for small horizontal axis wind turbines. ASME J Sol Energy Eng 120:108–114

Hasselman K et al (1973) Measurements of wind-wave growth and swell decay during the joint north sea wave project (JONSWAP). Deutschen Hydrographischen Zeitschrift, Supplement A, 8(12):95ff

Hodges DH (1980) Torsion of pretwisted beams due to axial loading. J App Mech 47:393–397

Hohm RME (2003) Comparative study of maximum power point tracking algorithms. Prog Photovolt: Res Appl 47–62

Houbolt JC, Brooks GW (1957) Differential equations of motion for combined flapwise bending, chordwise bending, and torsion of twisted nonuniform rotor blades. NACA-TN-3905

Jou H-L, Wu K-D, Wu J-C, Shen J-M (2008) Simplified maximum power point tracking method for the grid-connected wind power generation system. Electr Power Compon Syst 36(11):1208–1217

Kawabe I, Morimoto S, Sanada M (2007) Output maximization control of wind generation system applying square-wave operation and sensorless control. In: Proceedings of power conversion conference, vol 1. Nagoya, Japan, pp 203–209

Kaza KRV, Kvaternik RG (1977) Nonlinear flap-lag axial equations of a rotating beam. AIAA J 15(6):871–874

Kaza KRV, Kvaternik RG (1979) Examination of the flap-lag stability of rigid articulated rotor blades. J Aircr 16(12):79–4150

Kosmatka JB (1992) Extension-bend-twist coupling behaviour of nonhomogeneous anisotropic beams with initial twist. AIAA J 30(2):519–527

Koutroulis E, Kalaitzakis K (2006) Design of a maximum power tracking system for wind-energy-conversion applications. IEEE Trans Ind Electron 53(2):486–494

Krenk S (1983) The torsion-extension coupling in pretwisted elastic beams. Int J Solids Struct 19(1):67–72

Lee C-Y, Shen Y-X, Cheng J-C, Chang C-W, Li Y–Y (2009) Optimization method based MPPT for wind power generators. World Acad Sci Eng Technol 60:169–172

Liu K-C, Friend J, Yeo L (2009) The axial–torsional vibration of pretwisted beams. J Sound Vib 321:115–136

Manwell JF, McGowan JG, Rogers AL (2010) Wind energy explained, theory, design and application, 2nd edn. Wiley, UK

Miley SJ (1974) On the design of airfoils for low Reynolds numbers. In: Proceedings of AIAA/MIT/SSA 2nd international symposium of the technology and science of low-speed and motorless flight, soaring society of America. Los Angeles, pp 82–96 also AIAA Paper No. 74-1017

Munteanu I, Bratcu AI, Cutululis N-A, Ceanga E (2008) Optimal control of wind energy systems: towards a global approach, 1st edn. Springer, pp 109–128, chapter 5 (In the series on advances in industrial control)

Munteanu I, Bratcu AI, Ceanga E (2009) Wind turbulence used as searching signal for MPPT in variable-speed wind energy conversion systems. Renew Energy 34:322–327

Newman JN (1977) Marine Hydrodynamics. The MIT Press, Cambridge

Ochi M, Hubble E (1976) On six parameter wave spectra. In: Proceedings of 15th international conference on coastal engineering, ICCE, vol 1. Pp 301–328

Örs M (2009) Maximum power point tracking for small scale wind turbine with self-excited induction generator. Control Eng Appl Inform (CEAI) 11(2):30–34

Padfield GD (1996) Helicopter flight dynamics. Blackwell Science Ltd, Oxford

Pierson WJ, Moskowitz L (1964) A proposed spectral form of fully developed wind seas based upon similarity theory of S. A. Kitaigorodskii. J Geophys Res 69

Pitt DM, Peters DA (1981) Theoretical predictions of dynamic-inflow derivatives. Vertica 5:21–34

Qiao W, Zhou W, Aller JM, Harley RG (2008) Wind speed estimation based sensorless output maximization control for a wind turbine driving a DFIG. IEEE Trans Power Electron 23(3):1156–1169

Qiao W, Gong X, Qu L (2009) Output maximization control for DFIG wind turbines without using wind and shaft speed measurements. IEEE Energy Convers Congress Exposition, ECCE 2009, DOI: 10.1109/ECCE.2009.531613:404-410

Rosas P (2003) Dynamic influences of wind power on the power system. Ph.D. dissertation, Ørsted Institute, Section of electric power engineering, Technical University of Denmark

Rosen A (1983) Theoretical and experimental investigation of the nonlinear torsion and extension of initially twisted bars. J App Mech 50:321–326

Rosen A, Friedmann PP (1977) Nonlinear equations of equilibrium for elastic helicopter and wind turbine blades undergoing moderate deformation UCLA-ENG-7718 (revised June 1977), also published as NASA CR-159478, Dec 1978

Rosen A, Friedmann PP (1979) The nonlinear behavior of elastic slender straight beams, undergoing small strains and moderate rotations. ASME J App Mech 46:161–168

Slootweg JG, de Haan SWH, Polinder H, Kling WL (2003) General model for representing variable speed wind turbines in power systems dynamics simulations. IEEE Trans Power Syst 18(1):144–151

Sokolnikoff IS (1956) Mathematical theory of elasticity. McGraw-Hill Book Company, New York

Tan K, Islam S (2004) Optimum control strategies in energy conversion of PMSG wind turbine system without mechanical sensors. IEEE Trans Energy Convers 19(2):392–398

Taraft S, Rekioua D, Aouzellag D (2008) Performances study of the DFIG associated to the variable speed wind turbine connected to the grid. Int J Electr Power Eng 2(5):353–364

Tsuiji T (1985) Free vibrations of thin-walled pretwisted beams. Bull JSME 28(239):894–898

Van den Heever PD, Oberholzer S, Enslin JHR (1989) High-efficient solar panel/wind turbine converter with maximal power control. Proc Eur Conf Power Electron Appl 663–668

Vepa R (2011) Nonlinear, optimal control of a wind turbine generator. IEEE Trans Energy Convers 26(2):468–478

Wang Q, Chang L (2004) An intelligent maximum power extraction algorithm for inverter-based variable speed wind turbine systems. IEEE Trans Power Electron 19(5):1242–1248

Wehausen JV, Laitone EV (1960) Surface waves. Springer, Berlin

Wortmann FX (1974) The quest for high lift. In: Proceedings of AIAA/MIT/SSA 2nd international symposium of the technology and science of low-speed and motorless flight, soaring society of America. Los Angeles, pp 97–101, also AIAA paper no. 74-1018

Xu XS, Zhong WX, Zhang HW (1997) The Saint-Venant problem and principle in elasticity. Int J Solids Struct 34(22):2815–2827

Yu GJ, Jung YS, Choi JY, Kim GS (2003) A novel two-mode MPPT control algorithm based on comparative study of existing algorithms. Sol Energy 455–463

Chapter 5
Dynamic Modeling of Gas Turbines and Compressors

5.1 Gas Turbines: Typical Components and Dynamic Modeling

Gas turbines are undoubtedly the most popular energy-generating machines that have been developed by man. Their use is practically ubiquitous as they are used in aircraft, ships, trains, and automobiles. A gas turbine is also the primary subsystem of a jet engine. The gas turbine is complex machine put together by combine a number components which have a number of mechanically rotating and stationary parts. Each of the component subsystems is characterized by its own performance characteristics. The performance of each subsystem is dependent on either the aerodynamics or the thermodynamics of the fluids flowing in and out of the subsystem and in some cases on the interaction of both the aerodynamics and the thermodynamics of the fluids within a control volume. Modeling, simulation, and control of a gas turbine involve a thorough understanding of the dynamics and interaction of all of the component subsystems. For this reason, it is important to have an understanding of the modeling and simulation of each of the component subsystems that make up a gas turbine. Typically, a low-pressure compressor or a fan situated on the air intake side of the engine is driven by a low-pressure turbine while a high-pressure compressor is driven by a high-pressure turbine. The total mass flow rate splits between the primary flow and the by pass flow. The fuel is injected into the combustion chamber and the combustion takes place within the chamber. The low-pressure compressor and turbine may be mounted on one spool while the high-pressure compressor and turbine are mounted on a second spool. The rotation rates of both spools are the result of the difference between the power produced by each of the turbines and the power consumed by its associated compressor. The total mass flow rate and the primary mass flow rate are driven by choking conditions ahead of the high-pressure turbine and aft of the low-pressure turbine. Finally, the engine is controlled by the amount of fuel injected in the combustion chamber and the sectional area of the exhaust. The primary components of a typical multistage gas turbine and their key features of gas turbine

R. Vepa, *Dynamic Modeling, Simulation and Control of Energy Generation*, Lecture Notes in Energy 20,
DOI: 10.1007/978-1-4471-5400-6_5,

engine have been discussed at length by Mattingly (2005) and are only briefly enunciated in the following:

1. One or more axial or in some cases radial compressors: The primary purpose of the compressors is to deliver compressed air to the next stage which is usually a combustor. Compressors in gas turbine jet engines achieve pressurization of the flow by momentum transfer. Typically in a compressor of this type, the entering fluid is accelerated by the kinetic energy imparted to it in the rotor. The kinetic energy of the flow and is converted into potential energy by decelerating the fluid in a diverging channel. During the initial step, the total pressure of the fluid increases due to the momentum imparted to the flow by the rotor blades. In the final step of the pressurization, the deceleration of the fluid results in a pressure rise as the dynamic pressure is converted to static pressure while conserving the total energy. The performance of a compressor is often specified in a compressor map by curves that relate the rotation speed, the pressure rise across the compressor, and the mass flow through the compressor. Steady-state operating points with constant rotational speed are indicated by compressor characteristics. The load or throttle line represents the pressure requirements of the system. Consequently, the steady-state operating point of a compression system is given by intersection point of the compressor characteristic and the load line. Steady-state operating points must be sufficiently stable in the face of perturbations in a compressor that is used in a gas turbine engine. Stable operation can be guaranteed by operating the compressor at a safe distance from the region of the characteristics where the compressor is prone to instability. Rotating stall and surge are two types of instabilities associated with the flows through axial flow compressors that are used in gas turbine engines. Rotating stall refers to a non-axially symmetric flow perturbation that travels around the annulus of the compressor although it is itself limited to small region of flow separation over a blade where the flow is stalled and is known as a stall cell. Rotating stall is initiated when the flow's kinetic and potential energies are inadequate to overcome the required differential pressure across a blade passage to maintain the flow without separating and consequently the flow locally stalls. Surge refers to large amplitude axial oscillation of the flow caused by the alternating storage and release of the compressed air in the ducting downstream of the compressor. While the compressibility of the flow acts as non-linear spring, the local non-linear compressor characteristics provides a reduced or even a negative damping force, thus inducing large amplitude vibrations. Rotating stall and surge are both dynamic instabilities and static instability is avoided by design in compressors by ensuring that the non-linear equivalent spring stiffness of the flow is always positive. Rotating stall is generally a precursor to the onset of surge in many engines. Typically, effects of rotating stall and surge range from stress and wear of the compressor blades to perturbations of the operating point of the compressor which significantly affects the conditions downstream of the compressor in the vicinity of the combustion chamber resulting in some adverse consequences. One of the key variables that

influence the performance of a gas turbine engine is the compression efficiency. Compression ratio and compression isentropic or polytropic efficiency are functions of the rotation rate and the axial mass flow rate. Rotating stall and surge are two factors that can severely limit the compressors performance. However, they can be controlled either passively or actively are modern compressors are designed to avoid both of them.

Compressor performance is dependent on (among other parameters) the ratio of axial velocity to tangential velocity in the stages of the compressor. The tangential velocity is proportional to shaft speed. Performance can be improved across the stages of a compressor by limiting the variation of the ratio of axial velocity to tangential velocity. "Splitting" the compressor into two separate machines allows improved control over the range of the ratio within each compressor. This results in improved compression performance in machines with two.

2. The combustor serves to convert fuel chemical energy into thermal energy by the process of combustion. It is desirable to perform this conversion with minimum loss of the pressure that has been generated by the compression system, so pressure loss is a major figure of merit for the combustor. Another important factor is the stability of the whole process. A combustion chamber designed to obtain a desired outlet temperature and flow rate, so it is capable of driving one or more turbines. A gas turbine combustor is a device for raising the temperature of the incoming air stream from compressor by the addition and combustion of fuel. To this end, the combustor must satisfy many conflicting requirement. It must be capable of initiating ignition easily while it operates stably over a wide range of conditions. At any operating point, it must essentially provide for complete combustion of the fuel while minimizing the formation and emission of undesirable pollutants. To avoid damaging the turbine, sufficient mixing must be achieved in the combustor to obtain an acceptable exit gas temperature distribution. Simultaneously, the combustor must also operate with as minimum a pressure loss as practical to maintain high overall performance. Finally, all of these functions must be performed in a configuration, which has minimum size, weight, and cost and which is sufficiently durable to achieve an acceptable operating life. If one assumes stoichiometric or theoretical combustion, which is the ideal combustion process where fuel is burned completely, the combustion process involves the molecular processes of mixing, combustion, and thermal equilibration while obeying thermodynamic mixing rules based on mass and energy conservation. Typically, a stoichiometric mixture of reactants is formed, followed by combustion, which corresponds to a material transformation and then, thermal equilibration between the combustion products and diluents. All along the thermodynamic equilibrium conditions must be met by the fluid components while the flow field is governed by the multicomponent gas-dynamic conservation equations.
3. High- and/or low-pressure turbines which convert the thermal energy of the gases flowing out of a combustor to mechanical energy by expansion. The high-temperature gas from the exit of the combustor enters the turbine where it

undergoes isentropic expansion while the pressure falls to the ambient pressure. During the expansion, the energy is transferred from the flow to a mechanical impeller or rotor. The exhaust gases then leave the turbine and into the atmosphere thus completing a thermodynamic cycle, typically known as the Brayton cycle or one its variants. The expansion process within the turbine being isentropic, the work delivered to the turbine impeller is given by the product of the temperature difference across the turbine and specific heat coefficient of the gases in the turbine at constant pressure. Because the temperature ratio is a direct function of the pressure ratio, the specific work output of the turbine is function of the compression pressure ratio, the ambient temperature and the adiabatic index. The high- and/or low-pressure turbines are generally mounted on the same shaft as the corresponding high- and/or low-pressure compressors and can therefore provide the motive power to drive the latter. Most turbines used in gas turbine engines are of the axial type. Turbine can broadly be classified into two types, namely, impulse and reaction turbines. In the impulse turbine, the gases expand to high velocity in a nozzle and pass over the moving blades. The moving blades simply deflect the flow and this change in the flow direction enables the energy to be converted to work. There is no expansion or pressure drop over the blades themselves. In the reaction turbine, the expansion or pressure drop takes place partly in the stator blades and in the moving blades. It can be shown that the maximum efficiency of a turbine is obtained when the blade speed is half that of the gas stream velocity. In order to operate at the maximum efficiency while also delivering the maximum work output requires that the turbines operate in multiple stages. Consequently, either the velocity drop or the pressure drop can be achieved in several stages. Multistage velocity drop compounded or pressure drop compounded turbines are both used in conjunction with other components in gas turbine engines. The desired work output and operating speed of a turbine are fixed by the application and so are the turbine inlet pressure and inlet temperature. The turbine designer is only left with the option of choosing the type of the machine and the primary geometrical parameters like the blade height to rotor diameter ratio, the blade and nozzle angles, the ratio of the rotor clearances to the rotor diameter, and the outlet swirl angle which affect the losses in a turbine stage. With constant turbine efficiency, choked flow at the turbine inlet and nozzle exit means that turbine expansion ratio is a function of area ratio. The turbine designer must choose an optimum set of the geometrical parameters to obtain the highest efficiencies under given operating conditions.

4. In the case of gas turbine jet engines, they are additionally equipped with a nozzle and/or an afterburner to generate the required thrust. Normally, the exhaust flow from a turbine is through a nozzle and the primary role of the nozzle is to generate thrust. The nozzle produces thrust by expanding (accelerating) the flow. Static pressure and temperature decrease as the flow accelerates (thermal energy is converted to kinetic energy). If we assume the process is isentropic, then total pressure and total temperature are constant throughout the process. The nozzle is a key element as it determines the turbine entry

temperature as well as the mass flow of the engine and the exit velocity and pressure. Thus, the nozzle not only affects the work done by the turbine but also determines the magnitude of the thrust. Typically, the nozzle is a variable area duct where the flow velocity has reached a limit at the choked state and the pressure is increased beyond the choked state so as to generate the desired thrust. Modern nozzles are designed by the converging and then diverging the cross-sectional area to increase the velocity beyond the choked state and increase the thrust. Yet, there is limit to the maximum thrust generated by a nozzle.

Three basic types of jet engines are in current use and these are turbojet engines, turbofan engines, and turboprop engines. Historically, the turbojet was the first type of jet engine built and is the simplest of the three. Its major components include a compressor, a combustor, a turbine which drives the compressor, and an exhaust nozzle and is illustrated in Fig. 5.1. It produces a relatively high specific thrust defined as thrust per unit mass flow. It can be used both for high subsonic and supersonic flight speeds. The turbofan also has a compressor, combustor, and a turbine which drives the compressor. In addition, it has a fan in front of the main compressor and a second power turbine behind the first turbine to drive the fan. Turbofans are currently being designed and built in a variety of different configurations. There are high-bypass or low-bypass turbofans, afterburning turbofans, axial compressor or centrifugal compressor turbofans, single-spool, two-spool, or three-spool turbofans and aft fan or geared fan turbofans. High and low bypass refers to the amount of air which is ducted around the core of the engine. A low-bypass turbofan generally has a high specific thrust. A typical modern turbofan engine is illustrated in Fig. 5.2. The flow capacity of the fan is designed to be substantially larger than the compressor, and the excess air is bypassed around the combustor and exhausted through a separate nozzle. The bypass approach reduces engine-specific thrust but increases propulsion efficiency thereby reducing fuel consumption and is the engine that is used in most subsonic commercial aircraft. A component-model-based block diagram of a modern turbofan engine is illustrated in Fig. 5.3. The turboprop or turboshaft engine, illustrated in Fig. 5.4, includes a turbojet engine as its core and a power turbine but has no fan. Its power turbine can drive an external propeller or helicopter rotor through a gear reduction unit. The rotor or propeller further increases total engine air flow, decreases the specific

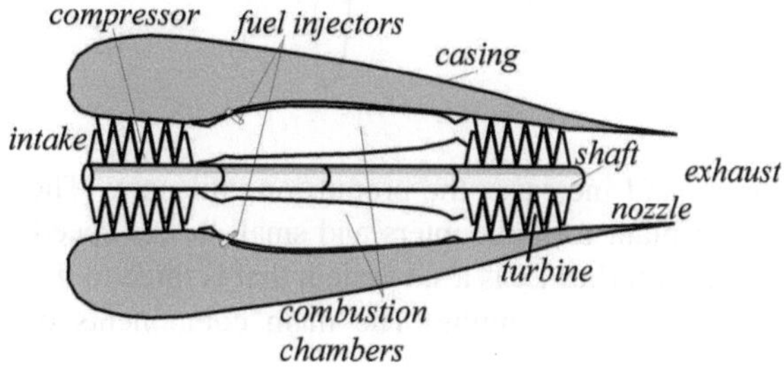

Fig. 5.1 Illustration of the components of basic turbojet engine

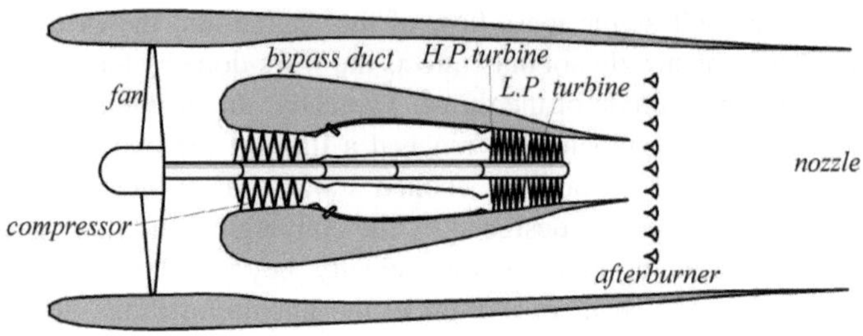

Fig. 5.2 Illustration of the components of a modern turbofan engine

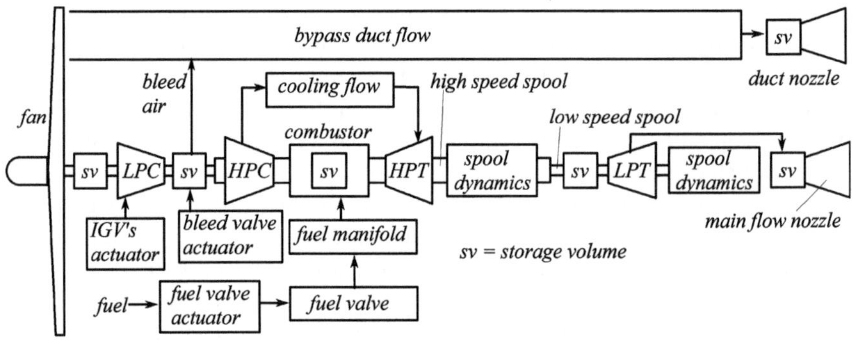

Fig. 5.3 Component-model-based block diagram of a modern turbofan engine

Fig. 5.4 Illustration of the concept of a turboprop engine

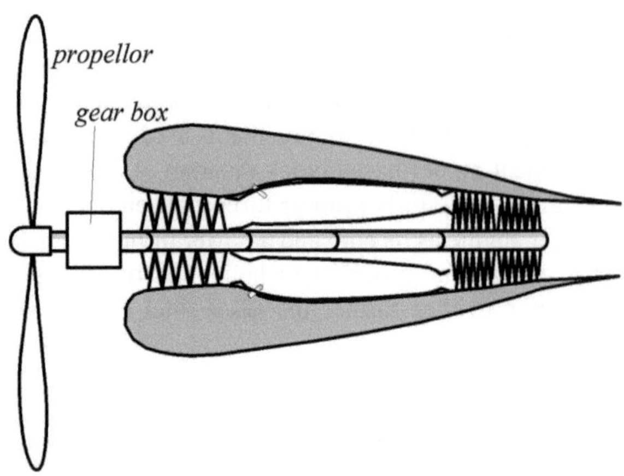

thrust, and increases the propulsion efficiency. The turboshaft is the best type of power plant for helicopters and small, lower-speed commercial aircraft.

The afterburner is a subsystem that is fitted to the exit of low-pressure turbine in the gas turbine engine. The main components of afterburner are diffuser that reduces the flow velocity, screech liner to attenuate any transverse oscillations that may be present, fuel manifolds to provide proper distribution of the fuel and a flame stabilizer that anchors the flame in the recirculation zone. Afterburning is a method of increasing the basic thrust produced by a nozzle in a jet engine over a short period of time in order to increase the thrust generated during aircraft take off, climb, or during combat. The fuel injected into a combustion chamber generally burns in an excess amount of air. Thus, the fuel–air mixture exhaust from a combustion chamber has sufficient amounts of oxygen to support further combustion, but not enough fuel. Because of this, it is possible to inject additional fuel for burning downstream of the turbine, so as to increase the overall thrust generated by the jet engine. Afterburning of injected fuel into a jet pipe prior to the flow entering the nozzle provides the best method of thrust augmentation for short periods of time. An afterburning jet pipe is larger than a normal jet pipe would be so as to obtain a reduced velocity gas stream. Therefore, an afterburning jet pipe is terminated by either a two-position or a variable area nozzle to provide for operation under all conditions. The nozzle will open or close to give an exit area suitable for the volume of the gas stream, which prevents any pressure increase from occurring in the jet pipe. Afterburning is primarily achieved on low-bypass engines by mixing the bypass and turbine streams before the fuel injection into the afterburner. In such a situation, combustion takes place in the mixed exhaust stream. Alternately, the fuel is injected and the flame stabilized independently in the bypass and turbine streams. Following combustion, the gases reach a common exit temperature at the nozzle's exit.

5. In the case of electricity generating gas turbines, a high-power generator is also mounted on the same shaft as one of the turbine and compressor pairs.

5.2 Axial Flow Compressor Systems: Modeling as a One-Dimensional Duct

The modeling of typical axial flow compressor begins with a typical compressor map describing its steady-state performance. This a family of curves for different rotor speeds, where each curve is plot of the compressor pressure ratio for increasing values of the mass flow rate. The compressor map is best obtained experimentally. By choosing the steady operating point in the map, which must lie on one of the curves, as the reference and defining a non-dimensional pressure ratio and mass flow variables, all the curves in the map may be represented by a single cubic polynomial of the form,

$$\Psi_{\text{comp}}(\Phi) = \Psi_{C0} + H\left(1 + \frac{3}{2}\left(\frac{\Phi - \Phi_0}{F} - 1\right) - \frac{1}{2}\left(\frac{\Phi - \Phi_0}{F} - 1\right)^3\right) \tag{5.2.1}$$

where $\Psi_{\text{comp}}(\Phi)$ is the non-dimensional compressor pressure ratio which is a function of the non-dimensional mass flow rate variable, Φ, Φ_0 and Ψ_{C0} are the non-dimensional mass flow rate and the non-dimensional compressor pressure ratio at the steady operating point, and H and F are two parameters which define the shape of the compressor characteristic. When $\Phi = \Phi_0$, $\Psi_{\text{comp}}(\Phi_0) = \Psi_{C0}$, which implies that the characteristic defined with reference to an origin located at the steady operating point. The mass flow rate and the pressure jump are both assumed to be one-dimensional. In general in duct, this implies that the variables are circumferentially averaged.

5.2.1 Actuator Disk Theory

For most performance assessments, the static compressor characteristic is adequate. Often the characteristic is supplemented by a flow model referred to as the actuator disk model. The basic idea of the actuator disk model, developed in detail by Hawthorne and Horlock (1962) in connection with compressors, is to replace the compressor by a permeable disk of equivalent area where the forces from the compressor blades are uniformly distributed on the circular disk. In fact, in the actuator disk approach, the compressor is represented as an infinitely thin disk with given distribution of external forces on the disk surface. The distributed forces on the actuator disk depend on the local velocities through the disk and in general the entire flow field around the rotor disk. The actuator disk is assumed to be a limiting case in which the number of compressor blades is infinite. The mass flux across the disk is considered to be continuous. The dynamic pressure jump across the disk is assumed to be given the pressure jump defined by the compressor map. The actuator disk theory essentially ignores all radial dynamics. Given the flow velocity on the upstream side of the actuator disk and given the dynamic pressure rise, it is now possible to compute the flow velocity on the downstream side of the disk. Moreover, it can be shown that the velocity at the disk itself is the mean of the velocities at the inlet and outlet sides.

5.3 The Moore–Greitzer Model

Moore and Greitzer (1986) developed a phenomenological model for rotating stall and surge. The model, which is illustrated in Fig. 5.5, is obtained by serially connecting four components: The first component is the inlet duct that allows infinitesimally small disturbances at the duct entrance to grow until they reach an appreciable magnitude at the compressor face. The second component is the compressor itself, physically modeled as an actuator disk associated with a non-dimensional compressor map, which raises the pressure ratio by doing work on the

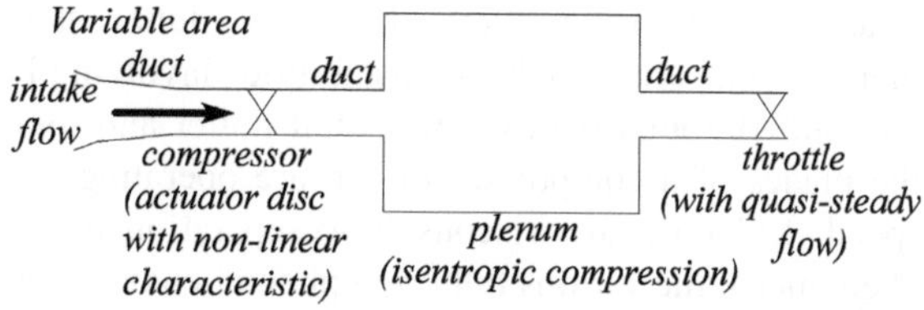

Fig. 5.5 The Moore–Greitzer compressor model

fluid. The third component is the plenum chamber (or diffuser) downstream, which acts as a large reservoir or as a Helmholtz resonator and responds to fluctuations in mass flow with fluctuations in pressure behind the actuator disk. The fourth and the final component is a throttle valve representing the load on the compressor.

The Moore–Greitzer model gives rise to a finite set of ordinary differential equations which can be solved for three important flow variables. The first of these variables is the non-dimensional, circumferentially averaged one-dimensional mass flow rate through the compressor, and the second for the non-dimensional total-to-static pressure rise across the compression system, and the third is the amplitude of the non-axisymmetric mass flow rate fluctuations which is indicative of rotating stall. The underlying assumptions of the Moore–Greitzer model are that (1) it is valid under small perturbations of the amplitude of non-axisymmetric mass flow rate fluctuations and (2) the time scale of the dynamics governing the amplitude of non-axisymmetric mass flow rate fluctuations caused by rotating stall is much faster than the time scale of the dynamics governing perturbations to the circumferentially averaged axisymmetric mass flow rate. The Moore–Greitzer model is a small perturbation model that cannot explain the underlying non-linear mechanism that triggers the instabilities in a compressor. However, the model can be used to predict the conditions under which the disturbances will grow and allow the designer to synthesize parametric alterations or feedback controls to control and suppress the instabilities. Furthermore, since rotating stall is generally a precursor to surge, prediction of the incidence of rotating stall is also an indication of the initiation and development of surge. Moreover, one can design two sets of controls, first one could adjust the steady characteristics defined by the compressor and throttle maps to guarantee stability while second one could design a physical unsteady feedback control designed to inject air into the system either in front of the duct or into the plenum itself which in turn could inhibit rotating stall. Altering the dynamic characteristics of the Helmholtz resonator or the plenum would be another novel and practical way of controlling the instabilities.

5.3.1 Compressor Surge and Rotating Stall

Compressor surge and rotating stall vibrations place fundamental limitations on aircraft engine performance and remain persistent problems in the development of axial compressor and fan stages. Compressor surge and rotating stall are purely fluid mechanic instabilities, while blade flutter, stall flutter, and surge flutter and their variants are aeroelastic instabilities involving both blade vibrations and fluid motion. Although both rotating stall flutter and rotating stall tend to occur when the blades of a compressor or fan are operating at high incidence angles and/or speed, and unsteady viscous flow separation plays a key roll in both of these phenomena, the various fluttering phenomenon are precursors to compressor surge.

Surge is characterized by large amplitude fluctuations of the pressure in unsteady, circumferentially uniform, annulus-averaged mass flow. It is a one-dimensional instability that spreads through the compression system as a whole and culminates in a limit cycle oscillation in the compressor map. In most situations, surge is initiated in a compressor when the compressor mass flow is obstructed and throttled. The frequency of surge oscillations is relatively in a low-frequency band (<25–30 Hz) which could couple with the aeroelastic modes of vibration. The performance of the compressor in surge is characterized by a loss in efficiency leading to high aeroelastic vibrations in the blade as well as influences the stress levels in the casing. In jet engines, surge can lead to the so-called flame-out of the combustor which could involve reverse flow and chaotic vibrations.

Based on the amplitude of mass flow and pressure fluctuations, surge was classified into four distinct categories: mild surge, classical surge, modified surge, and deep surge by de Jager (1995). This classification is now widely accepted and is used to differentiate between different forms of surge and rotating stall vibrations. During mild surge, the frequency of oscillations is around the Helmholtz frequency associated with the resonance within a cavity, i.e., the resonance frequency of the compressor duct and the plenum volume connected to the compressor. This frequency is typically over an order of magnitude smaller than the maximal rotating stall frequency which is normally of the same order as the rotor frequency. Classical surge is a non-linear phenomenon such as bifurcation and chaos with larger oscillations and at a lower frequency than mild surge, but the mass flow fluctuations remain positive. Modified surge is a mix of both classical surge and rotating stall. Deep surge, which is associated with reverse flow over part of the cycle, is associated with a frequency of oscillation well below the Helmholtz frequency and is induced by transient non-linear processes within the plenum. Mild surge may be considered as the first stage of a complex non-linear phenomenon which bifurcates into other types of surge by throttling the flow to compressor to lower mean mass flows. Mild surge is generally a relatively low-frequency phenomenon ($\approx$5–10 Hz) while rotating stall is a relatively higher-frequency phenomenon ($\approx$25–30 Hz).

There two modes of stable control of a compressor, the first is based on surge avoidance which involves by operating the compressor in instability free domain (Epstein et al. 1989; Gu et al. 1996). Most control systems currently used in industry are based on this control strategy. In this simple strategy, a control point is defined in parameter space with a redefined stability margin from the conditions for instability defined in terms of stall point. This stability margin is defined by (1) typical uncertainties in the location of the stall point, (2) typical disturbances including load variations, inlet distortions, and combustion noise, and (3) a consideration of the available sensors and actuators and their limitations. Generally, a bleed valve or another form of bleeding or recycling of the flow is used to negate the effect of throttling the flow. The control is either the valve position or if one employs an on/off approach as in pulse width modulation, the relative full opening times of the bleed valve in a cycle. Such an approach achieves stability at the expense of performance and the approach is not particularly suitable when the flow is compressible. In short, the surge avoidance approach is not performance optimal. There are also problems associated with the detection of instability. The second mode of control involves continuous feedback control of the mass flow by introducing a control valve or an independently controlled fan. This method involves stability augmentation as the changes in the mass flow will effectively change the conditions for instability and thus increase the stability margin. Rather than operating away from the domain of instability, the domain is pushed further away from the operating point. Based on the experiments performed by a number of earlier researchers (see for example Greitzer 2009), a 20 % increase in mass flow is deemed achievable by this means of stability augmentation.

Several attempts have been made to incorporate the influence of blade dynamics into model for stall prediction. Compressor surge by itself places a fundamental limitation on performance. Hence, active control methods that tend to suppress the various forms of stall will allow the system to be effectively employed over the parameter space prior to the occurrence of surge. Moreover, it is important to consider the various forms of stall in a holistic and integrated fashion as it would be quite impossible to design individual control systems to eliminate each of the individual instabilities. To this end, it is also important to develop a holistic and integrated dynamic model. The model developed by Moore and Greitzer (1986) based on the assumptions that the system is incompressible except in a plenum which is assumed to enclose the compressor and turbine stages, and that radial variations are unimportant, represents compressor surge as a Helmholtz-type hydrodynamic instability. In the original Moore–Greitzer model, an empirical, semi-actuator disk representation of the compressor was used, incorporating Hawthorne and Horlock's (1962) original actuator disk model of an axial compressor and it served as the basic model incorporating rotating stall. By introducing a semi-empirical actuator disk theory into the model, Moore and Greitzer were able to predict rotating stall and surge. The advantage of the Moore–Greitzer model is the analyst ability to incorporate a host of hysteresis models into the compressor characteristics that permit the prediction of a variety of limit cycle response characteristics. Gravdahl and Egeland (1997) extended the Moore–Greitzer model by including the spool dynamics and the input torque into the same framework as the original model, thus permitting the inclusion of the control inputs into the dynamics. The models may be derived by the application of finite volume type analysis and may also be extended to the case of rotating stall instability and rotating stall induced flutter. In the Moore–Greitzer model, the downstream flow field is assumed to be a linearized flow with vorticity, so a solution of a form similar to the upstream solution can be found. The plenum chamber is assumed to be an isentropic compressible chamber in which the flow is negligibly small and perturbations are completely mixed and distributed. Thus, the plenum acts merely as a "fluid spring." The throttle is modeled as a simple quasi-steady device across which the drop in pressure is only a function of the mass flow rate. Flow variations across the compressor are subject to fluid-inertia lags in both the rotor and the stator, and these lags determine the rotation rate of rotating stall. Stability of rotating stall is determined by the slope of the compressor total-to-static pressure rise map. Greitzer (2009) discussed the possibility of the active control of both stall and rotating stall by controlling the relevant Helmholtz cavity resonance frequencies which could be achieved by structural feedback.

Apart from the numerous methods of synthesizing control laws that have been proposed by the application of linear control law synthesis methods, which are only suitable for the guaranteed stabilization of mild surge, a few non-linear control law synthesis methods have also been proposed. In order to design an active feedback controller that can control deep surge, an inherently non-linear surge-control model is essential. A number of non-linear models have been proposed (Chen et al. 1988; Krstic et al. 1995; Nayfeh and Abed 1999; Paduano et al. 1993; Young et al. 1998), but almost all of these are oriented toward rotating stall control synthesis and include the dynamics of the amplitude of the leading circumferential mode. Many of these models (Gu et al. 1997; Hõs et al. 2003) have been employed to perform a bifurcation analysis to explore the behavior of the post-instability dynamics.

In this section, an unsteady non-linear and extended version of the Moore–Greitzer model is developed to facilitate the synthesis of a surge and stall controller. The motivation is the need for a comprehensive and yet low-order model to describe the various forms of stall as well as the need to independently represent the transient disturbance and control inputs in the compressor pressure rise dynamics. Furthermore the extended version of the Moore–Greitzer model is developed by reducing the number of independent model parameters to a minimum. Our preliminary studies indicate that model can effectively capture the dynamics of the phenomenon of compressor surge and that its post-stall instability behavior is well representative of the observed behaviors in real axial flow compressors. The controller is synthesized in two steps. In the first step, the desired equilibrium throttle position and the desired equilibrium value of the ratio of the non-dimensional pressure rise at minimum flow to a quarter of the peak to peak variation of the pressure fluctuation at the compressor exit are established. This defines the equilibrium point and ensures that the desired equilibrium point is stable. In the second step, the margin of stability at the equilibrium point is tuned or increased by an appropriate feedback of change in the mass flow rate about the steady mass flow rate at the compressor exit. The first step may be considered to be an equilibrium point controller while the second corresponds to stability augmentation. Such a two-step process then ensures that both the desired equilibrium solution is reachable and that any perturbations about the equilibrium point are sufficiently stable.

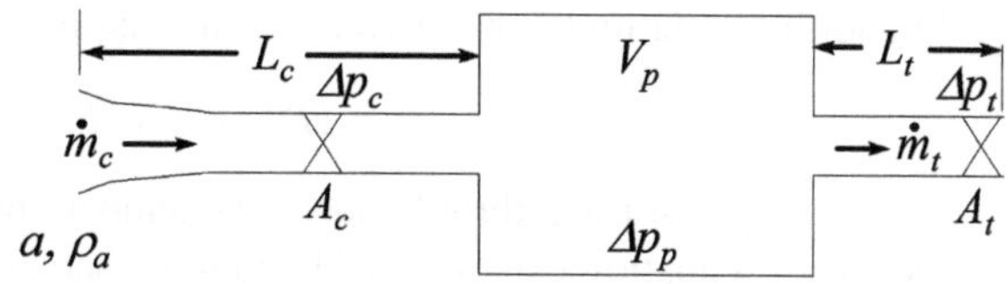

Fig. 5.6 Primary variables in the Moore–Greitzer compressor model

5.3.2 Derivation of the Moore–Greitzer Model Equations

The primary Moore–Greitzer model parameters are defined in Fig. 5.6. The plenum is connected to the atmosphere by a duct, in which the compressor works. If the length of the interconnecting duct is relatively small compared to the wavelength of the pressure oscillations, the gas in the duct can be assumed to be incompressible. The compressibility effects are only associated with the compression in the plenum while inertia effects are lumped on the acceleration of the gas in the compressor duct. The compressor is replaced by an actuator disk and a straight duct of length L_c. The geometry of the compressor duct is assumed to be such that a given mass flow change leads to the same pressure difference in the actual duct and as in the model. The throttle behavior is modeled as a simple valve with a known flow characteristic.

The following primary assumptions are made in the derivation of the model:

1. The flow in the ducts is one-dimensional and incompressible.
2. The behavior of the fluid in the plenum can be represented by an isentropic compression process.
3. In the plenum, the velocity is negligible while pressure is uniformly distributed.
4. The compressor and throttle behavior are assumed to be quasi-steady.
5. The influence of rotor speed variations is assumed to negligible.
6. The temperature across the system is assumed to be almost the same.

To describe the one-dimensional flow of the gas in the ducts and in the plenum, principles of conservation of mass, momentum, and energy are applied. The general unsteady, equilibrium, dynamics of quasi-one-dimensional flow with no heat addition or frictional losses may be expressed by the continuity, conservation of momentum and conservation of energy equations defined in Chap. 1 as,

$$\frac{\partial}{\partial t}(\rho A) + \frac{\partial}{\partial x}(\rho A u) = 0, \tag{5.3.1}$$

$$\frac{\partial}{\partial t}(\rho A u) + \frac{\partial}{\partial x}(\rho A u^2) = -A\left(\frac{\partial p}{\partial x} + \rho g_u\right), \tag{5.3.2}$$

$$\frac{\partial}{\partial t}(\rho A h) + \frac{\partial}{\partial x}(\rho A u h) = \frac{\partial}{\partial t}(pA) + \frac{\partial}{\partial x}(pAu) - \frac{\partial q}{\partial x}. \tag{5.3.3}$$

In addition, relations for a perfect gas are applied:

$$p = \rho RT,\ h = c_p T. \tag{5.3.4}$$

In the compressor and throttle ducts, the temperature is assumed to be uniform, so the energy equation can be considered to be satisfied. Furthermore, the flow in the ducts is assumed to be incompressible. Consequently, the continuity or mass conservation equation is satisfied. Moreover, the velocity fluctuations associated with surge are supposed to have small magnitudes, so the second-order terms in velocity may be ignored. Thus, applying the one-dimensional momentum equation to the incompressible flow in the ducts gives:

$$\rho A \frac{\partial u}{\partial t} = -A \frac{\partial p}{\partial x}. \tag{5.3.5}$$

Integrated over the compressor duct volume, this equation becomes:

$$\frac{L_c}{A_c} \frac{\mathrm{d}\dot{m}_c}{\mathrm{d}t} = -\left(p_{\text{plenum}} - p_{\text{atmos}} - \Delta p_c\right) \tag{5.3.6}$$

where A_c is the average cross-sectional area of the compressor duct, $\dot{m}_c = \rho A_c u$ is the mass flow rate through the compressor, p_{plenum} is the pressure in the plenum chamber, p_{atmos} is the atmospheric pressure at the intake, and Δp_c is the discrete pressure rise across the compressor actuator disk.

Consider fluid flowing through a one-entrance, one-exit throttle. The fluid experiences a pressure drop as it flows through the throttle while no net work is done by the fluid. If one assumes the process is adiabatic and that the kinetic and potential energies are neglected, the conservation of mass and energy equations are satisfied. This process is called a throttling process, and when throttling an ideal gas, the temperature does not change. Similar to the compressor duct, the flow in the throttle duct can be described by:

$$\frac{L_t}{A_t} \frac{\mathrm{d}\dot{m}_t}{\mathrm{d}t} = \left(p_{\text{plenum}} - p_{\text{atmos}} - \Delta p_t\right) \tag{5.3.7}$$

where L_t is the length of the throttle duct, A_t is the average cross-sectional area of the throttle duct, $\dot{m}_t = \rho A_t u$ is the mass flow rate through the throttle, and Δp_t is the discrete pressure drop across the throttle.

In the plenum chamber where the flow velocity can be neglected and the temperature is uniform, only the continuity equation must be satisfied and it follows that

$$\frac{\partial}{\partial t}(\rho A) = -\frac{\partial}{\partial x}(\rho A u). \tag{5.3.8}$$

Integrating over one one-dimension, within the plenum chamber one has,

$$V_p \frac{\mathrm{d}\rho_p}{\mathrm{d}t} = -(\dot{m}_t - \dot{m}_c) \tag{5.3.9}$$

where V_p is the volume of the plenum chamber and ρ_p is the density within the plenum chamber. Since the fluid is isentropic within the chamber, it can be shown that

$$\frac{d\rho_p}{dt} = \frac{1}{\gamma} \frac{\rho_p}{p_{\text{plenum}}} \frac{dp_{\text{plenum}}}{dt}. \tag{5.3.10}$$

Consequently in the plenum chamber,

$$\frac{1}{\gamma} \frac{V_p \rho_p}{p_{\text{plenum}}} \frac{dp_{\text{plenum}}}{dt} = -(\dot{m}_t - \dot{m}_c). \tag{5.3.11}$$

Using the equation of state, $p = \rho RT$ and the fact that the speed of sound satisfies $a = \sqrt{\gamma RT} = \sqrt{\gamma p/\rho}$,

$$\frac{V_p}{a_p^2} \frac{dp_{\text{plenum}}}{dt} = -(\dot{m}_t - \dot{m}_c). \tag{5.3.12}$$

Scaling the mass flow rates by $\rho_c A_c U_t$ and the pressures by $(1/2)\rho_c U_t^2$ the duct flow and plenum pressure equations can be expressed as,

$$\frac{d(\dot{m}_c/\rho_c A_c U_t)}{dt} = -\frac{2(p_{\text{plenum}} - p_{\text{atmos}} - \Delta p_c)/\rho_c U_t^2}{2L_c/U_t}, \tag{5.3.13}$$

$$\frac{d(\dot{m}_t/\rho_c A_c U_t)}{dt} = \frac{2(p_{\text{plenum}} - p_{\text{atmos}} - \Delta p_t/\rho_c U_t^2)}{(2L_c/U_t)(L_t A_c/L_c A_t)}, \tag{5.3.14}$$

$$\frac{d(2p_{\text{plenum}}/\rho_c U_t^2)}{dt} = -\frac{(\dot{m}_t - \dot{m}_c/\rho_c A_c U_t)}{U_t V_p/2A_c a_p^2}. \tag{5.3.15}$$

In the above equations, one may assume that the speed of sound in the plenum is equal to the speed of sound in ambient conditions, $a_p = a$. Considering the plenum chamber, the Helmholtz natural frequency for the compressor duct and cavity combination is given by,

$$\omega_H = a\sqrt{\frac{A_c}{V_p L_c}} = \frac{a}{L_c}\sqrt{\frac{A_c L_c}{V_p}}. \tag{5.3.16}$$

Hence, the duct flow and plenum pressure equations can be expressed as,

$$\frac{d(\dot{m}_c/\rho_c A_c U_t)}{\omega_H dt} = -\frac{2(p_{\text{plenum}} - p_{\text{atmos}} - \Delta p_c)/\rho_c U_t^2}{2\omega_H L_c/U_t}, \tag{5.3.17}$$

$$\frac{d(\dot{m}_t/\rho_c A_c U_t)}{\omega_H dt} = \frac{2(p_{\text{plenum}} - p_{\text{atmos}} - \Delta p_t)/\rho_c U_t^2}{(2\omega_H L_c/U_t)(L_t A_c/L_c A_t)}, \tag{5.3.18}$$

$$\frac{\mathrm{d}\left(2p_{\text{plenum}}/\rho_c U_t^2\right)}{\omega_H \mathrm{d}t} = -\frac{(\dot{m}_t - \dot{m}_c)/\rho_c A_c U_t}{U_t/(2\omega_H L_c)}. \tag{5.3.19}$$

5.3.3 The Moore–Greitzer Model Equations

The unsteady and steady fluid mechanics of the flow upstream and downstream of the compressor is considered while the viscous effects are limited to within the actuator disk of the compressor which allows one to define non-dimensional total-to-static pressure rise map. Compressibility is assumed to be confined to the plenum chamber downstream of the compressor where the compression is assumed to be uniform and isentropic. The throttle map sets the mass flow through the system and is function of the plenum pressure and the throttle opening. It is essential in defining the flow characteristics of the compressor. The rate of change of the plenum pressure is determined from the one-dimensional continuity conditions and is a function of difference in the compressor flow averaged over the face of the compressor and the throttle flow. The second equation is defined by the one-dimensional rate of change of momentum which relates to the dynamic pressure. Two other equations complete the definition of the complete dynamics of the Moore–Greitzer model; the first relates to the rate of change of the throttle flow and the second defines the compressor dynamics and is based on an unsteady adaptation of the actuator disk model. These equations were first proposed by Greitzer (1976).

The dimensionless compressor mass flow is assumed to be ϕ_c and ψ is the dimensionless plenum pressure rise. Furthermore, $\Psi_{c,\text{ss}}$ is the dimensionless steady-state compressor pressure rise given in the compressor map, whereas Ψ_c is the dimensionless dynamic compressor pressure rise. The dimensionless throttle mass flow is ϕ_t and dimensionless pressure drop across the throttle is Ψ_t.

$$\frac{1}{B}\frac{\mathrm{d}}{\mathrm{d}\tau}\phi_c = \Psi_c - \psi, \tag{5.3.20}$$

$$\frac{G}{B}\frac{\mathrm{d}}{\mathrm{d}\tau}\phi_t = \psi - \Psi_t, \tag{5.3.21}$$

$$B\frac{\mathrm{d}}{\mathrm{d}\tau}\psi = \phi_c - \phi_t, \tag{5.3.22}$$

$$\tau_c\frac{\mathrm{d}}{\mathrm{d}\tau}\Psi_c = \Psi_{c,\text{ss}} - \Psi_c, \tag{5.3.23}$$

where $\phi_c = \dot{m}_c/(\rho_a A_c U_t)$, $\phi_t = \dot{m}_t/(\rho_a A_c U_t)$, $\Psi_c = 2\Delta p_c/\left(\rho_a U_t^2\right)$, $\Psi_t = 2\Delta p_t/\left(\rho_a U_t^2\right)$, B is the Greitzer parameter given by, $B = U_t/2\omega_H L_c$, $\omega_H = a\sqrt{A_c/\left(V_p L_c\right)}$ is the Helmholtz cavity resonance frequency for the plenum, τ is the non-dimensional time defined in terms of the Helmholtz frequency and the time t, in seconds as $\tau = \omega_H t$, G is the geometry ratio parameter of the throttle duct and control volume given by $G = (L_t/A_t)/(L_c/A_c)$ and τ_c is the time constant of the compression system that would be different for stall and for rotating stall. In the preceding definitions of the model parameters, $\dot{m}_c$ is the mass flow rate through the compressor, $\dot{m}_t$ is the mass flow rate through the throttle, Δp_c is the pressure rise across the compressor, Δp_t is the pressure drop across the throttle, ρ_a is the ambient air density, a is the speed of sound corresponding to ambient conditions, A_c is the cross-sectional area of the control volume, L_c is the length of the control volume, A_t is the cross-sectional area of the throttle duct, L_t is the length of the throttle duct, V_p is the volume of the plenum chamber, and U_t is the rotor tip speed.

The compressor map in steady flow is a plot of the non-dimensional pressure with the non-dimensional mass flow rate through the compressor for each rotation speed. However, the plots are self-similar and can be reduced to single plot by scaling the non-dimensional mass flow rate and the non-dimensional dynamic pressure rise. The compressor surge line is obtained simply by linking the maximum point on each compressor characteristic for a particular rotational speed. Representing the compressor characteristics in a non-dimensional manner for each of rotation speeds and appropriately scaling the axes, simply reduces the "surge line" to a single point which is the maximum point on the characteristic. Following, Hõs et al. (2003) the scaled compressor map in steady flow when $\phi_c = \phi_{\text{cs}}$ is assumed to be

$$\Psi_{c,\text{ss}}(\phi_{\text{cs}}) = \Psi_{c0} + \frac{H}{2}\left(2 + 3\left(\frac{\phi_{\text{cs}}}{F} - 1\right) - \left(\frac{\phi_{\text{cs}}}{F} - 1\right)^3\right). \tag{5.3.24}$$

In (2), H defines half the peak to peak variation of the pressure fluctuation at the compressor exit or the amplitude of the pressure fluctuation while F is half the change in the steady mass flow rate, ϕ_{cs} required for the pressure to change from the minimum to the maximum. The definitions of the parameters H and F are illustrated in Fig. 5.7.

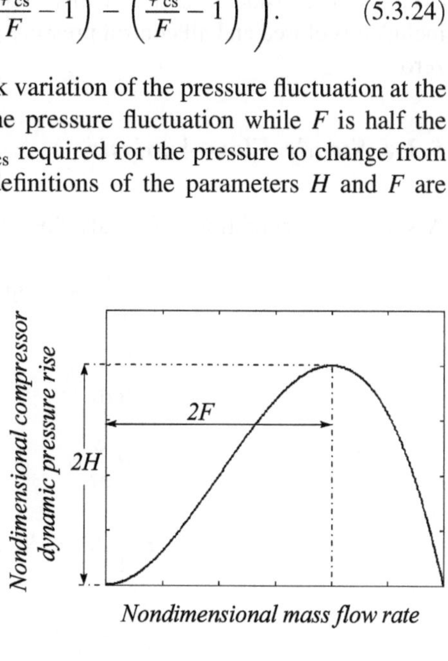

Fig. 5.7 Definitions of the compressor characteristic parameters H and F

The throttle map in steady flow when $\phi_t = \phi_{\text{ts}}$ is taken to be:

$$\Psi_{t,\text{ss}} = (\phi_{\text{ts}}/C_t\gamma)^2, \tag{5.3.25}$$

where the dimensionless throttle parameter C_t is a coefficient defining the capacity of the fully opened throttle and γ is the dimensionless throttle position.

Following Gravdahl and Egeland (1997), the input torque to the compressor may be included and the dynamics of the spool as another state equation given by,

$$I(\mathrm{d}\omega/\mathrm{d}t) = T_{\text{ext}} - T_c, \tag{5.3.26}$$

where I is the mass moment of inertia of the compressor rotor, ω the angular velocity which may be expressed in terms of the Greitzer parameter and tip radius as $\omega = U_t/R_t = 2\omega_H L_c B/R_t$, T_{ext} is the external torque input and T_c is the torque necessary to drive the compressor which may be expressed in terms of the slip ratio σ, as $T_c = \rho_a A_c U_t^2 R_t \phi_c \sigma$. The slip ratio σ can be defined as the ratio of the tangential velocity of the fluid at the compressor exit guide vanes and the tip speed. The external torque may be expressed in non-dimensional form as $\Gamma_{\text{ext}} = T_{\text{ext}}/\rho_a A_c U_t^2 R_t$. Hence, (4) may be expressed in non-dimensional form as,

$$\mathrm{d}B/\mathrm{d}\tau = B^2(\Gamma_{\text{ext}} - \phi_c\sigma)/\mu, \tag{5.3.27}$$

where $\mu = I/2\rho_a R_t^2 A_c L_c$ is the non-dimensional inertia parameter, and Γ_{ext} is the non-dimensional torque input.

In this analysis, all controls are initially assumed to be fixed as the uncontrolled dynamics is considered first. For this reason, any bleed valve that may have been included is closed and all control pressure perturbations are assumed to be equal to zero.

5.3.4 Steady Flow Analysis

Assuming the conditions of steady flow, the equations are as follows:

$$\frac{1}{B}\frac{\mathrm{d}}{\mathrm{d}\tau}\phi_c = \Psi_c - \psi = 0, \tag{5.3.28}$$

$$\frac{G}{B}\frac{\mathrm{d}}{\mathrm{d}\tau}\phi_t = \psi - \Psi_t = 0, \tag{5.3.29}$$

$$B\frac{\mathrm{d}}{\mathrm{d}\tau}\psi = \phi_c - \phi_t = 0, \tag{5.3.30}$$

$$\tau_c\frac{\mathrm{d}}{\mathrm{d}\tau}\Psi_c = \Psi_{c,\text{ss}} - \Psi_c = 0, \tag{5.3.31}$$

$$\frac{\mathrm{d}B}{\mathrm{d}\tau} = \frac{B^2}{\mu}(\Gamma_{\text{ext}} - \phi_c\sigma) = 0. \tag{5.3.32}$$

From the third of the above equations, Eq. 5.3.30, in steady flow let,

$$\phi_{cs} = \phi_{ts} = \phi_{s0}. \tag{5.3.33}$$

The steady flow conditions are obtained from the first two of the above Eqs. 5.3.28 and 5.3.29 and are given by $\Psi_{c,ss} = \Psi_{t,ss}$; i.e.,

$$\Psi_{c,ss}(\phi_{s0}) = \Psi_{c0} + \frac{H}{2}\left(2 + 3\left(\frac{\phi_{s0}}{F} - 1\right) - \left(\frac{\phi_{s0}}{F} - 1\right)^3\right) = (\phi_{s0}/C_t\gamma)^2 \tag{5.3.34}$$

A parameter p, is defined as,

$$p = (2/H)(F/C_t\gamma_n)^2, \tag{5.3.35}$$

where p is the throttle non-dimensional pressure rise at minimum flow and a parameter p_0,

$$p_0 = \frac{2}{H}\Psi_{c0} + 2 \tag{5.3.36}$$

which is the ratio of the non-dimensional pressure rise at minimum flow to a quarter of the peak to peak variation of the pressure fluctuation at the compressor exit; then, Eq. 5.3.24 reduces to,

$$\Psi_{c,ss}(\phi_{s0}) = \frac{H}{2}\left(p_0 + 3x - x^3\right) = p\frac{H}{2}(1 + x)^2, \tag{5.3.37}$$

where the variable x is,

$$x = (\phi_{s0}/F) - 1. \tag{5.3.38}$$

If one assumes that with the minimum flow through the compressor and the throttle, the flow is always steady, then with $\phi_{s0}/F = 1$, and one obtains from Eq. 5.3.35,

$$p_0 = p. \tag{5.3.39}$$

Assuming that the position of the throttle γ is set to a nominal value $\gamma = \gamma_n$ when Eqs. 5.3.37–5.3.39 are satisfied, Eq. 5.3.37 may be rearranged and written as,

$$\Psi_{c0} = \frac{H}{2}(p - 2). \tag{5.3.40}$$

Eliminating Ψ_{c0}, the steady flow characteristics may be defined entirely in terms of the compressor and throttle map parameters, H, F, and the product $\gamma_n C_t$ and is:

$$3x - x^3 = px(x + 2). \tag{5.3.41}$$

and Eq. 5.3.41 may be expressed as,

$$x\left(x^2 + px + 2p - 3\right) = 0. \tag{5.3.42}$$

From the first factor of Eq. 5.3.42, the assumed solution $\phi_{s0}F = 1$ is recovered. Assuming $x \neq 0$ and solving for p,

$$p = \left(3 - (\varphi_{s0}/F - 1)^2\right)\Big/(\varphi_{s0}/F + 1). \tag{5.3.43}$$

If one assumes that with the flow through the compressor and the throttle either minimum or below minimum, it is always steady, then with $x = x_0$, when,

$$\frac{H}{2}\left(p_0 + 3x_0 - x_0^3\right) = p\frac{H}{2}(1 + x_0)^2. \tag{5.3.44}$$

Eliminating p_0 one obtains,

$$\left(3(x - x_0) - \left(x^3 - x_0^3\right)\right) = 2p(x - x_0) + p\left(x^2 - x_0^2\right). \tag{5.3.45}$$

Solving for p, one obtains,

$$p = \frac{\left(3 - \left(x^2 + xx_0 + x_0^2\right)\right)}{(x + x_0 + 2)}. \tag{5.3.46}$$

When $x_0 = 0$, Eq. 5.3.46 reduces to 5.3.43.

5.3.5 Unsteady Non-Linear Extended Moore–Greitzer Model

Rather than combine the quasi-steady and transient components of compressor pressure rise, the independent contributions from these two components of the pressure rise are separately identified. If one defines, $\Delta\Psi_c = \Psi_c - \Psi_{c,\mathrm{qs}}$, as the transient disturbance and control pressure component of the compressor pressure rise, the first three unsteady equations may be expressed as,

$$\mathrm{d}\phi_c/B\mathrm{d}\tau = \Psi_{c,\mathrm{qs}} - \psi + \Delta\Psi_c, \tag{5.3.47}$$

$$G\mathrm{d}\phi_t/B\mathrm{d}\tau = \psi - \Psi_{t,\mathrm{qs}}, \tag{5.3.48}$$

$$B\mathrm{d}\psi/\mathrm{d}\tau = \phi_c - \phi_t. \tag{5.3.49}$$

The compressor transient disturbance and control dynamics, in the absence of a control pressure input, are defined entirely in terms of, $\Delta\Psi_c$ as,

$$\tau_c\mathrm{d}\Delta\Psi_c/\mathrm{d}\tau = -\Delta\Psi_c + \Psi_{c,\mathrm{ss}} - \Psi_{c,\mathrm{qs}}, \tag{5.3.50}$$

where the unsteady compressor characteristics, $\Psi_{c,\mathrm{qs}}$ and the unsteady throttle map $\Psi_{t,qs}$, are assumed to satisfy the quasi-steady model equations given by,

$$\Psi_{c,\mathrm{qs}}(\phi_c) = \Psi_{c0} + \frac{H}{2}\left(2 + 3\left(\frac{\phi_c}{F} - 1\right) - \left(\frac{\phi_c}{F} - 1\right)^3\right), \tag{5.3.51}$$

$$\Psi_{c0} = \frac{H}{2}(p-2),\ \Psi_{t,\mathrm{qs}} = \left(\frac{\phi_t}{C_t\gamma}\right)^2. \tag{5.3.52}$$

Furthermore,

$$\Psi_{c,\mathrm{ss}}(\phi_{s0}) = \frac{H}{2}\left(p + 3\left(\frac{\phi_{s0}}{F} - 1\right) - \left(\frac{\phi_{s0}}{F} - 1\right)^3\right). \tag{5.3.53}$$

Further from the definition of the parameter, p, one may write,

$$C_t^2\gamma_n^2 = 2F^2/pH. \tag{5.3.54}$$

In Eq. 5.3.54, one considers the throttle's non-dimensional nominal position, $\gamma = \gamma_n$ to be fixed and any perturbations to it must be considered as a deviation. If $\Delta\gamma$ is the deviation of the throttle position from the nominal position, $\gamma = \gamma_n$ then in the general case, Eq. 5.3.54 may be written as,

$$C_t^2\gamma^2 = \left(\sqrt{2F^2/pH} + C_t\Delta\gamma\right)^2. \tag{5.3.55}$$

Considering the last equation for the dynamics of the compressor spool one assumes that the non-dimensional torque input, Γ_{ext} is provided by a non-dimensional power input and can be defined by $\Gamma_{\mathrm{ext}} = \Pi_{\mathrm{ext}}/B$. The equation for the spool dynamics is

$$\frac{\mathrm{d}B}{\mathrm{d}\tau} = \frac{B}{\mu}(\Pi_{\mathrm{ext}} - B\phi_c\sigma). \tag{5.3.56}$$

where the non-dimensional power input is related to the real power P_{ext} by the equation

$$\Pi_{\mathrm{ext}} = P_{\mathrm{ext}}/2\rho_a U_t^2 A_c \omega_H L_c. \tag{5.3.57}$$

In most practical situations involving jet engines, it is power that is delivered to a turbine driving the compressor by a combustor and this can be modeled independently.

Using Eqs. 5.3.47–5.3.56, the complete unsteady non-linear equations not including the control inputs may be expressed in terms of the five states ϕ_c, ϕ_t, ψ, $\Delta\Psi_c$ and B, as,

$$\frac{\mathrm{d}\phi_c}{\mathrm{d}\tau} = B\Psi_{c,\mathrm{qs}} + B(\Delta\Psi_c - \psi) \tag{5.3.58}$$

$$\frac{\mathrm{d}}{\mathrm{d}\tau}\phi_t = -\frac{B}{G}\frac{\phi_t^2}{\left(\sqrt{2F^2/pH} + C_t\Delta\gamma\right)^2} + \frac{B\psi}{G} \tag{5.3.59}$$

$$\frac{\mathrm{d}}{\mathrm{d}\tau}\psi = \frac{(\phi_c - \phi_t)}{B} \tag{5.3.60}$$

$$\frac{\mathrm{d}\Delta\Psi_c}{\mathrm{d}\tau} + \frac{\Delta\Psi_c}{\tau_c} = \frac{\left(\Psi_{c,\mathrm{ss}} - \Psi_{c,\mathrm{qs}}\right)}{\tau_c} \tag{5.3.61}$$

and

$$\frac{\mathrm{d}B}{\mathrm{d}\tau} = \frac{B}{\mu}(\Pi_{\mathrm{ext}} - B\phi_c\sigma), \tag{5.3.62}$$

with

$$p = \left(3 - (\varphi_{s0}/F - 1)^2\right)\Big/(\varphi_{s0}/F + 1). \tag{5.3.63}$$

The eight model parameters are $\phi_{s0}/F,\ H, G, \tau_c, F, C_t\Delta\gamma, \mu$ and σ. The input to the model is defined by Π_{ext}, the non-dimensional power input to the compressor.

5.3.6 Application to Rotating Stall Vibrations

The Eqs. 5.3.58–5.3.63 describe surge in our one-dimensional model but does not include rotating stall. The extension needed is derived and explained in detail by Moore and Greitzer (1986) by Galerkin projection and presents here the essence of the method. The Galerkin projection procedure represents the reduction of the differential equation by a set of basic or coordinate functions to capture the behavior in the circumferential direction with a finite set of modes. One-mode truncation via Galerkin projection results in an additional equation in terms of a new variable J that must be included with Eqs. 5.3.58–5.3.63. The square of the new variable J represents the amplitude of the asymmetric mass flow rate fluctuations. Following, Hõs, Champneys and Kullmann (2003), the dynamics of J is described by:

$$\tau_J\frac{\mathrm{d}J}{\mathrm{d}\tau} = \frac{H}{F}J\left(1 - \left(\frac{\phi_c}{F} - 1\right)^2 - \frac{1}{4}J\right) \tag{5.3.65}$$

where the time constant τ_J is related to the time constant of an N-stage compressor τ_c, and the slope of the compressor duct flow parameter m, by the relations, $\tau_J = \omega_H R(1 + ma)/3aU_t$ with $a = R/\tau_c U_t$.

The presence of rotating stall influences the compressor characteristic given by Eq. 5.2.1, and following Hõs et al. (2003), it is modified as,

$$\Psi_{c,\mathrm{ss}}(\phi_{\mathrm{cs}}) = \Psi_{c0} + \frac{H}{2}\left(2 + 3\left(\frac{\phi_{\mathrm{cs}}}{F} - 1\right)\left(1 - \frac{J}{2}\right) - \left(\frac{\phi_{\mathrm{cs}}}{F} - 1\right)^3\right). \quad (5.3.65)$$

Conditions for steady flow now require additionally that either $J = J_s = 0$, corresponding to an equilibrium with no rotating stall disturbance, or $J = J_s = 4(1 - x^2)$ corresponding to an equilibrium with a rotating stall disturbance. Since J represents the amplitude of rotating stall amplitude, to avoid rotating stall J must tend to zero. If it tends to any other finite value the rotating stall amplitude is non-zero, implying that rotating stall exists. In the case when the rotating stall amplitude is non-zero, Eqs. 5.3.35 and 5.3.36 are unchanged, but Eqs. 5.3.42 and 5.3.43 are, respectively, modified, in the case J is given by the latter non-zero equilibrium point as,

$$5x^2 - xp - 3 - 2p = 0, \quad (5.3.66)$$

$$p = \left(5(\phi_{s0}F/ - 1)^2 - 3\right)/(\phi_{s0}/F + 1). \quad (5.3.67)$$

where the definition of the parameter p is unchanged. In the model, it should be noted that the Greitzer parameter B is no longer a parameter but a slowly varying state. In this respect, our analysis is different from that of Moore and Greitzer (1986) who treated it as a parameter and stated the conditions for surge in terms of this parameter. For control applications, particularly when the external control input is due to a control torque, it is most appropriate to allow the Greitzer parameter B to vary. However, when the Greitzer parameter B is assumed to be variable, it is essential that both the compressor steady characteristic parameters, H and F, are not constants but functions of B. Based on a set of typical characteristics, the parameters, H and F, are assumed to be linear functions of the Greitzer parameter B and given by

$$H = H_0 + H_B B,\ F = F_B B \quad (5.3.68)$$

where H_0, H_B and F_B are assumed to be constants. Thus, in steady state, when $B = B_0$, H and F are given by

$$H_s = H_0 + H_B B_0,\ F_s = F_B B_0. \quad (5.3.69)$$

If one defines the change in J by $\Delta J = J - J_s$ in the unsteady case, Eqs. 5.3.58–5.3.63 are now modified as,

$$\frac{\mathrm{d}\phi_c}{\mathrm{d}\tau} = B\Psi_{c,\mathrm{qs}} + B(\Delta\Psi_c - \psi), \quad (5.3.70)$$

$$\frac{\mathrm{d}}{\mathrm{d}\tau}\phi_t = -\frac{B}{G}\frac{\phi_t^2}{\left(\sqrt{2F^2/pH} + C_t\Delta\gamma\right)^2} + \frac{B\psi}{G}, \quad (5.3.71)$$

$$\frac{\mathrm{d}}{\mathrm{d}\tau}\psi = \frac{(\phi_c - \phi_t)}{B}, \tag{5.3.72}$$

$$\frac{d\Delta\Psi_c}{d\tau} + \frac{\Delta\Psi_c}{\tau_c} = \frac{\left(\Psi_{c,\mathrm{ss}} - \Psi_{c,\mathrm{qs}}\right)}{\tau_c}, \tag{5.3.73}$$

$$\frac{\mathrm{d}B}{\mathrm{d}\tau} = \frac{B}{\mu}(\Pi_{\mathrm{ext}} - B\phi_c\sigma), \tag{5.3.74}$$

and

$$\tau_J\frac{\mathrm{d}\Delta J}{\mathrm{d}\tau} = \frac{H}{F}(J_s + \Delta J)\left(1 - \left(\frac{\phi_c}{F} - 1\right)^2 - \frac{1}{4}(J_s + \Delta J)\right), \tag{5.3.75}$$

with

$$J_s = 0, \tag{5.3.76}$$

or

$$J_s = 4\left(1 - ((\phi_{s0}/F) - 1)^2\right), \tag{5.3.77}$$

where the parameter F is evaluated under steady conditions. Only the former is used and it also required the equilibrium point to be stable. Moreover, there is now an additional parameter τ_J, which may be related to τ_c as,

$$\tau_J = \tau_c\omega_H(1 + mR\tau_c/U_t)/3, \tag{5.3.78}$$

but will be treated as an independent parameter. Equations 5.3.70–5.3.78 represent a six-state dynamic model of the dynamics of the compressor system.

5.3.7 Model Response and Instability

Although our primary interest is in establishing a non-linear model for synthesizing an active surge controller, one needs to understand the dynamic response of the uncontrolled model not only in the vicinity of the domain of instability but also in the post-instability domains in the parameter space. For this reason, the dynamic response of the model proposed in the preceding section is considered, without including any controls which could include a bleed valve or a feedback controller that influences the transient dynamics of the compressor. The rotating stall dynamics is ignored in the first instance.

Table 5.1 lists the nominal typical values of the parameters, initial values of the states and the inputs used in the simulation of the dynamic response, for which the system was stable. The parameter p is not shown in Table 5.1 as it is computed from the parameters in the table. It is, however, an important parameter as a high

Table 5.1 Typical parameter and initial state values for simulation

Parameter	Primary Value	State/ Input	Initial Value
ϕ_{s0}	0.375	ϕ_c	0.4
F_B	0.625	ϕ_t	0.3
H_0	0.06	ψ	0.0
H_B	0.3	$\Delta\Psi_c$	1.0
G	2	B	0.4
σ	0.9	ΔJ	3.1 or 0.1
$C_t\Delta\gamma$	0.0	J_s	0
μ	40	Π_{ext}	0.17
τ_c	0.05	τ_J	0.5

value represents greater levels of throttling and a reduced mass flow rate through the throttle. The system was not unstable unless either H was negative or $\gamma < \gamma_n$. A typical example of a stable response is shown in Fig. 5.8 from Vepa (2010).

The first case considered was with $C_t\Delta\gamma = 0$. In this case, no chaotic behavior was observed although both stable and unstable behaviors were observed. When the compressor was stable, the behavior was always lightly damped and oscillatory. Choosing the parameter $\gamma = \gamma_n$ represents a case of tuning or matching the throttle to the compressor. In most cases, the instability could be eliminated by proper tuning of the parameters and no active stabilization was deemed necessary.

When H is locally negative, it corresponds to the case of negative slope in the characteristic that was considered by Hõs et al. (2003). When H is negative and the parameter, $\gamma > \gamma_n$, the throttle mass flow is not matched to the compressor mass flow. Although the system was unstable, no chaos was observed. When H is negative and $\gamma < \gamma_n$, there was a clear incidence of chaos in the flow through the compressor, which was identified by a one-dimensional Poincaré map. The chaotic response with a negative H is significant as it represents the case of flame-out in jet engines. However, this case is not of much practical importance for controller synthesis as the compressor becomes unstable before it becomes chaotic.

The responses of B and J, when H is negative and $\gamma > \gamma_n$ in the rotating stall case, are illustrated in Fig. 5.9. Apparently, the Greitzer "parameter" is itself stable in this case, but the sustained response in J away from the trivial equilibrium solution ($J = 0$) represents the presence of rotating stall disturbances.

Considering the case of rotating stall with H positive and $C_t\Delta\gamma = 0$, the system always exhibited stability in the sense that the response converged to a steady state. With $\gamma \neq \gamma_n$ or $\gamma = \gamma_n$, H positive and $\phi_{s0}/F < 2$ the equilibrium solution jumps from one with $J_s = 0$ to one with $J_s = 1$ and this is followed by the pressure in the plenum chamber falling to zero. The state responses in this case are illustrated in Fig. 5.10a. The corresponding unsteady compressor map and the operating point on the map are shown in Fig. 5.10b.

Although, when the compressor flow and throttle flow were matched, i.e., with $\gamma = \gamma_n$, the system is stable, and it is also important to maintain J at zero, as it represents the amplitude of the rotating stall disturbance amplitude. It can be

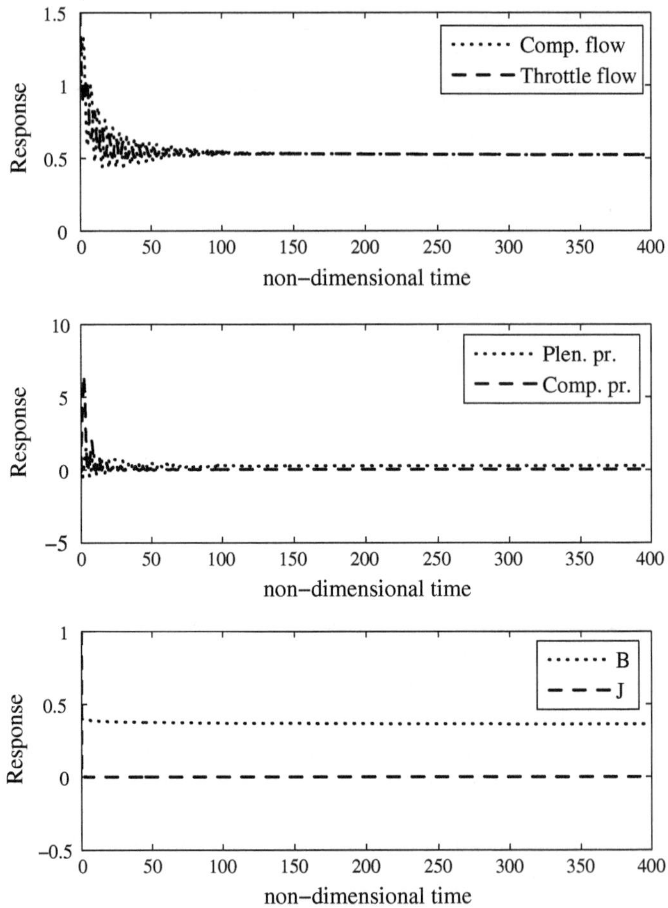

Fig. 5.8 Stable responses of states to disturbance, for the nominal typical values of the parameters

concluded that open-loop stability is not enough to drive the operating point to $\gamma = \gamma_n$ and also suppress rotating stall disturbances, by using a controller such as an automatically controlled bleed valve. The bleed valve by itself is not always adequate to maintain J at zero, and additional feedback is essential to suppress the rotating stall disturbance by changing the operating equilibrium point. Some authors (Gu et al. 1996) have referred to this requirement as "bifurcation control."

5.3.8 Control Law for Throttle Setting

To design the throttle controller, one rewrites Eq. 5.3.71 as,

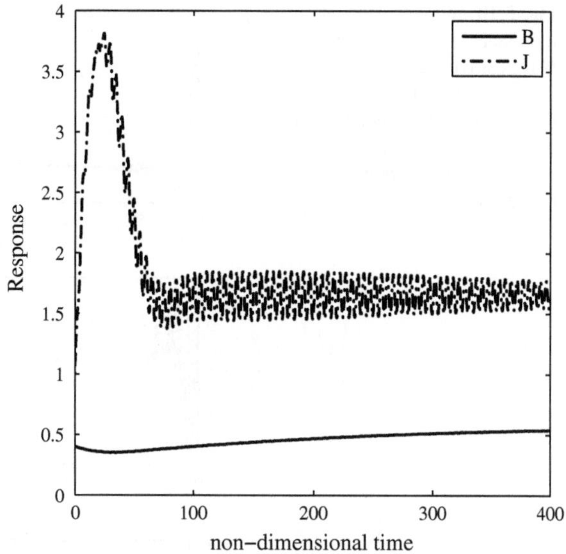

Fig. 5.9 Typical responses of B and J when H is negative and $\gamma < \gamma_n$ in the rotating stall case

$$\frac{\mathrm{d}}{\mathrm{d}\tau}\phi_t = -\frac{B}{G}\frac{\phi_t^2}{\left(\sqrt{2F^2/p_d H} + C_t\Delta\gamma\right)^2} + \frac{B\psi}{G}, \tag{5.3.79}$$

where p_d is the desired set value for p. The first step in designing a controller is to choose an appropriate value for p_d. The next step is to gradually wash out $\Delta\gamma$ according to some dynamic law such as

$$\tau_u\frac{\mathrm{d}\Delta\gamma}{\mathrm{d}\tau} = -\Delta\gamma, \tag{5.3.80}$$

where τ_u is an appropriate time constant, so the washout does not interfere with the plant dynamics.

If one further chooses $x > 1$, the equilibrium with $J = J_s = 0$ is stable. To establish the controller parameter p_d, a suitable choice may be made by first choosing x_0 and the operating point x and using 5.3.46. A typical choice could be $x_0 = 0$ and $x > 1$ giving a value for $p_d < 0.666$. If the initial value of p is p_0 and is greater than this value, then the steady-state value of $\Delta\Psi_c$ must be increased by

$$\Delta\Psi_{c,\mathrm{ss}} = \frac{H}{2}(p_d - p_0) = \frac{H\Delta p}{2}. \tag{5.3.81}$$

The corresponding initial condition for $\Delta\gamma$ is then given by,

$$\Delta\gamma(t)|_{t=0} = \gamma - \gamma_n = \gamma - \frac{\sqrt{2F^2/p_d H}}{C_t}. \tag{5.3.82}$$

Fig. 5.10 **a** Typical open-loop state responses when H is positive and $\phi_{s0}/F < 2$ in the rotating stall case. **b** Unsteady, quasi-steady and steady characteristics of the compressor corresponding to Fig. 5.10a

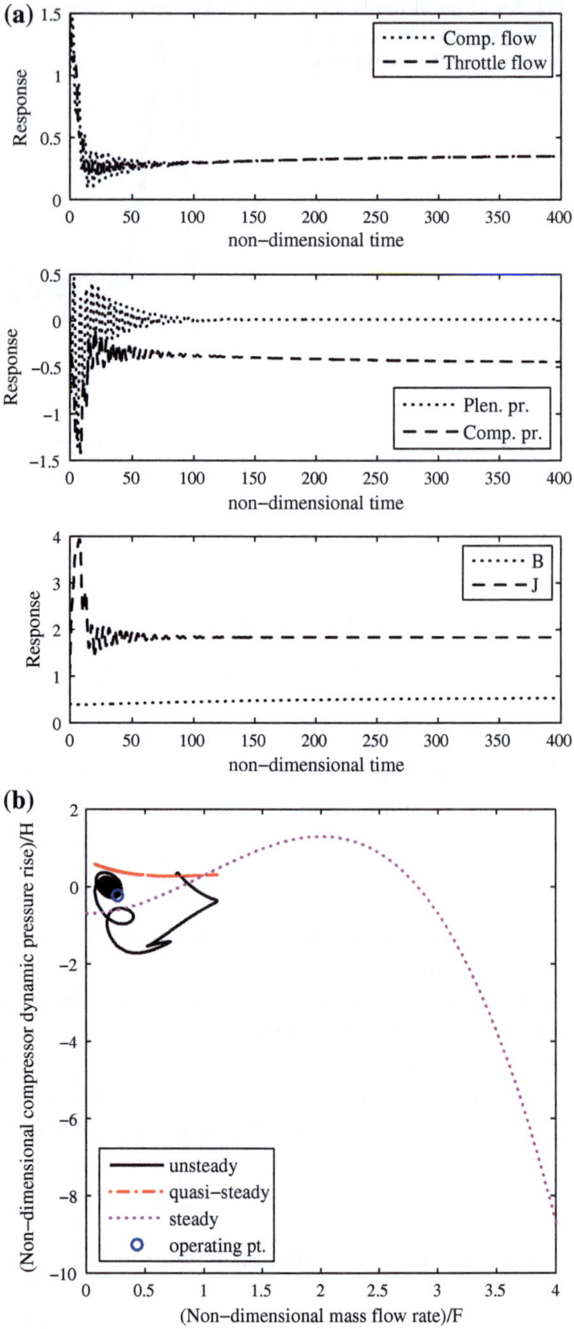

5.3.9 Control of the Rotating Stall Vibration Amplitude

To increase the steady-state value of $\Delta\Psi_c$ it is important to increase the steady flow delivered by the compressor. This can be done by increasing the input to the compressor. To incorporate such a feature in our model, one assumes a distribution of pressure sources at the inlet to the compressor and writes the compressor unsteady pressure dynamics equation with a source control term included as,

$$\frac{d\Delta\Psi_c}{d\tau} + \frac{\Delta\Psi_c}{\tau_c} = \frac{(\Psi_{c,ss} - \Psi_{c,qs})}{\tau_c} + \frac{\Delta u_0}{\tau_c}, \tag{5.3.83}$$

where, the control input is a distribution of pressure sources which are integrated over the inlet area of the compressor and chosen according to the control law,

$$\Delta u_0 = \frac{H}{2}(p_d - p_0) + \Delta u = \frac{H\Delta p}{2} + \Delta u, \tag{5.3.84}$$

where Δu is the control input perturbation to provide feedback. The complete model Eqs. 5.3.70–5.3.77 including the controller may be expressed as

$$\frac{d\phi_c}{d\tau} = B\Psi_c(p_0, \phi_c) + B(\Delta\Psi_c - \psi), \tag{5.3.85}$$

$$\frac{d}{d\tau}\phi_t = -\frac{B}{G}\frac{\phi_t^2}{\left(\sqrt{2F^2/p_d H} + C_t\Delta\gamma\right)^2} + \frac{B\psi}{G}, \tag{5.3.86}$$

$$\frac{d}{d\tau}\psi = \frac{(\phi_c - \phi_t)}{B}, \tag{5.3.87}$$

$$\frac{d\Delta\Psi_c}{d\tau} + \frac{\Delta\Psi_c}{\tau_c} = \frac{(\Psi_{c,ss}(p_0, \phi_{cs}) - \Psi_c(p_0, \phi_c))}{\tau_c} + \frac{\Delta u_0}{\tau_c}, \tag{5.3.88}$$

$$\frac{dB}{d\tau} = \frac{B}{\mu}(\Pi_{ext} - B\phi_c\sigma), \tag{5.3.89}$$

and

$$\tau_J\frac{d\Delta J}{d\tau} = \frac{H}{F}(J_s + \Delta J)\left(1 - \left(\frac{\phi_c}{F} - 1\right)^2 - \frac{1}{4}(J_s + \Delta J)\right) \tag{5.3.90}$$

$$\tau_u\frac{dC_t\Delta\gamma}{d\tau} = -C_t\Delta\gamma, \tag{5.3.91}$$

where

$$\Psi_c(p_0,\phi_c) = \frac{H}{2}\left(p_0 + \left(\frac{\phi_c}{F}-1\right)\left(1-\frac{J}{2}\right) - \left(\frac{\phi_c}{F}-1\right)^3\right), \tag{5.3.92}$$

$$\Psi_{c,\mathrm{ss}}(p_0,\phi_{\mathrm{cs}}) = \Psi_c(p_0,\phi_c)|_{t\to\infty}, \tag{5.3.93}$$

and

$$\Delta u_0 = \frac{H}{2}(p_d - p_0) + \Delta u = \frac{H\Delta p}{2} + \Delta u. \tag{5.3.94}$$

To implement such a controller, the parameter p_0 must be known. This parameter must therefore be identified off-line a priori or adaptively, so the control input can be synthesized.

5.3.10 Stability of Controlled Equilibrium

An important step in the validation of the controller is the assessment of the stability of the closed-loop equilibrium. To determine the stability of the controlled equilibrium, one first linearizes Eqs. 5.3.85–5.3.91, about the controlled equilibrium solution which is characterized by $p = p_d$, $\phi_c = \phi_t = \phi_{\mathrm{sd}}$. Perturbing the state vector and the control input and linearizing Eqs. 5.3.85–5.3.91 about the equilibrium states result in

$$\begin{aligned}\frac{\mathrm{d}\Delta\phi_c}{\mathrm{d}\tau} &= \left(\Psi_c(p_d,\phi_{\mathrm{sd}}) + B_0\frac{\mathrm{d}\Psi_c(p_d,\phi_{\mathrm{sd}})}{\mathrm{d}B}\right)\Delta B \\ &\quad + B_0\left(\frac{\mathrm{d}\Psi_c(p_d,\phi_{\mathrm{sd}})}{\mathrm{d}\phi_c}\Delta\phi_c + \frac{\mathrm{d}\Psi_c(p_d,\phi_{\mathrm{sd}})}{\mathrm{d}J}\Delta J + \Delta\Psi_c - \Delta\psi\right)\end{aligned} \tag{5.3.95}$$

$$\frac{\mathrm{d}\Delta\phi_t}{\mathrm{d}\tau} = \frac{B_0\Delta\psi}{G}$$

$$\begin{aligned}\frac{\mathrm{d}\Delta\phi_t}{\mathrm{d}\tau} &= \frac{B_0\Delta\psi}{G} - \frac{B_0}{G}\frac{p_d H_s \phi_{\mathrm{sd}}}{2F_s^2} \\ &\quad \left(2\Delta\phi_t - \frac{2\phi_{\mathrm{sd}}C_t\Delta\gamma}{\sqrt{2F_s^2/p_d H_s}} - \left(\frac{F_B}{F_s} - \frac{H_B}{2H_s}\right)\frac{\phi_{\mathrm{sd}}}{\sqrt{2F_s^2/p_d H_s}}\Delta B\right),\end{aligned} \tag{5.3.96}$$

$$\frac{\mathrm{d}}{\mathrm{d}\tau}\Delta\psi = \frac{(\Delta\phi_c - \Delta\phi_t)}{B_0}, \tag{5.3.97}$$

$$\begin{aligned}\frac{\mathrm{d}\Delta\Psi_c}{\mathrm{d}\tau} + \frac{\Delta\Psi_c}{\tau_c} &= -\frac{1}{\tau_c}\left(\frac{\mathrm{d}\Psi_c(p_d,\phi_{\mathrm{sd}})}{\mathrm{d}\phi_c}\Delta\phi_c + \frac{\mathrm{d}\Psi_c(p_d,\phi_{\mathrm{sd}})}{\mathrm{d}J}\Delta J + \frac{\mathrm{d}\Psi_c(p_d,\phi_{\mathrm{sd}})}{\mathrm{d}B}\Delta B\right) \\ &\quad + \frac{\Delta u}{\tau_c},\end{aligned} \tag{5.3.98}$$

$$\frac{\mathrm{d}\Delta B}{\mathrm{d}\tau} + \frac{\Delta B}{\mu}(2B_0\phi_{\mathrm{sd}}\sigma - \Pi_{\mathrm{ext}}) = -\frac{B_0^2 F_s \sigma}{\mu}\frac{\Delta\phi_c}{F_s}, \tag{5.3.99}$$

$$\begin{aligned} \tau_J\frac{\mathrm{d}\Delta J}{\mathrm{d}\tau} = & -\frac{H_s}{F_s}\left(2J_s\left(\frac{\phi_{\mathrm{cs}}}{F_s} - 1\right)\frac{\Delta\phi_c}{F_s}\right) \\ & -\frac{H_s}{F_s}\left(\left(\left(\frac{\phi_{\mathrm{cs}}}{F_s} - 1\right)^2 - 1 + \frac{J_s}{4}\right)J_s\left(\frac{H_B}{H_s} - \frac{F_B}{F_s}\right)\Delta B + \left(\left(\frac{\phi_{\mathrm{cs}}}{F_s} - 1\right)^2 - 1 + \frac{J_s}{2}\right)\Delta J\right) \end{aligned} \tag{5.3.100}$$

and

$$\tau_u\frac{\mathrm{d}C_t\Delta\gamma}{\mathrm{d}\tau} = -C_t\Delta\gamma, \tag{5.3.101}$$

where Δu is the control input perturbation and $\Delta\phi_c$, $\Delta\phi_t$, $\Delta\psi$, $\Delta\Psi_c$, ΔJ, ΔB are the perturbations to the corresponding states.

From the above, observe that the last three of the linearized perturbation equations Eqs. 5.3.99–5.3.101 are only weakly coupled with the earlier four, Eqs. 5.3.95–5.3.98. An analysis of the stability indicates that the controlled system is stable. Assume that the compressor perturbation mass flow ($\Delta\phi_c$) is measured, the root locus plot is obtained and shown in Fig. 5.11. The two lightly damped poles correspond to modes associated primarily with $\Delta\phi_c$ and $\Delta\psi$. To increase the stability margins, one could include and stability augmentation negative feedback (gain = 3.3), and this is implemented in calculating the closed-loop response in the next section. The chosen value of the gain corresponds to the maximum stability margin based on root locus plot.

The controller can now be tested by simulating it and the complete non-linear plant. The case of a compressor with the parameters as listed in Table 5.1 is considered. The desired compressor flow ratio is chosen to be $\phi_{\mathrm{sd}}/F = 2.1$. The desired value of the parameter $p = p_d$ is then estimated from Eq. 5.3.67. The initial value for $\Delta\gamma$ is chosen to be -0.2. The results of the closed-loop simulation including negative feedback are illustrated in Fig. 5.12a, which corresponds to the same case as the one shown in Fig. 5.10a without feedback.

Figure 5.12b illustrates the unsteady characteristics of the closed-loop compressor which are compared with the steady-state characteristics. Also shown in the figure is the steady-state closed-loop operating point. The results clearly indicate that the compressor now operates with the equilibrium $J = J_s = 0$ being stable. Thus, the rotating stall disturbance is eliminated.

5.3.11 Control of Compressor Rotating Stall

The dynamics of compressor stall has been re-parameterized in a form that would facilitate the construction of a non-linear control law for the active non-linear control of compressor stall. The regions of stable performance in parameter space

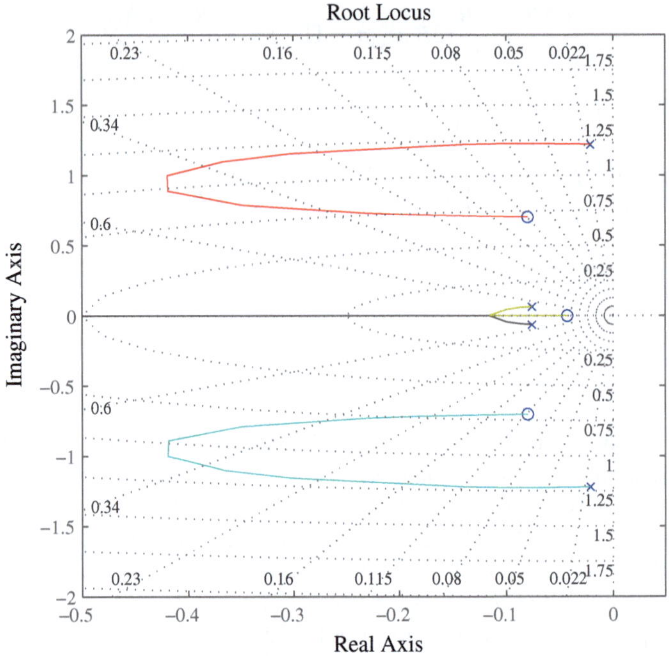

Fig. 5.11 Root locus plot illustrating the effect of the negative feedback of $\Delta\phi_c$

$(\gamma = \gamma_n, H > 0, J = J_s = 0)$, unstable performance $(\gamma \neq \gamma_n \,\text{or}\, H < 0, J \neq 0)$ were identified. This has led to the belief that a control law that maintains both $\gamma = \gamma_n, H > 0$ and $J = J_s = 0$ would actively stabilize the compressor. One observes that by merely setting the throttle at its optimum equilibrium position does not maintain, $J = J_s = 0$. An additional control input must aim to manipulate the transient and control pressure dynamics defined by Eq. 5.3.88 which would involve control inputs to the compressors inlet guide vanes or some other means of feedback control. That in turn points to a need for a better compressor pressure rise model incorporating the control input dynamics. Yet the relatively simple and systematic approach adopted in this section clearly highlights the main features on the controller that is capable of inhibiting compressor surge and rotating stall. Moreover, the method can be adopted for any axial compressor provided its steady-state compressor, and throttle maps are known. Furthermore, the linear perturbation controller synthesized in the previous section could be substituted by a non-linear controller synthesized by applying the backstepping approach as demonstrated by Krstic et al. (1995). Preliminary implementations of such a controller have supported the view that there is a need for an improved, matching, non-linear compressor pressure rise model including disturbance and uncertainty effects, and the results of this latter study involving a robust complimentary non-linear H_∞ optimal control law will be reported elsewhere. Coupled with the views

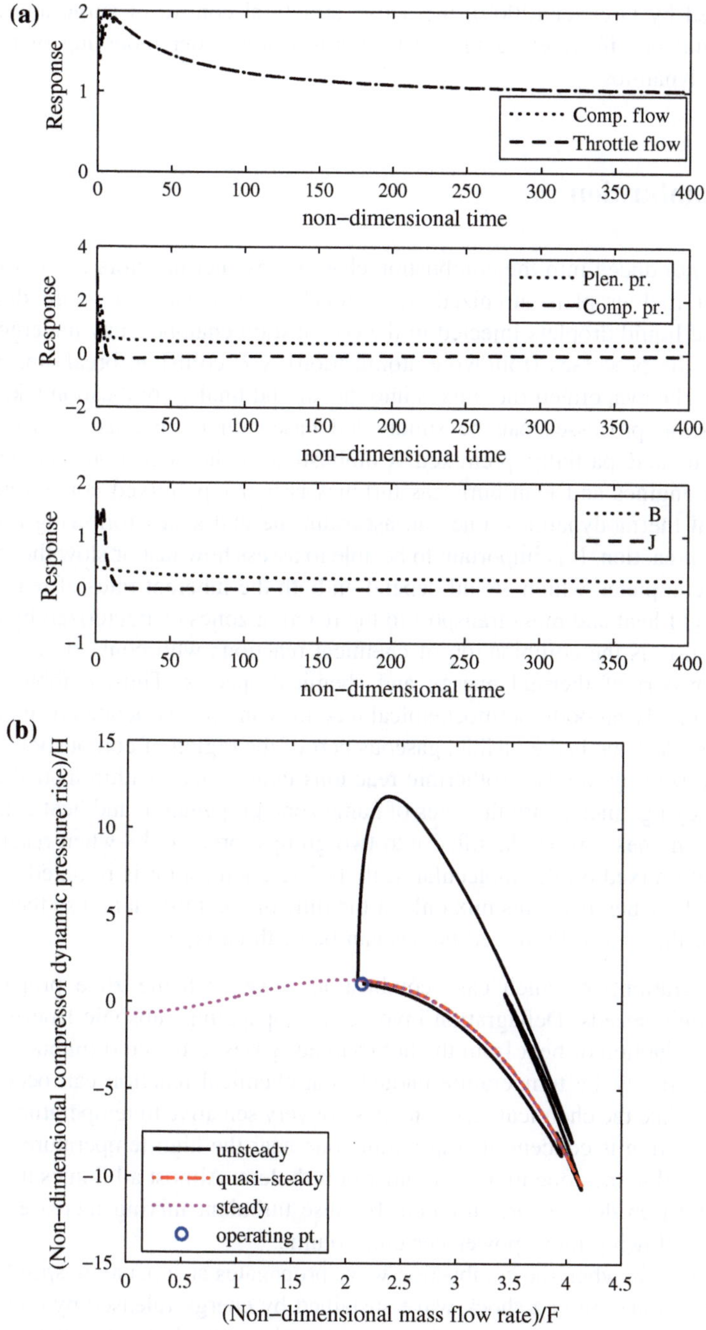

Fig. 5.12 **a** Typical closed-loop state responses when H is positive and $\phi_{sd}/F > 2$ in the rotating stall case. **b** Unsteady, quasi-steady and steady characteristics of the closed-loop compressor

expressed by Greitzer (2009), the active structural control of surge and rotating stall could be effectively achieved by realistic low-order modeling of the compressor dynamics.

5.4 Combustion

Fuel is introduced into the combustion chamber by fuel injectors in the form of a spray of fuel droplets atomized by a nozzle with a large pressure drop. The atomized liquid droplets injected in the combustion chamber may undergo one or more of the processes (following atomization), viz. collision, break-up, or evaporation. The evaporated fuel mixes into the air and finally combustion takes place. Combustion processes can be subdivided based on mixing as premixed, non-premixed, and partially premixed. Combustion in homogeneous-charge spark-ignition engines and lean-burn gas turbines is under premixed conditions. From chemical thermodynamics, one can ascertain the end states following a thermochemical reaction. It is important to be able to assess how fast or slow the reactions will be completed which would depend on both the inherent rates of reaction and the rates of heat and mass transport to the reaction zones characterized by a flame. Combustion is the combination of chemical reactions with convective and diffusive transport of thermal energy and chemical species. Thus, combustion is a process involving both thermochemical reactions and the associated heat and mass transport. A flame is the visible, gaseous part of the region of combustion which is characterized by highly exothermic reactions taking place within a smaller zone. So the key parameters are the rates of flame zone propagation and heat generation. Broadly, flames can be classified into two groups: premixed—when reactants are intimately mixed on the molecular scale before combustion is initiated; and non-premixed—when reactants mix only at the time of combustion and so they have to mix first then burn. Premixed flames can be of three types:

1. Deflagration, in which case combustion wave or flame zone propagates at subsonic speeds. Deflagration involves a propagating subsonic front sustained by conduction of heat from the hot (burned) gases to the cold (unburned) gases which raises the temperature enough that chemical reaction can occur. However, since the chemical reaction rates are very sensitive to temperature, most of the reaction is concentrated in a thin zone near the high-temperature side. The flow within the zone may be laminar or turbulent. Almost all flames in practical combustion devices are turbulent because turbulent mixing increases burning rates, allowing more power per unit volume.
2. Detonation, when the combustion wave propagates at supersonic speeds. In this case, detonation is a shock wave sustained by energy released by combustion. The combustion process, in turn, is initiated by shock wave compression and results in high temperatures. Detonations involve interaction between fluid mechanic processes (shock waves) and thermochemical processes (combustion).

In this case, there are qualitative differences between upstream and downstream properties across the detonation wave which are similar to property differences across a normal shock although there also some significant differences between a normal shock wave and detonation wave. Detonation and deflagration velocities may be found by an analysis similar to normal shock analyses in Chap. 1.

3. Homogeneous flame, when one has a fixed mass (control mass) with uniform spatial distributions of temperature and pressure as well as a uniform composition. Furthermore, there is no "propagation" in space but is dynamic in time.

Chemical reactions relevant to combustion are generally very complicated and are approximated by a one-step overall reaction. To obtain quantitative model of the reaction rate, one can apply the law of mass action which states that the rate of reaction is proportional to the number of collisions between the reactant molecules, which in turn is proportional to the concentration of each reactant. Consider a reaction of the form,

$$n_A A + n_B B \rightarrow n_C C + n_D D. \tag{5.4.1}$$

The law of mass action states that the rate of reaction,

$$\frac{\mathrm{d}}{\mathrm{d}t}\left\{\frac{[A]}{n_A}\right\} = \frac{\mathrm{d}}{\mathrm{d}t}\left\{\frac{[B]}{n_B}\right\} = -\frac{\mathrm{d}}{\mathrm{d}t}\left\{\frac{[C]}{n_C}\right\} = -\frac{\mathrm{d}}{\mathrm{d}t}\left\{\frac{[D]}{n_D}\right\} = -k_f[A]^{n_A}[B]^{n_B} \tag{5.4.2}$$

where $[S]$ is the concentration of the species S in moles per unit volume. The proportionality constant k_f is known as the forward reaction rate constant. The total number of moles of gas per unit volume for all reactants is calculated from the universal gas law, $p = n\rho R_u T$, where R_u is the universal gas constant. The concentration of a particular species is then found by multiplying the total concentration by the ratio of the number moles of the species in the reaction n_S to the total number of moles n. The reaction rate constant k_f is usually of the Arrhenius form

$$k_f = k_{f0} T^a \exp(-E/R_u T) \tag{5.4.3}$$

where E is the activation energy in calories/mole, and k_{f0} and a are constants which are determined experimentally. Boltzman showed in the 1800s that the fraction of molecules in a gas with translational kinetic energy greater than some value E is proportional to $\exp(-E/R_u T)$. Thus, E represents the "energy barrier" that must be overcome for reaction to occur.

5.4.1 Combustion Chambers

In a combustion chamber, the propellants are injected through a number of coaxial injectors, with oxidant injected usually through the center units and the fuel through the others to facilitate rapid premixing. The injectors are usually of the shear coaxial type, the swirl-coaxial type, or the impinging type. The shear coaxial injectors are supposedly effective in mixing two propellants with substantial density differences, while impinging and swirl-coaxial injectors are used for two liquid propellants with similar densities. In swirl-coaxial injectors, one of the two propellants is given an initial angular velocity. There are several variations in the geometry of injectors which all have a significant influence on combustion performance characteristics and stability. Modern gas turbine designs are increasingly based on the premixed mode of combustor operation. This operational mode and accompanying design implications have been primarily driven by the strict emission guidelines on pollutants like NOx and CO. One of the principal strategies adopted to meet the requirements of emission standards is to burn the fuel–air mixture at lean equivalence ratios. At lean conditions, the production of thermal NOx is greatly reduced due to the lowered temperatures. However, one shortcoming of lean, premixed mode of operation is that these combustion chamber designs are especially prone to self-excited oscillations. Humming was reported in ring burners in one particular gas turbine with associated flickering gas flames, as well as serious vibration and shaking of loose parts. The dynamic phenomenon of self-excited thermoacoustic oscillations generally occurs when the unsteady combustion process couples with the natural aeroacoustic modes of the combustion chamber. Combustion instabilities result from complex dynamic interactions between acoustics, heat release, and vortex dynamics within the combustion zone. These instabilities are destructive to combustion chambers and adversely affect performance and emissions.

The condition under which acoustic oscillations are amplified by a fluctuating source of heat release was first postulated by Lord Rayleigh in 1878 [Rayleigh (1878a, b)]. In a thermodynamic cycle, the addition of heat at high pressure leads to a net energy input, which is then available to do work. If the energy gained from the unsteady combustion exceeds the rate of work done, the acoustic waves grow in amplitude. A description of this phenomenon is given in Rayleigh (1945). Rayleigh's energy criterion for self-excited thermoacoustic oscillations essentially states that the instabilities in a combustor are feasible when the unsteady combustion processes add energy in phase to the acoustic oscillations. In general, this instability is facilitated and subsequently sustained if the rate at which the energy is added to the acoustic oscillations is greater than the rate at which the acoustic energy is absorbed. Essentially, a perturbation in a flow variable can cause a change in the flame heat release rate. The unsteadiness in the flame heat release results in the generation of acoustic waves. The acoustic waves cause further changes in the flow variables, and this process is akin to a positive feedback loop which is inherently unstable. In a typical combustor, this feedback mechanism, illustrated in Fig. 5.13, can be initiated by a variety of disturbance perturbations like oscillations in the equivalence ratio, oscillations in the velocity field, extinction of the flame, vortex dynamics, flame–wall interaction as well as several other mechanisms. Depending on the operating conditions, one or more of these mechanisms can sustain the oscillations.

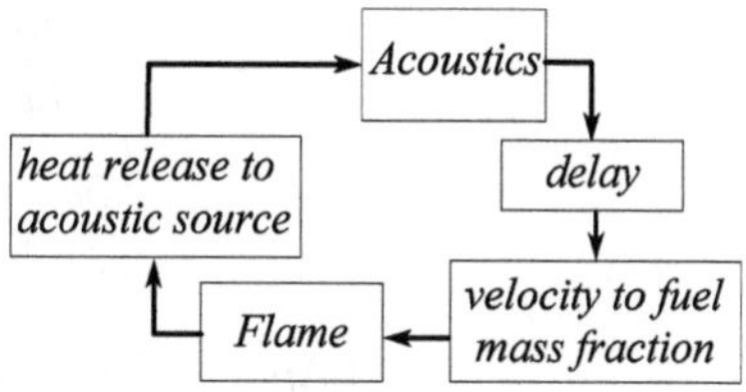

Fig. 5.13 Feedback loop nature of thermoacoustic coupled oscillations

Mathematically for small amplitude disturbances in a combustion chamber of volume V and bounded by the surface area S, with unsteady heat addition q, the condition for the growth of acoustic waves is,

$$\frac{(\gamma-1)}{\rho a^2}\int_V \left(\overline{\Delta p\, q}\right) dV > \int_S \overline{\Delta p \mathbf{u}} \cdot d\mathbf{S} \tag{5.4.4}$$

where Δp denotes the unsteady pressure (the overbar denotes the mean value of the products covered by it), $\mathbf{u}$ the flow velocity vector, ρ the density of the flow, a the speed of sound and γ the ratio of specific heats. When the inequality in Eq. 5.4.4 is satisfied, the energy gained from the unsteady combustion exceeds the rate at which work is done on the surroundings. The acoustic energy within the combustor increases till the amplitude increases and the fluctuations in the heat input saturate. Subsequent increases are less rapid with amplitude than the energy loss term. A constant amplitude limit cycle is obtained when the left- and right-hand sides of Eq. 5.4.4 are equal to each other.

5.4.2 Aeroacoustics of Combustion Chambers

When the perturbations are infinitesimally small, the system dynamics can be analyzed using linear aeroacoustics. Apart from predicting the frequency and growth rate of the oscillations, the linearized analysis also provides information about the operating conditions under which the system may be unstable. From a practical standpoint, gas turbine designers are not only interested in the unstable frequencies but also in the conditions for the existence and stability of the limit cycle amplitude of the oscillations. There exists an "acceptable/unacceptable threshold" of the instability amplitude (based on the potential to damage hardware or affect performance) which is dependent on the frequency of the limit cycle oscillations.

Since the flame zone is thin enough, we can approximate the heat source spatially as a point source which is represented by a Dirac delta function. Assuming combustion in a duct and a point source of heat, a one-dimensional model comprising the mass, energy, and linear momentum conservation equations can be simplified, by ignoring the body force and viscosity, as,

$$\frac{\partial \rho}{\partial t} + \frac{\partial}{\partial x}(\rho u) = 0, \tag{5.4.5}$$

$$\frac{\partial u}{\partial t} + u\frac{\partial u}{\partial x} = -\frac{1}{\rho}\frac{\partial \Delta p}{\partial x}, \tag{5.4.6}$$

$$\frac{\partial \Delta p}{\partial t} + \gamma p\frac{\partial u}{\partial x} + u\frac{\partial \Delta p}{\partial x} = (\gamma - 1)q(t)\delta(x - x_f). \tag{5.4.7}$$

Thus, following linearization and neglecting second-order terms of the fluctuating components, with the assumption that for low Mach number flows, the effect of the step change in the pressure is negligible compared to the change in the mean velocity or mean density, and the spatial gradient of the mean pressure as well as the mean velocity can be considered small, one obtains the governing linear equations for the velocity and pressure perturbations as,

$$\frac{\partial u}{\partial t} + \frac{1}{\bar{\rho}}\frac{\partial \Delta p}{\partial x} = 0, \tag{5.4.8}$$

$$\frac{\partial \Delta p}{\partial t} + \gamma \bar{p}\frac{\partial u}{\partial x} = (\gamma - 1)q(t)\delta(x - x_f). \tag{5.4.9}$$

Differentiating the first with respect to the spatial variable and the second with respect to time, the unsteady pressure Δp can be shown to satisfy the inhomogeneous wave equation,

$$\frac{\partial^2 \Delta p}{\partial t^2} - a^2\nabla^2(\Delta p) = (\gamma - 1)\frac{\partial q}{\mathrm{d}t}\delta(x - x_f). \tag{5.4.10}$$

In the case of distributed rate of heat release sources, the right-hand side of the inhomogeneous wave equation is a non-uniformly and spatially distributed function. A much more general expression is given by Cullick (1988). Controlling the spatially distributed rate of heat release source driving the acoustic modes would be one method of controlling the instabilities in a combustion chamber. The term on the right-hand side of Eq. (1001) describes how the unsteady addition of heat generates a acoustic pressure source. The oscillatory pressure is expressed as a series of modes satisfying the relevant boundary conditions at the ends of the cavity. Thus, typically one could assume that,

$$\Delta p = \sqrt{\frac{2}{\bar{\rho} L_{\mathrm{eq}}}} \sum_{i=1}^{\infty} \cos\left(\frac{2i-1}{2L_{\mathrm{eq}}}\pi x\right)\eta_i(t), \tag{5.4.11}$$

where L_{eq} is the acoustic equivalent length of the pipe. The pressure mode amplitudes $\eta_i(t)$ satisfy

$$\frac{\mathrm{d}^2\eta_i(t)}{\mathrm{d}t^2} + \omega_i^2\eta_i(t) = -\bar{\rho}(\gamma - 1)\sqrt{\frac{2}{\bar{\rho} L_{\mathrm{eq}}}}\cos\left(\frac{2i-1}{2L_{\mathrm{eq}}}\pi x_f\right)\frac{\mathrm{d}q(t)}{\mathrm{d}t}. \tag{5.4.12}$$

The natural frequencies ω_i correspond to a pipe closed at one end which is a special case of a cavity and duct resonator known as the Helmholtz resonator. They are,

$$\omega_i = 2\pi\left(\frac{2i-1}{4}\right)\left(\frac{a}{L_{\text{eq}}}\right). \tag{5.4.13}$$

When $i = 1$ it corresponds to the first natural frequency of a pipe closed at one end which is also known as Rijke tube and is considered to be a model of a combustion chamber. Rijke (1859) demonstrated the phenomenon experimentally using a simple tube originally in the year of 1859. The higher natural frequencies are the resonant overtones of the pipe given by the odd harmonics of the first natural frequency. The action of a Rijke tube can be visualized most simply as a mass–spring–dashpot system. The mass is the volume mass trapped in the duct of length L_{duct} and the spring stiffness is provided by the compressibility of the gas in the volume of the cavity. Hence, one should expect that the natural frequency should decrease as the volume of the cavity increases and should also decrease as the mass of the air in the duct increases. For a relatively long duct, the formula for the lowest resonant frequency of the Helmholtz resonator is,

$$\omega_H = \frac{a}{L_{\text{duct}}}\sqrt{\frac{V_{\text{duct}}}{V_{\text{cavity}}}}. \tag{5.4.14}$$

The term within the square root is the ratio of acoustical capacitance of the duct to that of the cavity. For shorter ducts, the formula for the lowest resonant frequency of the Helmholtz resonator is modified and expressed as,

$$\omega_H = \frac{a}{L_{\text{duct}}}\sqrt{\frac{V_{\text{duct}}}{V_{\text{cavity}}}\left(\frac{L_{\text{duct}}}{L_{\text{duct}}+\sqrt{A_{\text{duct}}}}\right)}. \tag{5.4.15}$$

The physical interpretation of the Helmholtz resonance frequency allows one to model any combustion chamber irrespective of its geometry. The development of a physically based model for the rate of heat release begins with simplified equations governing a one-dimensional laminar premixed flame. A single-step global reaction mechanism for the fuel is considered. The rate of heat release is the sum of contributions from heat conduction (Fourier's Law) or temperature diffusion and enthalpy diffusion arising from the burning fuel. Assuming that the fuel is the limiting species, the temperature distribution and the mass fraction distribution of the fuel are expressed in terms of a source term that is a function the unsteady pressure. The rate of heat release to the fluid is then given in terms of the temperature difference and the fuel flow rate. The unsteady heat source is generally related to the time rate of change of the of the pressure mode amplitudes $\eta_i(t)$ by a nonlinear function of $\eta_i(t)$ and $\dot{\eta}_i(t)$ given by,

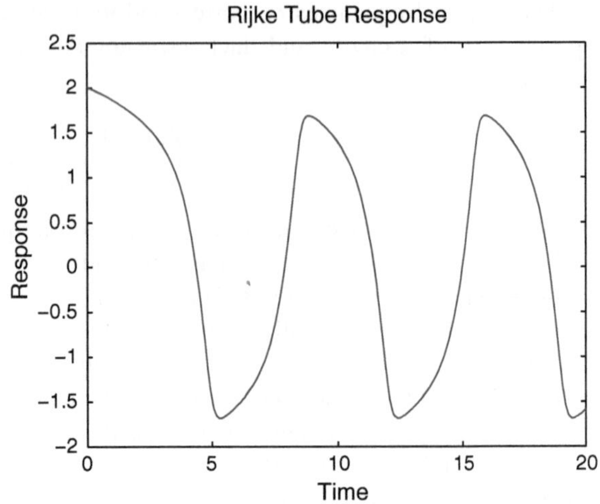

Fig. 5.14 Typical unstable response of a Rijke tube

$$\frac{dq}{dt} = -\sum_{i=1}^{\infty} \left(b_{i1}\left(1 - \eta_i^2\right) + b_{i2}\left(1 - \dot{\eta}_i^2\right) + b_{i3}\left(1 - \eta_i^2 - \dot{\eta}_i^2\right) + b_{i4}\left(1 - \eta_i^4\right)\dot{\eta}_i(t) \right). \tag{5.4.16}$$

where b_{ij} are coefficients that generally determined either experimentally or numerically and are functions of both the temperature and the fuel flow rate. A typical unstable response of a Rijke tube with

$$\begin{bmatrix} b_{11} & b_{12} & b_{13} & b_{14} \end{bmatrix} = \begin{bmatrix} 0.7584 & 0.0969 & 0.0588 & 0.9845 \end{bmatrix} \tag{5.4.17}$$

is shown in Fig. 5.14.

As rightly pointed out by Dowling (2003) and Dowling and Morgan (2005), the dynamics of amplitudes of the pressure modes are coupled with the vortex dynamics of the fuel flow and temperature diffusion dynamics within the combustion chamber both of which are primarily unsteady and diffusive. Time delays have been ignored in the above model nut should be considered as they add an interesting additional dimension to the instability problem. Eliminating the heat source, the complete nonlinear dynamics of the pressure amplitude is obtained which can shown to exhibit limit cycle oscillations for certain ranges of values of the parameters b_{ij}. Linear models of the heat release rate are constructed by relating it to the fuel mass flow rate by a "flame transfer function." The *flame transfer function* is usually constructed in the frequency domain and the parameters of the model determined by experiment.

5.4.3 *Flow-Coupled and Thermoacoustic Instabilities: POGO, Buzz, Chugging and Screech*

Flow-coupled and thermoacoustic instabilities may be classified depending on their typical bandwidth or on the basis of where they occur (Cullick 1988). Strong coupling between chamber pressure oscillations, low-frequency structural vibrations and the propellant feed system produces the common POGO instability in liquid-fueled rocket engines. The POGO instability generally occurs in a relatively low-frequency band. Supersonic inlets are characterized by shock-induced oscillations commonly referred to as "buzz". The "buzz" type of thermoacoustic instability has been discussed by Bloxside et al. (1988). Resonating cavities have been used in small supersonic inlets to reduce these shock-induced oscillations which occur in mid-frequency bands (100 Hz–1 kHz). Axial oscillations in Ramjet engines are known to be troublesome, and their influence on the shock-induced oscillations in the inlet diffuser can result in a reduction in the inlet stability margin. Bulk oscillations known as "chugging" are also observed in ramjets and are low-frequency instabilities which could be stabilized by controlling the supply of liquid propellant. These low-frequency oscillations correspond to rigid spatial modes and are also observed in liquid-fueled rocket engine. Because of their light construction, thrust augmentors are susceptible to failure of flame holders or of the basic structure when high-frequency ($\sim$ 10 kHz) combustion instabilities known as "screech" become severe.

The thermoacoustic instabilities and the ensuing vibrations can be effectively controlled, and the design of these controllers is facilitated by the construction of reduced-order models that succinctly capture the interactions between acoustics, heat release, vortex dynamics, inlet dynamics, and fuel delivery for a variety of geometries and flame-anchoring mechanisms. One approach to the control of such instabilities is to couple a Helmholtz resonator to the system (Bellucci et al. (2001)) so it absorbs the oscillations in very much the same manner as Stockbridge vibration isolator absorbs the galloping oscillations of suspension bridge cables. To apply feedback control principles to the active control of combustion instabilities a range of physical quantities (Docquier and Candel 2002) such as the air–fuel ratio, the heat release and the heat release rate, the mass flux, the pressure, temperature, and the homogeneity of the mixing of air and fuel must be measured within the combustion chamber. Using a range of actuators such as gas injectors and atomizers, control valves, microphones, and heating/cooling elements, a variety of controllers have been proposed and experimented with (DeLaat and Chang 2003; Kopasakis 2003; Le et al. 2003; Auer et al. 2004; Campos-Delgado et al. 2003a, b; Evesque et al. 2004; Riley et al. 2004).

Every air-interacting component of a jet engine is a potential noise source by disturbing the ambient air and causing pressure fluctuations. The primary sources of engine noise apart from the combustion chamber are from the fan, compressor, turbine, and jet. Jet noise is the noise created when high-temperature, high-velocity exhaust jet gases mix with ambient air. The three main mechanisms that contribute

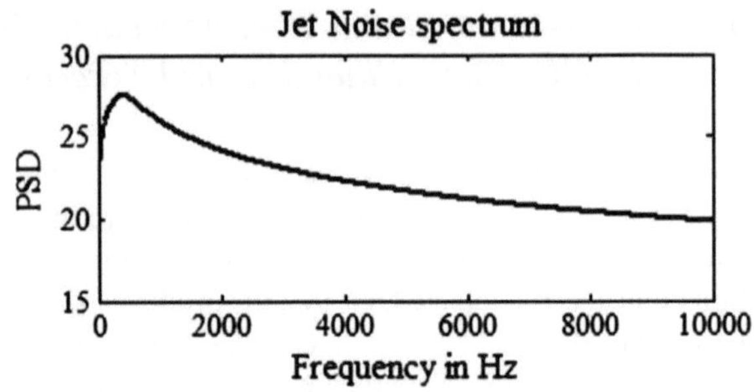

Fig. 5.15 Typical simulated power spectrum of jet noise

to jet noise are the turbulent mixing of the exhaust gases with the ambient air, the shocks that form in the vicinity of the exhaust plane which interact with the turbulence, producing broadband noise, and the high temperature of the jet. Large eddies in the mixing region are responsible for low-frequency noise, and small eddies in high-frequency noise in the broadband spectrum arising from the jet mixing region. These three sources are normally referred to as jet mixing noise, jet shock noise, and jet entropy noise. A typical simulated power spectrum of jet noise is illustrated in Fig. 5.15.

Fan noise which is made up of three main contributions is characteristic of turbofans. The three main sources are a broadband noise that arises from the interaction between rotating blade and incoming air, blade-passing noise due to the forward movement of the blade pressure distribution at the fan rotational speed frequency and its higher harmonics and fan "buzz" or "buzz-saw" noise which originates from small geometrical differences and asymmetric vibrations of successive fan blades.

5.5 Overall Jet Engine Volume Dynamics Modeling

A modern gas turbine jet engine is a complex system that is a multidimensional and highly parametric characterized by non-linear dynamic behavior with stochastic properties. It is generally required to perform optimally in spite of a wide spectrum of changes in its operating environment (e.g., temperatures from −60 to +40 °C, widely varying humidity and pressures, etc.). It is most desirable to be able to control the thrust delivered by aircraft gas turbine jet engine. Yet when a certain desired thrust is commanded, the engine must necessarily deliver this thrust while meeting the operating limits on the engine. The auxiliary aim of an engine controller, after delivering the desired thrust or maximum thrust is to guarantee that the engine will be maintained within its operating limits at all times, regardless of how the operator may move the throttle lever or of inlet conditions (e.g., altitude and Mach number). The engine's operating limits may be of a mechanical, thermal, or aerodynamic nature (e.g., stall margin). In addition to these constraints, the overall engine performance is also subject to both regulatory standards and the requirements of the airframe manufacturer. In many engines, the controlled variable is not the actual thrust itself but some other variable on which the directly dependent on. In turbofan engines, for example, it may be the fan speed.

The first of the limited variables is the temperature within the combustion chamber. The second requirement is the surge and stall margin of the compressor. The stall margin is the distance of the compressors operating point from the boundary separating the region of instability from the operating region. The inlet to the jet engine is the first subsystem exposed to the incoming flow field and is therefore of some importance to the overall propulsion system. The purpose of the inlet is to supply the engine with the required mass flow with the highest pressure and least distortion. The larger the inlet flow distortion, the larger is the stall margin that is required. Engines that power high-performance commercial aircraft are currently operated with a stall margin that is large enough to accommodate the worst-case combinations of inlet flow distortion, throttle transients, and engine capacity. Operating with a large-stall margin restricts the rate of change of engine thrust, and for this reason, the commanded value of the engine thrust cannot be achieved instantaneously but after a certain time delay. Thus, it is not only important to predict this delay but to operate all the subsystems in modes that will ensure that the delay in achieving the commanded thrust is a minimum. Another variable that must meet strict performance limits is the engine pressure ratio which is roughly equivalent to the ratio of the turbine outlet and compressor inlet pressures. In order achieve the overall control function of an aircraft jet engine, key engine variables are monitored and these are the spool speeds, the inlet and outlet compressor pressures, and the turbine inlet temperature. Several environmental variables are also measured and continuously monitored such as the altitude, the flight Mach number, the outside air temperature. The compressor control inputs are a bleed valve on the outlet side of the compressor and angles of the inlet guide vanes in the compressor stator while the engine control input is the rate of fuel flow. It is also essential to build an integrated dynamic model of the jet engine that facilitates the simulation of the measured variables, the control inputs, and the controlled output variables. It is essential that the model includes only those dynamical subsystems that have significant influence on the overall performance of the jet engine, the controlled outputs and captures the essential dynamics of the measured variables and outputs and their relationships to the control inputs.

The modeling approach is derived from work by Seldner et al. (1972), Seldner and Cwynar (1978) and refined by Stuber and Melcher (2004) and Kopasakis (2007) for a turbojet engine and by Kopasakis et al. (2008) and Connolly (2009) for a turbofan engine. The equations are derived here for completeness and clarity when ultimately integrating into the model for each component. In addition, the equations here are defined for a turbofan with each component being modeled by a lumped-volume-based system of equations. Each of the fluid flow components is modeled using a set of derived conservation equations modified from the standard form and written in total condition form, where the basic equations are the continuity, momentum, and energy equations, respectively, defined in Chap. 1 as,

$$\frac{\partial}{\partial t}(\rho A) + \frac{\partial}{\partial x}(\rho A u) = 0, \tag{5.5.1}$$

$$\frac{\partial}{\partial t}(\rho A u) + \frac{\partial}{\partial x}\left(\rho A u^2\right) = -A\left(\frac{\partial p}{\partial x} + \rho g_u\right), \tag{5.5.2}$$

$$\frac{\partial}{\partial t}(\rho A h) + \frac{\partial}{\partial x}(\rho A u h) = \frac{\partial}{\partial t}(p A) + \frac{\partial}{\partial x}(p A u) - \frac{\partial q}{\partial x}. \tag{5.5.3}$$

In addition,

$$h = c_p T. \tag{5.5.4}$$

The Eqs. 5.5.1–5.5.3 are linearized and expressed as,

$$\frac{\partial}{\partial t}(\rho) = -\frac{1}{A}\frac{\partial}{\partial x}(\rho A u) = 0, \tag{5.5.5}$$

$$\frac{\partial}{\partial t}(\rho A u) = -A\frac{\partial p}{\partial x}, \tag{5.5.6}$$

$$\frac{\partial}{\partial t}(\rho T) = -\frac{\partial}{\partial x}(\rho A u(h - p/\rho)) = -\frac{1}{A}\frac{c_v}{c_p}\frac{\partial}{\partial x}(\rho A u T) = -\frac{\gamma}{A}\frac{\partial}{\partial x}(\rho A u T). \tag{5.5.7}$$

Since $\rho A u = \dot{m}$ the mass flow rate, discretizing the spatial variable as,

$$\frac{\partial}{\partial x} = \frac{\Delta}{l}, \tag{5.5.8}$$

the equation for the density, mass flow rate, and temperature in the stage volume are,

$$\frac{\mathrm{d}}{\mathrm{d}t}(\rho) = -\frac{\Delta \dot{m}}{V}, \tag{5.5.9}$$

$$\frac{\mathrm{d}}{\mathrm{d}t}\dot{m} = -\frac{A}{l}\Delta p, \tag{5.5.10}$$

$$\frac{\partial}{\partial t}(\rho T) = -\frac{\gamma}{V}\Delta(\dot{m}T). \tag{5.5.11}$$

The engine model is based on component volume dynamics. Isentropic processes are assumed for the diffuser and the converging nozzle. Actual performance of the components in the form of performance maps for the compressor and turbine is used. Every component is described by the steady-state characteristics, followed by an equivalent volume where mass and energy storage take place. Figure 5.16 shows the station numbering of the mathematical model for the engine.

The engine considered is a hypothetical engine very similar to the GE J85 single-spool turbojet engine with a multistage compressor and two-stage turbine. The modeling approach follows Connolly (2009) and Connolly et al. (2010)

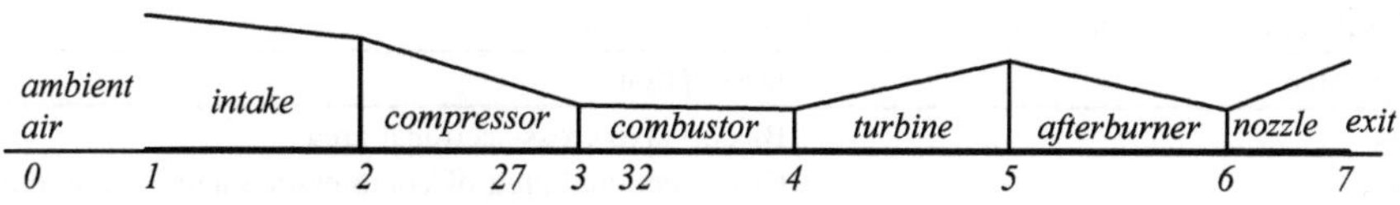

Fig. 5.16 Stage numbering for engine modeling

although there are some significant differences particularly in the definition of the compressor and turbine maps.

5.5.1 Compressor Modeling

The compressor is modeled as a single lumped-volume subsystem and a non-dimensional characteristic performance map. All the variables used in the compressor subsystem are defined in Table 5.2. The three-state variables chosen for the modeling of each component are the total pressure, total temperature, and mass flow rate. The compressor is modeled as a subsonic system, so the disturbances travel with finite velocities. Mass flow rate disturbances travel upstream while the temperature and pressure disturbances travel downstream. The compressor's non-dimensional performance map is in the form of table lookups to obtain a given pressure ratio and efficiency, which is then used to get the characteristic pressure and temperature. The table lookups are based on a non-dimensional specific mass flow rate and a specific engine speed which are, respectively, defined in terms of the total inlet temperature T_{inlet} and the total inlet pressure P_{inlet} at the compressor inlet, the reference temperature T_{ref} and the reference pressure P_{ref}, the compressor inlet flow $\dot{m}_{\text{cp}}$, the design mass flow rate $\dot{m}_d$ and the reference engine design speed N_{ref}, as:

$$\dot{m}_{\text{cm}} = \frac{\sqrt{T_{\text{inlet}}/T_{\text{ref}}}\,\dot{m}_{\text{cp}}}{P_{\text{inlet}}/P_{\text{ref}}\ \ \dot{m}_d},\ N_{\text{cm}} = \sqrt{\frac{T_{\text{ref}}}{T_{\text{inlet}}}}\frac{N}{N_{\text{ref}}}. \tag{5.5.12}$$

The generic performance map used to model the engine compressor is illustrated in Fig. 5.17. The performance properties of the compressor, the pressure ratio $P_{\text{cp\_ratio}}$ and the efficiency η_{cp}, are obtained from the non-dimensional performance map. The actual compressor outlet static pressure and static temperature are estimated using the following two equations:

$$P_{\text{cp}} = P_{\text{cp\_ratio}} P_{\text{inlet}}, \tag{5.5.13}$$

and

$$T_{\text{cp}} = \left(1 + \frac{\left(P_{\text{cp\_ratio}}\right)^{\frac{\gamma_{\text{cp}}-1}{\gamma_{\text{cp}}}} - 1}{\eta_{\text{cp}}}\right) T_{\text{inlet}}. \tag{5.5.14}$$

Table 5.2 Definition of compressor subsystem variables

Variable	Description
A_{bv}	Bleed valve cross-sectional area
A_{cp}	Cross-sectional area of compressor stage volume duct
c_{p_cp}	Specific heat at constant pressure for air
K_{bv}	Bleed valve coefficient
l_{cp}	Length of compressor stage volume duct
$\dot{m}_{\mathrm{bv}}$	The bleed valve mass flow rate
$\dot{m}_{\mathrm{cb}}$	The combustion chamber inlet mass flow rate
$\dot{m}_{\mathrm{cm}}$	Compressor non-dimensional specific mass flow rate
$\dot{m}_{\mathrm{cp}}$	The compressor inlet mass flow rate
$\dot{m}_{d}$	The compressor inlet design mass flow rate
M_{cp}	Mach number in compressor stage volume
N	Actual spool speed (rpm)
N_{cm}	Compressor non-dimensional specific spool speed
N_{ref}	The compressor (engine) reference spool speed
P_{cp}	Compressor characteristic pressure
P_{cpt}	Compressor outlet total pressure
$P_{\mathrm{cp\_ratio}}$	Compressor map pressure ratio
P_{inlet}	The total inlet pressure
P_{ref}	The compressor map reference pressure
R_{air}	Gas constant for air
T_{cp}	Compressor characteristic temperature
T_{cps}	Compressor outlet static temperature
T_{cpt}	Compressor outlet total temperature
T_{inlet}	The total inlet temperature
T_{ref}	The compressor map reference temperature
V_{cp}	Volume of compressor stage volume duct
γ_{cp}	Ratio of specific heats in compressor stage volume
η_{cp}	Compressor map efficiency
$\rho_{\mathrm{cp\_sv}}$	Compressor stage volume density

Once the characteristic properties of the compressor outlet static temperature and static pressure are determined from the map, they are used in the state equations for the compressor stage volume dynamics which are:

$$\frac{\mathrm{d}}{\mathrm{d}t}\rho_{\mathrm{cp\_sv}} = \frac{\left(\dot{m}_{\mathrm{cp}} - \dot{m}_{\mathrm{cb}} - \dot{m}_{\mathrm{bv}}\right)}{V_{\mathrm{cp}}}, \tag{5.5.15}$$

$$\frac{\mathrm{d}}{\mathrm{d}t}\dot{m}_{\mathrm{cp}} = \frac{A_{\mathrm{cp}}}{l_{\mathrm{cp}}}\left(P_{\mathrm{cp}} - P_{\mathrm{cpt}}\right)\left(1 + \frac{\gamma_{\mathrm{cp}} - 1}{2}M_{\mathrm{cp}}^2\right)^{\frac{1}{\gamma_{\mathrm{cp}}-1}}, \tag{5.5.16}$$

$$\frac{\partial}{\partial t}T_{\mathrm{cpt}} = \frac{\gamma_{\mathrm{cp}}}{\rho_{\mathrm{cp\_sv}}V_{\mathrm{cp}}}\left(T_{\mathrm{cp}}\dot{m}_{\mathrm{cp}} - T_{\mathrm{cpt}}\left(\dot{m}_{\mathrm{cb}} + \dot{m}_{\mathrm{bv}}\right)\right). \tag{5.5.17}$$

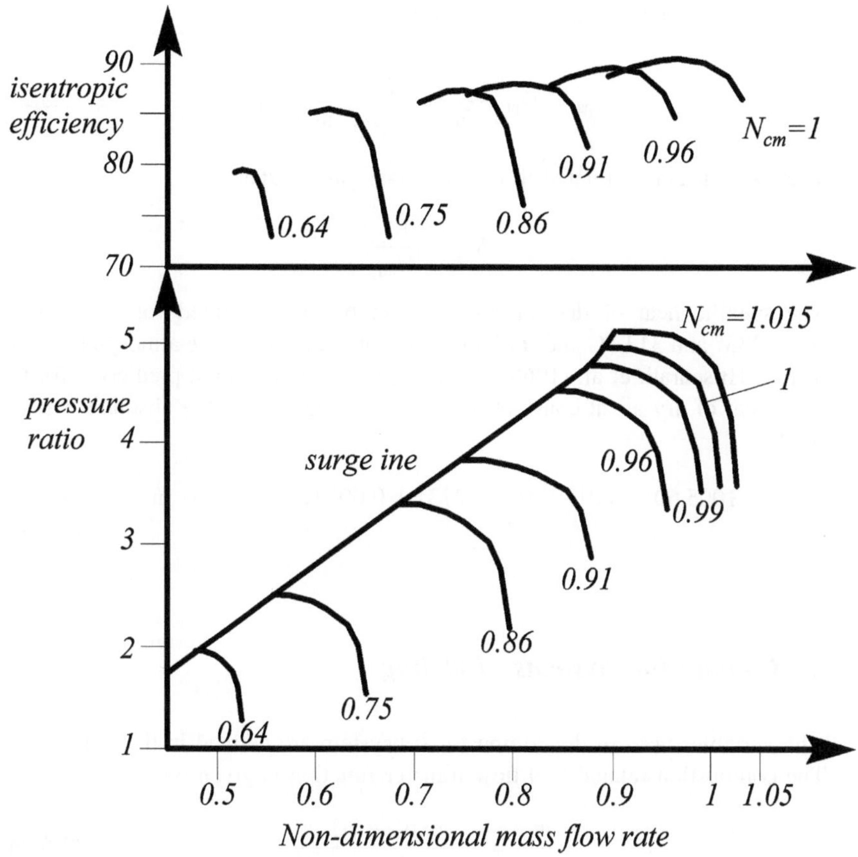

Fig. 5.17 Typical compressor map

The compressor outlet total pressure is related to the compressor outlet total temperature by,

$$P_{\text{cpt}} = \rho_{\text{cp\_sv}} R_{\text{air}} T_{\text{cpt}} \left(1 + \frac{\gamma_{\text{cp}} - 1}{2} M^2 \right)^{\frac{1}{\gamma_{\text{cp}} - 1}}. \tag{5.5.18}$$

The bleed valve flow rate is given by the equation,

$$\dot{m}_{\text{bv}} = K_{\text{bv}} A_{\text{bv}} \frac{P_{\text{cpt}}}{\sqrt{T_{\text{cpt}}}}. \tag{5.5.19}$$

The local Mach number is determined from,

$$M_{\text{cp}} = \frac{\dot{m}_{\text{cp}}}{\rho_{\text{cp\_sv}} A_{\text{cp}} \sqrt{\gamma_{\text{cp}} R_{\text{air}} T_{\text{cpt}}}}. \tag{5.5.20}$$

The compressor outlet static temperature is obtained from,

$$T_{\mathrm{cps}} = T_{\mathrm{cpt}} - \frac{1}{2c_{p\_\mathrm{cp}}}\left(\frac{\dot{m}_{\mathrm{cp}}}{\rho_{\mathrm{cp\_sv}}A_{\mathrm{cp}}}\right)^2. \tag{5.5.21}$$

The gas constant is related to the universal gas constant by,

$$R_{\mathrm{air}} = \frac{R_u}{M_{\mathrm{air}}}. \tag{5.5.22}$$

The specific heat of dry air appears can be approximated for temperature between 233° and 313 °C and for barometric pressures near one atmosphere using data from Hilsenrath et al. (1960) by the following polynomial prediction for the specific heat of dry air at constant pressure as a function of the absolute temperature, T:

$$c_{p\_\mathrm{air}} = 1005.60 + 0.017211(T - 273) + 0.000392(T - 273)^2 \mathrm{in\ Jkg}^{-1°}\mathrm{K}^{-1}. \tag{5.5.23}$$

5.5.2 Combustion Systems Modeling

All the variables used in the combustor subsystem are defined in the Table 5.3.

The combustion-related fuel flow transfer function is given by

$$\frac{\dot{m}_f}{\dot{m}_{f\,\mathrm{in}}} = \frac{\exp(-st_{\mathrm{td}})}{(st_{\mathrm{fl}} + 1)(st_a + 1)} \tag{5.5.24}$$

where t_{td} is the transport delay, t_{fl} is the flame dynamics lag, and t_a is the acoustic-related lag.

The state equations for the combustor stage volume dynamics may be expressed as,

$$\frac{\mathrm{d}}{\mathrm{d}t}\left(\rho_{\mathrm{cb\_sv}}\right) = \frac{\left(\dot{m}_{\mathrm{cb}} + \dot{m}_f - \dot{m}_{\mathrm{tb}}\right)}{V_{\mathrm{cb}}}, \tag{5.5.25}$$

$$\frac{\mathrm{d}}{\mathrm{d}t}\dot{m}_{\mathrm{cb}} = \frac{A_{\mathrm{cb}}}{l_{\mathrm{cb}}}\left(P_{\mathrm{cpt}} - P_{\mathrm{cbt}} - \Delta P_{\mathrm{cbt}}\right), \tag{5.5.26}$$

$$\frac{\partial}{\partial t}T_{\mathrm{cbt}} = \frac{\gamma_{\mathrm{cb}}}{\rho_{\mathrm{cb\_sv}}V_{\mathrm{cb}}}\left(T_{\mathrm{cpt}}\dot{m}_{\mathrm{cb}} + \frac{\left(c_{p\_\mathrm{comb}}T_{\mathrm{cbt}} + \eta_{\mathrm{cb}}LHV\right)}{c_{p\_\mathrm{comb}}}\dot{m}_f - T_{\mathrm{cbt}}\dot{m}_{\mathrm{tb}}\right). \tag{5.5.27}$$

The combustor outlet total pressure is related to the combustor outlet total temperature by,

$$P_{\mathrm{cbt}} = \rho_{\mathrm{cb\_sv}}R_{\mathrm{comb}}T_{\mathrm{cbt}}. \tag{5.5.28}$$

Table 5.3 Definition of combustor subsystem variables

Variable	Description
A_{cb}	Combustor stage volume cross-sectional area
c_{p_comb0}	Value of c_{p_comb} at temperature $T = T_0$
c_{p_comb}	Specific heat at constant pressure for combustor
c_{pg}	Temperature gradient of c_{p_comb} at $T = T_0$
K_b	Constant in expression for combustor pressure loss
K_c	Constant in expression for combustor pressure loss
K_p	Constant in expression for combustor pressure loss
l_{cb}	Length of combustor stage volume duct
LHV	Lower heating value of fuel
$\dot{m}_{\text{cb}}$	The combustion chamber inlet mass flow rate
$\dot{m}_d$	The compressor inlet design mass flow rate
$\dot{m}_f$	Combustor fuel mass flow rate
$\dot{m}_{f\,\text{in}}$	Combustor inlet fuel mass flow rate
$\dot{m}_{\text{tb}}$	The turbine inlet mass flow rate
M_{air}	Molecular weight of air
M_{cp}	Mach number in compressor stage volume
M_f	Molecular weight of fuel
P_{cbt}	Combustor outlet total pressure
P_{cpt}	Compressor outlet total pressure
ΔP_{cbt}	Combustor pressure loss
P_{inlet}	The total inlet pressure
P_{ref}	The combustor map reference pressure
R_{comb}	Gas constant for gases in combustor
R_u	Universal Gas constant
s	Laplace transform variable
t_a	Acoustic time lag in transfer function
t_{fl}	Flame dynamics time lag in transfer function
t_{td}	Transport delay in transfer function
T_{cbs}	Combustor outlet static temperature
T_{cbt}	Combustor outlet total temperature
V_{cb}	Volume of combustor stage volume duct
γ_{cb}	Ratio of specific heats in combustor stage volume
ϕ	Fuel air ratio
$\rho_{\text{cb\_sv}}$	Combustor stage volume density

The specific heat coefficient c_{p_comb} at constant pressure in the combustion chamber at any temperature T is assumed to be given by a linear law,

$$c_{p\_\text{comb}}(T) = c_{p\_\text{comb0}} + (T - T_0)c_{pg}. \tag{5.5.29}$$

The gas constant is given by,

$$R_{\text{comb}} = \frac{R_u}{(1-\phi)M_{\text{air}} + \phi M_f}. \tag{5.5.13}$$

The pressure loss in the combustor is empirically modeled by the relation,

$$\frac{\Delta P_{\mathrm{cbt}}}{P_{\mathrm{cbt}}} = K_c \left(\frac{\dot{m}_{\mathrm{cb}}}{A_{\mathrm{cb}}}\right)^2 \frac{\left(K_p T_{\mathrm{cpt}} - K_b T_{\mathrm{cbt}}\right)}{P_{\mathrm{cbt}} P_{\mathrm{cpt}}}, \tag{5.5.14}$$

where K_p, K_b and K_c are experimentally determined constants.

5.5.3 Turbine Dynamics Modeling

The turbine consists of alternating stators where each stator is stationary arrangement of blades on a hub, with each blade cross section is a typical aerofoil section followed by rotors where each rotor is rotating arrangement of blades on a hub with each blade cross section is similar to the corresponding stator blade cross section. Each row decreases the pressure of the incoming fluid by a certain factor, converting the incoming pressure energy to kinetic energy ("expanding") by turning the flow. The turbine extracts the tangential velocity from the flow to power the compressor. The dimensional analysis of the turbine reveals the similarity with the compressor. Consequently, the component maps for a turbine and compressor are fundamentally similar. A typical turbine map is shown in Fig. 5.18.

All the additional variables used in the turbine subsystem are defined in the Table 5.4.

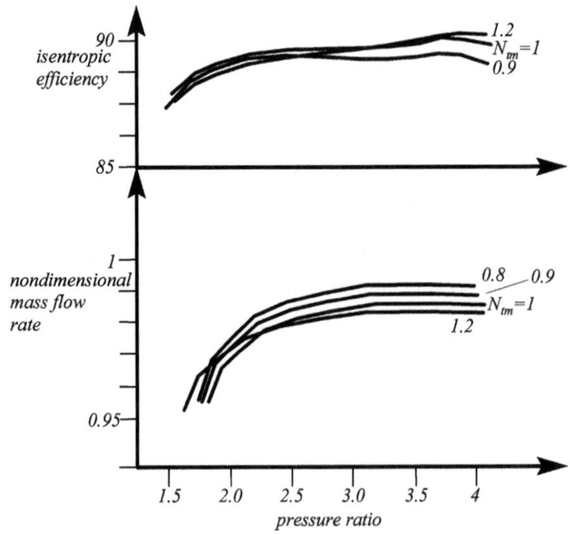

Fig. 5.18 Typical turbine map

Table 5.4 Definition of turbine subsystem variables

Variable	Description
A_{tb}	Cross-sectional area of turbine stage volume duct
c_{p_tb}	Specific heat at constant pressure for turbine gases
h_{in}	Total sp. enthalpy of working fluid into turbine
h_{out}	Total sp. Enthalpy of working fluid out of turbine
K_{tm}	Turbine map-related constant
l_{tb}	Length of turbine stage volume duct
$\dot{m}_{\text{ab}}$	The afterburner inlet mass flow rate
$\dot{m}_{\text{bv}}$	The bleed valve mass flow rate
$\dot{m}_{\text{ca}}$	Mass flow rate of cooling air
$\dot{m}_{\text{tb}}$	The turbine inlet mass flow rate
$\dot{m}_{\text{tref}}$	The turbine map mass flow rate constant
M_{air}	Molecular weight of air
$M_{\text{cool\_air}}$	Molecular weight of cool air
M_f	Molecular weight of fuel
M_{tb}	Mach number in turbine stage volume
N_{tm}	Turbine non-dimensional specific spool speed
N_{tref}	The turbine (engine) reference spool speed
P_{tref}	Turbine map constant
P_{tbc}	Turbine map characteristic pressure
P_{tbt}	The turbine total outlet pressure
R_{exhaust}	Gas constant for combustion chamber exhaust
T_{cat}	The total cool air temperature
T_{tbc}	Turbine characteristic temperature
T_{tbs}	Turbine outlet static temperature
T_{tbt}	Turbine outlet total temperature
T_{tref}	The turbine map reference temperature
V_{cb}	Volume of combustor stage volume duct
γ_{cb}	Ratio of specific heats in combustor stage volume
η_{tb}	Turbine isentropic efficiency
$\rho_{\text{cb\_sv}}$	Combustor stage volume density
χ	Cool air to hot air ratio
ω	Turbine shaft speed in rad/s
P_{turbine}	Turbine power output
T_{turbine}	Turbine torque output

The turbine map is taken to be in the form

$$\dot{m}_{\text{tb}} = \frac{\dot{m}_{\text{tref}}}{1 + K_{\text{tm}}(N_{\text{tm}} - 1)} \left(1 - e^{-\frac{P_{\text{tbc}}}{P_{\text{tref}}}}\right) \tag{5.5.15}$$

where $\dot{m}_{\text{tref}}, P_{\text{tref}}$ and K_{tm} are functions of N_{tm} and N_{tm} is the non-dimensional turbine map specific speed,

$$N_{\text{tm}} = \sqrt{\frac{T_{\text{tref}}}{T_{\text{cbt}}}} \frac{N}{N_{\text{tref}}}. \tag{5.5.16}$$

The turbine efficiency and $\dot{m}_{\text{tref}}, P_{\text{tref}}$ and K_{tm} are obtained from a lookup table.

The turbine outlet characteristic static temperature is found from the turbine characteristic static pressureP_{tbc},

$$T_{\text{tbc}} = \left(1 - \eta_{\text{tb}}\left(1 - \left(\frac{P_{\text{cbt}}}{P_{\text{tbc}}}\right)^{-\frac{\gamma_{\text{cp}}-1}{\gamma_{\text{cp}}}}\right)\right)T_{\text{cbt}}. \tag{5.5.17}$$

The state equations for the turbine stage volume dynamics may be expressed as

$$\frac{\mathrm{d}}{\mathrm{d}t}\left(\rho_{\text{tb\_sv}}\right) = \frac{\left(\dot{m}_{\text{tb}} + \dot{m}_{\text{bv}} + \dot{m}_{\text{ca}} - \dot{m}_{\text{ab}}\right)}{V_{\text{tb}}}, \tag{5.5.18}$$

$$\frac{\mathrm{d}}{\mathrm{d}t}\dot{m}_{\text{tb}} = \frac{A_{\text{tb}}}{l_{\text{tb}}}\left(P_{\text{tbc}} - P_{\text{tbt}}\right)\left(1 + \frac{\gamma_{\text{tb}} - 1}{2}M_{\text{tb}}^2\right)^{\frac{\gamma_{\text{tb}}}{\gamma_{\text{tb}}-1}}, \tag{5.5.19}$$

$$\frac{\partial}{\partial t}T_{\text{tbt}} = \frac{\gamma_{\text{tb}}}{\rho_{\text{tb\_sv}}V_{\text{tb}}}\left(T_{\text{tbc}}\dot{m}_{\text{tb}} + T_{\text{cpt}}\dot{m}_{\text{bv}} + T_{\text{cat}}\dot{m}_{\text{ca}} - T_{\text{tbt}}\dot{m}_{\text{ab}}\right). \tag{5.5.20}$$

The turbine outlet total pressure is related to the turbine outlet total temperature by,

$$P_{\text{tbt}} = \rho_{\text{tb\_sv}}R_{\text{exhaust}}T_{\text{tbt}}\left(1 + \frac{\gamma_{\text{tb}} - 1}{2}M_{\text{tb}}^2\right)^{\frac{1}{\gamma_{\text{tb}}-1}} \tag{5.5.21}$$

and the corresponding Mach number is,

$$M_{\text{tb}} = \frac{\dot{m}_{\text{tb}}}{\rho_{\text{tb\_sv}}A_{\text{tb}}\sqrt{\gamma_{\text{tb}}R_{\text{exhaust}}T_{\text{tbt}}}}. \tag{5.5.22}$$

The gas constant is given by,

$$R_{\text{exhaust}} = \frac{R_u}{(1-\phi)((1-\chi)M_{\text{air}} + \chi M_{\text{cool\_air}}) + \phi M_f}. \tag{5.5.23}$$

The turbine outlet static temperature is obtained from,

$$T_{\text{tbs}} = T_{\text{tbt}} - \frac{1}{2c_{p\_\text{tb}}}\left(\frac{\dot{m}_{\text{tb}}}{\rho_{\text{tb\_sv}}A_{\text{tb}}}\right)^2. \tag{5.5.24}$$

5.5.4 Turbine Power and Torque Outputs

The power developed by the turbine is calculated using the change in enthalpy in the working fluid. This enthalpy is based on the linearized enthalpy temperature relationship of the air and fuel entering the combustion chamber and that of the products of combustion.

$$P_{\text{turbine}} = \eta_{\text{tb}}\dot{m}_{\text{tb}}(h_{\text{in}} - h_{\text{out}}) \tag{5.5.25}$$

where $P_{turbine}$ is the total power developed by the turbine, $\dot{m}_{tb}$ is the total mass flow rate into the turbine, η_{tb} is the isentropic efficiency of turbine, h_{in} is the total specific enthalpy of the working fluid into the turbine, h_{out} is the total specific enthalpy of the working fluid out of the turbine based on room temperature. To estimate h_{in} it is essential to model all the components contributing to the turbine output.

The torque developed by the turbine can be calculated using the power developed by the turbine and the angular velocity of the rotor from the compressor model.

$$T_{turbine} = P_{turbine}/\omega \tag{5.5.26}$$

where $T_{turbine}$ is the total torque developed by the turbine. The net torque out available of the entire gas turbine is the difference between $T_{turbine}$ and the torque need to run the compressor.

5.5.5 One-Dimensional Variable Area Ducts: The Afterburner and Nozzle

The afterburner and nozzle are modeled as a single component. The afterburner can be used in one of two modes, either including the combustor or by in a mode when no combustion takes place. Typically in supersonic flight, if an afterburner is used, it is only ignited at the time of takeoff and powered though the transonic flight regime. In a cruise condition, it usually operates in a non-combusting mode. Thus, if one is interested in simulating the flight in a cruise condition, a non-combusting mode of operation of the afterburner meets the requirements. The afterburner essentially acts as a large volume that is capable of attenuating disturbances. Furthermore, simulating the dynamics of the mass flow rate at the nozzle with a variable nozzle area must satisfy the exit boundary condition. The exit boundary condition is a typical choked flow. Even when the nozzle varies the exit area based on the corrected speed of the engine to match steady-state operating conditions, the choked flow must be maintained through the nozzle. All the additional variables used in the afterburner and nozzle subsystem are defined in the Table 5.5. The nozzle mass flow rate is given by,

$$\dot{m}_{nz} = \frac{K_{nz}A_{nz\_ref}P_{abt}}{\sqrt{T_{abt}}}\left(\frac{P_{ambient}}{P_{abt}}\right)^{\frac{1}{\gamma_{ab}}}\sqrt{1-\left(\frac{P_{ambient}}{P_{abt}}\right)^{\frac{\gamma_{ab}-1}{\gamma_{ab}}}}. \tag{5.5.27}$$

The variable nozzle area and flow coefficient are represented by K_{nz} while A_{nz_ref} is a reference nozzle throat area. This variable can be used as a tuning factor or as control variable to obtain the expected steady-state results.

The state equations for the combined nozzle and afterburner stage volume dynamics may be expressed as,

$$\frac{\mathrm{d}}{\mathrm{d}t}\left(\rho_{\mathrm{ab\_sv}}\right) = \frac{\left(\dot{m}_{\mathrm{ab}} - \dot{m}_{\mathrm{nz}}\right)}{V_{\mathrm{ab}}}, \tag{5.5.28}$$

$$\frac{\mathrm{d}}{\mathrm{d}t}\dot{m}_{\mathrm{ab}} = \frac{A_{\mathrm{ab}}}{l_{\mathrm{ab}}}\left(P_{\mathrm{tbt}} - P_{\mathrm{abt}}\right)\left(1 + \frac{\gamma_{\mathrm{ab}} - 1}{2}M_{\mathrm{ab}}^2\right)^{\frac{\gamma_{\mathrm{ab}}}{\gamma_{\mathrm{ab}}-1}}, \tag{5.5.29}$$

$$\frac{\partial}{\partial t}T_{\mathrm{abt}} = \frac{\gamma_{\mathrm{ab}}}{\rho_{\mathrm{ab\_sv}}V_{\mathrm{ab}}}\left(T_{\mathrm{tbt}}\dot{m}_{\mathrm{ab}} - T_{\mathrm{abt}}\dot{m}_{\mathrm{nz}}\right). \tag{5.5.30}$$

The afterburner nozzle outlet total pressure is related to the afterburner nozzle outlet total temperature by,

$$P_{\mathrm{abt}} = \rho_{\mathrm{ab\_sv}}R_{\mathrm{comb}}T_{\mathrm{abt}}\left(1 + \frac{\gamma_{\mathrm{ab}} - 1}{2}M_{\mathrm{ab}}^2\right)^{\frac{1}{\gamma_{\mathrm{ab}}-1}} \tag{5.5.31}$$

and the corresponding Mach number is,

$$M_{\mathrm{ab}} = \frac{\dot{m}_{\mathrm{ab}}}{\rho_{\mathrm{ab\_sv}}A_{\mathrm{ab}}\sqrt{\gamma_{\mathrm{ab}}R_{\mathrm{comb}}T_{\mathrm{abt}}}}. \tag{5.5.32}$$

Table 5.5 Definition of afterburner and nozzle subsystem variables

Variable	Description
A_{ab}	Cross-sectional area of afterburner stage volume duct
A_{ex}	Nozzle exhaust area
$A_{\mathrm{nz\_ref}}$	Reference nozzle throat area
c_{p_ab}	Specific heat at constant pressure in the afterburner
K_{nz}	Nozzle coefficient
l_{ab}	Length of the afterburner stage volume duct
$\dot{m}_{\mathrm{ab}}$	The afterburner inlet mass flow rate
$\dot{m}_{\mathrm{nz}}$	Nozzle mass flow rate
M_{ab}	Mach number in the afterburner stage volume
P_{abt}	Afterburner outlet total pressure
P_{ambient}	Ambient pressure
P_{exhaust}	Pressure in the nozzle exhaust
T_{abt}	Afterburner outlet total temperature
U_e	Velocity of exhaust gases
U_0	Velocity of jet engine
V_{ab}	Volume of afterburner stage volume duct
γ_{ab}	Ratio of specific heats in afterburner stage volume
γ_{nz}	Ratio of specific heats in the nozzle
η_{nz}	Nozzle efficiency
$\rho_{\mathrm{ab\_sv}}$	Afterburner stage volume density

The net thrust generated by the nozzle is the sum of the thrust generated due to the change in momentum and due to the pressure difference acting over the cross-sectional area of the exhaust and is,

$$\mathbf{T} = (U_e - U_0)\dot{m}_{\text{nz}} + (P_{\text{exhaust}} - P_{\text{ambient}})A_{\text{ex}} \tag{5.5.33}$$

where

$$P_{\text{exhaust}} = P_{\text{abt}}\left(1 - \frac{\gamma_{\text{nz}} - 1}{\eta_{\text{nz}}(\gamma_{\text{nz}} + 1)}\right)^{\frac{\gamma_{\text{nz}}}{\gamma_{\text{nz}}-1}}. \tag{5.5.34}$$

5.5.6 The Spool Dynamics Modeling

All the additional variables used in the spool shaft subsystem are defined in Table 5.6. The dynamics of the spool shaft is given by,

$$I_{\text{spool}}\frac{\mathrm{d}\omega}{\mathrm{d}t} = \left(\frac{P_{\text{turbine}} - P_{\text{compressor}}}{\omega}\right), \tag{5.5.35}$$

where the power generated by the turbine is,

$$P_{\text{turbine}} = \dot{m}_{\text{cb}}c_{p\_\text{comb}}T_{\text{cbt}} + \dot{m}_{\text{ca}}c_{p\_\text{air}}T_{\text{cat}} - \dot{m}_{\text{ab}}c_{p\_\text{ab}}T_{\text{abt}} \tag{5.5.36}$$

and the power consumed by the compressor

$$P_{\text{compressor}} = (\dot{m}_{\text{cb}} + \dot{m}_{\text{bv}})c_{p\_\text{cp}}T_{\text{cpt}} - \dot{m}_{\text{cp}}c_{p\_\text{air}}T_{\text{inlet}}. \tag{5.5.37}$$

5.5.7 Typical Simulation Results

Connolly (2009) and Connolly et al. (2010) have presented a number of typical results of simulation studies of a typical turbofan engine and the reader is referred to them for the details.

Table 5.6 Definition of spool shaft subsystem variables

Variable	Description
I_{spool}	Mass moment of inertia of spool shaft
$P_{\text{compressor}}$	Power consumed by compressor
P_{turbine}	Power delivered by the turbine
c_{p_air}	Specific heat at constant pressure for air at the inlet

5.6 Full-Authority Digital Engine Control Systems

Electronic controllers for jet engines were developed in the 1980s. Currently, the jet engine controllers use highly customized computers to synthesize the control laws with a number of sensor inputs and actuator outputs. Such electronic controllers result in higher engine operating efficiencies by allowing for precise engine control through the use of multiple control loops and improved control laws to reduce transient overshoot or undershoot. It also allows implementation of control algorithms which would be difficult to implement using traditional wired circuitry. Current engine control systems are generally implemented digitally and cannot be overruled by the pilot and for this reason are referred to as full-authority digital engine controllers (FADEC). The primary objective of a FADEC is to operate the engine while avoiding any instabilities over the entire engine while delivering the commanded performance in terms of the thrust generated. Several modes of instability can occur in a jet engine: surge, which is a longitudinal flow oscillation over the length of the compressor and turbine, and rotating stall, which is the lack of pressure rise between the compressor blades. Often rotating stall occurs at low rotor speeds and surge at high rotor speeds. Both surge and rotating stall generate violent axial oscillations of the internal air column which can cause substantial damage to both the compressor and the engine. Fan stalls can be caused by operation with too small a fan duct nozzle area, booster stalls by a throttle reduction to a lower engine rotational speed, and compressor stalls by a rapid throttle increase. The engine and control system must be designed to avoid surge/stall with sufficient design margin to offset the effects of altitude, increased tip clearances, component deterioration, and engine/airflow operation at high angles of attack. Too much fuel can result in a blowout, where soaking the flame with fuel displaces the oxygen and lowers the temperature enough to extinguish the flame. Depending on the combustor pressure, the combustor can operate in certain regions where they tend to "blow out." There are two regions of blowout associated with most jet engine combustors. The high-temperature blowout region is referred to as "rich blowout" and the low-temperature region is referred to as "lean blowout." Flame-outs which refer to the extinguishing of the flame, although rare, must also be avoided in the combustor as the consequences of flame-outs coupled with other potential instabilities can be quite disastrous.

While higher turbine inlet temperatures can lead to improved specific thrust, a lighter engine but requires expensive turbine materials and a complex turbine cooling system which reduces cruise performance. A proper balance of temperature must be maintained in the engine, and it will depend on the relative importance of specific thrust which sets engine size and weight and fuel requirements in cruise.

Typically, a FADEC must ensure that the following limits are not exceeded:

1. maximum fan speed and compressor speeds,
2. maximum turbine temperature,
3. avoid fan and compressor stall,

4. meet maximum and minimum compressor discharge pressure requirements,
5. Avoid lean and rich blowouts.

To meet these requirements, a minimum set of measurements must be made and these are as follows:

1. High- and low-pressure spool speeds.
2. Engine pressure ratio.
3. Ratio of the fuel flow rate to compressor exit pressure.
4. Low-pressure turbine inlet temperature and
5. The control inputs which could be the fuel–air ratio, the compressor inlet guide vanes, and/or the variable nozzle throat area.

All the states of the system are generally estimated from the measurements, and for this reason, a much larger set of measurements is generally desirable. Jet engine dynamic models may involve as many as 35–40 state variables and four times as many measurements are made in practice.

Primary control strategies are to control the spool shaft speeds, usually of the low-pressure spool so as to indirectly be able to control the thrust. Direct control of thrust, although rare, is also possible in some jet engines. For a more detailed description of the issues involved in the design of the controller, the reader is referred to Spang and Brown (1999).

References

Auer MP, Gebauer C, Mösl KG, Hirsch C, Sattelmayer T (2004) Active instability control: feedback of combustion instabilities on the injection of gaseous fuel. In: Proceedings of the ASME Turbo Expo 2004, power for land, sea and air, Vienna, Austria

Bellucci V, Paschereit CO, Flohr P, Fulvio M (2001) On the use of Helmholtz resonators for damping acoustic pulsations in industrial gas turbines. In: Proceedings of the ASME TURBO EXPO 2001, New Orleans, USA

Bloxsidge GJ, Langhorne PJ, Dowling AP (1988) Reheat buzz: an acoustically coupled combustion instability. II. Theory. J Fluid Mech 193:445–473

Campos-Delgado DU, Schuermans BBH, Zhou K, Paschereit CO, Gallestey EA, Poncet A (2003) Thermoacoustic instabilities: modeling and control. In: IEEE transactions on control systems technology, vol. 11, pp 429–447

Campos-Delgado DU, Zhou K, Allgood D, Acharya S (2003b) Active control of combustion instabilities using model-based controllers. Combust Sci Technol 175:27–53

Chen L, Smith R, Dullerud G (1988) A linear perturbation model for a non-linear system linearized at an equilibrium neighbourhood. In: Proceedings of the 37th IEEE conference on decision and control, Tampa, Florida USA, pp 4105–4106

Connolly JW (2009) Turbo fan engine dynamic modeling and control. Masters thesis, Department of Electrical Engineering and Computer Science, Case Western Reserve University

Connolly JW, Kopasakis G, Lemon KA (2010) Turbofan volume dynamics model for investigations of Aero-Propulso-Servo-Elastic effects in a supersonic commercial transport. NASA/TM—2010-216069

Cullick FEC (1988) Combustion instabilities in liquid fuelled propulsion systems. AGARD-CP-450, 1–1:1–72

de Jager B (1995) Rotating stall and surge control: a survey. In: Proceedings of 34th IEEE conference on decision and control, New Orleans, LA, USA, pp 1857–1862

DeLaat JC, Chang CT (2003) Active control of high frequency combustion instability in aircraft gas-turbine engines. In: Proceedings of 16th international symposium on airbreathing engines, sponsored by the international society for airbreathing engines, NASA/TM—2003-212611, Cleveland, Ohio

Docquier N, Candel S (2002) Combustion control and sensors: a review. Prog Energy Combust Sci 28:107–150

Dowling AP (2003) The challenges of lean premixed combustion. In: Proceedings of the international gas turbine congress, IGTC2003Tokyo KS-5, Tokyo

Dowling AP, Morgans AS (2005) Feedback control of combustion oscillations. Ann Rev Fluid Mech 37:151–182

Epstein AH, Williams JEF, Greitzer EM (1989) Active suppression of aerodynamic instabilities in turbomachinery. J Propul 5:204–211

Evesque S, Park S, Riley AJ, Annaswamy AM, Dowling AP (2004) Adaptive combustion instability control with saturation: theory and validation. J Propul Power 20:1086–1095

Gravdahl JT, Egeland O (1997) A Moore-Greitzer axial compressor model with spool dynamics. In: Proceedings of the 36th conference on decision and control, San Diego, CA, USA, pp 4714–4715

Greitzer EM (1976) Surge and rotating stall in axial flow compressors, Part I, II. J Eng Power 98(2):190–217 (ASME Transactions)

Greitzer EM (2009) Some aerodynamic problems of aircraft engines: fifty years after-the 2007 IGTI scholar lecture. ASME J Turbomach 121(031101):1–13

Gu G, Banda S, Sparks A (1996) An overview of rotating stall and surge control for axial flow compressors. In: Proceedings of IEEE conference on decision and control, Kobe, Japan, pp 2786–2791

Gu G, Sparks A, Banda S (1997) Bifurcation based non-linear feedback control for rotating stall in axial flow compressors. Int J Contr 6:1241–1257

Hawthorne WR, Horlock JH (1962) Actuator disc theory of the incompressible flow in axial compressor. Proc Inst Mech Eng London 196(30):798

Hilsenrath J et al (1960) Tables of thermodynamic and transport properties. Pergamon Press, New York

Hõs C, Champneys A, Kullmann L (2003) Bifurcation analysis of surge and rotating stall in the Moore-Greitzer compression system. IMA J Appl Math 68(2):205–228

Kopasakis G (2003) High-frequency instability suppression controls in a liquid-fueled combustor. In: Proceedings of 39th joint propulsion conference and exhibit, AIAA–2003–1458, Huntsville, AL

Kopasakis G (2007) Feedback control systems loop shaping design with practical considerations, NASA TM-2007-215007

Kopasakis G, Connolly JW, Paxson DE, Ma P (2008) Volume dynamics propulsion system modeling for supersonics vehicle research, NASA.TM- 2008-215172

Krstic M, Protz JM, Paduano JD, Koktovic PV (1995) Backstepping designs for jet-engine stall and surge control. In: Proceedings of IEEE conference on decision and control, pp 3049–3055

Le D, DeLaat J, Chang C (2003) Control of thermo-acoustic instabilities: the multi-scale extended Kalman approach. In: Proceedings of 39th joint propulsion conference and exhibit, AIAA–2003–4934, Huntsville, AL

Mattingly JD (2005) Elements of gas turbine propulsion. McGraw-Hill Education (India) Pvt. Ltd., Noida

Moore FK, Greitzer EM (1986) A theory of post-stall transients in axial compressors: part I—development of the equations. ASME J Eng Gas Turbines Power 8:68–76

Nayfeh MA, Abed EH (1999) High-gain feedback control of surge and rotating stall in axial flow compressors. In: Proceedings of the American control conference, Ca. 2663-2667, San Diego

Paduano JD, Epstein AH, Valavani L, Longley JP, Greitzer EM, Gunette GR (1993) Active control of rotating stall in a low-speed axial compressor. ASME J Turbomach 115:48–56

Rayleigh JWS (1878a) The explanation of certain acoustical phenomena. Nature 18:319

Rayleigh JWS (1878b) The explanation of certain acoustical phenomena. Royal Inst Proc 8:536–542

Rayleigh JWS (1945) The theory of sound. Dover, New York

Rijke PL (1859) Notiz über eine neue Art, die in einer an beiden Enden offenen Röhre enthaltene Luft in Schwingungen zu versetzen. Annalen der Physik und Chemie 183(2–107):339–343

Riley AJ, Park S, Dowling AP, Evesque S, Annaswamy AM (2004) Advanced closed-loop control on an atmospheric gaseous lean-premixed combustor. ASME J Eng Gas Turbines Power 126:708–716

Seldner K, Cwynar DS (1978) Procedures for generation and reduction of linear models of a turbofan engine, NASA TP- 1261

Seldner K, Mihaloew JR, Blaha RJ (1972) Generalized simulation technique for turbojet engine system analysis, NASA TN-D-6610

Spang HA III, Brown H (1999) Control of jet engines. Contr Eng Pract 7:1043–1059

Stueber TJ, Melcher KJ (2004) J85-13 turbojet engine model for propulsion system control studies, NASA TP-2004-212223

Vepa R (2010) Modelling and quasi-linear control of compressor surge and rotating stall vibrations. Math Prob Eng Article ID 314172 (Special Issue on Nonlinear Vibrations, Stability Analysis and Control)

Young S, Wang Y, Murray RM (1998) Evaluation of bleed valve rate requirements in non-linear control of rotating stall on axial flow compressors. Division of engineering and applied science, CALTECH CDS TR 98-001, California Institute of Technology, CA, USA

Chapter 6
Modelling and Simulation of Fuel Cells

6.1 Fuel Cell Systems

As already discussed, the fuel cell (FC) is an electrochemical energy conversion system in Chap. 1, where chemical energy is converted directly into electrical energy and heat. There are a number of advantages of fuel cell technologies which include higher efficiencies at partial loads, lower or no emissions, noiselessness operation due to the nonexistence of mechanical parts, controllability of electric and heat generation, and overall energy savings. The energy savings result from the higher conversion efficiency, typically 40 % or higher, depending on the type of fuel cell. When utilized in a cogeneration applications for recovering all the available thermal energy output, overall energy utilization efficiencies can be in the order of 85 % or more.

Physically, a FC plant consists of three major subsystems: (1) a fuel processing subsystem that preprocesses and conditions the fuel for power generation, (2) a power generation subsystem that consists of several stacks of fuel cells and mainly generates electric power, and (3) a power conditioning or post-processing subsystem that conditions the generated power, so it can be directly supplied to the electricity grid.

The five primary FC technologies are (1) low-temperature FCs operating at or below 80 °C based on proton exchange membrane and alkaline FCs, (2) medium temperature FCs operating at 200 °C and based on phosphoric acid as the electrolyte and using natural gases as fuels (PAFC), and (3) high-temperature FCs operating at or above 650 °C and based on molten carbonate FC (MCFC) and solid oxide FC (SOFC). This last category of high-temperature systems are preferred for cogeneration and in the development hybrid prime movers (hybrid gas turbines). The main advantage of SOFCs in comparison with MCFC is the higher temperatures, allowing internal reforming of hydrocarbons such as methane and butane. Moreover, the solid electrolyte of SOFC has an advantage over liquid electrolytes in an MCFC where special protection facilities which are necessary for the corrosive electrolytes. The primary fuel cell types are compared in Table 6.1. They are extensively discussed by O'Hayre et al. (2006).

R. Vepa, *Dynamic Modeling, Simulation and Control of Energy Generation*, Lecture Notes in Energy 20,
DOI: 10.1007/978-1-4471-5400-6_6, © Springer-Verlag London 2013

Table 6.1 Fuel cells compared

Fuel cell type→	PEMFC	PAFC	MCFC	SOFC
Electrolyte→	Nafion (perfluoro sulfonic acid based), Pt_3Ni-based polymer	H_3PO_4 in a Teflon-bonded silicon carbide matrix	Li_2CO_3/K_2CO_3	8 mol % yttria-stabilized zirconia
			(lithium or potassium carbonate) in a porous lithium aluminate matrix	Scandia-stabilized zirconia
				Gadolinium-doped ceria
Mobile ions→	H^+	H^+	CO_3^{2-}	O^{2-}
Anode→	Platinum(Pt) catalyzed, PTFE s(Teflon)-bound, thin-film (carbon, Pt)	Pt catalyzed with carbon black	Nickel (Ni) or its alloy with chromium or aluminum	Nickel ceramic cermets
Cathode→	Pt catalyzed, PTFE(Teflon)-bound, thin-film (carbon, Pt)	Pt catalyzed with carbon black	Nickel oxide, NiO	Lanthanum strontium manganite
Operating temperature→	<80°	~200°	>650°	>650°
Start-up time→	10 s	3–10 h	2 h	2 h
Power output, mW/cm^2→	400	150–300	150	150
Primary fuel→	Hydrogen	Hydrogen, natural gas, methanol, naptha, lighter oils	Hydrogen, natural and coal gas, methanol, naptha, petroleum	Hydrogen, natural and coal gas, methanol, naptha, petroleum
Reforming→	No	No	Yes	Yes
Efficiency→	40–50 %	40 %	50–60 %	50–60 %
Efficiency with cogeneration→	–	–	75 %	75 %
CO poisoning→	Severe	Limited	No	No
Applications→	Backup or portable power	Distributed generation	Electric utility	Auxiliary power, electric utility
	Distributed generation		Distributed generation	Distributed generation
	Transportation			

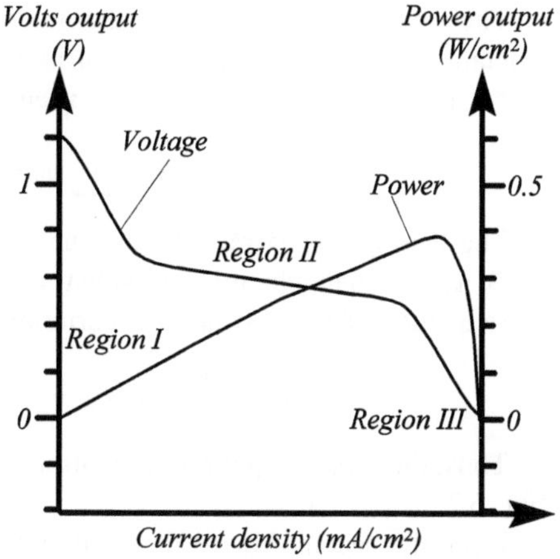

Fig. 6.1 Fuel cell output voltage and power versus current density *curve*

The performance of a fuel cell is represented by the output voltage and output power versus current density curve as shown in Fig. 6.1. The open-circuit voltage, as determined by the Gibbs potential or "free energy" property of reactants, is 1.23 V. In practice, the fuel cell voltage is below this theoretical value. At small current densities (Region 1), the activation energy associated with the chemical reaction results in a drop in the voltage. At higher current densities (Region 2), the voltage drop is dominated by the Ohmic losses in the electrode structure and the electrolyte. At even higher current densities (Region 3), the rate at which the reactants can diffuse to or the products to diffuse away from the reaction site determines the losses. Due to the limitation caused by the diffusion process, the current which can be drawn from the fuel cell has a maximum value called the limiting current. Better cells are those with flatter curves and higher limiting currents. Fuel cell performance can only in principle be increased by increasing the cell temperature and the reactant pressure. At higher temperature and pressure operating conditions, material and hardware limitations tend reduce the efficiencies.

6.2 Thermodynamics and Electrochemistry of Fuel Cells

6.2.1 Thermodynamics of Fuel Cells

A fuel cell (FC) is an electrochemical device that converts chemical energy of the fuel and oxidant directly and reversibly into electrical energy. Hydrogen and oxygen may be used as examples of the fuel and oxidant, respectively, to illustrate the thermodynamics of the fuel cell.

The first and second law of thermodynamics describe the reversible FC. The reactants (fuel and air) deliver total input enthalpy, and the total output enthalpy leaves the fuel cell, so the change in enthalpy, ΔH, is

$$\Delta H = \sum n_{\text{in}}\text{H}_{\text{in}} - \sum n_o\text{H}_o. \tag{6.2.1}$$

The heat q_{FC} is extracted from the fuel cell and simultaneously the reversible work w_{FC} is done which is equivalent to extracting $-w_{\text{FC}}$ in terms of energy. As a consequence of the first law of thermodynamics,

$$\Delta H = \sum n_{\text{in}}H_{\text{in}} - \sum n_o H_o = q_{\text{FC}} + w_{\text{FC}}. \tag{6.2.2}$$

Thus, the input/output representation of the fuel cell follows and is illustrated in Fig. 6.2.

From the second law of thermodynamics,

$$\oint \mathrm{d}S \geq 0. \tag{6.2.3}$$

In the ideal or reversible case,

$$\Delta S = q_{\text{FC}}/T_{\text{FC}}. \tag{6.2.4}$$

Thus, the reversible work may be obtained from the above equations as,

$$w_{\text{FC}} = \Delta H - T_{\text{FC}}\Delta S. \tag{6.2.5}$$

This is equal to the change in the Gibbs potential. Thus, for ideal gases and a reversible process,

$$\Delta G = w_{\text{FC}} = \Delta H - T_{\text{FC}}\Delta S \tag{6.2.6}$$

where

$$S(p,T) = S_0 + \int_{T_0}^{T} \frac{C_p(\tau)}{\tau} d\tau - R \cdot \ln\left(\frac{p}{p_0}\right) \tag{6.2.7}$$

Thus, the Gibbs potential may be expressed as,

$$\Delta G(p,T) = \Delta G_0 + RT \cdot \ln(K_{\text{eq}}) \tag{6.2.8}$$

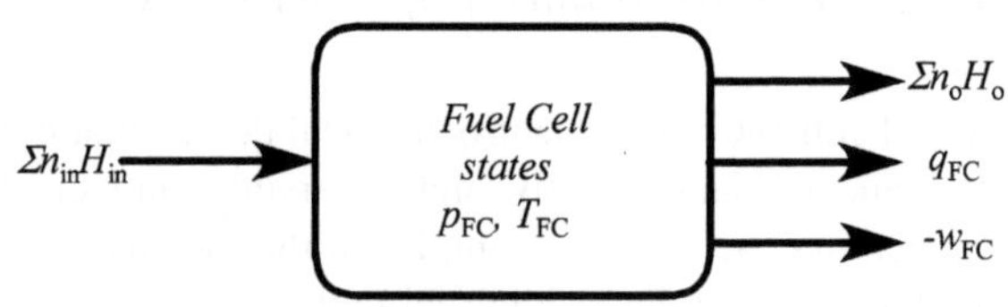

Fig. 6.2 Input/output representation of a fuel cell

where K_{eq} is the equilibrium constant. Thus, the Nernst potential may be expressed as:

$$\Delta V_N = -\frac{\Delta G(p,T)}{n_{el}F} = -\frac{\Delta G_0 + RT \cdot \ln(K_{eq})}{n_{el}F}. \tag{6.2.9}$$

6.2.2 Fuel Cell Electrochemistry and Electrocatalysis

The electrochemical reactions occurring in a FC at the anode are

$$H_2 \rightarrow 2H^+ + 2e^-,$$

where the corresponding anode thermodynamic potential is $E_{0a} = 0.00$ V under SHE (under standard hydrogen electrode conditions).

At the cathode, the corresponding electrochemical reactions are

$$O_2 + 4H^+ + 4e^- \rightarrow 2H_2O.$$

Again, the corresponding cathode potential is $E_{0c} = 1.229$ V under SHE (under standard hydrogen electrode conditions). The overall hydrogen fuel cell reaction is therefore

$$O_2 + 2H_2 \rightarrow 2H_2O + \text{Heat},$$

with the standard equilibrium electromotive force calculated to be 1.229 V.

The fuel cell generally operates on pure H_2, while CO is a poison when present in a concentration greater than 0.5 %. If the fuel is a hydrocarbon such as natural gas (primarily methane), reforming of the fuel and shift of the reformate are essential to regenerate the required fuel by the cell. Reforming is given by the reaction:

$$CH_4 + H_2O \rightarrow 3H_2 + CO,$$

and the shift of the reformate by the reaction:

$$CO + H_2O \rightarrow H_2 + CO_2.$$

Polymer electrolyte membrane fuel cells (PEMFCs) typically utilize platinum-based catalysts on the anode to split the hydrogen into positive ions (protons) and negative electrons. The ions pass through the membrane electrolyte to the cathode to combine with oxygen to produce water. The electrons pass through an external circuit creating a current to rejoin the H_2 ion at the cathode. The chemical reactions at the anode and cathode combine to give the overall cell reaction.

$$\text{Anode}: 2H_2 \rightarrow 4H^+ + 4e^-$$

$$\text{Cathode}: O_2 + 4H^+ + 4e^- \rightarrow 2H_2O$$

Phosphoric acid fuel cells (PAFCs) use liquid phosphoric acid as an electrolyte—the acid is contained in a Teflon-bonded silicon carbide matrix—and porous carbon electrodes containing a platinum catalyst. Molten carbonate fuel cells (MCFCs) are high-temperature fuel cells that use an electrolyte composed of a molten carbonate salt mixture suspended in a porous, chemically inert ceramic lithium aluminum oxide (LiAlO2) matrix. For these cells, the chemical reactions at the anode and cathode are

$$H_2 + CO_3^{2-} \rightarrow H_2O + CO_2 + 2e^-$$

$$O_2 + 2CO_2 + 4e^- \rightarrow 2CO_3^{2-}.$$

Solid oxide fuel cells (SOFCs) use a hard, non-porous ceramic compound as the electrolyte. SOFCs and MCFCs can reform fuels internally, which enables them to use a variety of fuels. Using a variety of fuels can reduce the cost associated with adding a reformer to the system. For SOFCs, the chemical reactions at the anode and cathode are

$$H_2 + O^{2-} \rightarrow H_2O + 2e^-$$

$$O_2 + 4e^- \rightarrow 2O^{2-}.$$

Electrocatalysis plays a key role in fuel cells, and its primary role is to split the hydrogen at the anode into protons and electrons. The aim of improving the performance of catalysts is to obtain higher catalytic activity than the standard carbon-supported platinum particle catalysts and other catalysts used in current fuel cells. At the same time, it is essential to reduce any poisoning or degradation of performance due to the presence of impurities in fuel cell catalysts. Examples of these two approaches are given in the following sections. The platinum-based catalysts are extremely sensitive to CO poisoning. Thus, there is need for developing alternate catalysts or using an additional reactor to reduce CO concentration in the fuel gas if the hydrogen is derived from an alcohol or hydrocarbon fuel. When CO is present in the fuel stream because of reforming, the platinum is alloyed with other materials such as ruthenium to reduce poisoning of the fuel cell and retain the performance. Electrocatalysts are commonly prepared by solution precipitation, which is followed by reduction in platinum salt in either the gas or liquid phases. The two most common techniques of coating the catalyst ink on either the electrodes in the form of a gas diffusion layer (GDL) or the membrane to form a catalyst-coated membrane (CCM). PAFCs use a mixture of the electrocatalysts supported on carbon black and a polymeric binder to bind the carbon black particles together forming an integral structure, thus minimizing the use of platinum. Sometime different catalysts are used as oxygen reduction electrocatalysts at the cathode and as hydrogen oxidation electrocatalysts at the anode. Whichever catalysts are used, they can have a profound effect on fuel cell performance, but do not alter the general design or fabrication of the electrodes.

6.3 Hydrogen Production, Storage, and Distribution

An obstacle to using fuel cells in vehicles is hydrogen storage. Hydrogen is produced in central processing plants where it can be produced from chemical reforming of natural gas or other fossil fuels. Fuel cell-powered vehicles must store the hydrogen in its purest form onboard as a compressed gas in pressurized tanks. It is necessary to store enough hydrogen onboard to allow vehicles to travel the same distance as fossil fuel-powered vehicles by refuelling just once every 300–350 miles. Due to the low-energy density of hydrogen, the volume of hydrogen required is quite substantial. Moreover, storing hydrogen in liquid form can be expensive. An alternative to storing fuel directly in the form of hydrogen is to use an onboard reformer. This is relatively costly due to the additional plant required for producing the hydrogen onboard using a reformer and the increased system complexity. Higher-density liquid fuels, such as methanol, ethanol and related bio-fuels, natural gas, liquefied petroleum gas, and gasoline, can be used for fuel, but the vehicles must have an onboard fuel reformer to reform the methanol into hydrogen. The reformer releases carbon dioxide (a greenhouse gas) increasing the carbon footprint, though less than current fossil fuel-powered engines. The use of the reformer tends to make the fuel cell power engines bulkier while also increasing the total system and maintenance costs.

The hydrogen generation unit must be a compact, efficient, and low cost unit that process a fuel, such as methane, to produce hydrogen with low CO content which is then consumed by the fuel cell. Hydrogen production on an industrial scale is a mature technology and is based on (1) steam reforming of low molecular weight hydrocarbon or (2) partial oxidation of high molecular weight hydrocarbon such as methane. The partial oxidation process is sometimes substituted by autothermal reforming by adding steam and air to the methane–air mixture. The hydrogen produced by these processes is also passed through a plant for removal of carbon monoxide, desulfurization, and purification of the hydrogen. Sulfur is a poison for steam reforming nickel-based catalysts and for the platinum anode catalyst in the fuel cell and should be less than 0.001–0.1 ppm, while CO should be less than 5–50 ppm depending on the type of fuel cell. Desulfurization is achieved either by partial adsorption or by catalytic transformation of sulfur. Removal of CO is achieved in a water–gas shift unit where steam is added to convert the CO to carbon dioxide. Membrane reformers based on ultra-permeable *Pd* membranes and membrane purifiers have also developed for the purpose of reforming and purifying the fuel gas, specifically for PEMFC applications.

6.4 Stack Configurations and Fuel Cell Systems

To increase the amount of power generated, a number of cells can be connected in series. This is known as stacking of cells. When several single fuel cells are working in parallel but are connected in series, they make up a fuel cell stack.

The voltage of each cell is added up to give a stack voltage. The current through each cell remains same. The fuel cell needs proper inlet conditioning of gases and cooling of stack. Therefore, auxiliary components are added up with the fuel cell stack to make it a power producing device. There are four main subsystems of the fuel cell stack:

1. Reactant gases flow subsystems
2. Humidification subsystem
3. Stack cooling subsystem
4. Power conditioning subsystem.

In a conventional fuel cell stack, the reactant gases flow subsystem comprise hydrogen and air flow stream channels from the inlet to all the way down to exit. The hydrogen channel supplies hydrogen to the anode, and the air channel supplies oxygen to the cathode. The air channel includes a compressor which compresses the air to the desired level. The compressor intakes air from the atmosphere. The exit of the compressor is connected to the stack via pipe called supply manifold. The supply manifold is responsible for supplying air to the cathode channels of the fuel cells in the stack. A humidifier and air cooler are introduced between the compressor and the stack. The exits of the fuel cell cathodes are connected to a pipe called the return manifold from where the exhaust air is released to the atmosphere.

The power management subsystem is responsible for delivering all the output power via an autonomous power grid to a range of varying loads, including resistive, capacitive, and inductive loads, constant operating current loads, constant power dissipation loads, and induction motor loads. The power management system is responsible for the conversion of DC current to AC current, if required, by the use of an inverter and a synchronous generator. A super capacitor or an electric battery is used to discharge electrical energy when more output power is required and to charge and store energy when less power is required. This can help the fuel cell stack to run at steady current.

In current practical fuel cell applications, 24–384 fuel cells are organized into stacks which are interconnected so that combined voltage adds up. The stack architecture is specifically designed and integrated so as to generate up to 1.5 amps/cm^2, so the stack can be operated at pressures slightly above standard atmospheric pressures (1.6 bar) without the need for any external humidification or cooling. Current architectures are feasible as a consequence of innovative ways of organizing fuel cells into stacks for parallel operation which have been developed to facilitate (1) easier air breathing through the stack via a circular outer inlet (2) the direct humidification of the fuel cell membrane electrode assemblies (MEAs) with liquid water, by directly conveying the liquid water from an anode flow-field plenum through the nominally hydrophobic gas diffusion layer directly to the membrane, and (3) operation of the fuel cell stack at very near ambient pressure.

6.5 Control-Oriented Modelling and Dynamics

Fuel cells need to be properly controlled during both the transient- and steady-state operations for optimum performance efficiency and long life. If not properly controlled, the cell efficiency will fall significantly during transient operations and the cell's useful life will be reduced drastically. The efficiency of the fuel cell is highly dependent on the design of fuel cell stack controller and its subsystems (Yang et al. 1998). The control of fuel cell is difficult because of the nonlinear behavior of the fuel cell and its subsystems (Methekar et al. 2010). Methekar et al. (2010) argued that for control of dynamic systems, empirical-based model should be used instead of first-principle-based model to make the computation faster. Kalhammer et al. (1998) discussed the importance of proper management and control of the auxiliary components of fuel cell system. Badrinarayanan et al. (2001) discussed the effect of water content in the stack and temperature management of the fuel cell power system. Fronk et al. (2000) discussed the relation between stack performance and temperature management of the stack.

Although there are many publications found in the literature on modelling of fuel cell, not many of them are suitable for control purposes. The models of fuel cells found in the literature can be divided into three different categories, that is, in-depth models of fuel cell developed using partial differential equations, steady-state models of the fuel cell system developed using empirical formulations or graphs or experimental data, and dynamic models of the fuel cell system based on lumped assumption.

Most of the fuel cell models found in the literature are formulated for a single cell. These models are developed using thermodynamics, fluid mechanics, and electrochemical foundations and equations. These models define the dynamic and the steady-state single fuel cell performance at different operating conditions. Only the dynamic models can be used for controller design and to define the optimum operating conditions of a single fuel cell for optimum performance. Amphlett et al. (1994) presented a resistance model for a single fuel cell at different working conditions. Springer et al. (1991) formulated a mathematical model for the net flow of water molecules carried out by each proton through different transport phenomena in the membrane and the effect membrane water content on the resistance to flow of protons through the membrane. Bernardi (1990) model identifies reactants and stack temperature conditions that gives the net balance of water content in the membrane. Wohr et al. (1998) considered the dynamic effects of fuel cell current variations and showed that the cell voltage is dependent on both the current density and the current rates. They showed that this occurs due to different water content in the membrane and that for the same current value reached at different current rates, the cell voltage is different even when current is the same. Besides single fuel cell, some stack models have also been analyzed. Lee and Lalk (1998) develop a stack-oriented model to determine the fuel cell stack operating parameters.

A commonly used dynamic model of fuel cell system is defined by Pukrushpan et al. (2004a, b, c); here, they defined the stack voltage model, static cooler and humidification model, compressor model, anode flow model, cathode flow model, and cathode supply and return manifold model. The dynamics of the stack temperature is ignored in this model. The model by Turner et al. (1999) is discussed about the transient effects due to rise of fuel cell temperature.

The performance parameters that must be controlled are the pressure difference across the electrodes (across the membrane in the case of a PEMFC), the power output, and the direct humidification of the fuel cell (of the membrane electrode assemblies in the case of a PEMFC) with liquid water. The controlled inputs are the flow rates of fuel at the anode and the oxidant at the cathode. Thus, the fuel cell control problem is a multi-input/multi-output control problem.

6.6 Reduced-Order Modelling of PEMFCs

A schematic diagram of a PEMFC is shown in Fig. 6.3. As the dynamic power output and efficiency of a PEMFC are critically influenced by the relatively rapid variation of the temperature and the slow variation of the relative humidity and carbon monoxide content within the fuel cell, it is vital that the effects of temperature and humidity on the proton transport in the membrane and the associated Ohmic losses are effectively modelled. Similarly the oxygen concentration in the GDL on the cathode side is continuously changing and significantly affects the performance of the cell. Springer et al. (1991) developed an isothermal steady-

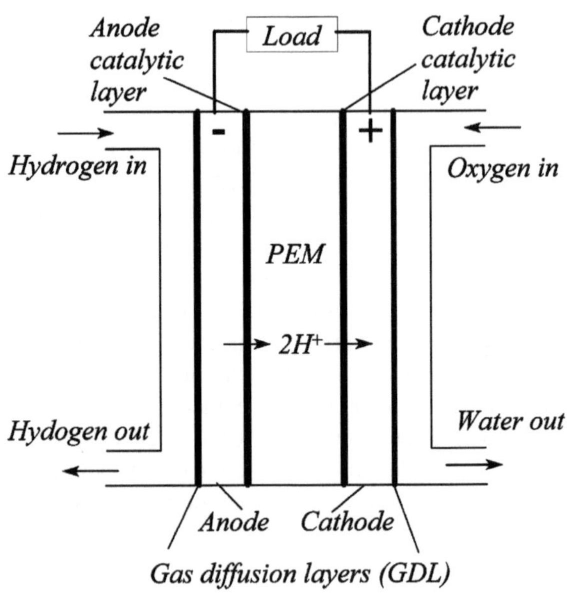

Fig. 6.3 A schematic diagram of a PEMFC

state model which concentrates on water transport mechanisms. The membrane resistance and concentration of oxygen are obtained by solving the proposed water balance equations which are in turn used to determine the cathode potential and V–I curve of the fuel cell. Amphlett et al. (1994) developed steady-state parametric model which gives cell voltage for a given set of operating conditions such as concentration of the reactant gases, pressure, and current. This model is developed using mass-transfer properties, thermodynamic equilibrium potentials, and over voltages. Amphlett et al. (1995) also obtained parametric equations using linear regression to determine the over voltages, i.e., the activation over voltage and Ohmic over voltage. Amphlett et al. (1996) also developed a transient model which predicts the cell voltage and temperature as function of time when perturbations such as step change in operating current or system shut down was imposed in the system. Mann et al. (2000) developed a generalized steady-state electrochemical model for a PEM fuel cell. Puranik et al. (2010), El-Sharkh et al. (2004), Uzunoglu and Alam (2006), Zhang et al. (2006), Yalcinoz and Alam (2008), Ziogou et al. (2011) have presented dynamic models which could in a sense be considered as reduced-order models for real-time filtering, control and estimation applications.

Rowe and Li (2001) developed a non-isothermal model to investigate the effects of design and operating conditions on the fuel cell performance, thermal response, and water management. The review by Wang (2004) underscored the importance of incorporating representative thermal dynamics into the dynamic model. Wang and Wang (2005), Ju et al. (2005), Stanton (2005), Zhang et al. (2006) have since considered thermal dynamics in some detail. Yu and Yuvarajan (2004) also developed a novel circuit model that could be used in power electronic applications where thermal considerations are significant.

Pasricha and Shaw (2006), Laffly et al. (2007), Garnier et al. (2004) considered detailed models of polarization dynamics in their work. There have also been a few studies involving experimental investigations of transient behavior by Chu and Jiang (1999), Morner and Klein (2001) and more recently by Adzakpa et al. (2008). Wang and Wang (2005) considered the role of the gas dynamics within the fuel cell and explained the current overshoot and undershoot exhibited by cells with step changes in output voltage. Ju et al.'s (2005) study identified the three parameters, GDL thermal conductivity, the incoming gas relative humidity, and the fuel cell current as main variables in controlling the thermal behavior of a fuel cell. Pukrushpan et al. (2004a, c) proposed a dynamic model with lumped parameters to predict the gas dynamics in a cathode electrode, but it does not consider the specific effects of oxygen in the GDL. In importance of water content, dynamics was highlighted by the studies of Grasser (2005), Ramousse et al. (2005), Grasser and Rufer (2007), Hinaje et al. (2008). The dynamics of membrane water content evolution and its dependence on the fuel cell current was investigated by Ceraolo et al. (2003) who were able to demonstrate the effects of the water flow dynamics time constant on the fuel cell output voltage. Ramousse et al. (2005), Hinaje et al. (2008) proposed a method of estimating the value of the water flux across the membrane as a function of the water content λ in the cathode-membrane and anode-membrane interfaces. Chia (2006) developed a mathematical model of the PEM fuel cell which captures the essential physics that control of the water balance within the cell. Shan and Choe (2005) consider the dynamics in temperature gradient across the fuel cell, the dynamics in water concentration redistribution in the membrane, the dynamics in proton concentration in the cathode catalyst layer, and the dynamics in reactant concentration redistribution in the cathode GDL. The physiochemical properties of fuel cell membranes, such as proton conductivity and water flux dynamics, were experimentally studied by Yang et al. (2004). Yang et al. (2004) provided useful data in modelling the membrane water content parameter as a function of the water activity parameter.

6.6.1 The PEMFC Model

Typical PEM fuel cell mathematical models consist of transport equations for the respective reactants and products and may include water transport and heat effects. Models for real-time monitoring and controller synthesis applications are required to capture the essence of the system dynamics ignoring unobservable variables that do not couple with the output. For our purposes, we only require the dynamics of variables contributing to the measurements and coupled to them, to ensure that the states are observable over the entire time frame. The dynamic equation of partial pressure of hydrogen, oxygen, and water can be written by differentiating the ideal gas law as $PV = nRT$, where R is the universal gas constant, and V is the constant volume of gas under consideration. Assuming pure feed operation, the gas phase pressure variation of hydrogen in the anode and oxygen in the cathode may be expressed in terms of three primary contributions from input and output flows and the flows during the reaction. For hydrogen, the relationship relating the rate of change of pressure to these three flows can be expressed as in terms of the fuel cell temperature, T_{fc}, as,

$$\frac{dP_{\text{H}_2}}{dt} = \frac{RT_{\text{fc}}}{V_{\text{an}}}\left(q_{\text{H}_2}^{\text{in}} - q_{\text{H}_2}^{\text{out}} - q_{\text{H}_2}^{r}\right) \tag{6.6.1}$$

Applying the electrochemical relationship to the reacted hydrogen flow within the cell and given the fuel cell system current I_{fc}, the flow rate of reacted hydrogen $q_{\text{H}_2}^{r}$ is given in terms of the number of cells N_s in the stack and the Faraday constant F by,

$$q_{\text{H}_2}^{r} = (N_s/2F)I_{\text{fc}}. \tag{6.6.2}$$

The inlet and outlets flows of hydrogen can, respectively, be expressed in terms of the anode channel inlet pressure P_A^{in} in atmospheres, and the ratio of the molar flow of water to the water pressure as,

$$q_{H_2}^{in} = \frac{(m_{H_2O})_{in}^a}{(P_{H_2O})_{in}^a} P_A^{in},\ q_{H_2}^{out} = q_{H_2}^{in} \frac{\left(2P_{H_2} - P_A^{in}\right)}{P_A^{in}}. \tag{6.6.3}$$

where the subscripts/superscripts "in" and "out" refer to the inlet and outlet channels, the superscripts "a" and "c", the subscripts "A" and "C", and the subscripts "an" and "ca" refer to the anode and cathode, respectively. Substituting Eqs. (6.6.2) and (6.6.3) in (6.6.1),

$$\frac{dP_{H_2}}{dt} = -\frac{RT_{fc}}{V_{an}} \left(\frac{2(m_{H_2O})_{in}^a}{(P_{H_2O})_{in}^a} P_{H_2} - \frac{2(m_{H_2O})_{in}^a}{(P_{H_2O})_{in}^a} P_A^{in} + \frac{N_s}{2F} I_{fc} \right). \tag{6.6.4}$$

Similarly for the oxygen flow,

$$\frac{dP_{O_2}}{dt} = -\frac{RT_{fc}}{V_{ca}} \left(\frac{2(m_{H_2O})_{in}^c}{(P_{H_2O})_{in}^c} P_{O_2} - \frac{2(m_{H_2O})_{in}^c}{(P_{H_2O})_{in}^c} P_C^{in} + \frac{N_s}{4F} I_{fc} \right). \tag{6.6.5}$$

Similarly for water, we have

$$\frac{dP_{H_2O}}{dt} = -\frac{RT_{fc}}{V_{ca}} \left(\frac{2(m_{H_2O})_{in}^c}{(P_{H_2O})_{in}^c} P_{H_2O} - \frac{2(m_{H_2O})_{in}^c}{(P_{H_2O})_{in}^c} P_{H_2O}^{sat} + \frac{N_s}{2F} I_{fc} \right). \tag{6.6.6}$$

Hydrogen enters the gas flow channels, where it is absorbed onto the anode surface and is catalytically oxidized into protons and electrons via a reaction. The chemical half-cell reaction at the anode is

$$H_2 \rightarrow 2H^+ + 2e^-. \tag{6.6.7}$$

Protons diffuse across the polymer electrolyte membrane, whereas electrons pass through an external load. Oxygen absorbed onto the cathode surface reacts with the crossover protons and electrons to produce water via the reaction at the cathode which is

$$O_2 + 4H^+ + 4e^- \rightarrow 2H_2O. \tag{6.6.8}$$

As a consequence, the overall chemical reaction of the PEMFC is

$$O_2 + 2H_2 \rightarrow 2H_2O \tag{6.6.9}$$

The reaction defined by Eq. (6.6.9) is exothermic, which means that the fuel cell produces heat. The composition of the polymer membrane is made in a way that it only allows protons to pass. Therefore, the electrons involved in the reaction have to flow in a circuit external to the fuel cell, doing electrical work. The temperature dynamics within the cell can be found by considerations of heat flow into and out of the cell. Given the total mass of the fuel cell stack, M_{fc}, the specific heat capacity, C_{fc}, the cell electrochemical heat generation rate, $\dot{Q}_{Chem}$, the rate at which electrical energy is converted into heat $\dot{Q}_{Elec}$ or the heat generated due to the flow of current and the rate of mechanical heat loss by conduction or convection $\dot{Q}_{Loss}$, the rate of change of the cell temperature is given by,

$$M_{\text{fc}} C_{\text{fc}} \frac{dT_{\text{fc}}}{dt} = \dot{Q}_{\text{Chem}} + \dot{Q}_{\text{Elec}} - \dot{Q}_{\text{Loss}}. \tag{6.6.10}$$

In Eq. (6.6.10), $\dot{Q}_{\text{Loss}}$ can be expressed in terms of an equivalent convective (h_{s_eq}) heat transfer and the coolant (h_{clt}) heat-transfer coefficients, the number of cells N_s, the coolant flow rate q_{cool}, and the difference between the fuel cell and the room temperatures (which is assumed to be the datum). $\dot{Q}_{\text{Elec}}$ is the product of the cell output voltage and the cell current, and $\dot{Q}_{\text{Chem}}$ is proportional to the enthalpy of combustion for hydrogen (286 kJ/mole), ΔH. Thus,

$$\begin{aligned}\dot{Q}_{\text{Loss}} &= \left(h_{s\_\text{eq}} N_s + q_{\text{cool}} h_{\text{clt}}\right)\left(T_{\text{fc}} - T_{\text{room}}\right),\ \dot{Q}_{\text{Elec}} = V_{\text{Cell}} I_{\text{fc}},\ \dot{Q}_{\text{Chem}} \\ &= \Delta H N_s I_{\text{fc}} / 2F.\end{aligned} \tag{6.6.11}$$

The enthalpy of combustion for hydrogen could also be expressed in terms of the Nernst potential which is discussed in a latter section. The equations discussed so far are almost standard and presented by several researchers. One final equation is needed to complete the reduced-order dynamic which we will use to estimate the relevant states from the measurements. This equation concerns the dynamics of membrane humidification.

6.6.2 Membrane Humidification Model

Membrane humidification is essential for the operation of a PEM fuel cell. The proton conductivity in a PEM is a function of both water activity and temperature. When the water content within the PEM is low, it tends to absorb water that ionizes the sulfonic acid groups. Ionization results in free protons that can hop between the fixed anions. With increasing water uptake, the distance between ionized sites decreases and consequently the hopping rate decreases exponentially. When the water content in the PEM is high, proton conductivity increases exponentially. The PEM also swells gradually decreasing the resistance to proton transport in the membrane. If the membrane is not adequately humidified, the proton conductivity will fall, which results in a lower current output. On the other hand, excess humidification will flood the electrodes and obstruct the transport of the gas to the membrane/electrode interface. Thus, the relative humidity content in the membrane must be maintained at an optimum level which is just below 100 %.

The dynamics of the water content may be established by considering the cell mass balance for water. In this section, while we explicitly model the dynamics of the membrane water content for simulation purposes, the membrane water content is assumed to be an independent unknown input to the reduced-order dynamics of the fuel cell, for estimation purposes. The mass balance equation is in terms of the Faraday's constant, the volumes of gas in the anode and cathode chambers, the accumulation of water, and the water removal rate. The accumulation of water in the membrane depends on the water uptake by the membrane itself and water in the gas phase. Water removal from the fuel cell depends on the exiting flow rates from each chamber. In this paper, we discretize and adapt this model so it can be cast in the form of a first-order noise-shaping filter and estimate the time constant of the filter. This results in a relatively simple dynamic model of the water content parameter.

Following Chia (2006) from considerations of cell mass balance,

$$\left(N_{\mathrm{SO_2}}\frac{\mathrm{d}\lambda}{\mathrm{d}a}+(V_{an}+V_{ca})\frac{P_{wv}(T_{\mathrm{fl}})}{RT_{\mathrm{fl}}}\right)\frac{\mathrm{d}a_w}{\mathrm{d}t}=\frac{i_{\mathrm{fc}}}{2F}-(q_a+q_c)\frac{P_{wv}(T_{\mathrm{fl}})}{RT_{\mathrm{fl}}}a_w, \quad (6.6.12)$$

where $P_{wv}(T_{fl})$ is the cell water vapor pressure at the cell temperature, $N_{\mathrm{SO_2}}$ is the number of moles of sulfonic acid groups in the membrane, and q_a and q_c are the anodic and cathodic volumetric flow rates. Following Chia (2006), Eq. (6.6.12) can be expressed in terms of lumped mass-transfer coefficients as,

$$N_{\mathrm{SO_2}}\frac{\mathrm{d}\lambda}{\mathrm{d}a_w}\frac{\mathrm{d}a_w}{\mathrm{d}t}=k_{\mathrm{GDL}}A(a_{\mathrm{GDL}}-a_w)-k_AA(a_w-a_a)=kA(\bar{a}_{\mathrm{GDL}}-a_w). \quad (6.6.13)$$

where A is the effective surface area of the membrane and

$$\bar{a}_{\mathrm{GDL}}=\frac{k_{\mathrm{GDL}}}{k}a_{\mathrm{GDL}}+\frac{k_A}{k}a_a,\ k=(k_{\mathrm{GDL}}+k_A) \quad (6.6.14)$$

In practice for a well-humidified fuel cell, one assumes that $\bar{a}_{\mathrm{GDL}}\approx 1$. In fact, the Eq. (6.6.13) is more appropriate for λ the membrane water content expressed as ratio of the moles of water to the moles in the sulfonic acid group. Thus,

$$\mathrm{d}\lambda/\mathrm{d}t=\omega_a(\lambda_{\bar{a}_{\mathrm{GDL}}=1}-\lambda). \quad (6.6.15)$$

where

$$\omega_a=kA/N_{\mathrm{SO_2}}\frac{\mathrm{d}\lambda}{\mathrm{d}a_w}. \quad (6.6.16)$$

The right-hand side of Eq. (6.6.15) is the overall flux of water transported through the membrane. The slope $\mathrm{d}\lambda/\mathrm{d}a_w$ is easily evaluated from the membrane water content λ expressed in terms of water molecules per membrane sulfonic acid group, as correlated experimentally by Yang et al. (2004). Using a simple polynomial fit to Yang's data to obtain the following expression for the water uptake,

$$\lambda=14.9a_w-44.7a_w^2+70.0a_w^3-26.5a_w^4-0.446a_w^5, \quad (6.6.17)$$

one finds that

$$\lambda|_{a_w=1}=13.354 \text{ and that, } d\lambda/da_w|_{a_w=1}=27.27. \quad (6.6.18)$$

Thus, an estimate of ω_a is given by,

$$\omega_a=kA/N_{\mathrm{SO_2}}\frac{\mathrm{d}\lambda}{\mathrm{d}a_w}=24.4689\times \mathrm{Am}^2. \quad (6.6.19)$$

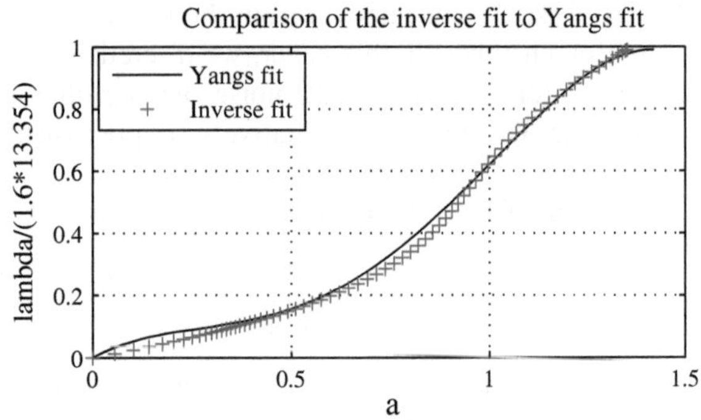

Fig. 6.4 Comparison of the inverse fit to Yang's fit

Moreover, we may numerically invert Yang's expression and write

$$a_w = 4.3491\lambda - 8.0880\lambda^2 + 7.2323\lambda^3 - 2.1269\lambda^4. \tag{6.6.20}$$

A comparison of the two curves is shown in Fig. 6.4.

6.6.3 The Cell Output Electrode Polarizations

The electromotive force generated by the electrochemical reaction under ideal open-circuit conditions is given the Nernst equation and is

$$E_{\text{Nernst}} = -\frac{\Delta G}{nF} + \frac{RT_{fc}}{nF}\log\left(\frac{P_{H_2}\sqrt{P_{O_2}}}{P_{H_2O}}\right) \tag{6.6.21}$$

where ΔG is the change in the Gibbs free energy potential defined as,

$$\Delta G = G_{f,H_2O} - G_{f,H_2} - G_{f,O_2}, \tag{6.6.22}$$

In Eq. (6.6.21), P_{H_2} is the partial pressure of hydrogen within the fuel cell, P_{O_2} is the partial pressure of oxygen, and P_{H_2O} is partial pressure of water in the cell, F is the Faraday constant (96,485 Coulombs/mole), and n represents the number of moles of electrons transferred per mole of hydrogen. At $T = 0$, E_{Nernst} is equal to the standard reference potential when the pressures are equal to one atmosphere. For this reason, all the pressures are measured in units of atmospheres.

The PEMFC exhibits different kinds of electrode polarizations. Polarizations refer to the reduction in the Nernst potential due to a particular physical reason. The first of these is the *activation* polarization or potential. The activation potential drop is due to the slowness of the chemical reactions taking place at electrode surfaces. It depends on the temperature and the operating pressures, the type of electrode, and the catalyst used. The activation over potential can be determined from the Tafel equation which is obtained from the Butler–Volmer equation for both the anodic and the cathodic polarizations, by ignoring the latter. The Tafel equation is

$$\Delta V_{\mathrm{act}} = \frac{\mathrm{RT}_{fc}}{\alpha nF}\log\left(\frac{I_{\mathrm{fc}}}{I_0}\right) \tag{6.6.23}$$

where α is the symmetry factor, and I_0 is the exchange current. On the other hand, Amphlett et al. (1995) have modified the activation loss equation by adding a constant η_0 to it and expressed it as,

$$\Delta V_{\mathrm{act}} = \eta_0 + \frac{RT_{\mathrm{fc}}}{\alpha nF}\log\left(\frac{I_{\mathrm{fc}}}{I_0}\right) \equiv \Delta V_{\mathrm{act0}} + \frac{\mathrm{RT}_{\mathrm{fc}}}{\alpha nF}\log(I_{\mathrm{fc}}). \tag{6.6.24}$$

The second polarization is due to the balance of the surface concentrations of the reactant species. This polarization can be expressed as a relationship between the rate of the electrode reaction and the supply of the reactants to the electrode, or the elimination of the species from the electrode due to diffusion. Therefore, this polarization is called the *concentration* polarization. This voltage drop results from the reduction in concentration of the reactants gases or, alternatively, from the transport of mass of oxygen and hydrogen and is also known as the *concentration* over potential. It is expressed as,

$$\Delta V_{\mathrm{conc}} = -\frac{RT_{\mathrm{fc}}}{nF}\log\left(1 - \frac{j}{j_{\mathrm{Limit}}}\right) = -\frac{RT_{\mathrm{fc}}}{nF}\log\left(1 - \frac{I_{\mathrm{fc}}}{I_{\mathrm{Limit}}}\right) \tag{6.6.25}$$

where j is the current density through the PEM, and j_{Limit} is a limiting value of it, which can be obtained by using the relationship with ideal gas equation and Fick's law.

In a PEMFC, the capacitive impedance due to double-layer charge effect is in parallel with impedance due to the current-dependent component of concentration and activation over potentials. The combined over potential due to the double-layer charge effect and current-dependent components of the concentration and activation over potentials satisfies the differential equation,

$$C_{\mathrm{cdl}}(R_{\mathrm{act}} + R_{\mathrm{conc}})\frac{d\Delta V_{\mathrm{cdl}}}{dt} + \Delta V_{\mathrm{cdl}} = I_{\mathrm{fc}}(R_{\mathrm{act}} + R_{\mathrm{conc}}) \tag{6.6.26}$$

where $R_{\mathrm{act}} \equiv (\Delta V_{\mathrm{act}} - \Delta V_{\mathrm{act0}})/I_{\mathrm{fc}}$ and $R_{\mathrm{conc}} \equiv (\Delta V_{\mathrm{conc}} - \Delta V_{\mathrm{conc0}})/I_{\mathrm{fc}}$ are nonlinear equivalent resistances and $\Delta V_{\mathrm{conc0}} = (RT_{\mathrm{fc}}/nF)\log I_{\mathrm{Limit}}$.

The voltage which is lost due to the resistance to the external flow of electrons through the electrodes and various interconnections and the internal resistance of the membrane (through the electrolyte) to the flow of ions is known as *Ohmic* over potential. Thus, the *Ohmic* potential drop is the sum of the resistance drop due to the electrons transfer through the collecting plates and carbon electrodes, and the resistance drop due to the protons transfer through the solid membrane. These over potentials are discussed by Amphlett et al. (1996). While the electron resistance may be modelled as constant (often neglected), the Ohmic potential loss in the membrane may be expressed in terms of a resistivity like parameter that is derived from a model of the proton conductivity of the membrane of the form,

$$\sigma_m = \sigma_{m0}\exp\left(4.18 \times 303 \times \left(\frac{1}{303} - \frac{1}{T}\right)\right). \tag{6.6.27a}$$

When the current density is zero,

$$\sigma_{m0} = (\lambda - 0.634)/181.6. \tag{6.6.27b}$$

The parameter σ_{m0} is a function of the membrane charge density j, the fuel cell temperature $T = T_{\rm fc}$ and λ. Thus, the Ohmic losses may be expressed as,

$$\Delta V_{\rm Ohmic} = I_{\rm fc}R_{\rm Ohmic},\ R_{\rm Ohmic} = R_C + \rho_m(l_m/A) \tag{6.6.28a}$$

and

$$R_C = 0.003\Omega,\ \rho_m = \frac{181.6\left(1 + 0.03j + 0.062(T/303)^2 j^{2.5}\right)}{(\lambda - 0.634 - 3j)\exp(4.18((T - 303)/T))}. \tag{6.6.28b}$$

In addition to these, there are a number of other potential losses due to contaminants such carbon monoxide which is discussed by Amphlett et al. (1993). The over potential due to carbon monoxide contamination is given by,

$$\Delta V_{\rm CO} = -\frac{RT_{\rm fc}}{F}\log(1 - \theta_{\rm CO}) \tag{6.6.29}$$

where $\theta_{\rm CO}$ represents the surface coverage of carbon monoxide contamination. Amphlett et al. (1993) provide data for converting carbon monoxide concentrations in parts per million (ppm) into $\theta_{\rm CO}$. By curve fitting these data, the function

$$\phi_{\rm co}(C_{\rm co}) = -\log(1 - \theta_{\rm CO}) \tag{6.6.30}$$

is expressed as a function of the carbon monoxide concentration, $C_{\rm co}$ in ppm. In fact, $\phi_{\rm co}(C_{\rm co})$ is approximated, with the maximum error under 2 %, as,

$$\phi_{\rm co}(C_{\rm co}) = \phi_{\rm co\_ss}\left\{1 - \exp\left(-\sum_{j=0}^{4} p_j (C_{\rm co})^j\right)\right\} \tag{6.6.31}$$

where the model parameters are listed in Table 6.2. Given the surface coverage of carbon monoxide $\theta_{\rm CO}$, the carbon monoxide concentration, $C_{\rm co}$, is found by inverse interpolation. Thus,

Table 6.2 Carbon monoxide over potential approximation coefficients

$\phi_{\rm co\_ss}$	18	p_2	−7.9065e-3
p_0	-8.22e-7	p_3	160.61529e-3
p_1	1.4716e-4	p_4	17.96306e-3

$$\Delta Y_{\mathrm{CO}} = \frac{RT_{\mathrm{fc}}}{F}\phi_{\mathrm{co\_ss}}\left\{1 - \exp\left(-\sum_{j=0}^{4} P_j (C_{\mathrm{co}})^j\right)\right\} \tag{6.6.32}$$

If the current-independent components of the concentration and activation over potentials are $\Delta V_{\mathrm{conc0}}$ and ΔV_{act0}, respectively, the cell output voltage delivered to the load is then given by,

$$V_{\mathrm{Cell}} = E_{\mathrm{Nernst}} - \Delta V_{\mathrm{act0}} - \Delta V_{\mathrm{conc0}} - \Delta V_{\mathrm{cdl}} - \Delta V_{\mathrm{Ohmic}} - \Delta V_{\mathrm{CO}}. \tag{6.6.33}$$

6.7 Flow-based Modelling of Proton Exchange Membrane Fuel Cells (PEMFCs)

In what follows we shall briefly and succinctly describe an alternate modelling approach from a different perspective. The modelling approach described in this section closely follows the flow modelling approach, to some extent, of Xue et al. (2006) and of del Real et al. (2007) within the electrode channels and for gaseous diffusion in the GDLs. The membrane is parameterized in terms of the electro-osmotic drag coefficient, the water content parameter, and the mass diffusivity of water vapor in the membrane. Although we adopt the same general approach as del Real et al. (2007) in defining the cell voltage, our model of the electrochemistry is representative of the voltage over potential due to the cell electrode charge double layer.

6.7.1 Modelling the Flow Dynamics within the PEM Fuel Cell

To begin with, it is assumed that a stack of polymer electrolyte membrane (PEM) fuel cells are supplied at the cathode with compressed air via a supply manifold. The compressor is assumed to draw air from the atmosphere and is assumed to be driven by a DC motor. At the cathode exit, the air enters a return manifold that is open to the atmosphere. The dynamics of the supply manifold, compressor, and return manifold may be expressed as,

$$\frac{\mathrm{d}m_{\mathrm{sm}}}{\mathrm{d}t} = \dot{m}_{\mathrm{cp},o} - \dot{m}_{\mathrm{sm},o}, \; \dot{m}_{\mathrm{sm},o} = K_{\mathrm{sm}}(p_{\mathrm{sm}} - p_{\mathrm{cach}}), \tag{6.7.1a}$$

$$\frac{M_A V_{\mathrm{sm}}}{\gamma R}\frac{\mathrm{d}p_{\mathrm{sm}}}{\mathrm{d}t} = \dot{m}_{\mathrm{cp,o}} T_{\mathrm{cp}} - \dot{m}_{\mathrm{sm},o} T_{\mathrm{sm}}, \tag{6.7.1b}$$

$$T_{\mathrm{sm}} = \frac{M_A V_{\mathrm{sm}}}{R m_{\mathrm{sm}}} \omega_{\mathrm{cp}} J_{\mathrm{cp}} \frac{\mathrm{d}\omega_{\mathrm{cp}}}{\mathrm{d}t} = P_{\mathrm{motor}} - P_{\mathrm{cp}}, \tag{6.7.1c}$$

$$\frac{\mathrm{d}m_{\mathrm{rm}}}{\mathrm{d}t} = \dot{m}_{\mathrm{ma,cach},o} - \dot{m}_{\mathrm{rm},o}, \tag{6.7.1d}$$

where m_{sm}, m_{rm} are the masses of air in the supply and return manifolds, respectively, $\dot{m}_{\mathrm{cp},o}$, $\dot{m}_{\mathrm{sm},o}$ are the mass flow rates out of the compressor and supply manifold, respectively, p_{sm}, p_{cach} are pressures in the supply manifold and cathode channels, respectively, K_{sm} is a typical gain relating the mass flow rate out of the supply manifold to the pressure differential across it, T_{cp} and T_{sm} are the temperatures in the compressor and supply manifold, respectively, M_A is the molar mass of dry air, V_{sm} is the volume of the supply manifold, R is the universal gas constant, and γ is the ratio of specific heats. Furthermore, ω_{cp}, J_{cp}, P_{motor}, and P_{cp} are the compressor shaft speed, shaft inertia, and the powers absorbed by the motor and compressor, respectively. The rate of change of mass in the return manifold is assumed to be zero as the cathode output flow ($\dot{m}_{\mathrm{ma,cach},o}$) reaches steady state and is equal to the flow out ($\dot{m}_{\mathrm{rm},o}$) of the return manifold. Thus, Eq. (6.7.1d) may be expressed as,

$$\frac{\mathrm{d}m_{\mathrm{rm}}}{\mathrm{d}t} = \dot{m}_{\mathrm{ma,cach},o} - \dot{m}_{\mathrm{rm},o} \approx 0. \tag{6.7.2}$$

The pressure in the return manifold is assumed to be equal to the standard atmospheric pressure. The variables P_{cp}, T_{cp}, and $\dot{m}_{\mathrm{cp},o}$ are obtained from a typical compressor map of the type adopted by Pukrushpan et al. (2004a, c).

The inlet flow into the cathode channel is assumed to be the out flow from the supply manifold, $\dot{m}_{\mathrm{cach,in}} = \dot{m}_{sm\_o}$, the temperature in the cathode channel to be the room temperature, the relative humidity in the cathode channel to be $\phi_{\mathrm{cach,in}}$, and the inlet flow into the anode channel to be $\dot{m}_{\mathrm{anch,in}} = \dot{m}_{H_2\mathrm{anch,in}}$. The first step is to determine the input species, i.e., oxygen, nitrogen, and water vapor, assuming the input gas is humidified (compressed) air. These are obtained as fractions of the inlet flow into the cathode channel, $\dot{m}_{\mathrm{cach,in}}$, and are given as,

$$\dot{m}_{\mathrm{O_2cach,in}} = \frac{0.21 M_{\mathrm{O_2}}}{1 + w_{r,\mathrm{cach-in}}} \dot{m}_{\mathrm{cach,in}}, \tag{6.7.3a}$$

$$\dot{m}_{\mathrm{N_2cach,in}} = \frac{0.79 M_{\mathrm{N_2}}}{1 + w_{r,\mathrm{cach-in}}} \dot{m}_{\mathrm{cach,in}}, \tag{6.7.3b}$$

and

$$\dot{m}_{\mathrm{wvcach,in}} = \frac{w_{r,\mathrm{cach-in}}}{1 + w_{r,\mathrm{cach-in}}} \dot{m}_{\mathrm{cach,in}}, \tag{6.7.3c}$$

where $M_{\mathrm{O_2}}$ is the molar mass of oxygen, $M_{\mathrm{N_2}}$ is the molar mass of nitrogen, and $w_{r,\mathrm{cach-in}}$ is

$$w_{r,\text{cach-in}} = \frac{M_{wv}}{M_A} \frac{\phi_{\text{cach-in}} \times p_{sat}(T_{\text{cach,in}})}{p_{\text{cach-in}} - \phi_{\text{cach-in}} \times p_{\text{sat}}(T_{cach,in})}, \tag{6.7.4}$$

with M_{wv} as the molar mass of water vapor, M_A is the molar mass of dry air, $p_{\text{cach-in}}$ is the pressure in cathode input channel, and p_{sat} is the saturated vapor pressure.

The flow state variables that must be updated each time step are $m_{\text{lw,cach}}$, $m_{wv,\text{cach}}$, $m_{\text{O}_2,\text{cach}}$, $m_{\text{N}_2,\text{cach}}$, the liquid water, the water vapor, the oxygen and nitrogen masses in the cathode channel, respectively, and $m_{\text{lw,anch}}$, $m_{wv,\text{anch}}$, $m_{\text{H}_2,\text{anch}}$, the liquid water, the water vapor and the hydrogen masses in the anode channel, respectively. The total mass in the cathode channel, $m_{\text{ma,cach}}$, is the sum of $m_{wv,\text{cach}}$, $m_{\text{O}_2,\text{cach}}$, and $m_{\text{N}_2,\text{cach}}$, while the total gas mass in the anode channel $m_{\text{ma,anch}}$, is the sum of $m_{wv,\text{anch}}$ and $m_{\text{H}_2,\text{anch}}$. The flows are assumed to be confined to certain "control volumes" defined the anode channel, the cathode channel, the membrane, and the gas diffusion layers. The various primary flow rates associated with different regions of the fuel cell (shown in Fig. 6.5) are listed in Table 6.3.

Considering the control volumes defined by the cathode and anode channels, the partial pressures (in units of atmospheres) in the cathode and anode channels are estimated from,

$$p_{\text{O}_2,\text{cach}} = \frac{RT_{\text{st}}}{p_{\text{atm}} V_{\text{cach}}} \frac{m_{\text{O}_2,\text{cach}}}{M_{\text{O}_2}}, \tag{6.7.5a}$$

$$p_{\text{N}_2,\text{cach}} = \frac{RT_{\text{st}}}{p_{\text{atm}} V_{\text{cach}}} \frac{m_{\text{N}_2,\text{cach}}}{M_{\text{N}_2}}, \tag{6.7.5b}$$

$$p_{wv,\text{cach}} = \frac{RT_{\text{st}}}{p_{\text{atm}} V_{\text{cach}}} \frac{m_{wv,\text{cach}}}{M_{wv}}, \tag{6.7.5c}$$

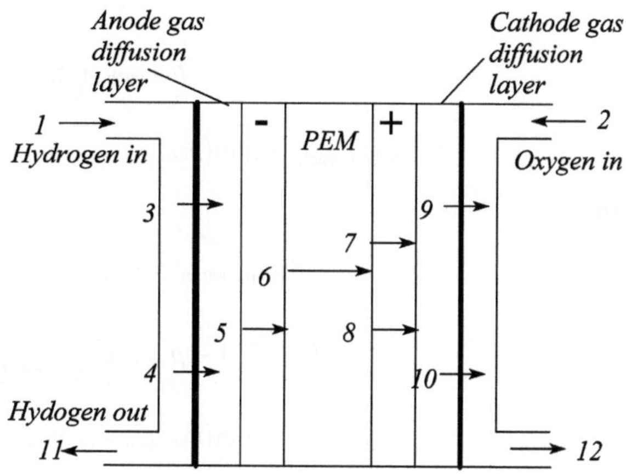

Fig. 6.5 Flow model of a PEMFC

Table 6.3 Primary flow variables associated with gas flow regions labelled in Fig. 6.5

Gas flow region	Mass flow rate variable name	Description
1	$\dot{m}_{\text{anch,in}}$	Anode channel in
2	$\dot{m}_{\text{cach,in}}$	Cathode channel in
3	$\dot{m}_{wv\text{AGDL2anch}}$	Water vapor, anode channel to GDL
4	$\dot{m}_{\text{H}_2\text{AGDL2anch}}$	Hydrogen, anode channel to GDL
5	$\dot{m}_{\text{H}_2\text{react}}$	Reacted hydrogen
6	$\dot{m}_{wv,\text{memb}}$	Membrane water vapor
7	$\dot{m}_{wv,\text{gen}}$	Water vapor generated
8	$\dot{m}_{\text{O}_2\text{react}}$	Reacted oxygen
9	$\dot{m}_{wv\text{CGDL2cach}}$	Water vapor, cathode channel to GDL
10	$\dot{m}_{\text{O}_2\text{CGDL2cach}}$	Oxygen, cathode channel to GDL
11	$\dot{m}_{\text{anch},o}$	Anode channel out
12	$\dot{m}_{\text{cach},o}$	Cathode channel out

$$p_{\text{H}_2,\text{anch}} = \frac{RT_{\text{st}}}{p_{\text{atm}} V_{\text{anch}}} \frac{m_{\text{H}_2,\text{anch}}}{M_{\text{H}_2}}, \tag{6.7.5d}$$

$$p_{wv,\text{anch}} = \frac{RT_{\text{st}}}{p_{\text{atm}} V_{\text{anch}}} \frac{m_{wv,\text{anch}}}{M_{wv}}, \tag{6.7.5e}$$

where R is the universal gas constant, T_{st} is the temperature of the fuel cell stack, V_{cach} and V_{anch} are the cathode and anode channel volumes, respectively. The total pressure in the cathode and anode is, respectively, given by,

$$p_{\text{cach}} = p_{\text{O}_2,\text{cach}} + p_{\text{N}_2,\text{cach}} + p_{wv,\text{cach}}, \tag{6.7.6a}$$

$$p_{\text{anch}} = p_{\text{H}_2,\text{anch}} + p_{wv,\text{anch}}. \tag{6.7.6b}$$

The water evaporation flow rates up to maximum of what is available from the GDLs are given at the cathode and anode sides as,

$$\dot{m}_{\text{evap,cach,in}} = A_{\text{fc}} p_{\text{sat}}(T_{\text{st}}) \times$$

$$\left(p_{\text{sat}}(T_{\text{st}}) - p_{wv,\text{cach}}\right) \sqrt{\frac{M_{wv}}{2\pi RT_{\text{st}}}}, \tag{6.7.7a}$$

$$\dot{m}_{\text{evap,cach}} = \min\left(\dot{m}_{\text{evap,cach,in}}, \dot{m}_{\text{lw,CGDL2cach}}\right) \tag{6.7.7b}$$

and

$$\dot{m}_{\text{evap,anch,in}} = A_{\text{fc}} p_{\text{sat}}(T_{\text{st}}) \times$$

$$\left(p_{\text{sat}}(T_{\text{st}}) - p_{wv,\text{anch}}\right) \sqrt{\frac{M_{wv}}{2\pi RT_{\text{st}}}}, \tag{6.7.8a}$$

$$\dot{m}_{\text{evap,anch}} = \min\left(\dot{m}_{\text{evap,anch,in}}, \dot{m}_{\text{lw,CGDL2cach}}\right). \tag{6.7.8b}$$

The flows on the cathode's outlet side are defined as,

$$\dot{m}_{\mathrm{ma,cach},o} = K_{\mathrm{cach},o} p_{\mathrm{atm}} (p_{\mathrm{cach}} - 1), \tag{6.7.9a}$$

$$\dot{m}_{O_2,\mathrm{cach},o} = \frac{\dot{m}_{O_2,\mathrm{cach}}}{\dot{m}_{\mathrm{ma,cach}}} \dot{m}_{\mathrm{ma,cach},o}, \tag{6.7.9b}$$

$$\dot{m}_{\mathrm{N_2,cach},o} = \frac{\dot{m}_{\mathrm{N_2,cach}}}{\dot{m}_{\mathrm{ma,cach}}} \dot{m}_{\mathrm{ma,cach},o}, \tag{6.7.9c}$$

$$\dot{m}_{wv,\mathrm{cach},o} = \frac{\dot{m}_{wv,\mathrm{cach}}}{\dot{m}_{\mathrm{ma,cach}}} \dot{m}_{\mathrm{ma,cach},o}, \tag{6.7.9d}$$

and

$$\dot{m}_{\mathrm{lw,cach},o} = \dot{m}_{\mathrm{lw,CGDL2cach}} - \dot{m}_{\mathrm{evap,anch}}. \tag{6.7.9e}$$

The flows on the anode's outlet side are defined as,

$$\dot{m}_{\mathrm{ma,anch},o} = K_{\mathrm{anch},o} p_{\mathrm{atm}} (p_{\mathrm{cach}} - 1) v_{o/c}, \tag{6.7.10a}$$

$$\dot{m}_{\mathrm{H_2,anch},o} = \frac{\dot{m}_{\mathrm{H_2,anch}}}{\dot{m}_{\mathrm{ma,anch}}} \dot{m}_{\mathrm{ma,anch},o}, \tag{6.7.10b}$$

$$\dot{m}_{wv,\mathrm{anch},o} = \frac{\dot{m}_{wv,\mathrm{anch}}}{\dot{m}_{\mathrm{ma,anch}}} \dot{m}_{\mathrm{ma,anch},o} \tag{6.7.10c}$$

and

$$\dot{m}_{\mathrm{lw,anch},o} = -\dot{m}_{\mathrm{lw,AGDL2anch}} - \dot{m}_{\mathrm{evap,anch}} \tag{6.7.10d}$$

In the expressions for flows on the anode's outlet side, the purge valve $v_{o/c}$ is 1 or 0 depending on whether it is open or closed.

The liquid water mass flow rates are assumed to be zero; thus,

$$\frac{\mathrm{d}}{\mathrm{d}t} m_{\mathrm{lw,cach}} = 0 =$$

$$-\dot{m}_{\mathrm{lw,cach},o} - \dot{m}_{\mathrm{evap,cach}} + \dot{m}_{\mathrm{lw,CGDL2cach}}, \tag{6.7.11a}$$

$$\frac{d}{dt} m_{\mathrm{lw,anch}} = 0 =$$

$$-\dot{m}_{\mathrm{lw,AGDL2anch}} - \dot{m}_{\mathrm{evap,anch}} - \dot{m}_{\mathrm{lw,anch},o}, \tag{6.7.11b}$$

The five component species mass update equations are given by,

$$\frac{\mathrm{d}}{\mathrm{d}t} m_{wv,\mathrm{cach}} = \dot{m}_{wv,\mathrm{cach,in}} - \dot{m}_{wv,\mathrm{cach},o} + \dot{m}_{wv,\mathrm{CGDL2cach}} + \dot{m}_{\mathrm{evap,cach}}, \tag{6.7.12a}$$

$$\frac{d}{dt}m_{O_2,cach} = \dot{m}_{O_2,cach,in} - \dot{m}_{O_2,cach,o} - \dot{m}_{O_2,CGL2cach}, \tag{6.7.12b}$$

$$\frac{d}{dt}m_{N_2,cach} = \dot{m}_{N_2,cach,in} - \dot{m}_{N_2,cach,o}, \tag{6.7.12c}$$

$$\frac{d}{dt}m_{wv,anch} = -\dot{m}_{wv,anch,o} - \dot{m}_{wv,aAGDL2nch} + \dot{m}_{evap,anch}, \tag{6.7.12d}$$

$$\frac{d}{dt}m_{H_2,anch} = \dot{m}_{H_2,anch,in} - \dot{m}_{H_2,anch,o} - \dot{m}_{H_2,AGDL2anch} \tag{6.7.12e}$$

The Eq. (6.7.10a) indicates that the flows are coupled to the flows from the cathode and anode channels into the GDLs. These interfacial flow rates must be defined considering both the flows of the gases and the flows of the water vapor, and liquid water separately from the electrode channels to the GDL. The interfacial flow rates of the gases and the water vapor ($\dot{m}_{wv,CGDL2cach}$, $\dot{m}_{O_2,CGL2cach}$, $\dot{m}_{wv,AGDL2anch}$, and $\dot{m}_{H_2,AGDL2anch}$) and the interfacial liquid water flows ($\dot{m}_{lw,CGDL2cach}$, $\dot{m}_{lw,AGDL2anch}$) are estimated in way similar to del Real et al. (2007). Thus, the interfacial flow rates of the gases and water vapor may be expressed in terms of the corresponding molar masses, cross-sectional areas, viscosity coefficient, and molar flux as,

$$\dot{m}_{wv,CGDL2cach} = A_{fc}\eta_{fc}M_{wv}N_{wv,ca}, \tag{6.7.13a}$$

$$\dot{m}_{O_2,CGDL2cach} = A_{fc}\eta_{fc}M_{O_2}N_{O_2,react}, \tag{6.7.13b}$$

$$\dot{m}_{wv,AGDL2anch} = A_{fc}\eta_{fc}M_{wv}N_{wv,an}, \tag{6.7.13c}$$

$$\dot{m}_{H_2,AGDL2anch} = A_{fc}\eta_{fc}M_{H_2}N_{H_2,react}. \tag{6.7.13d}$$

The capillary flows of liquid water are expressed additionally in terms of the absolute and relative permeability's μ and μ_r, the liquid water viscosity η_{lw}, the reduced liquid water saturation S_r, and the slope of the Leverett function J, which expresses the capillary pressure in terms of S_r at the anode and cathode GDLs and the thickness of the GDL, δ_{GDL}. The Leverett function J and the capillary pressure ρ_c are, respectively, defined as,

$$J(S_r) = S_r(1.417 - S_r(2.120 - 1.263S_r)),\ \rho_c = \frac{\sigma\cos\theta_c}{(K/\varepsilon)^{1/2}}J(S_r) \tag{6.7.14}$$

where σ is the surface tension which corresponds to water and air, θ_c is the contact angle, ε is the porosity of the equivalent capillary valve, and K is a valve coefficient. Then, capillary flows of liquid water are expressed as,

$$\dot{m}_{lw,CGDL2cach} = A_{fc}\eta_{fc}\mu\frac{\mu_{r,ca}}{\eta_{lw,ca}}\left|\frac{d\rho_c}{dS_{r,ca}}\right|\frac{S_{r,ca}}{\delta_{GDL}}, \tag{6.7.15a}$$

$$\dot{m}_{\mathrm{lw,AGDL2anch}} = A_{\mathrm{fc}}\eta_{\mathrm{fc}}\mu\frac{\mu_{r,\mathrm{an}}}{\eta_{\mathrm{lw,an}}}\left|\frac{d\rho_c}{dS_{r,an}}\right|\frac{S_{r,an}}{\delta_{\mathrm{GDL}}}. \tag{6.7.15b}$$

Molar fluxes are a function of the effective diffusivities, the concentration gradients, and the GDL thickness,

$$N_{wv,ca} = -\langle D_{wv,ca}\rangle\left(\frac{c_{wv,\mathrm{cach}} - c_{wv,\mathrm{CGDL}}}{\delta_{\mathrm{GDL}}}\right), \tag{6.7.16a}$$

$$N_{wv,\mathrm{an}} = -\langle D_{wv,\mathrm{an}}\rangle\left(\frac{c_{wv,\mathrm{anch}} - c_{wv,\mathrm{AGDL}}}{\delta_{\mathrm{GDL}}}\right). \tag{6.7.16b}$$

The oxygen and hydrogen molar fluxes and the amount of water steam generated in the electrochemical reaction are calculated using stoichiometric balances:

$$N_{wv,\mathrm{gen}} = \frac{1}{2}\frac{I_{\mathrm{st}}}{FA_{\mathrm{fc}}},\ N_{\mathrm{H_2,react}} = \frac{1}{2}\frac{I_{\mathrm{st}}}{FA_{\mathrm{fc}}}\ \text{and}\ N_{\mathrm{O_2,react}} = \frac{1}{4}\frac{I_{\mathrm{st}}}{FA_{\mathrm{fc}}}. \tag{6.7.17}$$

As the interfacial flow rates of the gases and water vapor depend on gaseous diffusion, the water vapor concentrations must be first estimated. These may be defined as,

$$c_{wv,\mathrm{CGDL}} = \frac{p_{\mathrm{atm}}}{RT_{\mathrm{st}}}p_{wv,\mathrm{CGDL}}, \tag{6.7.18a}$$

$$c_{wv,\mathrm{cach}} = \frac{p_{\mathrm{atm}}}{RT_{\mathrm{st}}}p_{wv,\mathrm{cach}}, \tag{6.7.18b}$$

$$c_{wv,\mathrm{AGDL}} = \frac{p_{\mathrm{atm}}}{RT_{\mathrm{st}}}p_{wv,\mathrm{AGDL}}, \tag{6.7.18c}$$

and

$$c_{wv,\mathrm{anch}} = \frac{p_{\mathrm{atm}}}{RT_{\mathrm{st}}}p_{wv,\mathrm{anch}}, \tag{6.7.18d}$$

where $p_{wv,\mathrm{CGDL}}$ and $p_{wv,\mathrm{AGDL}}$ are the water vapor partial pressures in the GDLs, which are independent state variables, that each satisfy the differential equation,

$$\frac{\mathrm{d}p_{wv,\mathrm{CGDL}}}{\mathrm{d}t} = RT_{\mathrm{st}}\left(\frac{N_{wv,\mathrm{gen}} - N_{wv,ca} + N_{wv,\mathrm{memb}}}{\delta_{\mathrm{GDL}}}\right) + \gamma\Delta p_{\mathrm{evap},ca}, \tag{6.7.19a}$$

$$\frac{\mathrm{d}p_{wv,\mathrm{AGDL}}}{\mathrm{d}t} = RT_{\mathrm{st}}\left(\frac{N_{wv,ca} - N_{wv,\mathrm{memb}}}{\delta_{\mathrm{GDL}}}\right) + \gamma\Delta p_{\mathrm{evap},an}. \tag{6.7.19b}$$

These are estimated in way similar to del Real et al. (2007), but the water vapor pressures in the GDL are assumed to have reached their steady-state values by running the simulations initially over long time. The quantities $\Delta p_{\mathrm{evap},j}$ $j = ca$ or an are the evaporation excess pressures which are assumed to be zero when no liquid water is available. When they are not zero, they are modelled as

$$\Delta p_{\text{evap},j} = \left(p_{\text{sat}}(T_{\text{st}}) - p_{wv,\text{jGDL}}\right). \tag{6.7.20}$$

The quantity $N_{wv,\text{memb}}$ is the membrane molar flow estimated in way similar to Springer et al. (1991) from the mass diffusivity of water vapor in the membrane D_w and the electroosmotic drag coefficient η_d. The quantity $N_{wv,\text{memb}}$ is given by,

$$N_{wv,\text{memb}} = \eta_d \frac{I_{\text{st}}}{FA_{\text{fc}}} - \alpha_w D_w \left(\frac{c_{wv,ca} - c_{wv,an}}{\delta_{\text{memb}}}\right). \tag{6.7.21}$$

The latter two characteristics of the membrane are estimated in the same manner as Shan and Choe (2005). The concentrations $c_{wv,\text{ca}}$ and $c_{wv,an}$ at the cathode and the anode are calculated as

$$\left(c_{wv,ca}, c_{wv,an}\right) = \frac{\rho_{\text{memb,dry}}}{M_{\text{memb,dry}}}(\lambda_{ca}, \lambda_{an}) \tag{6.7.22}$$

where $\rho_{\text{memb,dry}}$ is the dry membrane density, $M_{\text{memb,dry}}$ the dry membrane molecular weight, and λ_j, $j = ca$ or an , is calculated as

$$\lambda_j = 0.043 + a_j\left(17.81 - a_j\left(39.85 - 36.0a_j\right)\right) \tag{6.7.23}$$

a_j being the relative humidity of the gas inside the gas diffusion layers.

The effective diffusivities, $\langle D_{wv,ca}\rangle$, $\langle D_{wv,an}\rangle$ in the GDLs and the mass diffusivity of water vapor in the membrane D_w are estimated in way similar to del Real et al. (2007). The electroosmotic drag coefficient η_d is assumed to be,

$$\eta_d = (0.0029\lambda_{an} + 0.05)\lambda_{an} - 3.4 \times 10^{-19}. \tag{6.7.24}$$

Liquid water volumes in the GDLs, $V_{\text{lw,CGDL}}$ and $V_{\text{lw,AGDL}}$, are also assumed to have reached steady state and are assumed to be constant. They are estimated from,

$$\rho_{\text{lw}} \frac{dV_{\text{lw,CGDL}}}{dt} = -\dot{m}_{\text{lw,CGDL2cach}} - \varepsilon\gamma \frac{M_{wv}\Delta p_{\text{evap},ca} V_{\text{GDL}}}{RT_{\text{st}}} = 0, \tag{6.7.25a}$$

$$\rho_{\text{lw}} \frac{dV_{\text{lw,AGDL}}}{dt} = -\dot{m}_{\text{lw,AGDL2anch}} - \varepsilon\gamma \frac{M_{wv}\Delta p_{\text{evap},an} V_{\text{GDL}}}{RT_{\text{st}}} = 0, \tag{6.7.25b}$$

where V_{GDL} is the actual volume of the GDLs.

6.7.2 The Cell Output Equations

The approach adopted by del Real et al. (2007) is indeed most suitable for control applications. However, it is important to include the proper temperature dependence of the Nernst potential in the model and ensure that the dependence of the Ohmic resistance on the membrane water content is properly parameterized.

The cell voltage is expressed as,

$$V_{\text{Cell}} = E_{\text{Nernst}} - \Delta V_{\text{Ohmic}} - \Delta V_{\text{act0}} - \Delta V_{\text{cdl}}. \tag{6.7.26}$$

The *Ohmic* over potential may be expressed in terms of a resistivity like parameter that is function of the charge density the fuel cell temperature $T = T_{\text{fc}}$ and the membrane water content parameter λ, as,

$$\Delta V_{\text{Ohmic}} = I_{\text{fc}}\left(R_{p\_\text{Ohmic}} + R_e\right), \tag{6.7.27}$$

where

$$R_{p\_\text{Ohmic}} = \rho_m(l_m/A) \tag{6.7.28}$$

and

$$\rho_m = \frac{181.6 \cdot 10^{-4}\left(1 + 0.03 \cdot 10^{-4}j + 0.062 \cdot 10^{-10}(T_{st}/303)^2 j^{2.5}\right)}{(\lambda - 0.634 - 3j \cdot 10^{-4})\exp(4.18((T_{st} - 303)/T_{st}))} \tag{6.7.29}$$

where $\lambda = \lambda_{\text{mem}}$ in the membrane is assumed to be the average of the λ at the anode and cathode GDL boundaries. When

$$\lambda_{\text{mem}} < 0.634 + 3j \cdot 10^{-4} \equiv \lambda_{\text{crit}}, \tag{6.7.30}$$

the Ohmic resistance due to the decrease in proton conductivity is assumed to be very large. For a fully hydrated cell, $\lambda_{\text{mem}} = 13.5$, which will be assumed.

The Tafel equation for the activation over potential is

$$\Delta V_{\text{act}} = \frac{RT_{\text{st}}}{\alpha nF}\log\left(\frac{I_{\text{st}}}{I_0}\right), \tag{6.7.31}$$

where α is the symmetry factor, and I_0 is the exchange current. Separating the current-independent and current-dependent components, the activation over potential is expressed as,

$$\Delta V_{\text{act}} = \Delta V_{\text{act0}} + R_{\text{act}} I_{\text{st}}. \tag{6.7.32}$$

The concentration over potential is expressed as,

$$\Delta V_{\text{conc}} = R_{\text{conc}} I_{\text{st}} = -\frac{RT_{\text{fc}}}{nFI_{\text{st}}}\log\left(1 - \frac{j}{j_{\text{Limit}}}\right) \times I_{\text{st}}, \tag{6.7.33}$$

where j is the current density through the fuel cell, and j_{Limit} is a limiting value of it. Thus,

$$R_{\text{conc}} = -\frac{xc(10)T_{\text{fc}}}{I_{\text{st}}}\log\left(1 - \frac{j}{xc(11)}\right). \tag{6.7.34}$$

The combined over potential due to the double-layer charge effect and current-dependent components of the concentration and activation over potentials satisfies the differential equation,

$$C_{\text{cdl}}\frac{d\Delta V_{\text{cdl}}}{dt}+\frac{\Delta V_{\text{cdl}}}{(R_{\text{act}}+R_{\text{conc}})}=I_{\text{fc}}. \tag{6.7.35}$$

The cell voltage may then be expressed as,

$$\begin{aligned}V_{\text{Cell}} &= xe(1)-I_{\text{fc}}\left(xe(2)+R_{p\_\text{Ohmic}}\right)-\Delta V_{\text{cdl}}+xe(3)\\ &\times T_{\text{st}}\log\left(\frac{p_{\text{H}_2,\text{anch}}\sqrt{p_{\text{O}_2,\text{cach}}}}{p_{\text{atm}}\sqrt{p_{\text{atm}}}}\right),\end{aligned} \tag{6.7.36}$$

with

$$R_{p\_\text{Ohmic}}=\rho_m(l_m/A) \tag{6.7.37}$$

and

$$\rho_m=\frac{xe(4)\cdot 10^{-4}\left(1+0.03\cdot 10^{-4}j+0.062\cdot 10^{-10}(T_{st}/303)^2 j^{2.5}\right)}{(xe(5)-0.634-3j\cdot 10^{-4})\exp(4.18((T_{\text{st}}-303)/T_{\text{st}}))}. \tag{6.7.38}$$

The coefficients $xe(1)$ and $xe(5)$ are parameterized in terms of standard coefficients and stack temperature as follows:

$$xe(1)=0.72144+2.23e-03(T_{\text{st}}-273) \tag{6.7.39}$$

and

$$xe(5)=\lambda_{\text{mem}}=13.5. \tag{6.7.40}$$

The equivalent resistances, R_{act} and R_{conc}, may be expressed as,

$$R_{\text{act}}=xe(6)T_{\text{st}}(\log(I_{\text{st}})/I_{\text{st}}). \tag{6.7.41}$$

$$R_{\text{conc}}=-\frac{xe(7)T_{\text{fc}}}{I_{\text{st}}}\log\left(1-\frac{j}{xe(8)}\right). \tag{6.7.42}$$

The coefficients $xe(j)$, $j=2$, 3, 4, 6, 7 and 8 needed for evaluating the cell output voltage are listed in Table 6.4. The complete reduced-order model of the fuel cell can thus be described in terms of 9 first-order state equations which arise from Eqs. (6.7.1a)–(6.7.1c), (6.7.12), and (6.7.21). In addition to the stack current and stack temperature which are assumed to be measured accurately, it is assumed that three further noisy measurements of the stack voltage and the relative humidity in the anode and cathode channels and measurements of all the states

Table 6.4 The electrochemical constant parameters

Parameter	$xe(2)$	$xe(3)$	$xe(4)$	$xe(6)$	$xe(7)$	$xe(8)$
Value	0.0003	4.3085e-5	0.01816	9.724e-5	4.3085e-5	2781.28

relating to the supply manifold (mass, pressure, and compressor shaft speed) are available. These scalar measurements are expressed as,

$$z_j = h_j(\mathbf{x}), j = 1, \ldots, 6, \tag{6.7.43}$$

where $\mathbf{x}$ is the complete state vector describing the dynamics of the fuel cell. Expressing the dynamic equations and measurements in the standard forms, facilitates the application of control and filtering techniques based on optimal control and Kalman filtering. Optimal control techniques are discussed latter in the chapter, while Kalman filtering is discussed in the next chapter.

6.8 Solid Oxide Fuel Cells (SOFCs)

Solid oxide fuel cell (SOFC) is one of the most useful of the fuel cell types, and it works in high-temperature conditions and is characterized by higher performance and fuel utilization due to endothermic internal reforming occurring in the anodic porous layer. Tubular SOFC and planar SOFC are two basic types of SOFC currently being developed. The SOFC cell stack includes porous anode/cathode electrodes, the electrolyte, fuel and air channels, and solid connectors. The porous electrode structure of a SOFC allows the gases from the respective gas channels to penetrate through the porous layers, thus facilitating the reforming and electrochemical reactions near the interface between gas and solid phases. Mass and energy exchanges occur within the cell due to the reactions. While the high temperature improves the cell performance and allows the use of different fuels, it can also be responsible for excessive thermal stresses, catalyst degradation (carbon deposition), and thermal instability problems in materials. The SOFC primarily uses hydrogen as the fuel. The fuel reacts with oxygen to produce electricity. Fundamental to SOFC design is an understanding of reaction kinetics and its impact on power output and cell efficiency. The fundamental SOFC reactions and the corresponding enthalpies of formation $\Delta H_{298^\circ K}$, are:

$$\text{Anode}: 2\text{H}_2 + 2\text{O}^{2-} \rightarrow 2\text{H}_2\text{O} + 4e^-$$

$$\text{Cathode}: \text{O}_2 + 4e^- \rightarrow 2\text{O}^{2-},\ \Delta\text{H}_{298^\circ\,\text{K}} = 0,$$

$$\text{Overall}: 2\text{H}_2 + \text{O}_2 \rightarrow 2\text{H}_2\text{O},\ \Delta\text{H}_{298^\circ\,\text{K}} = -286\ \text{kJ/mol}$$

Methane steam reforming and electrochemical reactions occurring at the anode porous layer are the most important factors to affect the heat- and mass-transfer processes within the fuel cell. As internal reforming can occur in SOFC's anode, a mixture of hydrocarbons may be used as the fuel in the SOFC. Moreover, the reforming rate could be enhanced by a catalyst. However, ideally, the catalyst does not take any part in the overall reactions. H_2 and CO which are generated by methane and water gas steam reforming are the primary components of the electrochemical reactions in the anodic active layer. H_2O and CO_2 are then produced in the anodic active layer, and a certain proportion of the H_2O can be reused for methane steam reforming at the anode which in turn improves the fuel usage and overall efficiency. Oxygen which on the other hand is consumed in the cathode side does not contribute to the heat transfer within the cell. The equations corresponding to the reforming of hydrogen must be appended to the electrode electrochemical reaction equations. Hence, in addition, one has

$$\mathrm{CH_4\,partial\ oxidation}: CH_4 + H_2O \rightarrow CO + 3H_2,\ \Delta H_{298^\circ K} = 206\,\mathrm{kJ/mol}$$

$$\mathrm{CO\ oxidation}: +H_2O \rightarrow CO_2 + H_2,\ \Delta H_{298^\circ K} = -41\,\mathrm{kJ/mol}$$

$$\mathrm{CH_4\,oxidation}: CH_4 + 2H_2O \rightarrow CO_2 + 4H_2,\ \Delta H_{298^\circ K} = 165\ \mathrm{kJ/mol}.$$

There are two different approaches to reforming: external reforming, which is performed external to the cell before the fuel reaches it, and internal reforming, which takes place within the fuel cell anode. These are illustrated in Figs. 6.6 and 6.7.

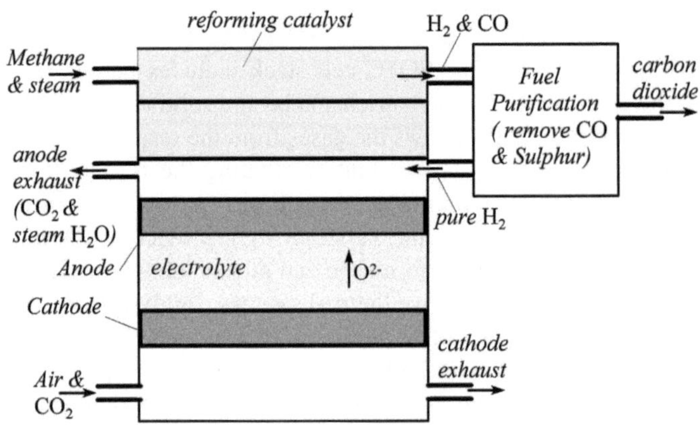

Fig. 6.6 SOFC with internal reforming

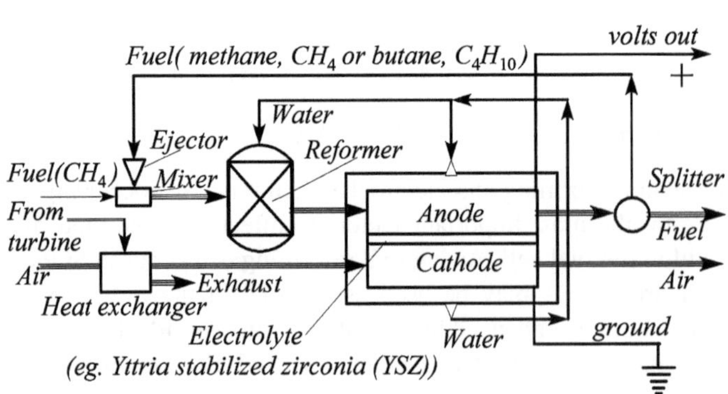

Fig. 6.7 SOFC with external reforming

In contrast to the steam reforming reactions, two of which, with the exception of the water gas shift reforming equation, are endothermic, the electrode electrochemical reactions are exothermic because of heat production within the fuel cell caused by the internal resistances. The reforming equations do not contribute to the electron flow but are only responsible for the reforming of hydrogen. Thus, for every two moles of hydrogen consumed, four moles of electrons are passed through the electric load. Electron mole flow is converted into charge units using Faradays constant ($F = 96,485$ coulombs/mole of electrons). Since the current is the rate of charge flow, the fuel cell delivers power to the load and it follows that

$$\text{Power} = \text{Current} \cdot \text{Voltage}. \ (\text{coulomb} \cdot \text{volt} = \text{joule and joule/s} = \text{watt}). \quad (6.8.1)$$

The fuel cell gains this power at the expense of the enthalpy released during the overall reaction, which is the oxidation of hydrogen to form water.

$$2H_2 + O_2 \rightarrow 2H_2O.$$

Yet, only a portion of this enthalpy can be converted into electric power, the remainder being accounted for by heat released by the reaction. This heat is normally transferred to the air and hydrogen fuel by heat exchangers. The performance of a fuel cell is typically expressed in terms of its total efficiency which is defined as energy delivered to the load divided by the total energy available from reaction.

In this section, a typical SOFC is considered with internal reforming only, although the model can be applied to an externally reformed cell. Assuming that the reactor is of a tubular section and narrow in width, the molar flow equations for the ith molar flow rate component N_i at the cathode reaction zone may be expressed as

$$\frac{d}{dt} N_{N_2} = \dot{N}_{N_2}^{\text{in}} - \dot{N}_{N_2}^{\text{out}}, \quad (6.8.2a)$$

$$\frac{d}{dt} N_{O_2} = \dot{N}_{O_2}^{\text{in}} - \dot{N}_{O_2}^{\text{out}} - 0.5 \frac{u_1}{2F}, \quad (6.8.2b)$$

At the anode reaction zone,

$$\frac{d}{dt} N_{F_i} = \dot{N}_{F_i}^{\text{in}} - \dot{N}_{F_i}^{\text{out}} + \sum_{k=1}^{3} a_{F_i,k} r_k + b_{F_i} \frac{u_1}{2F}, \ i = 1, \ldots, 2, \ b_{F_i} = \begin{bmatrix} -1 \\ 1 \end{bmatrix}, \quad (6.8.3)$$

$$\frac{d}{dt} N_{E_i} = \dot{N}_{E_i}^{\text{in}} - \dot{N}_{E_i}^{\text{out}} + \sum_{k=1}^{3} a_{E_i,k} r_k, \ i = 1, \ldots, 3, \quad (6.8.4)$$

$$a_{F_i,k} = \begin{bmatrix} 3 & 1 & 4 \\ -1 & -1 & -2 \end{bmatrix}, \ a_{E_i,k} = \begin{bmatrix} -1 & 0 & -1 \\ 1 & -1 & 0 \\ 0 & 1 & 1 \end{bmatrix}. \quad (6.8.5)$$

In the above equations, the first two terms correspond to the corresponding inlet and outlet flows of the component. Furthermore, the relation between reaction rate and fuel cell current is a proportional relationship. If we define r_i as the total rate of species i generated in the fuel cell (typically with units of mole/s) and I_{cell} is the total current leaving the cell, then

$$\frac{I_{\text{cell}}}{2F} = \frac{u}{2F} = -r_{\text{H}_2} = r_{\text{H}_2\text{O}} = -2r_{\text{O}_2}. \tag{6.8.6}$$

The reaction rates of the reforming equations may be defined on the basis of the reaction route network theory developed by Fishtik et al. (2004a, b, c) and applied by Master (2010) where coefficients of the rate equations found from experiments. A reaction route is defined as a linear combination of the elementary reactions, such that a certain number of species, either terminal or intermediate, are cancelled, thus producing a new reaction referred to as an overall reaction. By eliminating all the intermediate species, an overall reaction results and it is termed a full route. If all the species are eliminated in a reaction route, an empty route is formed where all the stoichiometric coefficients are equal to zero. Xu and Froment (1989a, b) determined intrinsic reaction rates for the reforming reactions in the form of the Langmuir–Hinshelwood kinetics model. The reaction rates in mols/s are defined by,

$$r_1 = \frac{v_{r1} V_r w_{ca} k_1}{\Delta^2 p_{\text{H}_2}^{5/2}} \left(p_{\text{CH}_4} p_{\text{H}_2\text{O}} - p_{\text{H}_2}^3 p_{\text{CO}} \middle/ K_{\text{eq}_1} \right), \tag{6.8.7a}$$

$$r_2 = \frac{v_{r2} V_r w_{ca} k_2}{\Delta^2 p_{\text{H}_2}} \left(p_{\text{CO}} p_{\text{H}_2\text{O}} - p_{\text{H}_2} p_{\text{CO}_2} / K_{\text{eq}_2} \right), \tag{6.8.7b}$$

$$r_3 = \frac{v_{r3} V_r w_{ca} k_3}{\Delta^2 p_{\text{H}_2}^{3/2}} \left(p_{\text{CH}_4} p_{\text{H}_2\text{O}}^2 - p_{\text{H}_2}^4 p_{\text{CO}_2} \middle/ K_{\text{eq}_3} \right), \tag{6.8.7c}$$

where w_{ca} is the density of the catalyst used, v_{ri} is the ratio of the actual volume over which the ith reaction is taking place to the reactor volume, V_r is the reactor physical volume,

$$\Delta = 1 + K_{\text{CO}} p_{\text{CO}} + K_{\text{H}_2} p_{\text{H}_2} + K_{\text{CH}_4} p_{\text{CH}_4} + K_{\text{H}_2\text{O}} p_{\text{H}_2\text{O}} / p_{\text{H}_2} \tag{6.8.8}$$

and p_{CO}, p_{H_2}, p_{CH_4}, p_{H_2O} are the partial pressures of CO, H_2, CH_4, and H_2O within the reaction volume. These partial pressures are calculated using the ideal gas equations,

$$p_j V_{\text{anode}} = N_j RT, \; j = \text{CO, H}_2\text{, CH}_4\text{, H}_2\text{O}. \tag{6.8.9}$$

The reaction rate constants are defined by Arrhenius-type relations given by,

$$k_i = A_i \exp(-E_i/RT), \; i = 1, 2, 3. \tag{6.8.10}$$

Table 6.5 Values of model constants

i	Units	1	2	3	
j		CO	H_2	CH_4	H_2O
A_i	mol × $MPa^{\chi_i}/g_{cat}s$	3.711×10^{11}	5.431×10^{3}	8.961×10^{10}	
χ_i	–	0.5	−1	0.5	
v_{ri}	–	~1	~1	~1	
$E_i \times 10^{-3}$	J/mol	240.1	67.13	243.9	
T_i	°K	−26830	4400	−22430	
T_{R_i}	°K	890.95	1090	860.11	
B_j	Pa^{-1}	8.23×10^{-10}	6.12×10^{-14}	6.65×10^{-9}	1.77×10^{5}
$\Delta H_j \times 10^{-3}$	J/mol	70.65	82.9	38.28	−88.68

K_{eq_1}, K_{eq_2}, and K_{eq_3} are the equilibrium constants for the reforming reactions, which are calculated from,

$$K_{eq_i} = \exp\left(T_i\left(\frac{1}{T} - \frac{1}{T_{R_i}}\right)\right),\ i = 1,\ 2,\ 3. \tag{6.8.11}$$

and

$$K_j = B_j \exp(-\Delta H_j/RT),\ j = \text{CO},\ \text{H}_2,\ \text{CH}_4,\ \text{H}_2\text{O}. \tag{6.8.12}$$

The constants A_i, E_i, T_i, T_{R_i}, B_j, and ΔH_j may be obtained from Xu and Froment (1989a, b) and are tabulated in Table 6.5.

The components F_i, in the component molar flow rate equations, are the components corresponding to the fuel (hydrogen) and the oxidized fuel (water steam). The components E_i are the components corresponding to the effluents from the reforming reactions, methane, carbon monoxide, and carbon dioxide. The coefficients $a_{F_i,k}$ and $a_{E_i,k}$ are the stoichiometric coefficients of the components F_i and E_i in the reaction k. The total pressure at the anode, which is also the input pressure at the anode, is given by,

$$p_{\text{anode}} = \sum_{j=1}^{4} p_j,\ j = \text{CO},\ \text{H}_2,\ \text{CH}_4,\ \text{H}_2\text{O}. \tag{6.8.13}$$

At the cathode, the partial pressure of Oxygen (O_2) is given by,

$$p_{O_2} V_{\text{cathode}} = N_{O_2} RT. \tag{6.8.14}$$

At the cathode, the total pressure at the cathode, which is also the input pressure at the cathode, is given by,

$$p_{\text{cathode}} V_{\text{cathode}} = N_{\text{air}} RT \tag{6.8.15}$$

where N_{air} is the total molar flow rate at the cathode. Finally, the mass flow rates at the anode and cathode channels are assumed to be given by,

$$\dot{m}_{\text{out},e} = \sqrt{k_e\left(p_e - p_{\text{out},e}\right)}, \ e = \text{anode, cathode}, \tag{6.8.16}$$

where k_e is a coefficient mass flow rate to the pressure difference.

The temperature within the fuel cell is given by, the energy balance equation,

$$mC_p\frac{\mathrm{d}T}{\mathrm{d}t} = P_{\text{anode}} + P_{\text{cathode}} - \sum_{j=1}^{4}\Delta h_j^r r_j - P_{DC} - P_{ht\_\text{loss}} \tag{6.8.17}$$

where P_{anode} is the total power input to cell at the anode, P_{cathode} is the total power input to the cell at the cathode, Δh_j^r is the molar specific enthalpy change due to the reaction j, P_{DC} is the electrical DC power loss across the output terminals, $P_{\text{ht\_loss}}$ represents the radiation and conduction heat loss to the surroundings from the SOFC.

The operating cell voltage is related to the open-circuit theoretical voltage and to the various losses as

$$E_{\text{cell}} = E_{\text{rev}} - \eta_a(i) - |\eta_c(i)| - \eta_{\text{conc}}(i) - \eta_{\text{ohm}}(i) \tag{6.8.18}$$

where $\eta_a(i)$ and $\eta_c(i)$ are the activation losses at the anode and cathode side, respectively, $\eta_{\text{conc}}(i)$ is the concentration over potential and $\eta_{\text{ohm}}(i)$ is the Ohmic over potential. For a solid oxide fuel cell, the Nernst equation gives the reversible cell potential which is,

$$E_{\text{rev}} = E_o + \frac{RT}{2F}\ln\left(\frac{p_{\text{H}_2}p_{\text{O}_2}^{1/2}}{p_{\text{H}_2\text{O}}}\right) \tag{6.8.19}$$

where $E_o = -\Delta G^\circ{}_{rxn}/2\text{ F}$, where $\Delta G^\circ{}_{rxn}$ is the Gibbs free energy of the overall reaction (evaluated at standard pressures and the operating temperature of the cell), and the p_i's are the partial pressures at the three-phase interface within the reaction region. The activation over potentials are obtained by using the Butler–Volmer equation given for each participating species s, i.e., H_2, O_2 separately. The functional relation between the activation loses and current density is described by the Butler–Volmer equations. For hydrogen oxidation, the Butler–Volmer equation takes the form,

$$i_{\text{H}_2} = i_{0,\text{H}_2}\left\{\exp\left(\frac{\alpha nF}{RT}\eta_a\right) - \exp\left(-\frac{(1-\alpha)nF}{RT}\eta_a\right)\right\}. \tag{6.8.20}$$

In the above equation, i_{0,H_2} is the exchange current given by,

$$i_{0,H_2} = c_1 \left(\frac{Y_{H_2}}{Y_{H_2,ref}} \right) \exp\left(\frac{E_{act,H_2}}{RT} \right). \tag{6.8.21}$$

The constants in the equation for the exchange current are obtained by fitting experimental data in the appropriate temperature range (750–850 °C for a SOFC). The activation energy barrier, E_{act,H_2}, is 110.3 kJ/mol. A modified version of the Butler–Volmer equation is used for oxygen reduction which takes the form,

$$i_{O_2} = i_{0,O_2} \left\{ \exp\left(\frac{\beta_a nF}{RT} \right) - \exp\left(\frac{\beta_c nF}{RT} \right) \right\} \tag{6.8.22}$$

where

$$i_{0,O_2} = i_{00,O_2} \exp\left(\frac{E_{act,O_2}}{RT} \right) \tag{6.8.23}$$

and i_{00,O_2} can be found from Zhu et al. (2005). The equations are then solved for the activation potentials numerically, assuming that $i_{H_2} = i_{O_2} = i_{cell} = i$. A good approximation is obtained by setting $\alpha = 0.5$and it follows that

$$i = i_0 \left\{ \exp\left(\frac{nF}{2RT} \eta_a \right) - \exp\left(-\frac{nF}{2RT} \eta_a \right) \right\} = 2 i_0 \sinh\left(\frac{nF}{2RT} \eta_a \right). \tag{6.8.24}$$

Hence,

$$\eta_{act} = \eta_a = \frac{2RT}{nF} \sinh^{-1}\left(\frac{i}{2 i_0} \right). \tag{6.8.25}$$

Moreover,

$$\frac{d\eta_{act}}{\mathrm{di}} = \frac{2RT}{\mathrm{nF}} \frac{1}{\sqrt{4 i_0^2 + i^2}} \tag{6.8.26}$$

Alternately, the activation loss may be obtained by fitting a Tafel-like equation to experimental data, in which case, the activation over potential may be expressed in the form,

$$\eta_{act} = \frac{RT}{\alpha nF} \ln\left(\frac{i}{i_0} \right). \tag{6.8.27}$$

The Ohmic loss may be expressed as,

$$\eta_{ohm} = R_{tot} j_{cell} \tag{6.8.28}$$

where R_{tot} is the total specific resistance of the cell (per unit area), and j_{cell} is the cell current density. The main contribution to the specific total resistance is from the electrolyte with the anode and cathode making a small contribution to it. Thus, it is expressed as,

$$R_{tot} = R_0 + t_{elec}/\sigma_{elec} \tag{6.8.29}$$

where R_0 is a small constant that accounts for the anode and cathode resistivities, t_{elec} is the thickness or width of the electrolyte, and σ_{elec} is the electrolyte conductivity. Yttria-stabilized zirconia (YSZ), which has a high ionic conductivity at high temperatures in the vicinity of 1,000 °C, is most common electrolyte used in SOFCs. For YSZ, the electrical conductivity is 6.67×10^{-2} S cm^{-1} at 1,000 °C. The relationship between electrical conductivity and temperature can be described by a thermally activated Arrhenius equation and expressed as,

$$\sigma_{elec} = \frac{\sigma_0}{T}\exp\left(\frac{E_{act,YSZ}}{RT}\right) \tag{6.8.30}$$

where the actuation potential $E_{act,YSZ} = 93\,\text{kJ/mol}$ for YSZ in the vicinity of 1,000 °C. Thus, the electrolyte conductivity may be expressed as,

$$\sigma_{elec} = \sigma_{1273}\frac{1273}{T}\exp\left(\frac{E_{act,YSZ}}{R}\left(\frac{1}{T} - \frac{1}{1273}\right)\right) \tag{6.8.31}$$

where T is in °K, and σ_{1273} is the electrical conductivity at 1,000 °C. In the Nernst equation, the species concentrations are taken as the reference value to calculate the open-circuit voltage. However, during the operation of the cell, the concentration at the reaction sites is not the one at the input channels, due to the depletion of the reactants and to mass-transfer resistances at the electrodes which act as porous media. Hence, Nernst equation must be corrected, both in the anode and in the cathode. Thus, the concentration over potentials in the anode and cathode are, respectively, given by,

$$\eta_{conc,an} = \frac{RT}{2F}\ln\left(\left(\frac{p_{H_2}}{p_{H_2O}}\right)_{in}\left(\frac{p_{H_2O}}{p_{H_2}}\right)_{re}\right), \tag{6.8.32a}$$

$$\eta_{conc,cath} = \frac{RT}{2F}\ln\left(\left(\sqrt{p_{O_2}}\right)_{in}\left(\frac{1}{\sqrt{p_{O_2}}}\right)_{re}\right), \tag{6.8.32b}$$

where the subscripts refer to the input channels and the reaction sites in the cell.

6.9 Cogeneration Applications of SO Fuel Cells

Hybrid power plants based on SOFCs are currently being developed that are expected to exceed values of net electrical efficiencies higher than 60 %. The analysis and design optimization of such systems can be done based on dynamic models of SOFCs and gas turbine systems to establish optimum design configurations of a hybrid power plant with the fuel cell integrated with a traditional Brayton cycle-based gas turbine illustrated in Fig. 6.8.

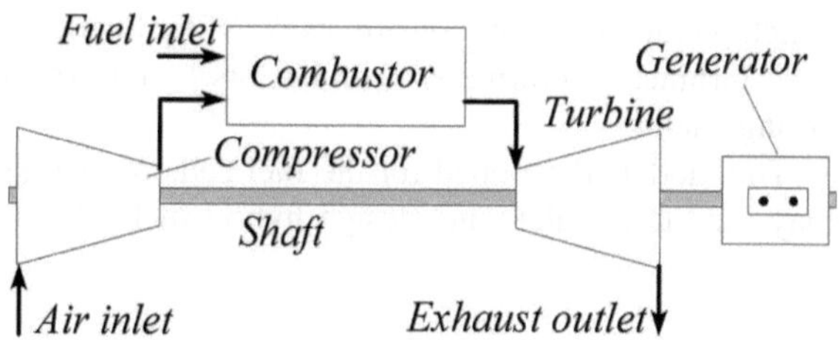

Fig. 6.8 Basic gas turbine system

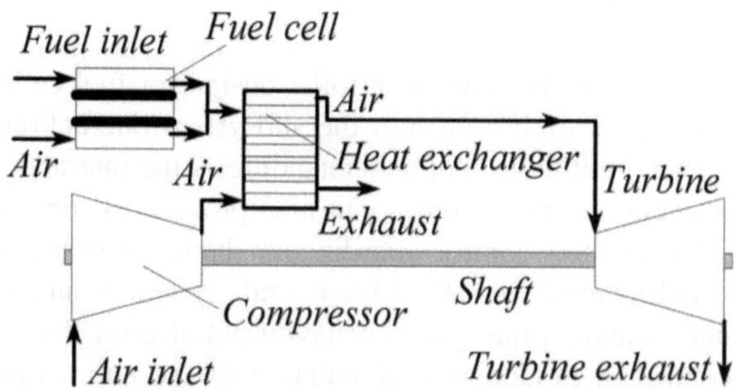

Fig. 6.9 Fuel cell-driven turbine compressor system

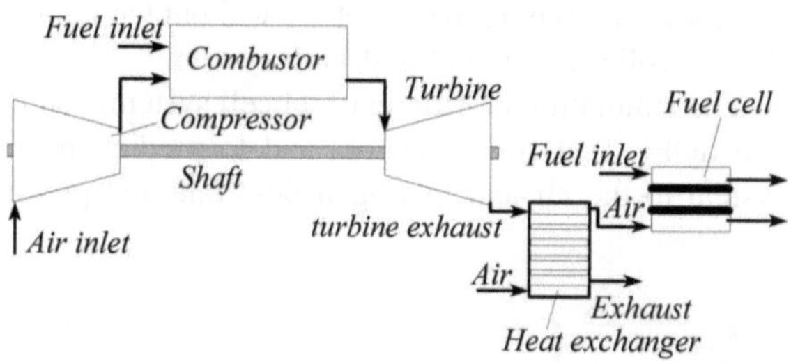

Fig. 6.10 Open-loop cogeneration with a SOFC

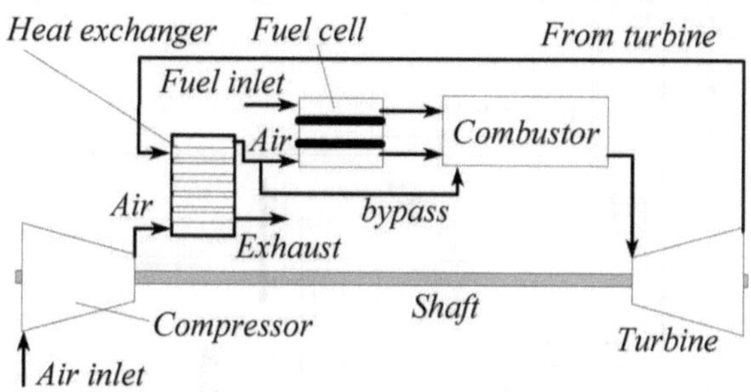

Fig. 6.11 Closed-loop cogeneration with a SOFC

A number of open-loop configurations (Figs. 6.9 and 6.10) and closed-loop configurations (Fig. 6.11) for hybrid SOFC—GT plants have been proposed in the literature (Calise et al. 2004; Bove and Ubertini 2006; Calise et al. 2006; Janardhanan and Deutschmann 2006; Kandepu et al. 2006; Pakalapati 2006; Pakalapati et al. 2006; Calise et al. 2007, Kandepu et al. 2007; Zhang et al. 2008; Chen et al. 2012), which include combinations of stacks of fuel cells, heat exchangers, compressors, gas turbines, reformers, combustors, and other subsystems in different arrangements.

The models developed for the fuel cells on the gas turbines system in Chap. 5 may used to simulate and design hybrid fuel cell—GT plants.

6.10 Simulation of Fuel Cells

Dynamic analysis and controller design for fuel cells are immensely facilitated if one has prior insight into the subsystem interactions and control design performance limitations. An understanding of the interactions and limitations of various feasible control schemes can also provide for developing guidelines for sensors selection and control coordination between subsystems. Creating a control-oriented dynamic model of the overall system is an essential first step not only for understanding the system behavior but also for the development and design of the control methodologies. Simulation of the entire open-loop and various feasible closed-loop systems using a tool such as MATLAB/SIMULINK then offers a rapid and easy prototyping methodology without having to actually build and implement the controller and the closed-loop system.

The simulation of different fuel cell systems facilitates the comparison of PEM and SOFC fuel cell behaviors and to predict the response of an integrated FC system to the air and hydrogen flow rate and pressure regulation. Fuel cells are

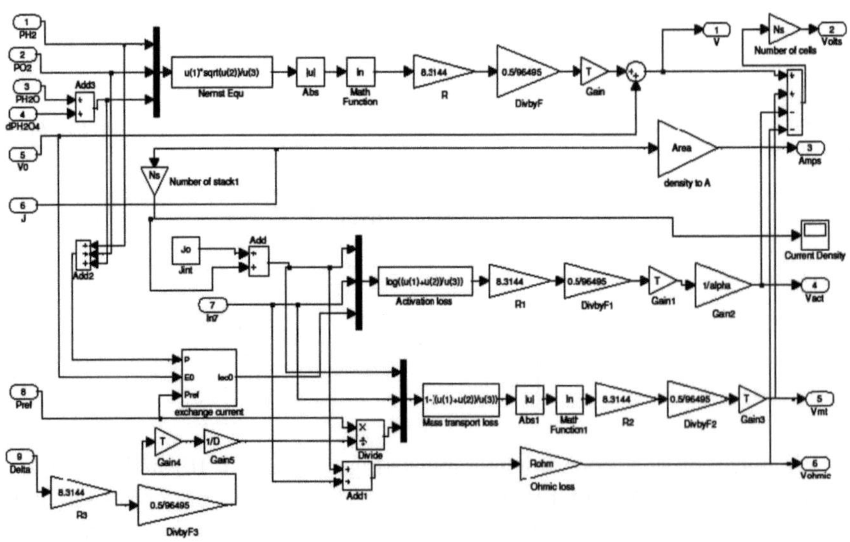

Fig. 6.12 SIMULINK diagram relating the input partial pressures and the currents to the output voltage of a PEM fuel cell

being studied extensively because of their potential in not only offering an alternate a new energy source for a large range of applications but also enhancing the efficiency of conventional energy generating systems based on fossil fuels which are a depleting resource. The benefits of energy production from fuel cells are the high efficiency and their environment-friendly by-products. In fuel cells, a chemical reaction converts hydrogen and oxygen into water, releasing electrons (current) in the process. To simulate the dynamics of fuel cell, it is indeed possible to simulate the dynamics of the flows and the partial pressures of the constituent gases but also the output voltages and the over voltages that limit their performance. To develop the simulation, it is necessary to establish a block diagram model of the complete process relating the inputs and the outputs of the plant. A typical example of a SIMULINK diagram relating the input partial pressures and the currents to the output voltage of a PEM fuel cell is shown in Fig. 6.12.

6.11 Estimation of the Parameters of a Fuel Cell

There have been only a few studies of the applications of filtering and estimation to fuel cells. McKay and Stefanopoulou (2004) developed a lumped parameter model for estimating relative humidity of the electrodes. A nonlinear estimator was developed for the estimation of membrane humidity and the pressure and inlet and outlet temperature at the electrodes. The estimation filter was also used to implement a feedback controller to regulate the excess oxygen ratio during changes in load (Pukrushpan et al. (2004a). Arcak et al. (2004) developed a nonlinear observer for fuel cell hydrogen estimation. Görgün et al. (2006) developed an algorithm for estimating the membrane water content in PEMFCs. Ingimundarson et al. (2008) adopted a model-based estimation approach to monitor hydrogen leaks in a fuel cell stack. Pan et al. (2010) have applied Kalman filtering to solve the state estimation problem of the fuel cell. Kandepu et al. (2008), Stiller et al. (2006), Kandepu et al. (2008, 2007, 2006, 2005) have applied state estimation using nonlinear Kalman filtering to a hybrid gas turbine and solid oxide fuel cell system. Haschka et al. (2006) have applied nonlinear estimation to solid oxide fuel cells whose impedance is represented by a fractional (non-integer) order impedance model. Tumuluri (2008) has applied nonlinear extended Kalman filtering (EKF) to the estimates the states of a PEMFC. All the nonlinear estimation algorithms used in these applications to fuel cells can be classified into three groups: deterministic nonlinear estimation, EKF based, and unscented Kalman filter (UKF) based. In the last two categories, the process and measurement noise covariance matrices are assumed to be constant and known. In the first category, nonlinear estimation algorithms were developed using deterministic model-based filtering. Real fuel cells are known to be extremely noisy, and for this reason, it is important to include the noise inputs both the process equations and the measurements. On the other hand, when covariance matrices are assumed to be constant and known, but are not representative of the true situation, the EKF is known to diverge in some situations, while the UKF is not stable and convergent. Moreover, for most control applications, the actual values of the covariance matrices, even if they are constant, are unknown or difficult to estimate.

Considering the state estimation of fuel cells, which is a problem that arises because of incomplete knowledge of the parameters and states of a real fuel stack, there is need to construct a reduced-order model of the system to facilitate state estimation and parameter identification. Most parameters are in reality dynamically evolving states, and they are treated as parameters when their slowly changing dynamics is ignored. The problem is that there are a number of states that are only weakly coupled to the main states, such as the limiting current density and the mean water content in the membrane, that contribute to one of the measureable outputs, the stack voltage, the stack current, the stack temperature, and the relative humidity at the anode and cathode channels. The weakly coupled states can lead to a problem of observability of the states or identifiability of the parameters. Hence, although there are a number of models of fuel cell dynamics in the literature, which we can mechanize in MATLAB, there is a need to carefully consider the model and extract the primary subsystem that contributes to the outputs. Furthermore, when dealing with a real practical fuel cell stack, difficulties have been experienced in matching the outputs of the simulations to the outputs of the real physical system. Thus, the only option available seems to be, to tune the states of the simulation, based on a real-time identification of key states that influence the output of the fuel cell stack.

One approach to avoid problems of observability and identifiability is to use circuit models where it is relatively easier to isolate decoupled subsystems.

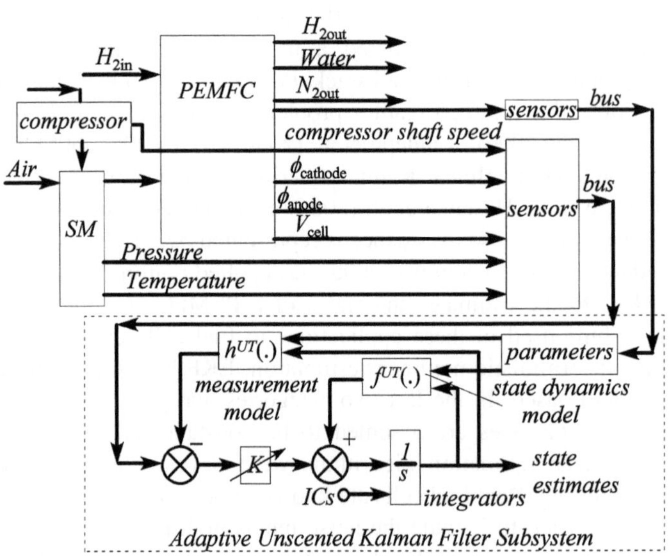

Fig. 6.13 Schematic diagram of the adaptive UKF (the adaptation loop is not shown)

Such an approach has been adopted by Fontes et al. (2010), Yu and Yuvarajan (2004), while Suares and Hoo (2000), Carnes and Djilali (2005), Pasricha et al. (2007) have opted to use a quasi-steady-state model which alleviates the problem to a certain extent. Another approach to avoid the problems of observability and identifiability is to a transfer function representation with unknown coefficients as has been done by Williams et al. (2007), Thanapalan et al. (2008). In these methods, while one does obtain a set of parameters for the assumed model, it is often quite impossible to relate these to the physical parameters used in a typical dynamic simulation.

In Fig. 6.13, a diagram of nonlinear state estimator is illustrated highlighting its role in the control of a fuel cell. In this implementation of the state estimation of the PEMFC by Vepa (2012), the adaptive method is coupled with the unscented Kalman filter developed by Julier and Uhlmann (2000), Julier et al. (2000), Julier (2002) and used to estimate the states of a state-space model of a fuel cell system. The principles of state estimation are discussed in Chap. 8.

6.12 Controller Design for Fuel Cells

The design of a controller for a fuel cell generally involves minimizing a certain performance functional. The control inputs are the flow rates of air and hydrogen, the compressor controls, and the relative humidity content. The cell voltage and power outputs, the relative humidity, and the temperature are the outputs of the system that must be regulated. It is important to regulate and control the stack temperature and the liquid water at both the anode and the cathode. The temperature and relative humidity (liquid water) are independently controlled, and consequently, the control loops for the regulating the temperature and the liquid water are only mildly coupled with the main control loop. First considering just the fuel cell, it is essential to maintain the oxygen excess ratio at the cathode close to a pre-established set-point in order to avoid lack of oxygen (oxygen starvation). The cathode inlet oxygen mass flow is often considered as a state variable, while the reacting oxygen mass flow, which is proportional to the fuel cell current is treated as an input variable. The oxygen excess ratio is the ratio of inlet oxygen mass flow to the reacting oxygen mass flow. The regulator of the oxygen excess ratio constitutes the main control loop, and it is the loop controlling the cathodic flow channel. It is also important to regulate and control the power loss and maximize efficiency. Furthermore, it is essential to regulate the differential pressure between anode and cathode and to maintain an acceptable level of fuel consumption and steam:carbon ratio (depending on the nature of the cell and fuel) for a given power output or electrical load. Generally, the regulation of oxygen supply is done by an air compressor which supplies air to all the fuel cells in a stack via a supply manifold. The air is collected after pass through the fuel cell in a return manifold and let out to the atmosphere. When this is done, one should also consider the consumption of power by the compressor. When the oxygen ratio is excessive, the compressor power required to maintain the flow of oxygen increases rapidly and there is fall in the efficiency. For the main control loop, the inputs are the compressor voltage and the fuel cell current, while the outputs are the fuel cell voltage and the oxygen excess ratio. Thus, the fuel cell system is multi-input/multi-output system, where some of the inputs/outputs are used to define the performance, while the others are used to implement the control loop. The methodology of optimal control and other methods related to it such as model predictive control (MPC) is most suited for implementing the main control loop. Pukrushpan et al. (2004b), Pukrushpan et al. (2004c), Vahidi et al. (2006) have discussed the control problem and its formulation in some detail.

Typically, the plant model is linearized and represented in state-space form as,

$$\frac{\mathrm{d}}{\mathrm{d}t}\mathbf{x}(t) = \mathbf{Ax} + \mathbf{B}u, \;\; \mathbf{y} = \mathbf{Cx} + \mathbf{Gu}. \tag{6.12.1}$$

The state-space description of a *linear time invariant* system is given by the above state equation where $\mathbf{x}$, $\mathbf{u}$, and $\mathbf{y}$ are the $n \times 1$ state vector, $m \times 1$ input vector, and the $1 \times k$ output vector, and $\mathbf{A}$, $\mathbf{B}$, $\mathbf{C}$, and $\mathbf{G}$ are $n \times n$, $n \times m$, $k \times n$, and $m \times m$ matrices. The first of Eq. (6.12.1) represents relationship between the inputs and the states, while the second represents the relationship of the states and the inputs to the outputs. Thus, in principle, the transfer function representation may be obtained by "eliminating" the states from the second of Eq. (6.12.1) using the first. For a single-input single-output (SISO) system, the number of inputs and number of outputs are equal to unity, i.e., $m = k = 1$.

The design of the control laws to optimize a performance criterion is done by minimizing cost function

$$J_{\mathrm{LQR}} = \frac{1}{2}\int_0^\infty \left(\mathbf{x}^T\mathbf{Qx} + \mathbf{u}^T\mathbf{Ru}\right)\mathrm{d}t = \frac{1}{2}\int_0^\infty \left(\|\mathbf{x}\|_{\mathbf{Q}} + \|\mathbf{u}\|_{\mathbf{R}}\right)\mathrm{d}t \tag{6.12.2}$$

The first term $\frac{1}{2}\int_0^\infty \|\mathbf{x}\|_{\mathbf{Q}}\mathrm{d}t$ represents the energy of the state response and provides the energy of a combination of the states to assess the performance. It is the H_2 norm of the state response. The second term in the cost function $\frac{1}{2}\int_0^\infty \|\mathbf{u}\|_{\mathbf{R}}\mathrm{d}t$ represents the energy contained in the control signal that is fed back into the plant by the controller. The linear quadratic regulator (LQR) uses a linear controller that minimizes the quadratic performance cost functional of the states and control inputs. Thus, the optimal gain matrix $\mathbf{K}$ is chosen such that for a given continuous-time state-space model, the state-feedback control law $\mathbf{u} = -\mathbf{Kx}$ minimizes the quadratic cost function

$$J_{\mathrm{LQR}} = \frac{1}{2}\int_0^\infty \left(\mathbf{x}^T\mathbf{Qx} + \mathbf{u}^T\mathbf{Ru}\right)\mathrm{d}t \tag{6.12.3}$$

subject to the state vector satisfying the model's dynamics equations $\dot{\mathbf{x}} = \mathbf{A}\mathbf{x} + \mathbf{B}\mathbf{u}$. A first choice for the matrices **Q** and **R** Bryson and Ho (1975) in the expression for the cost functional is given by *Bryson's rule*: select **Q** and **R** as diagonal matrices with elements,

$$q_{ii} = 1/\text{maximum expected value of } x_i^2, \tag{6.12.4a}$$

$$r_{ii} = 1/\text{maximum expected value of } u_i^2. \tag{6.12.4b}$$

The solution to the optimal control problem may be conveniently expressed in terms of the constant *Riccati* matrix **P** which is defined by,

$$\mathrm{d}\left(\mathbf{x}^T\mathbf{P}\mathbf{x}\right)/\mathrm{d}t = -\mathbf{x}^T\left(\mathbf{Q} + \mathbf{K}^T\mathbf{R}\mathbf{K}\right)\mathbf{x}. \tag{6.12.5}$$

Substituting the state-feedback control input and using the above equation,

$$J_{\mathrm{LQR}} = \frac{1}{2}\int_0^\infty \mathbf{x}^T\left(\mathbf{Q} + \mathbf{K}^T\mathbf{R}\mathbf{K}\right)\mathbf{x}\mathrm{d}t = -\frac{1}{2}\left(\mathbf{x}^T\mathbf{P}\mathbf{x}\right)\Big|_0^\infty = \frac{1}{2}\left(\mathbf{x}^T(0)\mathbf{P}\mathbf{x}(0)\right). \tag{6.12.6}$$

It then follows that J_{LQR} has a minimum which may be obtained by expanding (6.12.5)and solving the resulting algebraic matrix *Riccati* equation

$$\mathbf{A}^T\mathbf{P} + \mathbf{P}\mathbf{A}^T - \mathbf{P}\mathbf{B}\mathbf{R}^{-1}\mathbf{B}^T\mathbf{P} + \mathbf{Q} = 0, \tag{6.12.6}$$

for **P**, by eigenvalue decomposition of the equations representing the optimal solution. The optimal gain is given by,

$$\mathbf{K} = \mathbf{R}^{-1}\mathbf{B}^T\mathbf{P}. \tag{6.12.7}$$

A crucial property of an LQR controller is that this closed loop is asymptotically stable as long as the system is at least controllable. Furthermore, LQR controllers are inherently robust with respect to process uncertainty. Moreover, the gain margin of an LQR controller is infinite for gain increase and -6 dB for gain decrease, and the phase margin is at least 60°. Thus, the LQR controllers have some very desirable robustness properties and can be tuned to meet other requirements and thus extremely well suited for the design controllers for energy systems. For further details about the methods of optimal control, the reader is referred to the text by Ogata (2001).

The H_∞ norm minimizing controller minimizes the worst case of the ratio between the H_2 norm of the H_∞ performance assessment output $\mathbf{z}_\infty = \mathbf{T}_{\mathrm{H}_\infty}\mathbf{x}$ and the input **x** which is,

$$\left\|\mathbf{T}_{\mathrm{H}_\infty}(j\omega)\right\|_\infty = \sup_{\omega\in\mathrm{R}}\left[\max_{x\in C^n}\left[\frac{\left\|\mathbf{T}_{\mathrm{H}_\infty}(j\omega)\mathbf{x}\right\|_2}{\left\|\mathbf{x}\right\|_2}\right]\right] \tag{6.12.8}$$

where the H_2 and H_∞ norms are, respectively, defined Burl (1999) as:

$$\|\mathbf{T}\|_2 = \left(\frac{1}{2}\int_{-\infty}^{\infty} \text{trace}[\mathbf{T}^*(j\omega)\mathbf{T}(j\omega)]d\omega\right)^{\frac{1}{2}}, \quad \|\mathbf{T}\|_\infty = \sup_\omega[\sigma_{\max}(\mathbf{T}(j\omega))]. \quad (6.12.9)$$

The H_2 norm minimizing controller involves finding a control law, $\mathbf{u}(s) = \mathbf{K}(s)\mathbf{y}(s)$ over the set of all stabilizing controllers that minimizes the H_2 norm of the lower linear fractional transformation (LFT) $\mathbf{T_{zw}}$,

$$\mathbf{T_{zw}} = \mathbf{G_{zu}}(s)\mathbf{K}(s)\left(\mathbf{I} - \mathbf{G_{yu}}(s)\mathbf{K}(s)\right)^{-1}\mathbf{G_{yw}}(s) + \mathbf{G_{zw}}(s), \quad (6.12.10)$$

where the transfer function $\mathbf{G}(s)$ defines the relationship between the outputs $\mathbf{z}$ and $\mathbf{y}$ and inputs $\mathbf{w}$ and $\mathbf{u}$ as,

$$\begin{bmatrix}\mathbf{z}\\ \mathbf{y}\end{bmatrix} = \mathbf{G}(s)\begin{bmatrix}\mathbf{w}\\ \mathbf{u}\end{bmatrix} \equiv \begin{bmatrix}\mathbf{G_{zw}}(s) & \mathbf{G_{zu}}(s)\\ \mathbf{G_{yw}}(s) & \mathbf{G_{yu}}(s)\end{bmatrix}\begin{bmatrix}\mathbf{w}\\ \mathbf{u}\end{bmatrix}. \quad (6.12.11)$$

The H_∞ norm equal to $||\mathbf{T_{zw}}(s)||_\infty$ may be expressed as,

$$||\mathbf{T_{zw}}(s)||_\infty = \left\|\begin{matrix}W_s(s)S(s)\\ W_T(s)T(s)\end{matrix}\right\|_\infty < 1 \quad (6.12.12)$$

where $S(s)$ is the sensitivity-transfer function, and $T(s)$ is the complementary sensitivity-transfer function. The control problem requires that the H_∞ performance weights W_S and W_T are chosen appropriately. The H_∞ norm bounding but suboptimal controller may also be stated in the alternate form. Given the plant in a state-space form,

$$\dot{\mathbf{x}} = \mathbf{Ax} + \mathbf{B}_1\mathbf{w} + \mathbf{B}_2\mathbf{u} \quad (6.12.13a)$$

$$\mathbf{z} = \mathbf{C}_1\mathbf{x} + \mathbf{D}_{11}\mathbf{w} + \mathbf{D}_{12}\mathbf{u} \quad (6.12.13b)$$

$$\mathbf{y} = \mathbf{C}_2\mathbf{x} + \mathbf{D}_{21}\mathbf{w} + \mathbf{D}_{22}\mathbf{u}. \quad (6.12.13c)$$

Let the $\mathbf{w} \to \mathbf{z}$ transfer function be defined as, $\mathbf{T_{zw}}(s)$ with its H_∞ norm equal to $||\mathbf{T_{zw}}(s)||_\infty$. The problem is to find a control law, $\mathbf{u}(s) = \mathbf{K}(s)\mathbf{y}(s)$, such that the H_∞ norm is bounded as, $||\mathbf{T_{zw}}||_\infty < \gamma$ where $\gamma \in \mathrm{R}$. The solution may be conveniently stated in terms of the solutions of two algebraic Riccati equations (Glover and Doyle 1988; Doyle 1989).

Gahinet and Apkarian (1994) had shown that the same suboptimal problem may also be stated in terms of linear matrix inequalities (LMI), provided the controller is also defined as a dynamic controller. Khargonekar and Rotea (1991) had defined the mixed H_2/H_∞ norm optimization of a linear dynamic controller, and this again can be cast in LMI format. However, our experience indicated that finding a solution to this problem is not easy.

To define the H_2 or H_∞ norm optimization problems, one must define two independent outputs: one to assess the H_∞ performance and the other to assess the H_2 performance. If these two outputs are defined as,

$$\mathbf{z} = \begin{bmatrix} \mathbf{z}_\infty & \mathbf{z}_2 \end{bmatrix}^T \equiv \begin{bmatrix} \mathbf{z}_1 & \mathbf{z}_2 \end{bmatrix}^T, \tag{6.12.14}$$

and the corresponding transfer functions from $\mathbf{w} \to \mathbf{z}_1$ and $\mathbf{w} \to \mathbf{z}_2$ are $\mathbf{T}_1 = \mathbf{T}_{\mathbf{z}1,\mathbf{w}}(s)$ and $\mathbf{T}_2 = \mathbf{T}_{\mathbf{z}2,\mathbf{w}}(s)$; then to solve the H_2 and norm optimization problems, one must minimize either the H_∞ norm, $\|\mathbf{T}_1\|_\infty$ subject, or the H_2 norm satisfies, $\|\mathbf{T}_2\|_2$. The plant in a state-space form is assumed to be,

$$\dot{\mathbf{x}} = \mathbf{A}\mathbf{x} + \mathbf{B}_1\mathbf{w} + \mathbf{B}_2\mathbf{u} \tag{6.12.15a}$$

$$\mathbf{z}_1 = \mathbf{C}_1\mathbf{x} + \mathbf{D}_{11}\mathbf{w} + \mathbf{D}_{12}\mathbf{u} \tag{6.12.15b}$$

$$\mathbf{z}_2 = \mathbf{C}_2\mathbf{x} + \mathbf{D}_{21}\mathbf{w} + \mathbf{D}_{22}\mathbf{u} \tag{6.12.15c}$$

$$\mathbf{y} = \mathbf{C}_3\mathbf{x} + \mathbf{D}_{31}\mathbf{w} + \mathbf{D}_{32}\mathbf{u}. \tag{6.12.15d}$$

The key to finding a solution to the H_2 and H_∞ norm optimization problems is to define the appropriate performance assessing outputs defined by the equations for $\mathbf{z}_1$ and $\mathbf{z}_2$ and represent the model in a standard form.

If one assumes that the basic plant is modelled as,

$$\dot{\mathbf{x}}_p = \mathbf{A}_p\mathbf{x}_p + \mathbf{B}_p\mathbf{u},\ \mathbf{y}_\mathbf{o} = \begin{bmatrix} 1 & \mathbf{0} \end{bmatrix}\mathbf{x}_p, \tag{6.12.16}$$

The disturbance dynamics is modelled as,

$$\dot{\mathbf{x}}_d = \mathbf{A}_d\mathbf{x}_d + \mathbf{w}. \tag{6.12.17}$$

The plant dynamics including the disturbance states is

$$\dot{\mathbf{x}}_p = \mathbf{A}_p\mathbf{x}_p + \mathbf{A}_{\mathrm{pd}}\mathbf{x}_d + \mathbf{B}_p\mathbf{u} + \mathbf{B}_d\mathbf{w}. \tag{6.12.18}$$

The complete plant model including the state-space dynamics of the weighting filters is then expressed as,

$$\dot{\mathbf{x}}_c = \mathbf{A}_c\mathbf{x}_c + \mathbf{B}_c\mathbf{u} + \mathbf{B}_{\mathrm{cd}}\mathbf{w},\ \mathbf{y}_\mathbf{o} = \mathbf{C}_{oc}\mathbf{x}_c,\ \mathbf{y}_m = \mathbf{C}_c\mathbf{x}_c + \mathbf{D}_c\mathbf{w}, \tag{6.12.19}$$

The traditional block diagram representation of the plant, the controller, the measurement $\mathbf{y}_m$, and performance assessing outputs $\mathbf{z}_{11}$ and $\mathbf{z}_{12}$ is shown in Fig. 6.14.

An alternate representation leading to the lower LFT representation is shown in Figs. 6.15 and 6.16. In particular, the transfer function W_T could be incorporated in the plant as shown in Fig. 6.17, and the augmented plant may be expressed as,

$$\dot{\mathbf{x}}_a = \mathbf{A}_a\mathbf{x}_a + \mathbf{B}_a\mathbf{u} + \mathbf{B}_{\mathrm{ad}}\mathbf{w},\ \begin{bmatrix} \mathbf{z}_{12} & \mathbf{y}_m \end{bmatrix}^T = \mathbf{C}_a\mathbf{x}_a + \mathbf{D}_a\mathbf{w}. \tag{6.12.20}$$

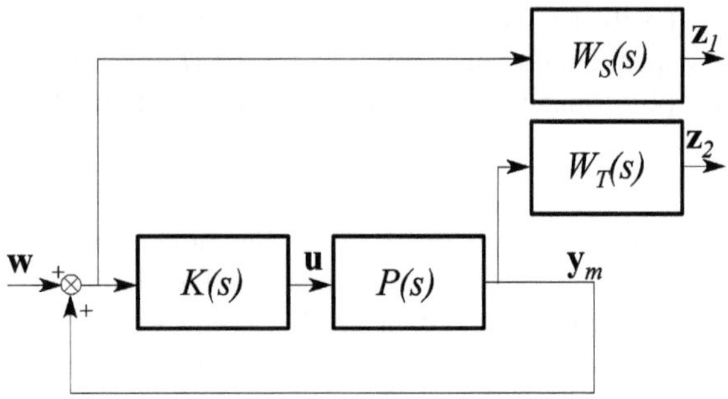

Fig. 6.14 Traditional representation of the plant, the controller, the measurement $\mathbf{y}_m$, and performance assessing outputs $\mathbf{z}_1$ and $\mathbf{z}_2$

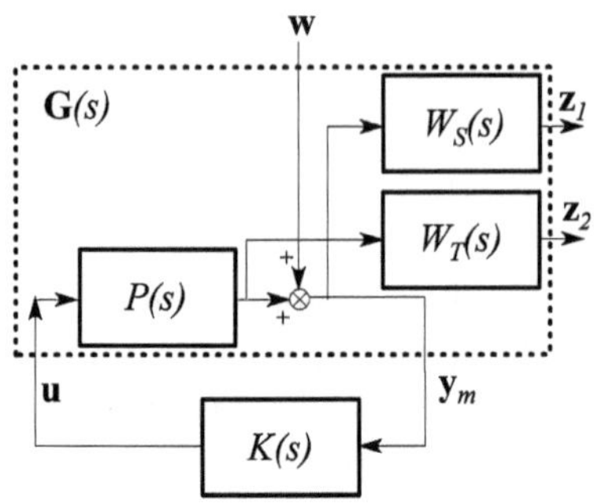

Fig. 6.15 Alternate representation of plant, measurement, and controller

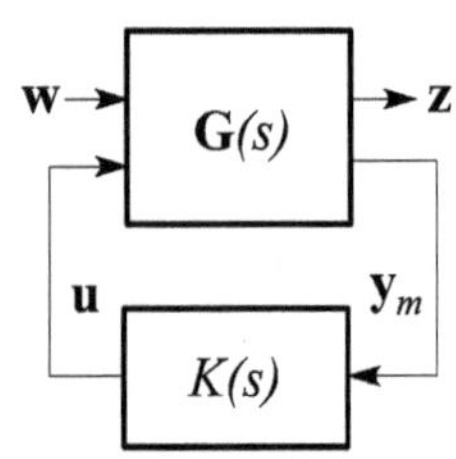

Fig. 6.16 Linear fractional transformation representation of feedback system

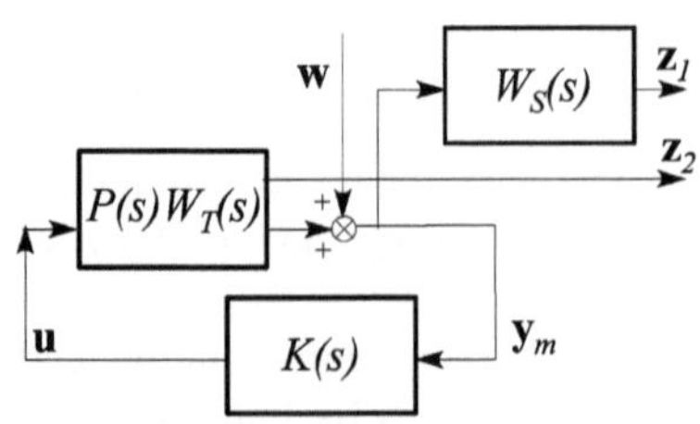

Fig. 6.17 Incorporating the weighting function in the plant

The optimal controller is then given by $\mathbf{u}(s) = \mathbf{K}(s)\mathbf{y}_m(s)$. Representing the dynamics in the form of Eq. (6.12.20) facilitates the use of optimal controller synthesis tools involving the use of full state and output feedback provided by the robust control toolbox in MATLAB. In the case of full state feedback, the states for feedback are replaced by their estimates. The estimation problem was discussed in the previous section. Pukrushpan et al. (2004b), Pukrushpan et al. (2004c), Vahidi et al. (2006) have discussed the application of the optimal design methods to the problem of detailed controller synthesis for a fuel cell, and the reader is referred to it for the details.

References

Adzakpa KP, Agbossous K, Dube Y, Dostie M, Fournier M, Poulin A (2008) PEM fuel cells modeling and analysis through current and voltage transients behaviors. IEEE Trans Energy Convers 23(2):581–591

Amphlett JC, Baumert RM, Mann RF, Peppley BA, Roberge RR, Rodrigues A (1993) The effect of carbon monoxide contamination on anode efficiency in PEM fuel cells. Am Chem Div Fuel Chem 38:1477–1482

Amphlett JC, Baumert RM, Mann RF, Peppley BA, Roberge RR, Rodrigues A (1994) Parametric modeling of the performance of a 5 kW proton exchange membrane fuel cell stack. J Power Sources 49:349–356

Amphlett JC, Baumert RM, Mann RF, Peppley BA, Roberge RR (1995) Performance modeling of the Ballard mark IV solid polymer electrolyte fuel cell: empirical model development. J Electrochem Soc 145:1–8

Amphlett JC, Baumert RM, Mann RF, Peppley BA, Roberge RR, Rodrigues A (1996) A model predicting transient responses of proton exchange membrane fuel cells. J Power Sources 61:183–188

Arcak M, Görgün H, Pedersen LM, Varigonda S (2004) A nonlinear observer design for fuel cell hydrogen estimation. IEEE Trans Control Syst Technol 12:101–110

Badrinarayanan P, Ramaswamy S, Eggert A, Moore R (2001) Fuel cell stack water and thermal management: impact of variable system power operation. SAE Paper 2001-01-0537

Bernardi D (1990) Water balance calculations for solid polymer electrolyte fuel cells. J Electrochem Soc 137(11):3344–3350

Bove R, Ubertini S (2006) Modeling solid oxide fuel cell operation: approaches, techniques and results. J Power Sources 159(1):543–559

Bryson AE, Ho Y-C (1975) Applied optimal control. Hemisphere Publishing Co., New York

Burl JB (1999) Linear optimal control: H_2 and H_∞ methods, Addison Wesley Longman, Inc

Calise F, d'Accadia MD, Palombo A, Vanoli L, Vanoli R (2004) Modelling, simulation and energy analysis of a hybrid SOFC-gas turbine system. 3rd international symposium on energy and environment, Sorrento

Calise F, d'Accadia MD, Palombo A, Vanoli L (2006) Simulation and energy analysis of a SOFC-gas turbine system. Energy Int J 31(15):3278–3299

Calise F, d'Accadia MD, Palombo A, Vanoli L (2007) A detailed one dimensional finite-volume simulation model of a tubular SOFC and a pre-reformer. Int J Thermodyn 10(3):87–96

Carnes B, Djilali N (2005) Systematic parameter estimation for PEM fuel cell models. J Power Sources 144:83–93

Ceraolo M, Miulli C, Pozio A (2003) Modeling static and dynamic behavior of proton exchange membrane fuel cells on the basis of electro-chemical description. J Power Sources 113:131–144

Chen S, Xue Z, Wang D, Xiang W (2012) An integrated system combining chemical looping hydrogen generation process and solid oxide fuel cell/gas turbine cycle for power production with CO_2 capture. J Power Sources 215:89–98

Chia E-S (2006) A chemical reaction engineering perspective of polymer electrolyte membrane fuel cells. Dissertation, Faculty of Chemical Engineering, Princeton University

Chu D, Jiang R (1999) Performance of polymer electrolyte membrane fuel cell (PEMFC) stacks. Part I. Evaluation and simulation of an air-breathing PEMFC stack. J Power Sources 83:128–133

del Real AJ, Arce A, Bordons C (2007) Development and experimental validation of a PEM fuel cell dynamic model. J Power Sources 173:310–324

Doyle JC, Glover K, Khargonekar PP, Francis B (1989) State-space solutions to the standard H_2 and H_∞ control problems. IEEE Trans Automat Contr 34:831–847

El-Sharkh MY, Rahman A, Alam MS, Byrne PC, Sakla AA, Thomas T (2004) A dynamic model for a stand-alone PEM fuel cell power plant for residential applications. J Power Sources 138:199–204

Fishtik I, Callaghan CA, Datta R (2004a) Reaction route graphs. I. Theory and algorithm. J Phys Chem B 108(18):5671–5682

Fishtik I, Callaghan CA, Datta R (2004b) Reaction route graphs. II. Examples of enzyme- and surface-catalyzed single overall reactions. J Phys Chem B 108(18):5683–5697

Fishtik I, Callaghan CA, Datta R (2004c) Reaction route graphs. III. Non-minimal kinetic mechanisms. J Phys Chem B 109(7):2710–2722

Fontes G, Turpin C, Astier S (2010) A large signal dynamic circuit model of a H_2/O_2 PEM fuel cell. IEEE Trans Ind Electron 57(6):1874–1881

Fronk M, Wetter D, Masten D, Bosco A (2000) PEM fuel cell system solutions for transportation. SAE paper 2000-01-0373

Gahinet P, Apkarian P (1994) A linear matrix inequality approach to H_∞ control. Int J Robust Nonlinear Control 4:421–448

Garnier S, Pera M-C, Hissel D, De Bernardinis A, Kauffmann J-M, Coquery G (2004) Dynamic behavior of a proton exchange membrane fuel cell under transportation cycle load. Proceeding of international symposium on industrial elect 1, pp 329–333

Glover K, Doyle JC (1988) State space formulae for all stabilizing controllers that satisfy an H_∞-norm bound and relations to risk sensitivity. Syst Control Lett 11:167–172

Görgün H, Arcak M, Barbir F (2006) An algorithm for estimation of membrane water content in PEM fuel cells. J Power Sources 157:389–394

Grasser F (2005) An analytical, control-oriented state-space model for a PEM fuel cell system, Ph.D. Dissertation, Ecole Polytechnique Fédérale de Lausanne, Lausanne, Switzerland

Grasser F, Rufer A (2007) A fully analytical PEM fuel cell system model for control applications. IEEE Trans Ind Appl 43(6):1499–1506

Haschka M, Weickert T, Krebs V (2006) Application of a sigma-point Kalman-filter for the online estimation of fractional order impedance *m* Fractional Differentiation and its Applications. Proceeding of the second IFAC workshop on fractional differentiation and its applications 2(1):194:199

Hinaje M, Nguyen D, Rael S, Davat B (2008) Modelling of the proton exchange membrane fuel cell in steady state. Proceeding of IEEE power electronics specialists conference 3550–3556

Ingimundarson A, Stefanopoulou A, McKay DA (2008) Model-based detection of hydrogen leaks in a fuel cell stack. IEEE Trans Control Syst Technol 16(5):1004–1012

Janardhanan VM, Deutschmann O (2006) CFD analysis of a solid oxide fuel cell with internal reforming: coupled interactions of transport, heterogeneous catalysis and electrochemical processes. J Power Sources 162:1192–1202

Ju H, Meng H, Wang C (2005) A single phase, non isothermal model for PEM fuel cells. Int J Heat Mass Transfer 48:1303–1315

Julier SJ (2002) The scaled unscented transformation. Proceeding of the American control conference, vol 6, pp 4555–4559

Julier SJ, Uhlmann J (2000) Unscented filtering and nonlinear estimation. Proc IEEE 92(3):401–422

Julier SJ, Uhlmann J, Durrant-Whyte HF (2000) A new method for the nonlinear transformation of means and covariances in filters and estimators. IEEE Trans Auto Cont 45(3):477–482

Kalhammer F, Prokopius P, Roan V, Voecks G (1998) Status and prospects of fuel cells as automotive engines. State of California Air Resources Board, Ca., USA

Kandepu R, Imsland L, Foss BA, Stiller C, Thorud B, Bolland O (2005) Control-relevant SOFC modeling and model evaluation. Proceedings of ECOS 5

Kandepu R, Foss BA, Imsland L (2006) Integrated modeling and control of a load-connected SOFC-GT autonomous power system. Proceeding of the American control conference 6:5

Kandepu R, Imsland L, Stiller C, Foss BA, Kariwala V (2006b) Control-relevant modeling and simulation of a SOFC-GT hybrid system. Model Ident Control 27(3):143–156

Kandepu R, Huang B, Imsland L, Foss BA (2007) Comparative study of state estimation of fuel cell hybrid system using UKF and EKF. IEEE international conference on control and automation, ICCA

Kandepu R, Imsland L, Foss BA, Stiller C, Thorud B, Bolland O (2007b) Modeling and control of a SOFC-GT-based autonomous power system. Energy 32(4):406–417

Kandepu R, Foss BA, Imsland L (2008a) Applying the unscented Kalman filter for nonlinear state estimation. J Process Control 18(7):753–768

Kandepu R, Imsland L, Foss BA (2008) Constrained state estimation using the unscented Kalman filter. 16th Mediterranean conference on control and automation, pp 1453–1458

Khargonekar PP, Rotea MA (1991) Mixed H_2/H_∞ Control: a convex optimization approach. IEEE Trans Auto Contr 36(7):824–837

Laffly E, Pera M-C, Hissel D (2007) Polymer electrolyte membrane fuel cell modeling and parameters estimation for ageing consideration. Proceeding IEEE international symposium on industrial elect, pp 180–185

Lee J, Lalk T (1998) Modeling fuel cell stack systems. J Power Sources 73:229–241

Mann RF, Amphlett JC, Hooper MAI, Jensen HM, Peppley BA, Roberge RR (2000) Development and application of a generalized steady state electrochemical model for a PEM fuel cell. J Power Sources 86:173–180

Master J (2010) Kinetics, catalysis and mechanism of methane steam reforming, Thesis Submitted to the Faculty of the Worcester Polytechnic Institute Department of Chemical Engineering

McKay D, Stefanopoulou AG (2004) Parameterization and validation of a lumped parameter diffusion model for fuel cell stack membrane humidity estimation. Proceeding of American control conferences 816–821

Methekar RN, Patwardhan SC, Rengaswamy R, Gudi RD, Prasad V (2010) Control of proton exchange membrane fuel cells using data driven state space models. Chem Eng Res Des 88(7):861–874

Morner SO, Klein SA (2001) Experimental evaluation of the dynamic behavior of an air-breathing fuel cell stack. J Solar Energy Eng 123:225–231

O'Hayre R, Cha SW, Colella W, Prinz FB (2006) Fuel cell fundamentals. Wiley, New York

Ogata K (2001) Modern control engineering, 4th edn. Prentice Hall, Chapters 11 and 12

Pakalapati SR (2006) A new reduced order model for solid oxide fuel cells, Ph.D. Thesis, West Virginia University, Morgantown, WV

Pakalapati SR, Yavuz I, Elizalde-Blancas F, Celik I, Shahnam M (2006) Comparison of a multidimensional model with a reduced order pseudo three-dimensional model for simulation of solid oxide fuel cells. Proceeding of the 4th international conference on fuel cell science, engineering and technology, Irvine, CA

Pan S, Su H, Wang H, Chu J (2010) The study of joint input and state estimation with Kalman filtering, Trans Inst Meas Control. ISSN: 01423312. doi:10.1177/0142331210361551

Pasricha S, Shaw SRD (2006) A dynamic PEM fuel cell model. IEEE Trans Energy Convers 21:484–490

Pasricha S, Keppler M, Shaw SR, Nehrir MH (2007) Comparison and identification of static electrical terminal fuel cell models. IEEE Trans Energy Convers 22(3):746–754

Pukrushpan JT, Peng H, Stefanopoulou A (2004a) Control-oriented modeling and analysis for automotive fuel cell systems. ASME J Dyn Syst Measure Control 126(1):14–25

Pukrushpan JT, Stefanopoulou AG, Peng H (2004b) Control of fuel cell power systems: principles, modeling and analysis and feedback design. In series on advances in industrial control. Springer

Pukrushpan JT, Stefanopoulou AG, Peng H (2004c) Control of fuel cell breathing. IEEE Control Syst Mag 30–46

Puranik SV, Keyhani A, Khorrami F (2010) State-space modeling of proton exchange membrane fuel cell. IEEE Trans Energy Convers 25(3):804–813

Ramousse J, Deseure J, Lottin O, Didierjean S, Maillet D (2005) Modelling of heat, mass and charge transfer in a PEMFC single cell. J Power Sources 145(6):416–427

Rowe A, Li X (2001) Mathematical modeling of proton exchange membrane fuel cells. J Power Sources 102:82–96

Shan Y, Choe S-Y (2005) A high dynamic PEM fuel cell model with temperature effects. J Power Sources 145:30–39

Springer T, Zawodzinski T, Gottesfeld S (1991) Polymer electrolyte fuel cell model. J Electrochem Soc 138(8):2334–2342

Stanton K (2005) A thermally dependent fuel cell model for power electronics design. Proceeding of IEEE power electronics specialists conference, pp 1647–1651

Stiller C, Thorud B, Bolland O, Kandepu R, Imsland L (2006) Control strategy for a solid oxide fuel cell and gas turbine hybrid system. J Power Sources 158(1):303–315

Suares GE, Hoo KA (2000) Parameter estimation of a proton-exchange-membrane fuel cell using voltage-current data. Chem Eng Sci 55:2237–2247

Thanapalan K, Wang B, Williams JG, Liu GP and Rees D (2008) Modeling, parameter estimation and validation of a 300 W PEM fuel cell system. Proceeding of UKACC'08, international conference on control 2008, Manchester, UK

Tumuluri U (2008) Nonlinear state estimation in polymer electrolyte membrane fuel cells. Master's Thesis, Chemical Engineering, Cleveland State University

Turner W, Parten M, Vines D, Jones J, Maxwell T (1999) Modeling a PEM fuel cell for use in a hybrid electric vehicle. IEEE 49th vehicular technology conference 2, pp 1385–1388

Uzunoglu M, Alam MS (2006) Dynamic modeling, design, and simulation of a combined PEM fuel cell and ultracapacitor system for stand-alone residential applications. IEEE Trans Energy Convers 21:767–775

Vahidi A, Stefanopoulou AG, Peng H (2006) Current management in a hybrid fuel cell power system: a model-predictive control approach. IEEE Trans Control Syst Technol 14(6):047–1057

Vepa R (2012) Adaptive state estimation of a PEM fuel cell. IEEE Trans Energy Convers 27:457–467

Wang C (2004) Fundamental models for fuel cell engineering. Chem Rev 104(10):4727–4766

Wang Y, Wang C (2005) Transient analysis of polymer electrolyte fuel cells. J Electrochemica Acta 50:1307–1315

Williams JG, Liu GP, Thanapalan K, Rees D (2007) Design and implementation of on-line self-tuning control for PEM fuel cells. Proceeding of EVS-23; sustainability- the future of transportation, Ca. USA, pp 359–375

Wohr M, Bolwin K, Schnurnberger W, Fischer M, Neubrand W, Eigenberger G (1998) Dynamic modeling and simulation of a polymer membrane fuel cell including mass transport limitation. Int J Hydrogen Energy 23(3):213–218

Xu JG, Froment GF (1989a) Methane steam reforming, methanation and water-gas shift. 1. Intrinsic kinetics. AIChE J 35(1):88–96

Xu JG, Froment GF (1989b) Methane steam reforming, 2. Diffusional limitations and reactor simulation. AIChE J 35(1):97–103

Xue X, Tang J, Sammes N, Ding Y (2006) Model-based condition monitoring of PEM fuel cell using Hotelling T^2 control limit. J Power Sources 162:388–399

Yalcinoz T, Alam MS (2008) Dynamic modeling and simulation of air-breathing proton exchange membrane fuel cell. J Power Sources 182:168–174

Yang W, Bates B, Fletcher N, Pow R (1998) Control challenges and methodologies in fuel cell vehicle development. SAE paper 98C054

Yang CR, Srinivasan S, Bocarsly AB, Tulyani S, Benziger JB (2004) A comparison of physical properties and fuel cell performance of Nafion and Zirconium Phosphate/Nafion composite membranes. J Membr Sci 237:145–161

Yu D, Yuvarajan S (2004) A novel circuit model for PEM fuel cells. Proceeding of IEEE APEC, Anaheim, Ca. 1:362–366

Zhang Z, Huang X, Jiang J, Wu B (2006) An improved dynamic model considering effects of temperature and equivalent internal resistance for PEM fuel cell power models. J Power Sources 161:1062–1068

Zhang H, Wang L, Weng S, Su M (2008) Performance research on the compact heat exchange reformer used for high temperature fuel cell systems. J Power Sources 183(1):282–294

Zhu H, Kee RJ, Janardhanan VM, Deutschmann O, Goodwin DG (2005) Modeling elementary heterogeneous chemistry and electrochemistry in solid-oxide fuel cells. J Electrochem Soc 152:A2427

Ziogou C, Voutetakis S, Papadopoulou S, Georgiadis MC (2011) Modeling, simulation and experimental validation of a PEM fuel cell system. Comput Chem Eng 35(9):1886–1900

Chapter 7
Batteries: Modeling and State of Charge Estimation

7.1 Introduction

In a battery, reduction and oxidation reactions occur at each of the two electrodes, which are separated by an ionically conductive electrolyte. Typically in a battery, several electrochemical cells are connected in series to provide fixed electromotive force or voltage. Unlike a fuel cell, an electrochemical cell used in a battery is one in which one or both of the reactants are permanently contained in the cell and are not continuously supplied from an external source, and the reaction products are not continuously removed. An electrochemical cell consists of two electrodes, an anode and a cathode, and an electrolyte inside a sealed container. When a passive electrical circuit is attached to the two electrodes, an electrical current is generated and the cell is discharged. The anode of a device is the terminal where the hole current flows in from outside and the electron current flows out. The cathode of a device is the terminal where the hole current flows out and the electron current flows in. In a discharging battery, the anode is the negative terminal since that is where the hole current flows into it and the electrons flow out. During discharge, electrons are supplied by the external circuit at the positive electrode or cathode, and the electrode is chemically reduced. Similarly, the negative electrode or anode is oxidized as it supplies electrons to the external circuit. Internally, the circuit is completed by positively charged ions flowing through the electrolyte, from the negative electrode to the positive electrode. Primary cells cannot be recharged as they are not designed for recharging and are discarded once they have produced all their electrical energy. On the other hand, a secondary cell is a cell capable of repeated use and can be recharged by connecting it in parallel to a circuit providing a higher voltage than the cell's terminal voltage. This allows the current to flow through the cell in the opposite direction, which recharges the cell. Its charge may be fully restored by passing an electric current through the cell in the opposite direction to that of discharge, which in turn reverses the reduction and oxidation reactions.

R. Vepa, *Dynamic Modeling, Simulation and Control of Energy Generation*, Lecture Notes in Energy 20,
DOI: 10.1007/978-1-4471-5400-6_7, © Springer-Verlag London 2013

7.2 Battery Electrochemistry

Following Reddy (2010), the reduction and oxidation at the positive and negative electrodes and the overall reactions may be expressed respectively as,

$$mB + nE^{+} + ne^{-} \to dD;$$

$$A \to cC + nE^{+} + ne^{-};$$

$$A + mB \leftrightarrow cC + dD,$$

where C and D are the energized species at each of the two electrodes, starting with a single mole of the species A. Assuming that the number of moles of the energized species and the diffusing ion species are both unity, the amount of C is directly proportional to the state of charge at positive electrode and the amount of E to the positive ionic charge that diffuses across the electrolyte to the negative electrode. In most batteries, a catalyst is also present, but it does not influence the overall reaction.

Figure 7.1 shows a schematic of the one-dimensional (1D) battery model. In a typical cell, during discharge, E^{+} ions diffuse to the surface of the negative electrode of material "A" where they react by oxidation and transfer to an electrolyte solution, while leaving behind the energized species "C". The positively charged E^{+} ions travel by diffusion and migrate via the electrolyte (K_{OH}), to the positive electrode where they react with the electrons entering the electrode by an external circuit by reduction and diffuse into active metal oxide material particles. The electrons themselves, produced in the negative electrode reaction, are blocked by an electronically insulating separator and must travel through an external circuit to the positive electrode.

Fig. 7.1 Schematic of a 1D battery model

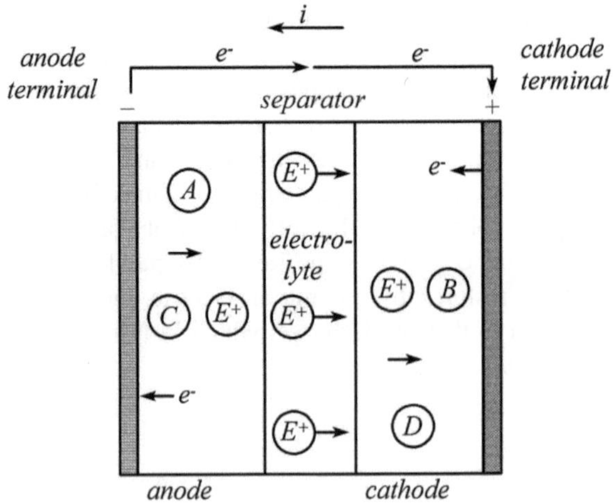

7.3 The Capacity and State of Charge

The capacity, Q, of a cell is the approximate amount of charge that can be stored in the cell when it is fully charged. Thus, the capacity is the amount of electric charge that a battery can store, which is stated in units of ampere-hour (A-h). The state of charge (SOC) is defined as the percentage of the capacity ($Q_{available}$) available at any time with respect to the maximum available capacity ($Q = Q_{max}$) of a battery. The SOC of an ideal cell can be expressed in terms of the initial state of charge, the capacity, and the charging current as,

$$\mathrm{SOC}(t) = \mathrm{SOC}(0) + \frac{1}{Q}\int_0^t i_c(\tau)\mathrm{d}\tau. \tag{7.3.1}$$

In terms of the discharging current it is,

$$\mathrm{SOC}(t) = \mathrm{SOC}(\infty) - \frac{1}{Q}\int_\infty^t i_d(\tau)\mathrm{d}\tau = 1 - \frac{1}{Q}\int_\infty^t i_d(\tau)d\tau. \tag{7.3.2}$$

For finite values of the SOC, the above equation is expressed as,

$$\mathrm{SOC}(t) = \mathrm{SOC}(\infty) - \frac{1}{Q}\int_\infty^t \eta(\tau)i_d(\tau)d\tau = 1 - \frac{1}{Q}\int_\infty^t \eta(\tau)i_d(\tau)d\tau \tag{7.3.3}$$

where $\eta(\tau)$ is the charge-transfer efficiency which is equal to unity for low values of the SOC.

The state of charge is an extremely important parameter in determining the remaining life of a battery at any point in time. A variety of battery SOC estimation methods Pop et al. (2008), Han et al. (2009) have been developed, which, in general, can be classified into four categories: methods based on Coulomb counting, methods based on techniques involving computational intelligence such as artificial neural nets, fuzzy logic, and support vector machines, model-based methods, and other mixed methods. The Coulomb counting methods are, in principle, simple and easy to implement in real-time systems. In these methods, the SOC is simply calculated by integrating the measured current over time with the information of the initial SOC in a fully charged state,

$$\mathrm{SOC}(t) = \frac{Q_{\mathrm{available}}}{Q} = \frac{Q\mathrm{SOC}(0) - \int_0^t i_d(\tau)d\tau}{Q} = \mathrm{SOC}(0) - \frac{1}{Q}\int_0^t i_d(\tau)d\tau. \tag{7.3.4}$$

Although simple in principle, the Coulomb counting methods have several disadvantages due to errors caused by factors such as a wrong initial SOC value and accumulation of estimation errors.

Moreover, the Coulomb counting methods cannot track the battery's non-linear capacity variations due to the rate capacity and recovery effects. Model-based SOC estimation methods use a state–space model derived from the equivalent electric circuit of the battery to design a state observer such as the Kalman filter (KF) for real-time SOC estimation. In fact, the SOC is continuously and dynamically estimated from-related measurements using sophisticated recursive non-linear estimators such as the extended KF Plett (2004a, b, c) and its many variants Bhangu et al. (2005), Plett (2006). A dynamic model describing how the output of the battery behaves with time is generally used in developing the SOC estimation algorithm. The SOC equation in terms of the discharge current for complete discharge is generally valid only when the discharge current is constant. Hence,

$$\mathrm{SOC}(T) = 1 - \frac{i_d T}{Q}. \tag{7.3.5}$$

Hence when,

$$\mathrm{SOC}(T) = 1 - \frac{i_d t_s}{Q} = 0, \tag{7.3.6}$$

$$Q = i_d t_s. \tag{7.3.7}$$

Given the state of charge and a constant discharge current, it is possible to estimate the number of hours the battery would be able to deliver the current. For example, if a fully charged cell has a capacity of 50 Ah (Amp-hours) and it is at 60 % SOC, then the cell is expected to produce a constant current of 3 A for 10 h before reaching full discharge.

7.3.1 Peukert's Equation

The actual capacity of a battery can vary significantly from the rated capacity due to the non-linear effects resulting from past charges and discharges, aging of the battery, charging or discharging history of the battery, ambient temperature variations, etc. Thus, it is customary to define a nominal capacity. The nominal capacity C of the cell is related to the discharge current and the total time of discharge by Peukert's equation (7.3.1). Peukert's equation is given as,

$$C = i_d^{\gamma} t_s = \left(\frac{Q}{t_s}\right)^{\gamma} t_s. \tag{7.3.8}$$

So, given the nominal capacity at some discharge rate and the time of discharge in hours H,

$$C = \left(\frac{Q}{H}\right)^{\gamma} H, \tag{7.3.9}$$

the corresponding capacity is,

$$Q = H\left(\frac{C}{H}\right)^{\frac{1}{\gamma}}. \tag{7.3.10}$$

The discharge time at a given discharge current I_d is then given by,

$$T_s = H\left(\frac{Q}{I_d H}\right)^{\gamma}. \tag{7.3.11}$$

It follows that the capacity for the given discharge current I_d is given by,

$$Q_{I_d} = I_d T_s = Q\left(\frac{Q}{I_d H}\right)^{\gamma-1}. \tag{7.3.12}$$

The exponent γ usually ranges from $1 \leq \gamma \leq \sqrt{2}$. For lead acid cells it is about, $\gamma = 1.35$.

The effect of temperature on Peukert's equation can be modeled by expressing the nominal capacity at any temperature T in terms of the nominal capacity at a certain given ambient temperature, T_0 as,

$$C_T = C_{T_0} \times (1 - \alpha(T_0 - T)). \tag{7.3.13}$$

The actual temperature within the battery follows a simple law and can be defined by,

$$mc_p \frac{\mathrm{d}T}{\mathrm{d}t} = i_{\mathrm{d}}^2 R_{\mathrm{int}} - h_c A(T - T_0) \tag{7.3.14}$$

where m is the battery equivalent mass, c_p is the specific heat capacity, R_{int} is the internal resistance, h_c is the mean heat-transfer coefficient for heat losses from the battery, and A is the external surface area.

7.3.2 Performance Plots

The performance characteristics of batteries and fuel cells can be compared with those of other energy storage and conversion systems by plotting the specific energy as a function of the specific power on log scales. Such a performance plot is also known as the Ragone plot. The resulting plots are generally straight lines with a positive slope and different intercepts. The intercepts on the vertical axis of the inclined lines indicate the relative time required by the system to extract and/or store energy. A Ragone plot demonstrates that both batteries and fuel cells have high theoretical specific energy, but lower power density when compared to conventional thermodynamic engines, indicating that batteries are excellent energy storage systems.

7.4 Battery Modelling

There are basically five types of battery models reported in the literature, specifically: distributed parameter electrochemical models, discrete dynamic electrochemical models, charge kinetics-based models, and electric circuit-based models. Battery modeling in general and electric circuit models in particular are extremely important both for the estimation of the SOC and for the simulation of the dynamic behavior of the battery.

Distributed parameter electrochemical models are not well suited to represent cell dynamics for the purpose of state-of-charge estimations of battery packs. The distributed parameter models arise due to the fact that the flow of ions in the electrolyte, like thermal diffusion, is also a diffusive process which is governed by a diffusion-related partial differential equation. The dynamic equations describe the evolution of the concentration of the electroactive species in the electrolyte in time and space and are based on Fick's first and second laws. Given the distribution of the flux of the electroactive species $J_s(x,t)$ at a distance x and at a time t, the species concentration $C_s(x,t)$ and the diffusion coefficient D, Fick's first and second laws are expressed as a pair of differential equations given by,

$$-J_s(x,t) = D\frac{\partial C_s(x,t)}{\partial x}, \quad \frac{\partial C_s(x,t)}{\partial t} = D\frac{\partial^2 C_s(x,t)}{\partial x^2}. \tag{7.4.1}$$

The equations are generally solved in Laplace transform domain. Based on these solutions one can show that the apparent charge lost at time, t is,

$$\Delta Q = \int_0^t \left(1 + 2\sum_{m=1}^{\infty} \exp\left(-\frac{m^2}{D^2}(t-\tau)\right)\right) i(\tau)\mathrm{d}\tau. \tag{7.4.2}$$

One method of approximating these partial differential equations is to replace spatial gradients by their central difference approximations. Discrete dynamic models may be developed by modeling the distributed parameter system models by finite state approximations such as the discrete transmission line models.

Another analytical model is the charge kinetic battery model or simply the kinetic battery model (KBM) proposed by Manwell and McGowan (1993, 1994). The KBM is derived intuitively and was originally developed to model chemical processes of large lead–acid batteries by a kinetic process.

The KBM models the chemical processes of a battery by the kinetics of flow between two wells. The KBM assumes that a battery has two charge wells, where the charge is stored with a capacity ratio c $(0 < c < 1)$ between the two wells, as shown in Fig. 7.2. The available charge well delivers charge directly to the load; while the bound-charge well supplies charge only to the available charge well through a valve with a discharge coefficient, k. The rate of charge flows from the bound-charge well to the available charge well depends on k and the difference in potential energy or height between the two wells, $h_1(t)$ and $h_2(t)$, where $h_1(t)$

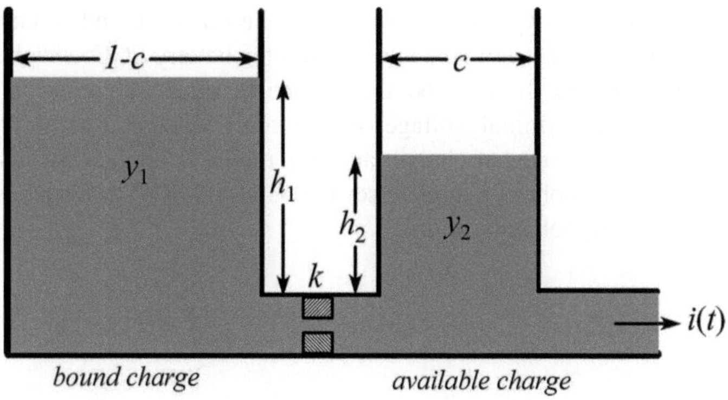

Fig. 7.2 The kinetic battery model

represents the SOC of the battery. The battery is fully discharged when $h_1(t)$ becomes zero. The changes of the charges in the two wells are expressed as,

$$\frac{dy_1(t)}{dt} = -i(t) + k\Delta h(t), \ \frac{dy_1(t)}{dt} = -k\Delta h(t), \ \Delta h(t) = h_2(t) - h_1(t), \quad (7.4.3)$$

where $y_1(t) = ch_1(t)$ is the total charge in the available charge well and $y_2(t) = (1-c)h_2(t)$ is the total charge in the bound-charge well. When battery is discharging and delivering a current $i(t)$, the rate of reduction of the available charge is greater than the rate of reduction of the bound charge and consequently the potential difference between the bound-charge well and the available charge well increases. When the current $i(t) = 0$, $\Delta h(t) = h_2(t) - h_1(t) \to 0$. Consequently, when the battery is idle, the rate of increase in the available charge in the corresponding charge well is greater than bound charge in the bound-charge well. This behavior is consistent with both recovery and rate effects observed in real batteries. In a fully charged state, $h_2(t) = h_1(t) = C_n$, the nominal battery capacity which can be used to establish the initial conditions for the state variables, $y_1(t)$, $y_2(t)$, which take the form,

$$y_i(t) = a_i \exp\left(-\frac{k}{c(1-c)}(t-t_0)\right) + b_i\left\{1 - \exp\left(-\frac{k}{c(1-c)}(t-t_0)\right)\right\} - y_{ip}(t), \quad (7.4.4)$$

where $y_{ip}(t)$ is a particular solution, and a_i and b_i are arbitrary constants. The charge depletes to zero when $y_1(t) = ch_1(t) = 0$ which corresponds to a zero SOC. The unavailable charge is given by,

$$u_c(t) = (1-c)|\Delta h(t)|. \quad (7.4.5)$$

Electric circuit-based models can also be used to represent electrical characteristics of batteries. The simplest of the electric circuit models consists of an ideal voltage source in series with an internal resistance (Hussain and Batarseh 2011). This model, however, does not take into account the battery SOC. Shepherd (1965) developed an equation to describe the electrochemical behavior of a battery directly in terms of terminal voltage, open-circuit voltage, internal resistance, discharge current, and state-of-charge, and this model is applied for discharge as well as for charge. Shepherd's model uses the battery's SOC as a state variable to represent the voltage behavior and is,

$$V_{oc} = V_B + i_d R_\Omega + \frac{K_S}{q_s}, q_s(t) = \mathrm{SOC}(t) = 1 - \frac{1}{Q}\int_{\infty}^{t} \eta(\tau)\, i_d(\tau)\mathrm{d}\tau. \tag{7.4.6}$$

Other approximations are the Unnewehr universal model (Nasar and Unnewehr 1993),

$$V_{oc} = V_B + i_d R_\Omega + K_U q_s \tag{7.4.7}$$

and the Nernst model (which is related to but not the Nernst equation),

$$V_{oc} = V_B + i_d R_\Omega + K_{N1}\ln(q_s) + K_{N2}\ln(1 - q_s) \tag{7.4.8}$$

where Q is the total charge capacity of the cell. Plett (2004a, b, c) combines these models to write,

$$V_{oc} = V_B + i_d R_\Omega + K_U q_s + \frac{K_s}{q_s} + K_{N1}\ln(q_s) + K_{N2}\ln(1 - q_s). \tag{7.4.9}$$

Assuming the charge-transfer efficiency $\eta(\tau)$ is equal to unity, the model enunciated by Plett (2004a) may be expressed in the state–space domain as,

$$\frac{\mathrm{d}q_s(t)}{\mathrm{d}t} = -\frac{i_d(t)}{Q}, \tag{7.4.10}$$

$$V_{oc} = V_B + i_d R_\Omega + K_U q_s + \frac{K_s}{q_s} + K_{N1}\ln(q_s) + K_{N2}\ln(1 - q_s). \tag{7.4.11}$$

The equations can be cast in the general state–space form in the continuous time domain,

$$\dot{\mathbf{x}} = \mathbf{A}\mathbf{x} + \mathbf{B}\mathbf{w} \tag{7.4.12a}$$

$$\mathbf{y} = \mathbf{H}\mathbf{x} + \mathbf{v}. \tag{7.4.12b}$$

The general solution of Eq. 7.4.12a can be expressed as,

$$\mathbf{x}(t) = \mathbf{\Phi}(t - t_0)\mathbf{x}(t_0) + \int_{t_0}^{t} \mathbf{\Phi}(t - \tau)\mathbf{B}(\tau)\mathbf{w}(\tau)\mathrm{d}\tau. \tag{7.4.13}$$

Assuming that $\mathbf{B}(\tau) = \mathbf{I}$, Eq. 7.4.13 may be discretized and expressed in recursive form as,

$$\mathbf{x}_k = \Phi_{k-1}\mathbf{x}_{k-1} + \mathbf{w}_{k-1}. \tag{7.4.14}$$

The above representation is the basis for several SOC estimation algorithms that will be discussed in the next section.

A slightly improved model may be derived from Thevenin's equivalent circuit as in Chan and Sutanto (2000). This model, in Fig. 7.3, is based on an open-circuit voltage in series with a resistance and parallel RC circuits with the so-called Warburg impedance which is the diffusion-related component of the charge-transfer impedance in parallel with the capacitance as defined by Buller et al. (2005). The model is modified by making the Thevenin resistances to be functions of the state–of-charge. The identification of all the parameters of this model is based on a rather complicated technique called impedance spectroscopy; Karden (2001); Mauracher and Karden (1997).

An extension of the Thevenin model is a linear electrical battery model, shown in Fig. 7.4. This model, which was developed specifically for a lead–acid battery by Appelbaum and Weiss (1982), uses a linear ohmic resistance to account for self-discharge (R_p) and a network of resistances and capacitances to represent the

Fig. 7.3 Equivalent circuit of a battery

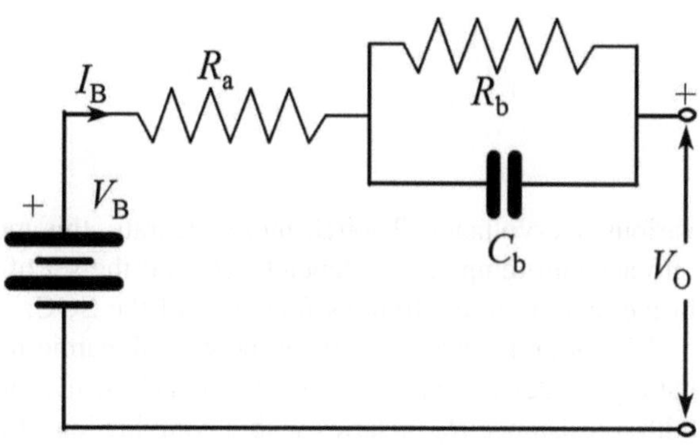

Fig. 7.4 Linear equivalent electrical model

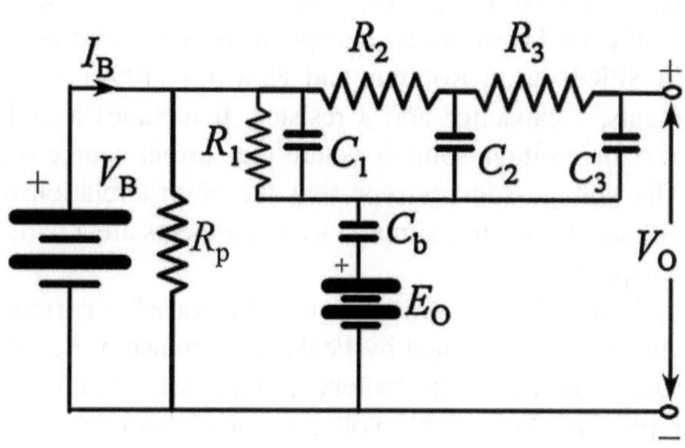

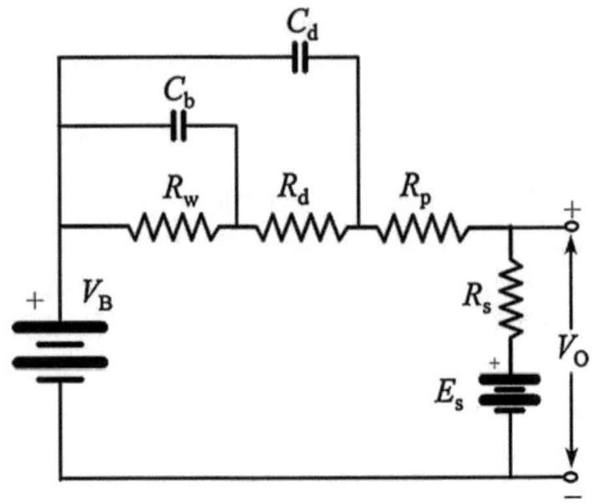

Fig. 7.5 Fourth-order dynamic model

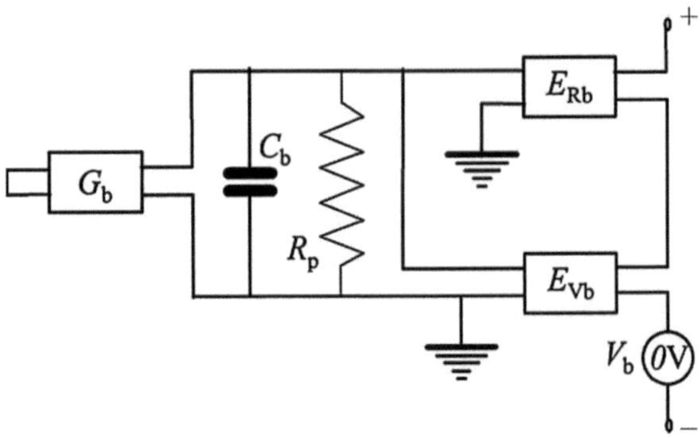

Fig. 7.6 Overcurrent dynamic model

various overvoltages. Though more accurate, this model, however, does not take into account temperature dependence, and the set of resistances and capacitances in the model are continuous functions of the SOC.

Giglioi and Cerolo (1990) proposed a dynamic model shown in Fig. 7.5. This battery model is comprised of two branches: (1) an electrolytic reaction branch with a resistance R_p in series with a combination of an ohmic component R_d, and its associated leakage capacitance C_d, and R_w representing energy dissipation with its associated leakage capacitance C_w; (2) a self-discharge branch through R_s.

Figure 7.6 shows a conceptual theoretical battery model that can be simulated in SPICE as in Robbins and Hawkins (1994) and uses three four-terminal elements, a capacitor and a resistor. It includes a variable current source and two variable voltage sources where the current source represents the Peukert relation. The voltage sources represent the Nernst relationship and the internal resistive voltage drop. The primary circuit elements are a variable resistor R in parallel with a capacitor C_b.

In the Fig. 7.6, G_b represents the variable current source to model the battery current and is defined by Peukert relationship, E_{Vb} represents the variable voltage source to model the battery voltage and is defined by the Nernst relation, E_{Rb} represents the variable voltage source characterizing the internal resistive voltage drop across the battery which is actually modeled as an internal resistance, R_{int}. Based on the previously developed battery models, a simple but representative battery model was proposed in the study by Salameh et al. (1992) and Casacca and Salameh (1992) which is shown in Fig. 7.7. It separately accounts for the non-linear battery element characteristics both during charging and during discharging, while retaining their dependence on the state-of-charge of the battery.

All the elements represented in this model are functions of the open-circuit voltage of battery, which in turn relates to SOC. The primary elements of this equivalent circuit are as follows:

1. Self-discharge resistance (R_p) which accounts for the electrolysis of water at high voltage and the slow leakage across the battery terminal at low voltage;
2. The charge and discharge (R_c and R_d) resistances which are the resistances associated with electrolyte resistance, plates resistance and fluid resistance, which are different in charging and discharging;
3. The overcharge and overdischarge resistances (R_{co} and R_{do}) which represent the increased internal resistances when the battery is overcharged or overdischarged due to an increase in the electrolyte diffusion;
4. Battery capacity (C_b) which represents a battery delivering or storing energy when it behaves as a large capacitor;
5. The diodes which represent the fact that the current travels only in one direction in each of the branches.

The model proposed in Fig. 7.7 may be generalized by including a large number of parallel connected *RC* elements, connected in series to model the fact that the electrolyte model is in fact a distributed parameter system like a transmission line. This results in a generic model as in Barsali and Ceraolo (2002) of the form shown in Fig. 7.8.

If only one of the parallel *RC* elements is retained in the circuit representation and the self-discharge element is ignored, the model reduces to the equivalent circuit in Fig. 7.3. Returning to this circuit, in the case of no current flowing, the

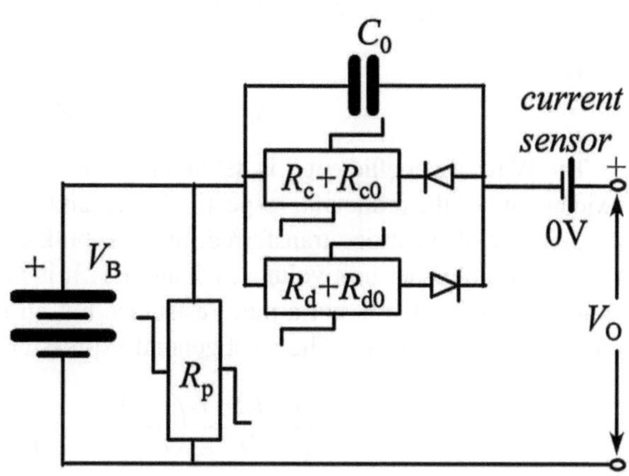

Fig. 7.7 Improved battery model

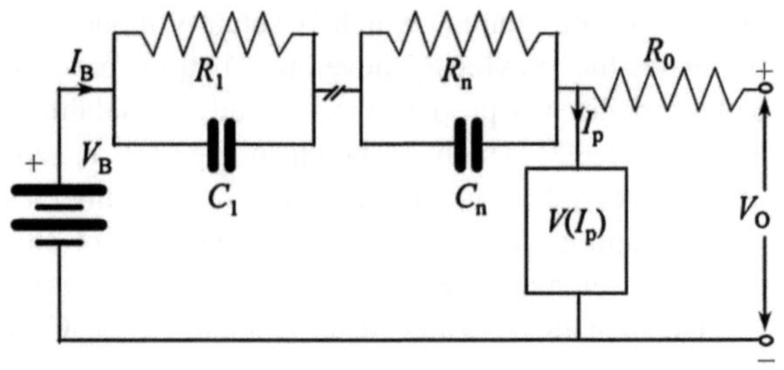

Fig. 7.8 Generic battery model

voltage across the circuit is the equilibrium voltage that can be described as an ideal voltage source which depends on the SOC. A constant equivalent resistance R_a is used for modeling all conductive media of the battery. The electrolytic double-layer capacitance is modeled by $C_b = C_{\mathrm{DL}}$. The resistive branch of the equivalent circuit loop contains the sum of the charge-transfer resistance and two Warburg impedances, which models the diffusion within the electrolyte. The complete equivalent circuit of the lead–acid battery is shown in Fig. 7.2. The circuit elements in the *RC loop* are non-linear. The double-layer capacitance can be shown to be a non-linear function of the voltage at the double layer V_{DL} and given by,

$$C_{\mathrm{DL}} = \alpha_{\mathrm{DL}} \frac{zF}{\sqrt{\mathrm{RT}}} \cosh\left(\frac{zFV_{\mathrm{DL}}}{2\mathrm{RT}}\right) \tag{7.4.15}$$

The constant α_{DL} is determined experimentally by impedance spectroscopy in Mauracher and Karden (1997). From Chap. 6,

$$R_{\mathrm{CT}} = \frac{\mathrm{d}\eta_{\mathrm{act}}}{\mathrm{d}i} = \frac{2RT}{nF} \frac{1}{\sqrt{4i_0^2 + i^2}}. \tag{7.4.16}$$

Traditionally the Warburg impedance is defined in terms of a Warburg coefficient σ and is given by,

$$Z_w = R_w + jC_w,\ R_w = \sigma/\omega^{1/2},\ C_w = 1/\sigma\omega^{1/2}. \tag{7.4.17}$$

Thus,

$$R_b \equiv Z_b = Z_w + R_{\mathrm{CT}}. \tag{7.4.18}$$

The Warburg coefficient σ is defined in terms of the diffusion coefficients of the oxidant and of the reductant, respectively, D_O and D_R, surface area of the electrode A_e, number of electrons transferred, and the bulk concentrations of the diffusing species in moles per unit volume, C_O^s and C_R^s. It is important to recognize that the Warburg impedance is not a pure resistance but an infinite combination of resistances and capacitances. The most general expression for Warburg coefficient σ is,

$$\sigma = \frac{1}{\sqrt{2}} \frac{RT}{n^2F^2} \frac{1}{A} \left(\frac{1}{C_O^s\sqrt{D_O}} + \frac{1}{C_R^s\sqrt{D_R}}\right). \tag{7.4.19}$$

The above equation of the Warburg impedance is valid only if the diffusion layer has an infinite thickness. Quite often this is not the case. If the diffusion layer is bounded and the thickness δ, the impedance at lower frequencies no longer obeys the equation above. Instead, one has:

$$Z_w = R_w + jC_w = \frac{\sigma}{\sqrt{\omega}}(1-j)\tanh\left(\delta\sqrt{\frac{2j\omega}{(D_O+D_R)}}\right). \qquad (7.4.20)$$

In this case, the electrolyte has a finite extension and is limited by a reservoir of constant concentration (ideal reservoir). However, when the electrolyte has the finite extension and is limited by a non-permeable wall (no transport of any substance through the wall),

$$Z_w = R_w + jC_w = \frac{\sigma}{\sqrt{\omega}}(1-j)\coth\left(\delta\sqrt{\frac{2j\omega}{(D_O+D_R)}}\right). \qquad (7.4.21)$$

In the limiting case of infinitely extended electrolyte, the two expressions for the Warburg impedances are equal. The actual impedance used for simulating the electrolyte is a series combination of two impedances where the actual choice of each of the impedances depends on the nature of the boundary conditions. The details are presented by Mauracher and Karden (1997).

7.5 SOC Estimation

SOC estimation is one of the most important steps in the route to the application of batteries in commercial electric vehicles. Anyone using a battery in a commercial vehicle needs to be aware at any time the current SOC Thus, the current SOC needs to be routinely and accurately estimated. One technique that has emerged is the main method of estimating the model parameters and SOC is based on the general subject of system and parameter identification and in particular on the Kalman Filter Vasebi et al. (2007). Because of its significance and importance in energy systems, a brief overview of system and parameter identification and Kalman filtering is presented in the next section.

7.6 System and Parameter Identification

Although it is not intended to provide a review of the developments in the field of non-linear system identification, a succinct description of the key developments that are particularly relevant to biological applications is essential. The field of system identification has fascinating history dating back to Gauss, Euler, Chebyshev, and two of his illustrious students, Aleksandr Lyapunov and Andrey Markov.

A complete review of the developments in linear system identification has been presented in the reviews by Kailath (1974) and Harris (1976). The textbooks by Ljung (1999) and Juang (1994) provide an excellent introduction to the subject of linear system identification. Serious research into non-linear system identification began in the 1970s but has matured into an independent field characterized by a number of focused reviews Ljung (2006), Kershen et al. (2006). A recent book by Nelles (2001) on the subject summarizes the subject. Dynamic non-linear system identification is a model structure and/or model parameter estimation process essential to establish the system dynamics using measured input–output data. Non-linear system identification is a very important prerequisite for non-linear controller design in most practical biological and energy systems which are patently non-linear. A good model structure representation facilitates the inclusion of an array of relevant system dynamics and provides for both the accuracy of modeling and compactness of structure. Therefore, obtaining an optimum fit for the measured data with just a few parameters is vital in the selection of model structure representation. Much of the literature has a range of non-parametric and parametric model representations for non-linear system identification problems. Among the parametric model representations, including black-box-type models, physical and semi-physical models, empirical models, block-oriented models, composite local models, and hybrid and linear parameter varying models (LPV). The black-box-type representations are particularly suitable for modeling neural networks using artificial neural net representations, wavelets, and neuro-fuzzy logic representations. The universal approximation properties of neural networks (see for example Janczak 2005) provide a powerful basis for modeling several biological non-linear systems. The block-oriented models are composed of dynamic linear blocks and static non-linear blocks and possess the flexibility of selecting blocks to represent the features of a given unknown system. The choices of different linear and non-linear blocks result in various structures. Two of the notable non-linear models are the Hammerstein and Wiener models, consisting of a dynamic linear part cascaded with an input or output static non-linear component, respectively. Discrete representations of component blocks almost invariably reduce to non-linear autoregressive moving average models with exogenous inputs (NARMAX) or a subset of a NARMAX model. In the final step on the identification process, one resorts to some form of regression, linear or non-linear or classification. Notable among the regression-like methods are blind signal identification, extended and adaptive KF-based approaches and approaches rooted in variational principles.

The KF, discussed in Sect. 7.7, is a recursive algorithm to optimally estimate and obtain a linear, unbiased, and minimum error variance response for the unknown state of a linear dynamic system from noisy data taken at discrete real-time intervals. If the process to be estimated and (or) the measurement relationship to the process are non-linear, the KF can be applied by linearizing the dynamics and the measurement's relationship to the states about the current mean and covariance and is the extended KF or EKF. There is however a flaw in the basis for the EKF in that the distributions (or densities in the continuous case) of the various random variables are no longer normal after undergoing their respective non-linear transformations. The EKF is simply an improvised state estimator that approximates the optimality of Bayes' rule by linearization. Julier and Uhlmann (2000) have developed a variation to the EKF, the unscented KF, which is discussed in Sect. 7.7.1, using methods that preserve the normal distributions up to a certain order, provided the non-linear transformation can be suitably approximated.

The plant disturbance and measurement noise statistics ought to be known a priori to be able to implement the optimum EKF. When these are unknown, one must necessarily adopt an adaptive approach as enunciated by Sage (1972) as far back as 1972. Moreover, the use of incorrect variance estimates for the plant disturbances could rend the filter unstable, leading to the classic problem of filter divergence. The solution to this problem is to adopt one of three strategies:

1. Error sensitivity analysis is to determine the performance sensitivity to uncertainties in the noise statistics to ensure that any performance degradation is within acceptable bounds;
2. Use a min-max approach such as a non-linear H^{∞}-type filter and determine the filter parameters to minimize the worst-case errors;
3. Adopt an adaptive approach and evaluate the noise statistics in real-time.

Four classes of methods have evolved to obtain the noise statistics in real-time which are (1) Bayesian methods, (2) Maximum likelihood based methods, (3) correlation methods, and (4) covariance matching method. The last two classes of methods are particularly suited to alleviate the problem of filter divergence due to improper knowledge of the variance of the process disturbance noise. Mehra (1970) developed a correlation method for estimating the plant disturbance and measurement noise covariances for time-invariant systems that are both controllable and observable.

The adaptive method may be coupled with the unscented Kalman filter (UKF) developed by Julier and Uhlmann (2000) and used to estimate the parameters of a model. The details of the adaptive KF implementation may be found in Vepa and Zhahir (2011). The identification is facilitated by the use of an UKF and adaptive UKF-based estimator where the non-linear process model is augmented by equations for parameters being estimated, which are assumed to be nominally constant. The identified, converged parameters are compared with an alternate KF implementation, the Ensemble Kalman filter (EnKF). The Ensemble Kalman Filter (EnKF) is another method primarily used to generate estimates of initial values when these are unknown. Thus, the EnKF may be used to estimate constant parameters which can be modeled as dynamic variables with initial values that must remain constant in principle. Like the UKF, the EnKF has the advantage of not requiring a linear model which makes it attractive for parameter identification. It was first proposed by Evensen (2003, 2007) for large non-linear systems associated with weather forecasting. The EnKF presents a solution to the optimal state estimation problem, which uses an ensemble of model states to approximate the mean state and covariance. The parameters are dynamically modeled by setting their derivatives equal to constant magnitude white noise processes.

7.7 The KF and the Extended KF

The discrete KF, outlined in several texts such as the one by Brown and Huang (1997) is the basis for developing the extended KF algorithm. Consider a linear discrete time model representing the dynamic states for a generic differential system given by,

$$\mathbf{x}_k = \Phi_{k-1}\mathbf{x}_{k-1} + \mathbf{w}_{k-1} \tag{7.7.1a}$$

$$\mathbf{z}_k = \mathbf{H}_k\mathbf{x}_k + \mathbf{v}_k, \tag{7.7.1b}$$

where $\mathbf{x}_k$ is a $(n \times 1)$ state vector, Φ_k is a $(n \times n)$ transition matrix, $\mathbf{z_k}$ is a $(m \times 1)$ measurement vector, and $\mathbf{H}_k$ is a $(m \times n)$ state to measurement distribution matrix.

Variables $\mathbf{w}_k$ and $\mathbf{v}_k$ are uncorrelated Gaussian white noise sequences with zero means

$$E\{\mathbf{w}_k\} = E\{\mathbf{v}_k\} = 0 \tag{7.7.2}$$

and covariance matrices defined by,

$$E\{\mathbf{w}_k\mathbf{v}_i^T\} = 0 \text{ and } E\{\mathbf{w}_k\ \mathbf{w}_i^T\} = 0, \tag{7.7.3a}$$

$$E\{\mathbf{w}_k\mathbf{v}_i^T\} = 0 \quad \text{for } i \neq k, \tag{7.7.3b}$$

and

$$E\{\mathbf{w}_k\mathbf{w}_k^T\} = \mathbf{Q}_k,\ E\{\mathbf{v}_k\mathbf{v}_k^T\} = \mathbf{R}_k \tag{7.7.3c}$$

where $E\{\cdot\}$ is the expectation operator. The parameters $\mathbf{Q}_k$ and $\mathbf{R}_k$ are the covariance matrices of the process noise sequence, $\mathbf{w}_k$ and the measurement noise sequence, $\mathbf{v}_k$, respectively. It can be shown that corresponding to the discrete state vector propagation Eq. (7.7.1a), the covariance of random state vector variable is propagated by the linear discrete matrix equation,

$$\mathbf{P}_k = \Phi_{k-1}\mathbf{P}_{k-1}\Phi_{k-1}^T + \mathbf{Q}_{k-1}, \tag{7.7.4}$$

where $\mathbf{P}_k = E\{\mathbf{x}_k\mathbf{x}_k^T\}$ is the state covariance matrix.

The state and covariance prediction equations defining the Kalman filter are:

$$\hat{\mathbf{x}}_k^- = \Phi_{k-1}\hat{\mathbf{x}}_{k-1} \tag{7.7.5a}$$

$$\widehat{\mathbf{P}}_k^- = \Phi_{k-1}\mathbf{P}_{k-1}\Phi_{k-1}^T + \mathbf{Q}_{k-1} \tag{7.7.5b}$$

where $\widehat{\mathbf{x}}_k^-$ is the state vector predicted from the corrected state vector, estimated at the end of the previous epoch, $\widehat{\mathbf{P}}_k^-$ is the corresponding predicted state covariance matrix, and $\mathbf{P}_{k-1}$ is the corresponding predicted state covariance matrix at the end of the previous epoch. The measurement correction or update equations defining the KF are,

$$\mathbf{K}_k = \widehat{\mathbf{P}}_k^- \mathbf{H}_k^T \left(\mathbf{H}_k \widehat{\mathbf{P}}_k^- \mathbf{H}_k^T + \mathbf{R}_k \right)^{-1} \tag{7.7.6a}$$

$$\widehat{\mathbf{x}}_k = \widehat{\mathbf{x}}_k^- + \mathbf{K}_k \left(\mathbf{z}_k - \mathrm{H}_k \widehat{\mathbf{x}}_k^- \right) \tag{7.7.6b}$$

$$\widehat{\mathbf{P}}_k = (\mathbf{I} - \mathbf{K}_k \mathbf{H}_k) \widehat{\mathbf{P}}_k^- \tag{7.7.6c}$$

where $\mathbf{K}_k$ is the optimal Kalman gain, which defines the correction that must be added to the predicted state vector in order to obtain the estimate.

The KF, discussed in this section, is a recursive algorithm to optimally estimate and obtain a linear, unbiased, and minimum error variance response for the unknown state of a linear dynamic system from noisy data taken at discrete real-time intervals. If the process to be estimated and (or) the measurement relationship to the process are non-linear, the KF can be applied by linearizing the dynamics and the measurement's relationship to the states about the current mean and covariance and is the extended KF or EKF. There is, however, a flaw in the basis for the EKF in that the distributions (or densities in the continuous case) of the various random variables are no longer normal after undergoing their respective non-linear transformations. The EKF is simply an improvised state estimator that approximates the optimality of Bayes' rule by linearization. Julier and Uhlmann (2000) have developed a variation to the EKF, the unscented KF, has been improved by Julier et. al. (2000) and Julier (2002) using improved approximation methods that preserve the normal distributions up to a certain order, provided the non-linear transformation can be suitably approximated.

Most dynamic models employed for purposes of state or parameter estimation or filtering of data related by dynamic models are generally not linear. To extend and overcome the limitations of linear models, a number of approaches such as the EKF have been proposed in the literature for non-linear estimation using a variety of approaches. Unlike the KF, the EKF may diverge, if the consecutive linearizations are not a good approximation of the linear model over the entire uncertainty domain. Yet the EKF provides a simple and practical approach to deal with essential non-linear dynamics. The model takes the form,

$$\mathbf{x}_k = \mathbf{f}_{k-1}(\mathbf{x}_{k-1}) + \mathbf{w}_{k-1} \tag{7.7.7a}$$

$$\mathbf{z}_k = \mathbf{h}_k(\mathbf{x}_k) + \mathbf{v}_k. \tag{7.7.7b}$$

Given the Jacobians,

$$\Phi_{k-1} = \nabla \mathbf{f}_{k-1}(\widehat{\mathbf{x}}_{k-1})|_{k-1}, \tag{7.7.8}$$

and

$$\mathbf{H}_k = \nabla \mathbf{h}_k \left(\widehat{\mathbf{x}}_k^- \right) \big|_k, \tag{7.7.9}$$

the state prediction equation defining the EKF is:

$$\widehat{\mathbf{x}}_k^- = \mathbf{f}_{k-1}(\widehat{\mathbf{x}}_{k-1}) \tag{7.7.10}$$

while the covariance prediction equation is,

$$\widehat{\mathbf{P}}_k^- = \Phi_{k-1}\mathbf{P}_{k-1}\Phi_{k-1}^T + \mathbf{Q}_{k-1}. \tag{7.7.11}$$

The measurement correction equations defining the EKF are,

$$\mathbf{K}_k = \widehat{\mathbf{P}}_k^- - \mathbf{H}_k^T\left(\mathbf{H}_k\widehat{\mathbf{P}}_k^-\mathbf{H}_k^T + \mathbf{R}_k\right)^{-1} \tag{7.7.12a}$$

$$\hat{\mathbf{x}}_k = \hat{\mathbf{x}}_k^- + \mathbf{K}_k\left[\mathbf{z}_k - \mathbf{h}_k\left(\hat{\mathbf{x}}_k^-\right)\right] \tag{7.7.12b}$$

$$\widehat{\mathbf{P}}_k = (\mathbf{I} - \mathbf{K}_k\mathbf{H}_k)\widehat{\mathbf{P}}_k^-. \tag{7.7.12c}$$

Equations (7.7.11, 7.7.12a), and (7.7.2) are identical to Eqs. (7.7.5b, 7.7.6a), and (7.7.6c), respectively. The main difficulty in applying the algorithm to problems related to the estimation of orbital ephemeris is in determining the proper Jacobian matrices. The UKF is a feasible alternative that has been proposed to overcome this difficulty, by Julier and Uhlman (2000) as an effective way of applying the KF to non-linear systems.

The UKF gets its name from the unscented transformation, which is a method of calculating the mean and covariance of a random variable undergoing non-linear transformation $\mathbf{y} = \mathbf{f}(\mathbf{w})$. Although it is a derivative-free approach, it does not really address the divergence problem. In essence, the method constructs a set of *sigma vectors* and propagates them through the same non-linear function. The mean and covariance of the transformed vector are approximated as a weighted sum of the transformed *sigma vectors* and their covariance matrices.

7.7.1 The Unscented KF

In the non-linear estimator the complete non-linear function $\mathbf{f}(\mathbf{x})$ is used rather than its approximation so as to generate a reasonably accurate estimate of the state. The non-linear estimator is implemented by employing the unscented transformation.

Consider a random variable $\mathbf{w}$ with dimension L which is going through the non-linear transformation, $\mathbf{y} = \mathbf{f}(\mathbf{w})$. The initial conditions are that $\mathbf{w}$ has a mean $\overline{\mathbf{w}}$ and a covariance $\mathbf{P}_{ww}$. To calculate the statistics of $\mathbf{y}$, a matrix χ of $2L + 1$ sigma vectors is formed. Sigma vector points are calculated according to the following conditions:

$$\chi_0 = \overline{\mathbf{w}} \tag{7.7.13a}$$

$$\chi_i = \overline{\mathbf{w}} + \left(\sqrt{(L+\lambda)\mathbf{P}_{ww}}\right)_i, \quad i = 1, 2, \ldots, L, \tag{7.7.13b}$$

$$\chi_i = \overline{\mathbf{w}} - \left(\sqrt{(L+\lambda)\mathbf{P}_{ww}}\right)_i, \quad i = L+1, L+2, \ldots, 2L, \tag{7.7.13c}$$

where $\lambda = \alpha^2(L + \kappa) - L$, α is a scaling parameter between 0 and 1 and κ is a secondary scaling parameter. $\left(\sqrt{(L+\lambda)\mathbf{P}_{ww}}\right)_i$ is the ith column of the matrix square root. This matrix square root can be obtained by Cholesky factorization. The weights associated with the sigma vectors are calculated from the following:

$$W_0^{(m)} = \lambda/(L+\lambda) \tag{7.7.14a}$$

$$W_0^{(c)} = (\lambda/(L+\lambda)) + 1 - \alpha^2 + \beta \tag{7.7.14b}$$

$$W_i^{(m)} = W_i^{(c)} = 1/2(L + \lambda),\ i = 1, 2, \ldots, 2L, \tag{7.7.14c}$$

where β is chosen as two for Gaussian distributed variables. The mean, covariance, and cross-covariance of $\mathbf{y}$ calculated using the unscented transformations are given by,

$$\mathbf{y}_i = \mathbf{f}(\chi_i) \tag{7.7.15a}$$

$$\overline{\mathbf{y}} \approx \sum_{i=0}^{2L} W_i^{(m)} \mathbf{y}_i \tag{7.7.15b}$$

$$\mathbf{P}_{yy} \approx \sum_{i=0}^{2L} W_i^{(c)} (\mathbf{y}_i - \overline{\mathbf{y}})(\mathbf{y}_i - \overline{\mathbf{y}})^T \tag{7.7.15c}$$

$$\mathbf{P}_{xy} \approx \sum_{i=0}^{2L} W_i^{(c)} (\chi_i - \overline{\chi})(\mathbf{y}_i - \overline{\mathbf{y}})^T \tag{7.7.15d}$$

where $W_i^{(m)}$ and $W_i^{(c)}$ are the set of weights defined in a manner so approximations of the mean and covariance are accurate up to third order for Gaussian inputs for all non-linearities and to at least second order for non-Gaussian inputs. The sigma points in the sigma vectors are updated using the non-linear model equations without any linearization.

Given a general discrete non-linear dynamic system in the form,

$$\mathbf{x}_{k+1} = \mathbf{f}(\mathbf{x}_k,\ \mathbf{u}_k) + \mathbf{w}_k,\ \mathbf{y}_k = \mathbf{h}(\mathbf{x}_k) + \mathbf{v}_k \tag{7.7.16}$$

where $\mathbf{x}_k \in R^n$ is the state vector, $\mathbf{u}_k \in R^r$ is the known input vector, $\mathbf{y}_k \in R^m$ is the output vector at time k. $\mathbf{w}_k$, and $\mathbf{v}_k$ are, respectively, the disturbance or process noise and sensor noise vectors, which are assumed to be Gaussian white noise with zero mean. Furthermore, $\mathbf{Q}_k$ and $\mathbf{R}_k = \mathbf{I}$ are assumed to be the covariance matrices of the process noise sequence, $\mathbf{w_k}$ and the measurement noise sequence, $\mathbf{v_k}$, respectively. The unscented transformations of the states are denoted as,

$$\mathbf{f}^{UT} = \mathbf{f}^{UT}(\mathbf{x}_k,\ \mathbf{u}_k),\ \mathbf{h}^{UT} = \mathbf{h}^{UT}(\mathbf{x}_k) \tag{7.7.17}$$

while the transformed covariance matrices and cross-covariance are respectively denoted as,

$$\mathbf{P}_k^f = \mathbf{P}_k^f(\widehat{\mathbf{x}}_k,\ \mathbf{u}_k),\ \mathbf{P}_k^h = \mathbf{P}_k^h(\widehat{\mathbf{x}}_k),\ \mathbf{P}_k^{fh} = \mathbf{P}_k^{fh}(\widehat{\mathbf{x}}_k,\ \mathbf{u}_k). \tag{7.7.18}$$

The unscented KF may then be expressed in a compact form. In the intended applications of the filter considered in this section, the measurement is assumed to be a linear combination of the states, and the unscented transformation of the states to the measurement may be evaluated equivalently by the linear combination of the state estimates and is given by,

$$\mathbf{h}_k^{UT}\left(\widehat{\mathbf{x}}_k^-\right) = \mathbf{C}_2\widehat{\mathbf{x}}_k^-. \tag{7.7.19}$$

The state time-update equation, the predicted covariance, the optimal gain, the corrected covariance, and the state estimate are respectively given by,

$$\widehat{\mathbf{x}}_k^- = \mathbf{f}_{k-1}^{UT}(\widehat{\mathbf{x}}_{k-1}) \tag{7.7.20a}$$

$$\widehat{\mathbf{P}}_k^- = \mathbf{P}_{k-1}^f + \mathbf{Q}_{k-1},\ \mathbf{Q}_{k-1} = \widehat{\mathbf{P}}_{k-1}\left(\mathbf{C}_1^T\mathbf{C}_1\Delta t/\gamma^2\right)\widehat{\mathbf{P}}_{k-1} + \mathbf{B}_1\mathbf{B}_1^T\Delta t \tag{7.7.20b}$$

$$\mathbf{L}_k = \widehat{\mathbf{P}}_k^{fh-}\left(\widehat{\mathbf{P}}_k^{h-} + \mathbf{I}\right)^{-1} \tag{7.7.20c}$$

$$\widehat{\mathbf{P}}_k = \alpha_H\widehat{\mathbf{P}}_{k-1} + (1-\alpha_H)\left(\widehat{\mathbf{P}}_k^- - \mathbf{L}_k\left(\widehat{\mathbf{P}}_k^{h-} + \mathbf{I}\right)^{-1}\mathbf{L}_k^T\right), \tag{7.7.20d}$$

$$\widehat{\mathbf{x}}_k = \widehat{\mathbf{x}}_k^- + \mathbf{L}_k\left[\mathbf{y}_k - \mathbf{h}_k^{UT}\left(\widehat{\mathbf{x}}_k^-\right)\right]. \tag{7.7.20e}$$

Equations (7.7.20a–7.7.20e) are referred to as unscented KF equations. In the Eq. (7.7.20a–7.7.20e), the parameter α_H is chosen as close to zero as possible that will ensure convergence of the state covariance matrix estimate, $\widehat{\mathbf{P}}_k$.

7.7.2 Adaptive Kalman Filters

Reconsider the KF correction Eq. (7.7.6b) and observe that the correction is a function of the innovation sequence which is,

$$\upsilon_k = \left(\mathbf{z}_k - \mathbf{H}_k\widehat{\mathbf{x}}_k^-\right). \tag{7.7.21}$$

The innovation sequence is a white Gaussian noise sequence with zero mean when the filter is optimal. Moreover, the observation error and state estimation error are orthogonal to each other. The innovation sequence is different from the residual which is defined as,

$$\mathbf{r}_k = (\mathbf{z}_k - \mathbf{H}_k\widehat{\mathbf{x}}_k) = \mathbf{v}_k + \mathbf{H}_\mathbf{k}(\mathbf{x}_k - \widehat{\mathbf{x}}_k). \tag{7.7.22}$$

Thus, employing Eq. (7.7.1b) one could express the measurement noise $\mathbf{v}_k$ as a linear combination of two independent components, the residual $\mathbf{r}_k$ and the optimal error in the estimate. Eliminating the measurements, the innovation sequence may be expressed as,

$$\upsilon_k = \mathbf{H_k}\left(\mathbf{x}_k - \widehat{\mathbf{x}}_k^-\right) + \mathbf{v}_k \tag{7.7.23}$$

and the covariance of the innovation is,

$$E\left\{\upsilon_k \upsilon_k^T\right\} = \mathbf{H}_k \widehat{\mathbf{P}}_k^- \mathbf{H}_k^T + \mathbf{R}_k. \tag{7.7.24}$$

Assuming that the models are linear but with predicted states and measurements corrupted by some additive Gaussian noise with known variance of the type described in Eq. (7.7.3a), it is known that the KF converges to the steady state regardless of the initial conditions. The adaptive KF therefore assumes that the magnitudes of the covariance matrices of the additive Gaussian noises are unknown and seeks to estimate the noise covariance matrices $\mathbf{Q}_k$ and $\mathbf{R}_k$ pertaining respectively to the process and the measurement noise models. The adaptive KF is thus a method of self-tuning for adapting the covariance matrices, $\mathbf{Q}_k$ and $\mathbf{R}_k$ of the process, and measurement noise model sequences. It is achieved by making the statistics of the KF innovation sequences consistent with their theoretical covariances. This principle was established by Mehra (1972) and can be employed to tune both $\mathbf{Q}_k$ and $\mathbf{R}_k$. An estimate of the covariance of the innovation is obtained by averaging the previous innovation sequence over a window length N:

$$\mathbf{C}_U^{k,N} = \frac{1}{N}\sum_{j=k-N+1}^{k} \upsilon_j \upsilon_j^T \tag{7.7.25}$$

and the covariance of the measurement noise sequence may be updated in principle by employing the relation,

$$\widehat{\mathbf{R}}_k = \mathbf{C}_U^{k,N} - \mathbf{H}_k \widehat{\mathbf{P}}_k^- \mathbf{H}_k^T. \tag{7.7.26}$$

Assuming a fixed window length, the covariance matrix may be recursively updated by employing the recursive relation,

$$\mathbf{C}_U^{k+1,N} = \mathbf{C}_U^{k,N} + \frac{\left(\upsilon_{k+1}\upsilon_{k+1}^T - \upsilon_{k-N+1}\upsilon_{k-N+1}^T\right)}{N}. \tag{7.7.27}$$

One could also directly estimate $\mathbf{R}_k$ from the measurement residual. In this case, it has been shown by Mohamed and Schwarz (1999) that one has,

$$\widehat{\mathbf{R}}_k = \mathbf{C}_r^{k,N} + \mathbf{H}_k \widehat{\mathbf{P}}_k \mathbf{H}_k^T \tag{7.7.28}$$

where,

$$\mathbf{C}_r^{k,N} = \frac{1}{N} \sum_{j=k-N+1}^{k} \mathbf{r}_j \mathbf{r}_j^T. \tag{7.7.29}$$

The covariance of the process noise satisfies the equation,

$$\mathbf{Q}_{k-1} = \widehat{\mathbf{P}}_k^- - \Phi_{k-1}\mathbf{P}_{k-1}\Phi_{k-1}^T = \mathbf{K}_k\mathbf{H}_k\widehat{\mathbf{P}}_k^- + \widehat{\mathbf{P}}_k - \Phi_{k-1}\mathbf{P}_{k-1}\Phi_{k-1}^T. \tag{7.7.30}$$

Recognizing that the state estimate is an optimal estimate and considering the covariance of the state correction,

$$\mathbf{C}_{\Delta x}^{k,N} = \frac{1}{N} \sum_{j=k-N+1}^{k} \left(\widehat{\mathbf{x}}_k - \widehat{\mathbf{x}}_k^-\right)\left(\widehat{\mathbf{x}}_k - \widehat{\mathbf{x}}_k^-\right)^{\mathrm{T}} = \frac{1}{N} \sum_{j=k-N+1}^{k} \Delta\mathbf{x}\Delta\mathbf{x}^{\mathrm{T}}, \tag{7.7.31}$$

where,

$$\Delta\mathbf{x} = \left(\mathbf{x}_k - \widehat{\mathbf{x}}_k^-\right) - \left(\mathbf{x}_k - \widehat{\mathbf{x}}_k\right), \tag{7.7.32}$$

it may be expressed as,

$$\mathbf{C}_{\Delta x}^{k,N} = \frac{1}{N} \sum_{j=k-N+1}^{k} \Delta\mathbf{x}\Delta\mathbf{x}^T \approx \widehat{\mathbf{P}}_k^- - \widehat{\mathbf{P}}_k = \mathbf{K}_k\mathbf{H}_k\widehat{\mathbf{P}}_k^=. \tag{7.7.33}$$

The covariance of the state correction, which is linearly related to the innovation may also be expressed as,

$$\mathbf{C}_{\Delta x}^{k,N} = \frac{1}{N} \sum_{j=k-N+1}^{k} \Delta\mathbf{x}\Delta\mathbf{x}^T \approx \mathbf{K}_k\mathbf{C}_U^{k,N}\mathbf{K}_k^T. \tag{7.7.34}$$

This relationship between the covariance matrices suggests that the update of $\mathbf{R}_k$ could be done by employing the covariance of the residual, while the update of $\mathbf{Q}_k$ could be done by employing the covariance of the state correction. Hence, the equation for updating the covariance of the process noise may be expressed in principle as,

$$\mathbf{Q}_{k-1} = \widehat{\mathbf{Q}}_{k-1} \equiv \mathbf{C}_{\Delta x}^{k,\,N} + \widehat{\mathbf{P}}_k - \Phi_{k-1}\mathbf{P}_{k-1}\Phi_{k-1}^T. \tag{7.7.35a}$$

In some references [see for example Myers and Tapley (1976), Blanchet et al. (1997)], an unbiased estimator is employed for the covariance of the state correction and Eq. (7.7.35a) is expressed as,

$$\mathbf{Q}_{k-1} = \frac{N}{N-1}\mathbf{C}_{\Delta x}^{k,N} + \widehat{\mathbf{P}}_k - \Phi_{k-1}\mathbf{P}_{k-1}\Phi_{k-1}^T. \tag{7.7.35b}$$

7.7.3 Adaptive UKF-Based Parameter Identification

In order to employ the UKF when precise statistics of the process and measurement noise vectors are not available, the adaptive filter method proposed by Song, Qi and Han (2006) is used to estimate the parameters and identify them. The covariance matrixes of measurement residuals are recursively updated in the UKF. The measurement and state noise covariance matrices, in the case of the UKF, may be expressed as:

$$\widehat{\mathbf{R}}_k \equiv \mathbf{C}_U^{k,N} - \widehat{\mathbf{P}}_k^{hh-}, \tag{7.7.36a}$$

$$\widehat{\mathbf{Q}}_{k-1} \equiv \mathbf{C}_{\Delta x}^{k,N} + \widehat{\mathbf{P}}_k - \widehat{\mathbf{P}}_{k-1}^{ff} \tag{7.7.36b}$$

which are analogous to Eq. (7.7.26) and the right hand side of Eq. (7.7.35a). Corresponding to Eq. (7.7.28), we may express the measurement noise covariance as,

$$\widehat{\mathbf{R}}_k \equiv \mathbf{C}_r^{k,N} + \widehat{\mathbf{P}}_k^{hh} \tag{7.7.37}$$

which involves the further computation of $\widehat{\mathbf{P}}_k^{hh}$, by applying the unscented non-linear transformation, $\mathbf{h}^{UT}(\widehat{\mathbf{x}}_k)$ to the state estimate, $\widehat{\mathbf{x}}_k$. The measurement noise covariance may be updated in principle by employing the Eq. (7.7.36a).

The non-linear relationships between the covariance matrices also suggest that the update of $\mathbf{R}_k$ could be done by employing the covariance of the residual (Eq. 7.7.37), while the update of $\mathbf{Q}_k$ could be done by employing the covariance of the state correction (Eq. 7.7.36b). However, the simultaneous adaptation of both $\mathbf{Q}_k$ and $\mathbf{R}_k$ is not considered robust, as discussed by Blanchet, Frankignoul and Cane (1997). For this reason, we restrict our attention to $\mathbf{Q}_k$ adaptation as it is the process statistics that is really unknown. Furthermore, it was observed that the magnitudes of the filter gains were relatively small, and for this reason, Eq. (7.7.36b) was approximated as,

$$\widehat{\mathbf{Q}}_{k-1} \approx \mathbf{C}_{\Delta x}^{k,N}. \tag{7.7.38}$$

The application of the Extended KF and the UKF to battery estimation problems have been discussed by Plett (2004a, b, c; 2006).

References

Appelbaum J, Weiss R (1982) An electrical model of the lead-acid battery. IEEE telecommunications energy conference, INTELEC82, pp 304–307

Barsali S, Ceraolo M (2002) Dynamical models of lead-acid batteries: implementation issues. IEEE Trans Energy Convers 17(1):16–23

Bhangu BS, Bentley P, Stone DA, Bingham CM (2005) Nonlinear observers for predicting state-of-charge and state-of-health of lead-acid batteries for hybrid electric vehicles. IEEE Trans Vehicular Tech 54:783–794

Blanchet I, Frankignoul C, Cane MA (1997) A comparison of adaptive Kalman filters for a tropical pacific ocean model. Monthly Weather Rev, Am Meteorol Soc 125(1):40–58

Brown RB, Hwang PYC (1997) Introduction to random signals and applied Kalman filtering, 3rd edn. John Wiley, New York

Buller S, Thele M, De Doncker RWAA, Karden E (2005) Impedance-based simulation models of super capacitors and li-ion batteries for power electronic applications. IEEE Trans Ind Appl 41(3):742–747

Casacca MA, Salameh ZM (1992) Determination of lead-acid battery capacity via mathematical modeling techniques. IEEE Trans on Energy Convers 7(3):442–446

Chan HL, Sutanto D (2000) A new battery model for use with battery energy storage systems and electric vehicles power systems. IEEE Power Eng Soc Winter Meeting, 1(IEEE 0-7803-5935-6/00), pp 470–475

Evensen G (2003) The ensemble Kalman filter: theoretical and practical implementation. J Ocean Dyn, Springer 53:343–367

Evensen G (2007) Data assimilation: the ensemble Kalman filter, Springer

Giglioli R, Cerolo P (1990) Charge and discharge fourth order dynamic model of the lead battery. 10th international electric vehicle symposium, Hong Kong, pp 1–9

Han J, Kim D, Sunwoo M (2009) State-of-charge estimation of lead-acid batteries using an adaptive extended Kalman filter. J Power Sources 188:606–612

Harris CJ (1976) Problems in system identification and control. Bulletin of the IMA 12(5):139–150

Hussain AA, Batarseh I (2011) An overview of generic battery models, IEEE power and energy society general meeting, pp 1–6

Janczak A (2005) Identification of nonlinear systems using neural networks and polynomial models. Springer-Verlag, New York

Juang JN (1994) Applied system identification. Prentice-Hall, New Jersey

Julier SJ (2002) The scaled unscented transformation. Proc Am Control Conf 6:4555–4559

Julier SJ, Uhlmann J (2000) Unscented filtering and nonlinear estimation. Proc IEEE 92(3):401–422

Julier SJ, Uhlmann J, Durrant-Whyte HF (2000) A new method for the nonlinear transformation of means and covariances in filters and estimator. IEEE Trans on Automat Contr 45(3):477–482

Kailath T (1974) A view of three decades of linear filtering theory. Trans IEEE, IT-20, pp 146–181

Karden E (2001) Using low-frequency impedance spectroscopy for characterization, monitoring, and modeling of industrial batteries, Ph.D. dissertation, ISEA. RWTH Aachen, Aachen, Germany

Kerschen G, Worden K, Vakakis AF, Golinval JC (2006) Past, present and future of nonlinear system identification in structural dynamics. Mech Syst Sign Process 20:505–592

Ljung L (1999) System identification: theory for the user, prentice hall information and system science series, 2nd edn. Prentice Hall, Upper Saddle River, New Jersey

Ljung L (2006) Identification of nonlinear systems. IEEE ICARV, 1–4244–0342–1/06

Manwell J, McGowan J (1993) Lead acid battery storage model for hybrid energy system. Sol Energ 50:399–405

Manwell J, McGowan J (1994) Extension of the kinetic battery model for wind/hybrid power system. Proc 5th Eur Wind Energy Assoc Conf, pp 1182–1187

Mauracher P, Karden E (1997) Dynamic modelling of lead/acid batteries using impedance spectroscopy for parameter identification. J Power Sources 67:69–84

Mehra RK (1970) On the identification of variances and adaptive Kalman filtering. IEEE transactions on automatic control AC-15(2), pp 175–184

Mehra RK (1972) Approaches to adaptive filtering. IEEE Trans Autom Control 17(5):693–698

Mohamed AH, Schwarz KP (1999) Adaptive Kalman filtering for INS/GPS. J Geodesy 73:193–203
Myers KA, Tapley BD (1976) Adaptive sequential estimation with unknown noise statistics. IEEE Trans Autom Control 21:520–523
Nasar SA, Unnewehr LE (1993) Electromechanics and electric machines, 2nd edn. Wiley, USA
Nelles O (2001) Nonlinear system identification. Springer-Verlag, New York
Plett G (2004a) Extended Kalman filtering for battery management systems of LiPB-based HEV battery packs: part 1. Background J Power Sources 134:252–261
Plett G (2004b) Extended Kalman filtering for battery management systems of LiPB-based HEV battery packs: part 2. Model Ident J Power Sources 134:262–276
Plett G (2004c) Extended Kalman filtering for battery management systems of LiPB-based HEV battery packs: part 3, state and parameter estimation. J Power Sources 134:277–292
Plett G (2006) Sigma-point Kalman filtering for battery management systems of LiPB-based HEV battery packs: Part 1: Introduction and state estimation, J Power Sources 161:1356–368
Pop V, Bergveld H, Danilov D, Regtien P, Notten P (2008) Battery management systems: accurate state-of-charge indication for battery-powered applications. philips research book series. Springer press, Berlin, German
Reddy T (2010) Linden's handbook of batteries. 4th ed, McGraw-Hill professional
Robbins T, Hawkins J (1994) Battery model for over current protection simulation of DC distribution systems, Sixteenth IEEE international telecommunications energy conference, INTELEC94, pp 307–314
Sage AP (1972) System identification-history, methodology, future prospects. In: Pilkey WD, Cohen R (eds) System identification of vibrating structures: mathematical models from test data. ASME, New York
Salameh ZM, Casacca MA, Lynch WA (1992) A mathematical model for lead-acid batteries. IEEE Trans Energ Convers 7(1):93–98
Shepherd C (1965) Design of primary and secondary cells. J Electrochem Soc 112(7):657–664
Song Q, Qi J, Han J (2006) An adaptive UKF algorithm and its application in mobile robot control. ROBIO '06, IEEE international conference on robotics and biomimetic, Kunming, China, pp 1117–1122
Vasebi A, Partovibakhsh M, Mohammad S, Bathaee T (2007) A novel combined battery model for state-of-charge estimation in lead-acid batteries based on extended Kalman filter for hybrid electric vehicle applications. J Power Sources 174:30–40
Vepa R, Zhahir A (2011) High-precision kinematic satellite and doppler aided inertial navigation system. Royal Inst Navig J Navig 64(01):91–108

References

Mohamed AH, Schwarz KP (1999) Adaptive Kalman filtering for INS/GPS. J Geod [illegible]

Myers KA, Tapley BD (1976) Adaptive sequential estimation with unknown noise statistics. IEEE Trans Autom Control 21:520–523

[illegible]

Chapter 8
Non-Conventional Energy Generation: Solar, Wave, and Tidal Energy Generation

8.1 Introduction

Energy is an important resource for humans, which has become essential for the very survival of mankind and which plays a significant role in our daily needs, for example in using mobile phones, heating our home, driving our car, in making our food, in other facets of our life. Broadly, energy may be classed into two groups: renewable sources that primarily include energy from the sun, wind, biofuels, wave energy, and tidal energy, and non-renewable sources such fossil fuels that are expected to run out in 50 years.

Renewable Sources of Energy—These are natural resources and are therefore renewable, for example solar, wind, hydro electrical power, tidal power, and geothermal power. They are renewable as they are unlikely to run out in the foreseeable future quite unlike non-renewable sources of energy. Examples of renewable sources are the following:

Sun—The Sun is a star largely composed of hydrogen and helium situated at the center of the Solar System, with a distance of 150 million kilometers away from Earth it is still one of the few major power supplies of many renewable energy technologies. It has a diameter of 1,390,000 km and a core temperature of 15.6 million degrees Kelvin delivering a temperature of 5,800 °K at the surface.

Solar power or sunlight is converted into a usable form of energy, such as electricity, by photovoltaic cells (PV), which is done directly, by virtue of photoelectric effect within the cells where the energy is converted into electric current. Often it is first concentrated, which is done indirectly by the use of mirrors and lenses. Some solar energy is absorbed by green plants and used to make food via photosynthesis, so ultimately the Sun is the ultimate source of most energy resources available including fossil fuels.

The advantage of this type of energy generation is that it is an open source which is available relatively cheaply and helps to lower energy costs while it produces no waste or pollution and does not need any additional fuel to convert energy into electricity. The disadvantages are that large scale solar energy converters are extremely expensive to build and install though the costs are

R. Vepa, *Dynamic Modeling, Simulation and Control of Energy Generation*, Lecture Notes in Energy 20,
DOI: 10.1007/978-1-4471-5400-6_8, © Springer-Verlag London 2013

continually on the decline due to rapid improvements in technology. Solar energy is widely available in the tropical countries where it is expected to be most cost effective to convert solar energy in a large scale.

Wind, wave, and tidal energies—Sources of energy such as wind, wave, or tidal are used to convert wind and wave energies into a useful form of energy to make electricity. This is done via wind turbines, wave turbines, and tidal barrages, respectively, where the pressure of each is used to turn its individual turbine which acts as a generator in order to produce electricity. There is usually more than one turbine in order to make the production of electricity greater which is known as a farm. The waves get their energy largely from the wind, and consequently, it is best to convert wind energy at high altitudes before it is tapped by the oceans.

These types of energy sources can only work in certain areas and need to be placed in open areas where they are more likely to function better. Wave and tidal energy generating turbines should be placed in coastal areas where the waves are large or where the tides are significant. Conversion of wave and tidal energy is not an easy task, and although some novel energy converters have been built, it is not still a mature technology.

The primary advantages of such sources of energy are that they are open sources of energy and there is no waste generated or greenhouse gasses produced, no secondary fuel is needed for electricity generation. They can produce a great deal of energy if weather conditions permit and are not expensive to operate and maintain. However, the disadvantages are that they must be able to withstand tough weather conditions are generally noisy and can only be placed in suitable sites. As the weather can be unpredictable and seasonal one cannot always guarantee to produce electricity throughout the year.

BioFuels—Fuels such as biomass or biogas work in a way similar to fossil fuels. They are accumulated from organic materials that are living or recently living organisms such as plants, wood, waste, residue, or gas. These materials can be used directly as fuel or to produce biofuels such as methanol, natural gas, and oil. Biofuels account for approximately 80 % of renewable energy sources with majority of the remaining 20 % coming from a hydro electrical power supply.

The advantages of biofuels are that the fuel tends to be cheap, more appropriate to use waste materials where and are often suitable alternatives to fossil fuels. They can be used within fuel cells, although they may not be as efficient as pure hydrogen. The disadvantages are that the growth and collection of sufficient quantities of these fuels can be difficult and time-consuming. Furthermore, when burnt in a combustion chamber, they also produce greenhouse gasses.

Non-Renewable Sources of Energy—Non-renewable sources of energy are natural resources of economic value that cannot be readily replaced by natural means on a level equal to its consumption. Such non-renewable resources exist in fixed amounts and are heavily used in everyday life; they are consumed much faster than nature can create. Such materials that non-renewable are substances such as nuclear fuel and fossil fuels like crude oil, coal, and natural gasses.

Fossil fuels are so-called because they have been formed from organic remains of prehistoric plants and animals that lived up to 300 million years ago, they are found in deposits beneath the Earth. The three main types of fossil fuels are coal, oil, and natural gas, which are all naturally made well below the Earth's surface in similar ways. The energy originally stored in these elements is from the Sun and when they are burnt in a controlled manner within a combustion chamber to generate steam or increase the air flow, the energy can be converted to electric power as the flow or steam propels the turbines which drive the generators to generate electricity. The advantages of such energy sources are that they are relatively cheap, most are easy to obtain and easy to transport through the use of pipes (oil and gasses), they can provide a high generation of electricity and also the power stations can be built anywhere. The disadvantages include that they are non-renewable, and so once they are finished is hard to replenish them. Moreover, the pollution they produce, by combustion, in the form of greenhouse gasses like carbon dioxide contribute highly to global warming. Furthermore, transportation of these fuels in large container ships or transporters can result in oil spillages. Oil spillages and coal mining are an extreme hazard as they can be quite damaging to the Earth's ecological balance.

8.2 Solar Cells: Overview of Solar Cell Properties

A solar cell is a device that converts light from the sun directly into electricity. They are also called photovoltaic devices. There are a wide variety of solar cells available in the market, and they are all differentiated by their different materials and structures. All of these have one target which is to produce maximum amount of energy while leaving minimum installation and production cost at a low. More than half of the solar cells available in the market are made of silicon as they are very cheap due to their huge availability in the earth's crust. As it is very easily accessible, silicon is the most researched, developed, and used element to date in terms of solar cells.

8.2.1 Types and Characteristics

A solar cell is fundamentally a semiconductor-based device. The basics of solar cells are discussed in Fahrenburch and Bube (1983) and Mazer (1997). For the basics of semiconductor electronics, the reader is referred to Streetman (1995) and Sze (2011). The semiconductor of the solar cell is doped so that free electrons in the semiconductor can be achieved. These doping results into *N*-type and *P*-type silicon where *N*-type contains free electrons and *P*-type contains free holes which is a vacancy of electrons. When the two junctions are in contact, free electrons in the boundary area jumps to the *P*-type to fill the holes leaving a positive charge, and the free holes do the same by leaving a negative charge at the boundary. The attraction of opposite charges at the boundary produces a barrier which stops

Table 8.1 Properties of A300 solar silicon cell (Weider et al. 2006)

Efficiency	21 %
Power (%/c)	0.38
Weight (kg)	0.003
Dimensions (m)	0.125*0.034
Width (microns)	270 ± 40
Voltage (mv/°C)	-1.9
Maximum Power Current(A)	1.2
Maximum Power Voltage(V)	0.59
Short Circuit Current(A)	1.7
Open Circuit Voltage(V)	0.067

diffusion to different junctions. The junction region then forces electrons and holes in opposite directions. When a photon from the sunlight enters the semiconductor, it releases an electron from the electron–hole pair. Electrons near the field of the junction will be diffused to the *N*-type and the free holes to the *P*-type, which will force the electrons toward the positively charged holes through a circuit at the end of the junction. Electrons flow to the other side through the external circuit producing a current in the opposite direction.

Silicon-based solar multijunction cells are most efficient but are also expensive. A typical silicon solar cell currently available in the market, the *A-300* Silicon solar cell, has an efficiency of around 21 % as reported by the U.S. National Renewable Energy Laboratory (Mulligan et al. 2004). It is relatively very efficient with a low manufacturing cost. One cell is said to generate 3 W of electricity and 3 KW in less than 17 m^2. Based on a unique rear-contact design, it maximizes the working cell area, hence producing more efficiency. Due to its low-voltage temperature coefficient, low-light performance, and high sensitivity to light across the entire solar spectrum, the *A-300* converts almost every photon that passes through into useable electricity, maximizing annual energy production. As the cell is a mono-crystalline, it makes it expensive and the cost is currently less than £2/watt for each cell. The basic properties of the A300 solar cell are listed in Table 8.1.

Solar cells made from other elements, other than silicon, can be used like gallium arsenide, cadmium telluride, which results in higher efficiency but not without a higher manufacturing cost. Concentrator photovoltaic systems have higher efficiencies and the sizes of the cells are smaller but need an expensive tracking system to track down the sun all day.

8.2.2 PhotoSensitive Devices: Principles of Operation

Optoelectronics is the branch of electronics that deals with devices used for the conversion of signals between electrical form and light. Photosensitive semiconductor devices play a significant role in optoelectronics may be classified on the

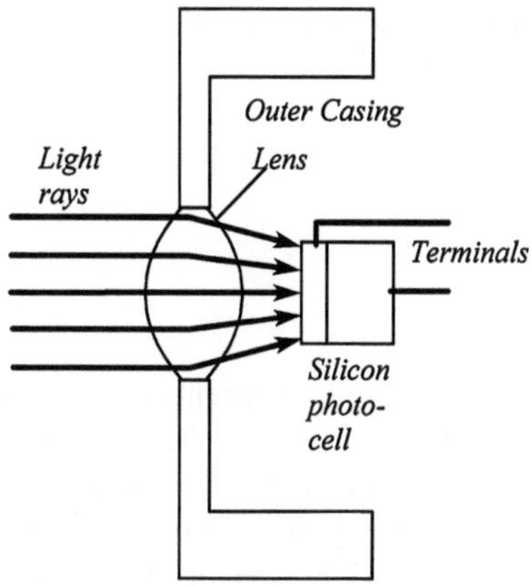

Fig. 8.1 Construction of a silicon photodiode

basis of these operations into three main classes: photodiodes, photoconductive cells, and photovoltaic cells. It is important to be able to recognize the differences between these three types of devices.

Silicon photodiodes are light-sensitive *p–n* junction diodes encased in a transparent material. The *p–n* junction is usually reverse-biased particularly when operating as a photoconductive cell. When operating as a photovoltaic cell, the photodiode is unbiased. The reverse leakage current present, when the diode is reverse-biased, is sensitive to light. The effect of light radiation is to supply enough energy to separate the valence electrons from the atom and reduce the thickness of the depletion layer. The effect of this reduction in thickness is to increase the reverse current by as much as 100 %. Figure 8.1 shows the construction of such a silicon photodiode of the *p–n* junction type. A plastic lens is usually incorporated into the glass housing to focus the light on to the active region of the junction.

A *p–n* junction bar, with light impinging on it, generates holes only on the n-side and only those within the hole diffusion length, L_p, of the space charge region contribute to the photocurrent. This hole current can be expressed by the relation:

$$I_p = q\mathrm{AL}_pG \tag{8.2.1}$$

where A is the junction cross-sectional area, G is the hole-electron generation rate, which is proportional to the incident light flux and the product AL_p represents the effective volume in which the relevant holes are generated.

Since only those photons that posses an energy exceeding the band gap can produce electron–hole pairs, the photocurrent is proportional to the number of these incident photons per unit time and can be written as follows:

$$I = I_0(e^{\frac{qV}{kT}} - 1) - q\mathrm{AG}(L_p + L_n) \tag{8.2.2}$$

where the first term on the right of the above equation represents the *p–n* junction dark current under the influence of a voltage V, and the second term represents the

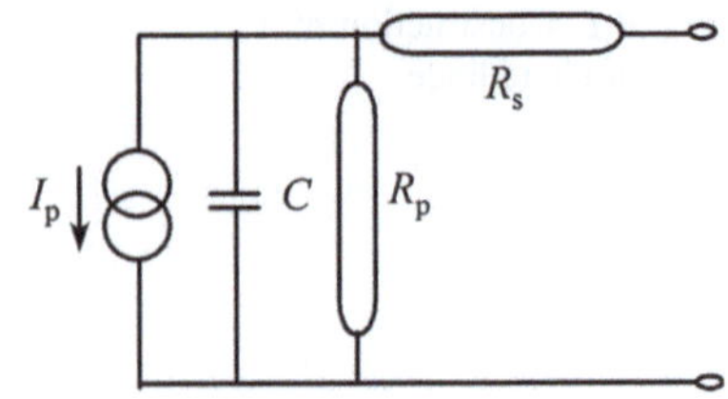

Fig. 8.2 Equivalent circuit of photodiode

combined photocurrent due to hole generation in the *n*-side and electron generation in the *p* side of the junction.

The equivalent circuit of a silicon photodiode is shown in Fig. 8.2. The effect of the current generated by the illumination is shown as a current source in parallel with the light-free or dark current characteristic. The typical current voltage characteristics of photodiode both in the dark and in the presence of light are shown in Fig. 8.3 for different values of the input power. When the diode is short circuited, $V = 0$ and $I = I_{sc}$. The optically generated current, I_p, is the current when the reverse bias voltage is very large.

Photoconductive cells or light-dependent resistors are based on the photoconductive effect. This is the change in the resistance of certain semiconductor materials when exposed to light. The basic process of photoconductivity is the absorption of photons by the semiconductor, resulting in the production of free carriers. Photoconductors exploit the resulting change of conductance. Both intrinsic and extrinsic photoconduction is possible. There are a number of photoconductive materials such as selenium, cadmium sulfide, lead sulfide, selenide, and indium antimonide. In a photoconductive material, incident light results in electron emission within the material. Bolometers also operate on a similar principle (change of resistance due to the incoming radiation, usually infrared) and the

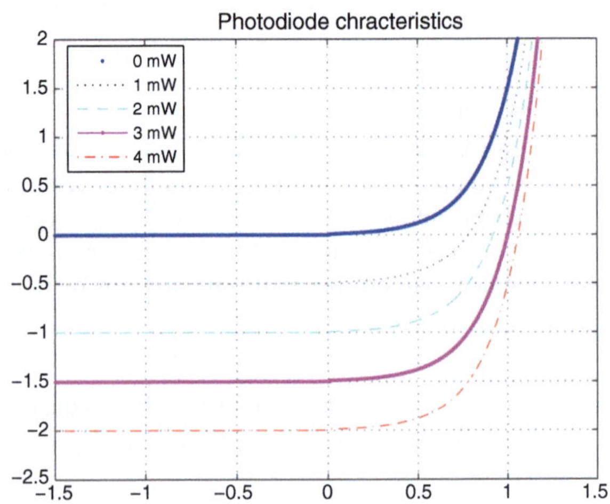

Fig. 8.3 Current voltage characteristics of a photodiode

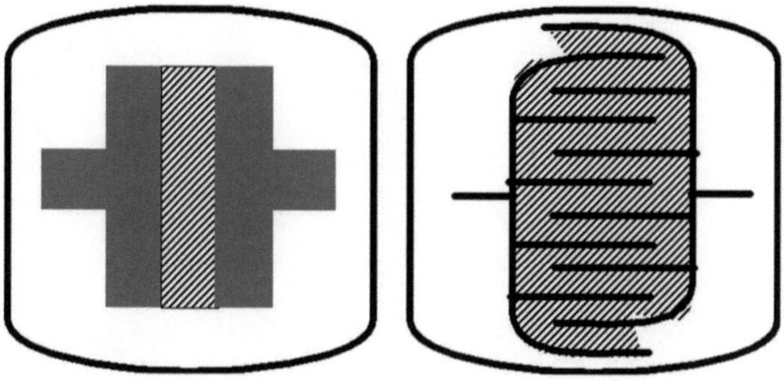

Fig. 8.4 Structure of typical photoconductive cells

sensitive elements may be metallic or semiconductive materials, whereas the elements of photoconductive cells comprise of semiconductive materials.

Under the influence of an external potential, the photoelectrons collide with the atoms in the semiconductors producing holes. For a prescribed light flux, equilibrium conduction is obtained when the rate of hole formation is equal to the rate of electron–hole recombination. The behavior of photoconductive materials can be summarized as follows:

(1) For short periods after exposure to light, the change in current after the exposure is proportional to the incident flux.
(2) With extended exposure, when equilibrium conditions are attained with a steady current, the change in current is proportional to the square root of the incident flux.
(3) When the incident flux is unsteady and is continually changing sinusoidally, the current is directly proportional to it but inversely proportional to the frequency of variation.

Figure 8.4 shows the structure of two forms of selenium-based photoconducting cells. The cell on the right which has a grid structure is formed by condensing vapor in the form of a very thin film on a double grid of gold or platinum mounted on a glass plate. By maintaining the temperature of the glass plate at a proper value, it can be ensured that the selenium has a crystalline structure. The selenium film is usually a few microns thick (1 micron = 0.001 mm), and the entire cell is encapsulated in a protective mount.

Illuminance is a measure of how much luminous flux is spread over a given area and is measured in lumens/square meter or lux in SI units. The corresponding unit in American traditional units is the foot-candle which is one lumen per square foot. One foot-candle is about 10.764 lux. Full sunlight outdoors is about 10,000–12,000 foot-candles (or an average of 160,000 lumens per square centimeter or 172 ftc). Photoconductive cells are characterized by an appreciable current even when the cell is in perfect darkness. The dark resistance of a photoconductive cell is the resistance offered by it when in the dark and it ranges from about 100 k to

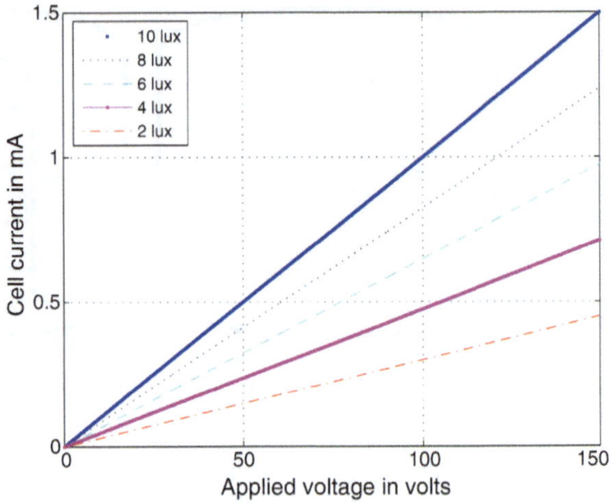

Fig. 8.5 Current voltage characteristics of photoconductive cells

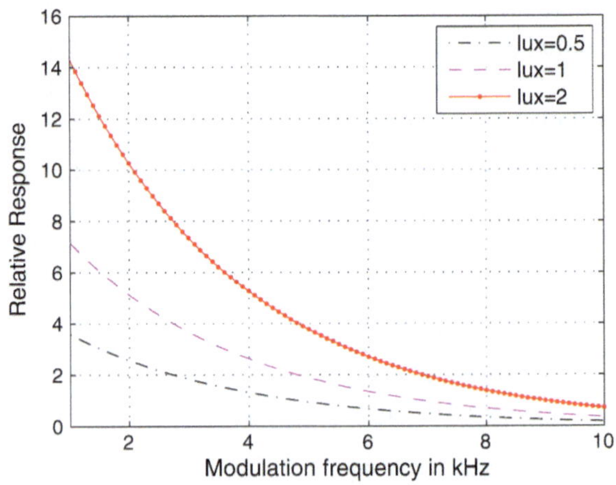

Fig. 8.6 Typical dynamic response curves for a typical photoconductive selenium cell

20,000 k. The ratio of the currents in light and in the dark is usually 8–10 and would be much higher in certain cases. The current voltage characteristics are linear and are shown in Fig. 8.5. The linearity indicates that the cell resistance at a given illumination is independent of the cell voltage. The current capacity of a photoconductive cell usually does not exceed a few milliamperes.

Photoconductive cells respond differently at different wavelengths. Selenium cells are particularly sensitive in the red and infra-red region and are therefore suitable as an infra-red detector as well. Figure 8.6 shows the typical dynamic response of a typical selenium cell where the curves show that the relative response is mainly proportional to the modulation frequency.

The photovoltaic effect is the generation of a potential difference across the junction of two semiconductor substrates. The main difference between the

Fig. 8.7 Example of a silicon photovoltaic cell

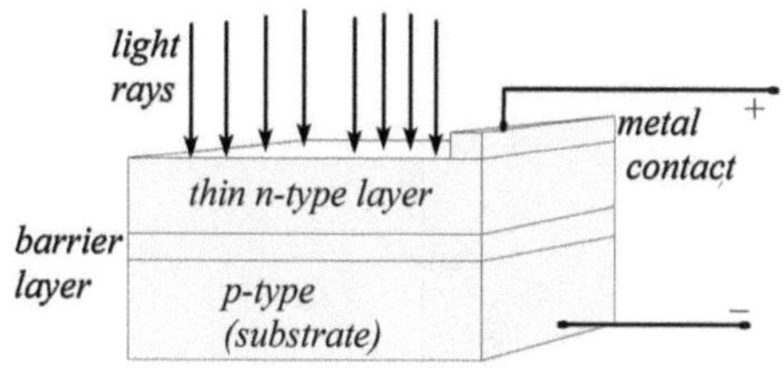

Fig. 8.8 Electron–hole pair generation in a silicon photocell (solar cell)

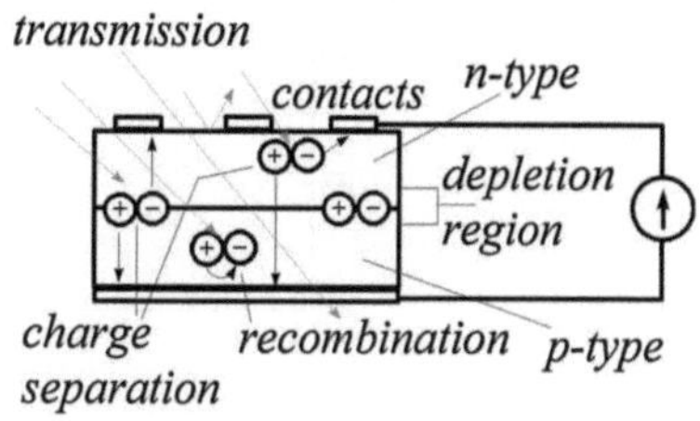

photovoltaic cell and the photodiode is the fact that the external voltage is used to bias the junction in a photodiode. These cells can be constructed from semiconductor *p*–*n* junctions as well as semiconductor metal junctions. In the former type, the effect of the reverse bias is generated by the incident light or thermal radiation. To be effective, the junction areas must be quite large compared with those of an ordinary rectifier. A typical silicon photovoltaic cell of this is illustrated in Fig. 8.7.

Silicon photocells cells are manufactured by the process of melting crystals of purified intrinsic silicon within a quartz crucible which is placed in a graphite oven. Minute traces of a pentavalent element such as arsenic or phosphorus are added, and the resulting crystal is cut into bars. The bars are then sliced, and each slice is then appropriately ground and lapped. The slices are then placed in a diffusion chamber filled with a vapor of boron trichloride, and boron is thus diffused into the *n*-type crystal. Thus, a layer of *p*-type material is formed and a *p*–*n* junction exists at the interface between the two layers. The silicon slice is then plated, and terminal wires are added.

At the *p*–*n* junction, a barrier layer is formed with very few charge carriers. The important feature of these barrier layers is that there is no need to reverse bias the junction as the potential barrier is almost as wide and there exists a potential hill. The incidence of radiant energy from sunlight creates an electron–hole pair by breaking the covalent bonds between atoms in the barrier layer.

The holes are swept to the *p*-side and the electrons to the *n*-side. This is illustrated in Fig. 8.8. If an external load is connected in the circuit, a current would flow and the energy from the incident radiation is dissipated in the load.

Only incident photons with an energy greater than the band gap generate electron–hole pairs and using Planck's Law, if follows that the photon frequency must satisfy the relation $hf > E_G$ in order to generate any photocurrent. The energy gap is itself a function of the voltage generated by the incident photons. Hence, different

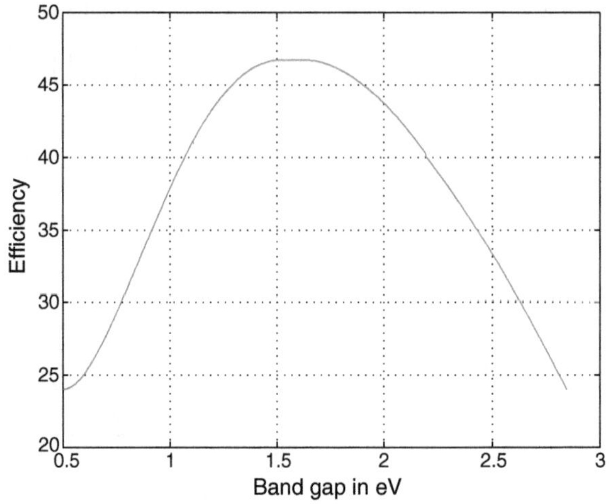

Fig. 8.9 Ideal efficiency of solar cells at 300 K

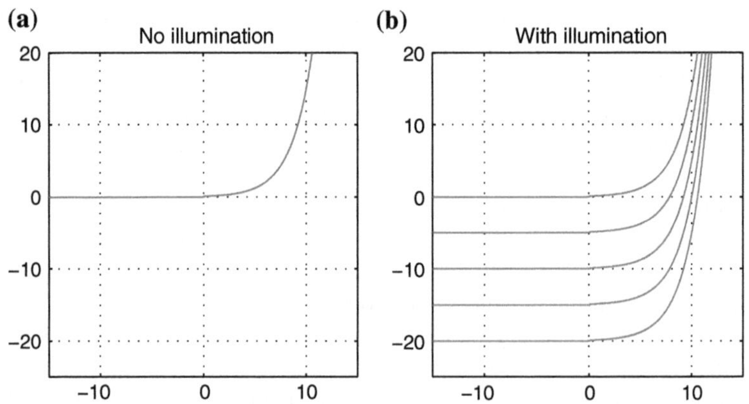

Fig. 8.10 (**a**) Current voltage characteristic of a silicon photovoltaic cell without illumination (**b**) compared with a silicon photodiode with different levels of illumination

semiconductor materials that are characterized by different band gaps have differing efficiencies in generating electron–hole pairs when exposed to sun light. Figure 8.9 illustrates the estimated ideal solar cell efficiency at 300 K when exposed to maximum radiation as a function of the cell material energy gap. Atmospheric absorption by the Earth's atmosphere corresponding to an air mass of 1.5 is assumed.

The current–voltage characteristic of a solar cell can be obtained from the bottom right hand part of the photodiode characteristic in Fig. 8.3 and is illustrated in Fig. 8.10. The objective is to obtain the maximum power from such a cell. Hence, the operating current and voltage must so chosen such that area enclosed by it is a maximum. The ratio of the product of the operating current and voltage,

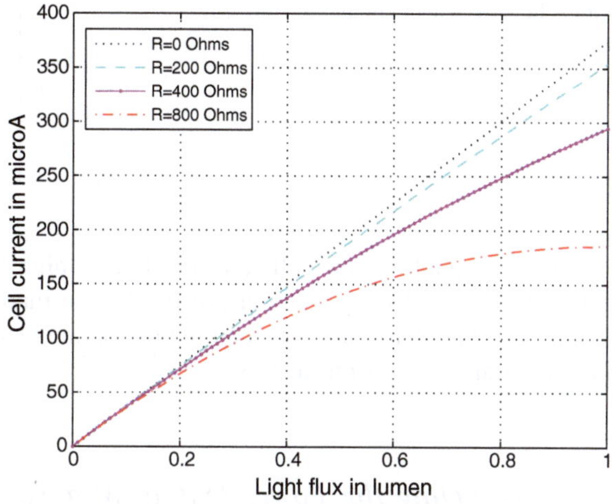

Fig. 8.11 Flux–current relationship for a barrier layer type photovoltaic cell for different values of the load resistance

$I_m V_m$, to the product of the short circuit current and the open circuit voltage, $I_{sc} V_{oc}$, is always less than one and is referred to as the fill factor ($I_{sc}V_{oc}$). Hence, the maximum power output of a silicon photocell may be written as follows:

$$P_{max} = FF\, I_{sc} V_{oc}. \tag{8.2.3}$$

Such solar cells are extensively used for low-power applications in unmanned satellites. The fact that they are extremely low in weight and yet can be connected in series–parallel combinations to deliver the necessary power in space is a major factor for their popularity for space applications. While silicon photocells are more common, GaAs cells, which can operate at higher temperatures than silicon, can be used more efficiently by using solar concentrators. However, solar sails use a large amount of semiconductor material, and the use of GaAs may not always be cost-effective.

The semiconductor metal type of barrier layer cells consists of a layer semiconductor on a metal base. Thus, this may consist of a copper oxide layer deposited on a metal base plate of copper, gold, or platinum. Alternatively, the cell may consist of a thin layer of ferrous selenide deposited on an iron plate. Illumination of the boundary of the barrier plane, between the semiconductor and the metal, results in a potential across the barrier. A current flows if an external circuit connects the semiconductor to the metal. The incident light energizes the electrons in the semiconductor beyond the band gap. They are then able to leave the boundary and flow into the metal and return to the external circuit. If the external circuit is open, a potential difference is set up until equilibrium is established and no electron flow is possible. The potential difference so generated is usually of the order of 4 mv per foot-candle of illumination.

The short circuit current of a barrier layer type cell is directly proportional to the incident light flux. When a resistance is introduced in the external circuit, the relationship between the current and the flux is no longer linear and this is illustrated in Fig. 8.11.

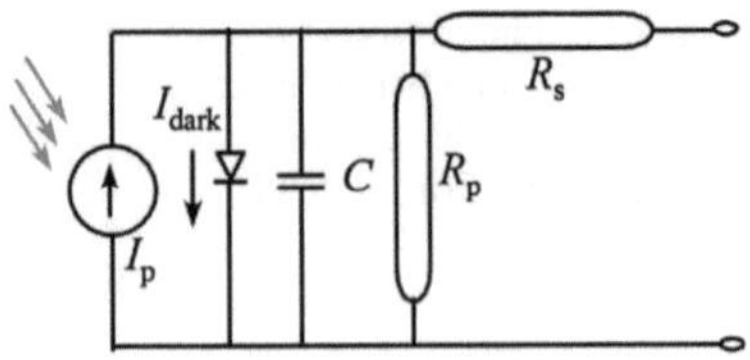

Fig. 8.12 Equivalent circuit of a solar cell

Although photovoltaic cells are sensitive to high temperature and the selenide cells get damaged if operated above 60 °C, the main advantage of these cells is that no external bias voltage is required to operate them. They are extensively used in photographic exposure meters.

8.2.3 Modeling the Power Output of a Solar Cell

The equivalent circuit of a solar cell is shown in Fig. 8.12. The two important characteristics of a solar cell are the open circuit voltage output with an infinite load across its terminals and the short circuit current with no load across its terminals. The short circuit current is directly proportional to the illumination ϕ. Since the short circuit current I_{sc} is approximately proportional to the area of the solar cell exposed to light, the short circuit current density, $J_{sc} = I_{sc}/A$, is the characteristic that is often measured. When the solar cell is connected to a load, a current will flow across its terminals and a voltage develops as there is a build-up of charge at the terminals. The current is a superposition of the short circuit current, caused by the absorption of photons, and a dark current, which is caused by the charge build-up at the terminals which flows to oppose the short circuit current. The dark current may be expressed in terms of the electronic charge q, the voltage across the terminals V, the absolute temperature T and a constant J_0 as,

$$J_{dark}(V) = J_0\left(e^{qV/kT} - 1\right). \tag{8.2.4}$$

Consequently, the total current is,

$$J = J_{sc} - J_0\left(e^{qV/kT} - 1\right). \tag{8.2.5}$$

To find the open circuit voltage, J is set equal to zero; i.e. $J = 0$. Solving for the corresponding voltage the,

$$V_{oc} = \frac{kT}{q}\ln\left(\frac{J_{sc}}{J_0} + 1\right). \tag{8.2.6}$$

The power density is simply the product of the load current density and the load voltage. Hence,

$$P_{den} = J_L V_L. \tag{8.2.7}$$

The efficiency of a solar cell is defined as the power (density) output divided by the power (density) output. Thus,

$$\eta = \frac{J_{\max} V_{\max}}{P_{\text{in}}}. \tag{8.2.8}$$

It is usual to express the power density in terms of the *fill factor* (FF) which is defined as,

$$\text{FF} = \frac{J_{\max} V_{\max}}{J_{\text{sc}} V_{\text{oc}}}. \tag{8.2.9}$$

In terms of the fill factor, the short circuit current density and the open circuit terminal voltage, the efficiency of the solar cell is,

$$\eta = \frac{J_{\text{sc}} V_{\text{oc}}}{P_{\text{in}}} \text{FF}. \tag{8.2.10}$$

When considering a practical equivalent circuit of a solar cell, it is customary to include a shunt resistance in parallel with the cell across its terminals and a series resistance. When this is done, the current density in the load may be expressed in terms of the load voltage V_L as,

$$J_L = J_{\text{sc}} - J_0\left(e^{q(V_L + J_L \text{AR}_s)/kT} - 1\right) - \frac{V_L + J_L \text{AR}_s}{\text{AR}_p}. \tag{8.2.11}$$

Linearizing this relation,

$$J_L \approx J_{\text{sc}} - J_0\left(e^{qV_L/kT}\left(1 + \frac{J_L \text{AR}_s}{V_L}\right) - 1\right) - \frac{V_L + J_L \text{AR}_s}{\text{AR}_p}, \tag{8.2.12}$$

and combining all the terms that are functions of charge density J_L on the right-hand side of the equation,

$$J_L \approx J_{sc} - J_L \frac{R_s}{R_p}\left(1 + \frac{R_p}{V_L} J_0 A e^{qV_L/kT}\right) - J_0\left(e^{qV_L/kT} - 1\right) - \frac{V_L}{\text{AR}_p}. \tag{8.2.13}$$

Solving for J_L,

$$J_L \approx \frac{J_{\text{sc}} - J_0\left(e^{qV_L/kT} - 1\right) - \frac{V_L}{\text{AR}_p}}{\left\{1 + \frac{R_s}{R_p}\left(1 + \frac{R_p}{V_L} J_0 A e^{qV_L/kT}\right)\right\}}. \tag{8.2.14}$$

Assuming that the dark current is constant one express the load current as,

$$I_L = I_{\text{sc}} - I_{\text{dark}} - \frac{V_L + I_L R_s}{R_p}. \tag{8.2.15}$$

Consequently,

$$I_L\left(1+\frac{R_s}{R_p}\right)=I_{sc}-I_{dark}-\frac{V_L}{R_p} \tag{8.2.16}$$

Solving for I_L,

$$I_L=\frac{I_{sc}-I_{dark}-\frac{V_L}{R_p}}{\left(1+\frac{R_s}{R_p}\right)}. \tag{8.2.17}$$

This expression is similar to the expression for J_L, with the exception of a correction term which accounts for the variation of the dark current. Hence, the expression for I_L may be written as,

$$I_L=\frac{I_{sc}-I_{dark}-\frac{V_L}{R_p}}{\left(1+\frac{R_s}{R_p}\left(1+\frac{R_p}{V_L}J_0Ae^{qV_L/kT}\right)\right)}. \tag{8.2.18}$$

Since the short circuit current is proportional to the illumination, if I_{scn} is the short circuit current when the illumination $\phi=\phi_n$, the equation for the load current is,

$$I_L=\frac{I_{scn}\left(\frac{\phi}{\phi_n}\right)-I_{dark}-\frac{V_L}{R_p}}{\left(1+\frac{R_s}{R_p}\left(1+\frac{R_p}{V_L}J_0Ae^{qV_L/kT}\right)\right)}. \tag{8.2.19}$$

The power delivered to the output is,

$$P_{deli}=V_LI_L=\frac{V_L\left(I_{scn}\left(\frac{\phi}{\phi_n}\right)-I_{dark}\right)-\frac{V_L^2}{R_p}}{\left(1+\frac{R_s}{R_p}\left(1+\frac{R_p}{V_L}J_0Ae^{qV_L/kT}\right)\right)} \tag{8.2.20}$$

which is a maximum when,

$$\begin{aligned}\frac{d}{dV_L}P_{deli}=&\frac{\left(1+\frac{R_s}{R_p}\left(1+\frac{R_p}{V_L}J_0Ae^{qV_L/kT}\right)\right)\left(\left(I_{scn}\left(\frac{\phi}{\phi_n}\right)-I_{dark}\right)-\frac{2V_L}{R_p}\right)}{\left(1+\frac{R_s}{R_p}\left(1+\frac{R_p}{V_L}J_0Ae^{qV_L/kT}\right)\right)^2}\\&-\frac{\left(R_sJ_0Ae^{qV_L/kT}\left(\frac{q}{kT}-\frac{1}{V_L}\right)\right)\left(\left(I_{scn}\left(\frac{\phi}{\phi_n}\right)-I_{dark}\right)-\frac{V_L}{R_p}\right)}{\left(1+\frac{R_s}{R_p}\left(1+\frac{R_p}{V_L}J_0Ae^{qV_L/kT}\right)\right)^2}=0.\end{aligned} \tag{8.2.21}$$

Solving for V_L,

$$V_L = \frac{R_p}{2}\left(I_{\text{scn}}\left(\frac{\phi}{\phi_n}\right) - I_{\text{dark}}\right) \times \left(\frac{1+\frac{R_s}{R_p}\left(1+\frac{R_p}{V_L}J_0Ae^{qV_L/kT}\right) - \left(R_sJ_0Ae^{qV_L/kT}\left(\frac{q}{kT}-\frac{1}{V_L}\right)\right)}{\left(1+\frac{R_s}{R_p}\left(1+\frac{R_p}{V_L}J_0Ae^{qV_L/kT}\right) - 0.5\left(R_sJ_0Ae^{qV_L/kT}\left(\frac{q}{kT}-\frac{1}{V_L}\right)\right)\right)}\right). \tag{8.2.22}$$

An approximate maximum is given by,

$$V_{L\max} \approx \frac{R_pk}{2}\left(I_{\text{scn}}\left(\frac{\phi}{\phi_n}\right) - I_{\text{dark}}\right), \tag{8.2.23a}$$

$$I_{L\max} = \frac{\frac{2-k}{2}\left(I_{\text{scn}}\left(\frac{\phi}{\phi_n}\right) - I_{\text{dark}}\right)}{\left(1+\frac{R_s}{R_p}\left(1+\frac{R_p}{V_{L\max}}J_0Ae^{qV_{L\max}/kT}\right)\right)}, \tag{8.2.23b}$$

where k may be considered to be a constant and is normally $1.2 \le k \le 1.4$.

Hence, the maximum power delivered by the solar cell is given by,

$$P_m = \frac{\frac{R_pk(2-k)}{4}\left(I_{\text{scn}}\left(\frac{\phi}{\phi_n}\right) - I_{\text{dark}}\right)^2}{\left(1+\frac{R_s}{R_p}\left(1+\frac{R_p}{V_{L\max}}J_0Ae^{qV_{L\max}/kT}\right)\right)}. \tag{8.2.24}$$

The current and power density outputs of a typical solar cell are shown in Fig. 8.13.

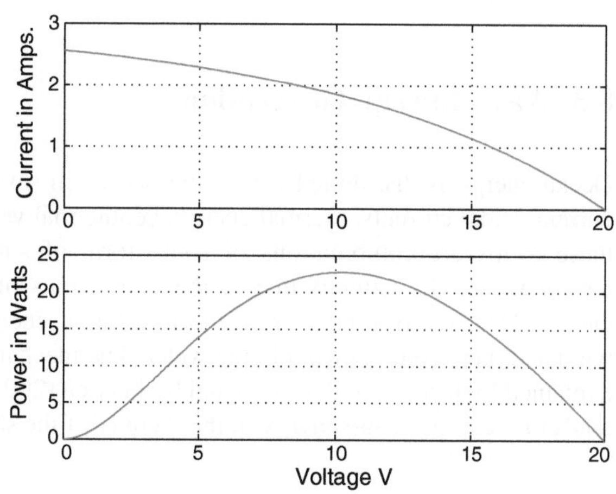

Fig. 8.13 Current and power density outputs of a typical solar cell

8.2.4 Maximum Power Point Tracking Control

Ability of a solar cell to produce power is limited. So it is necessary to force the cell to operate in conditions which match up with the cell's maximum power point (MPP). A MPP tracking (MPPT) controller which utilizes a MPPT algorithm can undertake this duty. When a solar cell operates in this condition, since the MPP is a unique and fixed operating point, the load power requirements are not taken into account and output voltage will vary with variations of load resistance. Thus, the MPPT controller works in the way that, the solar cell operates at its MPP while output voltage is kept constant. The solar cell is usually simulated in two steps, one for verifying performance of the MPPT controller and another for establishing the stability of voltage regulating controller.

A solar cell's output power depends non-linearly on the applied current or voltage, and there exists a unique MPP. Thus, it is possible to track the MPPs by a maximum seeking controller. The locus of MPPs varies non-linearly with the unpredictable variations in the solar cell's operation conditions. A MPPT controller is essential to continuously deliver the highest possible power to the load when variations in operation conditions occur. Usually, the MPPT controller is designed as a multiloop controller with at least two loops. The inner loop is designed to operate the solar cell at its MPP. A fast and efficient MPPT inner loop control scheme for solar cells is based on the methodology of the sliding-mode control approach. Typically, the closed loop system includes the solar cell, a boost chopper, a battery, and the controller. Sliding-mode control is used to control the duty cycle of the chopper in order to achieve MPPT. The outer loop uses a maximum seeking algorithm to estimate the MPP in real-time and then gives the estimated value to the inner loop as the set-point, at which the inner loop forces the solar cell to operate. The MPPT control system provides a simple control law that can keep the solar cell operating at the MPPs in real time. Simulation is used as tool to show the stable operation of the controller in the presence of variations in the solar cell operation conditions.

8.3 Wave Energy Generation

Ocean energy is distributed across the oceans in several forms including marine currents, tidal currents, thermal energy, geothermal vents, and ocean waves. All of these forms are combinations of some form of solar, gravitational, and wind energy where the latter is due to the interactions of Coriolis effects with solar energy. Ocean waves arise from the transfer of energy from the sun first to the wind and then from the wind to the water. The mechanics of this energy transfer is explained by Komen et al. (1996) and by Janssen (2004). However, since the ocean tends to concentrate the energy in the vicinity of the shores of land mass, there is a relatively high concentration of energy in the vicinity of the shallow water regions close to the coastline. The energy in the ocean waves is mainly in the form of kinetic energy which is continually converted to potential energy and vice versa. The basic characteristics of an ocean wave are listed in Table 4.1.

The total energy in a wave is the sum of the instantaneous potential and kinetic energies. The former is due to the height of the wave particle above the mean sea level while the latter is due to the velocity of the fluid particles in motion. The total energy is also equal to the maximum potential energy and to the maximum kinetic energy which may be found the energy over one wavelength. This process of integration over one wavelength yields the energy density of the wave. Thus, given the density of sea water $\rho = \rho_{\text{water}}$ and the local acceleration due to gravity g, for a single wave of height $H = 2A$, the energy density may be shown to be,

$$E_d = \frac{\rho g H^2}{8}. \tag{8.3.1}$$

To obtain the power of a single wave, consider the *wave period T which* is the time it takes two successive wave crests or troughs to pass a fixed point. Moreover, the *wave length* is the distance between two successive wave crests or troughs at a fixed time. Thus, the phase velocity of a wave is given by,

$$c = \frac{\lambda}{T} = \frac{2\pi/k}{2\pi/\omega} = \frac{\omega}{k}. \tag{8.3.2}$$

For ocean waves, the wave frequency ω is related to wave number k by the *dispersion relation* (Lamb 1932, p. 228):

$$\omega^2 = gk \tanh(kd), \tag{8.3.3}$$

where d is the water depth and g is the local acceleration of gravity. For deep water waves, i.e., for $kd > > 0$, $\tanh(kd) \approx 1$. Thus, it follows that for waves in deep waters,

$$\frac{\omega}{k} \approx \frac{g}{\omega}. \tag{8.3.4}$$

However, when kd is small but not equal to zero,

$$\left(\frac{\omega}{k}\right)^2 = gd\frac{\tanh(kd)}{kd} \text{ and } \frac{\omega}{k} = \sqrt{gd\frac{\tanh(kd)}{kd}}. \tag{8.3.5}$$

In the limit when $kd \to 0$,

$$\frac{\omega}{k} \approx \sqrt{gd}. \tag{8.3.6}$$

The power of a single wave is therefore given by,

$$P_w = \frac{\rho g H^2}{8} \frac{\omega}{k}. \tag{8.3.7}$$

For real ocean waves that are essentially random, the wave energy density in Joules per square meter is related to the variance of sea-surface displacement ζ by:

$$E_d = \rho g \langle \zeta^2 \rangle \tag{8.3.8}$$

where the brackets denote a time or ensemble average. In reality, ocean waves are composed of waves of various heights. A practical definition that is often used is the height of the highest one-third of the waves, which is known as *significant wave height* and is denoted as $H_{1/3}$. $H_{1/3}$ is related to the standard deviation of sea-surface displacement by the relation,

$$H_{1/3} = 4\sqrt{\langle \zeta^2 \rangle}. \tag{8.3.9}$$

The power of the ocean waves is therefore given by,

$$P_w = \frac{\rho g H_{1/3}^2}{16} \frac{\omega}{k}. \tag{8.3.10}$$

It is also useful to estimate the rate of transmission of energy as the waves are transmitted on the surface. Consider an element of height dz under the free surface of the wave. The total work done may be evaluated as,

$$W = \int_{-d}^{0} \text{pudz}. \tag{8.3.11}$$

Linearizing the Bernoulli equation, the pressure is given by,

$$p = \rho \frac{\partial \Phi}{\partial t} - \rho g z + p_{\text{atm}}, \tag{8.3.12}$$

where the velocity potential Φ is (see for example Newman 1977),

$$\Phi = \frac{gH}{\omega} \frac{\cosh(k(z+d))}{2\cosh kd} \cos(kx - \omega t), \tag{8.3.13}$$

and

$$u = \frac{\partial \Phi}{\partial x}. \tag{8.3.14}$$

Thus the mean rate at which work is done is given by,

$$\langle P_r \rangle = \frac{\rho g H^2}{8} \frac{c}{2} (1 + 2kd \times \text{cosech}(2kd)) \tag{8.3.15}$$

which may be expressed as,

$$\langle P_r \rangle = \frac{\rho g H^2}{8} c_g. \tag{8.3.16}$$

The velocity of energy flow is given by the *group velocity* which is defined by the relation,

$$c_g = \frac{c}{2}(1 + 2kd \times \text{cosech}(2kd)). \tag{8.3.17}$$

When $kd \to 0$, $c_g \approx c$ and when $kd > > 0$, for deep water waves, $c_g \approx c/2$.

8.3.1 Wave Power Transfer Devices

The actual wave power converted into useful power depends to a large extent the efficiency of conversion by the wave power transfer device employed for energy conversion. The wave power transfer devices currently in use fall broadly into two categories:

(1) Oscillating body type of wave power transfer devices

In this type of wave power transfer device, the waves are allowed to set a floating body into resonant oscillations and the motion is in turn converted into useful energy. A typical example is a Salter duck invented by Professor Stephen Salter (Salter 1974) in the United Kingdom. In this type of system, the major difficulty is the conversion of the oscillations of the body into useful energy. This would involve the use of a power take-off (PTO) system which must be chosen with matching characteristics in order to maximize the power transfer from the resonant device to an electricity generator

(2) Oscillating water column type of wave power transfer devices.

The oscillating water column also involves a floating body that is set in oscillation by the waves. Additionally, the floating body traps a volume of air in the body and above the wave. Consequently, the air is subjected to the oscillating pressure by the wave. The air is used to drive a special turbine that performs the role of a PTO and transfer power to a generator that in turn generates electricity. The development of wave energy utilization since the 1970s is extensive reviewed in Falcão (2010).

Several of the wave energy devices currently studied in the United Kingdom, Japan, Portugal, India, and other countries make use of the principle of the oscillating water column for converting wave energy to low pneumatic energy which in turn can be converted into mechanical energy. In this case, the development of a bidirectional air turbine has come up as an important problem. So far, several self-rectifying air turbines with different configurations have been proposed, including the Wells turbine (Gato and Falcao 1988; Raghunathan 1995;

Setoguchi et al. 1998), the Wells turbine with guide vanes (Gato and Falcao 1990), turbines with pitch-controlled blades (Sarmento et al. 1990; Salter 1993; Gato et al. 1991), impulse turbines with self-pitch control of blades (Kinoue et al. 2002), and other configurations. Among them the most promising turbine at present is the Wells turbine, although turbine efficiency is lower in comparison with that of the conventional turbines. However, installing guide vanes in front of and behind the rotor may be most effective in improving the turbine performance.

8.3.2 Wave Potentials Due to a Floating Body

The velocity potentials corresponding to the wave motion of the free surface are essential to compute the wave forces on a floating body. The velocity potential can be assume to be the sum of three components, the incident velocity potential which is responsible for the surface wave motion, the radiation potential arising from the body motion and representing the waves radiating outwards and the diffraction potential.

If the floating body has all six degrees of freedom, then the radiation potential is comprised of six components that are directly proportional to the magnitudes of the motions in the six directions, respectively. If the radiation potential of the body, floating in a quiescent fluid, due to a unit motion in the jth direction is ϕ_j, the boundary condition at the fluid–solid interface requires that the translational and rotational fluid velocity induced by the potential normal to the solid surface $\partial\phi_j/\partial n$ is equal to the respective translational and rotational velocity components due to the unit motion in the direction $n \cdot e_j$ where e_j is a unit vector in the jth direction. Furthermore, the waves must always radiate away from the body. The specific dynamics of floating bodies in each of the degrees of freedom is discussed in Sect. 4.6.4.

To compute the diffraction potential, one may assume that the body is stationary, and the radiation potential is absent. The sum of the incident and diffraction potentials $\phi_T = \phi_I + \phi_D$ must satisfy the appropriate boundary conditions on the body surface and at infinity. Thus,

$$\frac{\partial\phi_D}{\partial n} = -\frac{\partial\phi_I}{\partial n}. \tag{8.3.18}$$

The diffraction potential may be found by solving a boundary value problem, discussed in Sect. 4.6.3, as it must satisfy the Laplace equation along with the boundary conditions on the body surface and at infinity. The solution is generally expressed as a superposition of source potentials in the form of boundary integral equation which must be solved numerically by boundary element methods (Newman and Lee 2002). If we assume the body to be sufficiently small so as not to affect the pressure field due to an incident wave, then the diffraction effects can be ignored. This is based on the Froude–Krylov hypothesis and the resulting excitation force is:

$$F^{FK}(t) = \iint p(\phi_I)\mathbf{n} \cdot dS \tag{8.3.19}$$

where $p(\phi_I)$ is the pressure field exclusively due to the incident potential ϕ_I.

The pressure field generated by the wave motion of the fluid particles plays a key role in all forms of wave power transfer devices. The pressure field within the water wave may be obtained by substituting for the velocity potential $\Phi = \phi_I$ and the velocity components into the unsteady Bernoulli equation which is obtained integrating the Euler equation with respect to the time and is,

$$\frac{p}{\rho} + \frac{\partial \Phi}{\partial t} + \frac{u^2 + v^2 + w^2}{2} + g\Delta z = 0. \tag{8.3.20}$$

Using this equation, one can obtain the pressure distribution accurate to second order as,

$$p = \rho g H \frac{\cosh(k(z+d))}{2\cosh kd} \cos(kx - \omega t) - \rho g z + p_{\text{atm}} + \rho g k H^2 \frac{\tanh(kd)}{\sinh^2 kd} p_2. \tag{8.3.21}$$

The second-order correction to the first-order pressure distribution ($p_2 = 0$) is given by,

$$p_2 = \frac{3}{16}\left(\frac{\cosh(2k(z+d))}{\sinh^2 kd} - \frac{1}{3}\right)\cos(2(kx - \omega t)) - \frac{(\cosh(2k(z+d)) - 1)}{8}. \tag{8.3.22}$$

One can also predict power of a wave based on the second-order theory that can be shown to be,

$$P_w = \frac{\rho g H^2}{8} c \left(1 + \frac{9}{64}\frac{H^2}{k^4 d^6}\right). \tag{8.3.23}$$

8.3.3 Conditions for Optimum Power Take-Off

To illustrate the conditions that must be satisfied for optimum power transfer from the ocean to the generator, consider a wave power transfer device that consists of a body oscillating in heave. For such an oscillating body type of wave power transfer device, the heave equations motion may be expressed as follows:

$$(M + A_{33})\frac{d^2h}{dt^2} + B_{33}\frac{dh}{dt} + \rho g S_{wh} h = F_{wh} + F_{PTO,h}, \tag{8.3.24}$$

where M is the mass of the floater, A_{33} and B_{33} are the frequency dependent added mass and damping due to the body floating in water and radiating waves, h is the displacement in the heave degree of freedom, ρ is the density of water, g the acceleration due to gravity, S_{wh} the plane are enclosed by the waterline and F_{wh} is the external force exciting the heave motion due to wave forces acting on the floater, due to both the incident waves and diffraction effects and $F_{\mathrm{PTO},h}$ is the reaction force exerted by the power take-off system on the heaving body. The reaction force exerted by the power take-off system on the heaving body $F_{\mathrm{PTO},h}$ may be assumed to be in the form,

$$F_{\mathrm{PTO},h} = -B_{\mathrm{PTO}}\frac{dh}{dt} - K_{\mathrm{PTO}}h. \tag{8.3.25}$$

where B_{PTO} and K_{PTO} are the equivalent damping and stiffness of the power take-off system.

It is shown by Falnes (2002) that with harmonic motion in heave, the absorbed power from a heaving wave energy converter is given by

$$P_{\mathrm{TO}} = \frac{1}{2}\omega^2 B_{\mathrm{PTO}}|h_0|^2 \tag{8.3.26}$$

where h_0 is the heaving complex amplitude of the body, given by,

$$|h_0|^2 = \frac{A^2|\Gamma_{wh0}|^2}{(K_{\mathrm{PTO}} + \rho g S_{wh} - \omega^2(M + A_{33}(\omega)))^2 + \omega^2(B_{33}(\omega) + B_{\mathrm{PTO}})^2}, \tag{8.3.27}$$

where A is the amplitude of the incident wave, Γ_{wh0} is the hydrodynamic wave excitation force amplitude corresponding to wave of unit amplitude and ω is the frequency of harmonic oscillations. To maximize $|h_0|^2$, the frequency of harmonic oscillations must be equal to the resonance frequency,

$$\omega = \omega_r = \omega_n\sqrt{1 - 2\zeta_n^3} \approx \omega_n = \sqrt{\frac{K_{\mathrm{PTO}} + \rho g S_{wh}}{M + A_{33}(\omega_n)}} \; and \; B_{33}(\omega_n) = B_{\mathrm{PTO}}. \tag{8.3.28}$$

Assuming that these optimality conditions, which were first derived by Evans (1981), are satisfied,

$$|h_0|^2_{\max} \approx \frac{A^2|\Gamma_{wh0}|^2}{4\omega_n^2 B_{\mathrm{PTO}}^2} = \left(\frac{A|\Gamma_{wh0}|}{2\omega_n B_{PTO}}\right)^2 \tag{8.3.29}$$

and

$$P_{TO,\max} = \frac{A^2|\Gamma_{wh0}|^2}{8B_{\mathrm{PTO}}}. \tag{8.3.30}$$

Since $P_{\mathrm{TO,max}}$ is maximum when $B_{33}(\omega_n) = B_{\mathrm{PTO}}$ is a minimum, the shape of the heaving body must be chosen to minimize the body's hydrodynamic damping coefficient while also satisfying the condition for maximum power take-off and maximizing $|\Gamma_{wh0}|$. A similar analysis can be shown to hold for the oscillating water column type device, although in this case, the volume of air trapped is proportional to the heave amplitude of the body relative to the wave surface rather than relative to the mean sea water level. As shown by Evans and Porter (1995), the hydrodynamic forces acting on an oscillating water column are quite similar in nature to those acting on heaving body. Thus, in the case of the oscillating water column also, while the above conditions of optimal power transfer need to be met, the equivalent hydrodynamic coefficient $|\Gamma_{wh0}|$ can be simultaneously maximized by design.

8.3.4 The Wells Turbine

The special turbine used to generate electricity in an oscillating water column type of wave power transfer device is known as the Wells turbine. The Wells turbine, named after its inventor, Prof. Alan Wells, is a special type of self-rectifying air turbine, capable of maintaining constant direction of rotation, although the air flow passing through it is oscillating (see for example Raghunathan, Tan and Wells (1982)). The proposal was based on the phenomenon of lift and thrust generation in aerofoils which is well known to hydronautical and aeronautical engineers. When an aerofoil that is free to oscillate in both plunging and pitching or feathering modes, in a flow, then under certain conditions power is transferred from the flow to the aerofoil which is capable of performing sustained oscillations in the flow. In the case of an aircraft, this transfer of power to an oscillating wing can lead sustained and undesirable fluttering oscillations that can lead to disastrous structural failure. In a Wells turbine, this phenomenon is exploited to generate thrust and to draw power from an oscillating flow field with efficiencies that are comparable with conventional wind turbines.

8.4 Tidal Energy

During recent years, there has been considerable interest in the dynamics and control of tidal power systems for MPP operation. The potential exists for using such control schemes in the control of tidal barrages across estuaries, although the control of tidal power poses several problems to control theorists. While such systems employ approximate models of the hydrodynamics of the tidal flows, the characteristics of the generating machinery in the barrage, and the estimation of the power coefficient are based on standard electrical generators and the associated characteristics. Tidal power systems fall into two main groups, tidal barrage, or tidal stream. The operation of a tidal barrage involves the construction of a barrier to obstruct a natural tidal flow. The flow restriction creates a height difference on either side of the barrier and this pressure head is used to drive a low head hydroelectric system. Tidal stream devices rely on extracting the kinetic energy directly.

Typically, dynamics and control models have also fallen into one of two categories. The first involve controlling the turbine characteristics by changing features like the pitch of the blades while the second involve controlling the flows into the estuaries. In the latter case, the barrage or the stream is idealized, and the instantaneous rate of flow through the barrage or in the stream is controlled.

The power output of marine turbines, for their size, is considerably greater than wind turbines, because water is non-compressible and about 800-times denser than air. The turbines are to be deployed throughout the main estuary in sectors of where the depths are large and current velocities are greatest to maximize power output. Both conclusions are drawn from the Power equation, $P = \frac{1}{2} \rho AV^3 C_p$, where: ρ = density of seawater (1.025 kg/m^3); A = swept area of blades (m^2); V = velocity of currents; and C_p = power coefficient. Thus, current velocity and blade radius are the two major parameters determining turbine power output. While current velocity is determined by the tidal currents, current flow through each turbine would be enhanced substantially by the controlled Venturi shrouds, yaw angle, and the relative position of turbines to each other. Given the parameters determining marine turbine power output, optimal control theory is applied to maximize the current flow through each turbine to ensure optimal performance. Thus, there are two options for designing control systems for tidal power turbines: (1) active control of the Venturi shrouds or inflow control; (2) blade pitch control or turbine features control.

Most generators used in tidal power generation are induction generators. The most common method of controlling and maximizing the power output of these generators is by Field Orientation Control. To a large extent, these generators are controlled in much the same way as generators used in wind energy extraction. Consequently, the methods of MPPT are quite similar.

References

Evans DV (1981) Maximum wave-power absorption under motion constraints. Appl Ocean Res 3(4):200–203

Evans DV, Porter R (1995) Hydrodynamic characteristics of an oscillating water column device. Applied Ocean Res 17(3):155–164

Fahrenburch AL, Bube RH (1983) Fundamentals of solar cells: photovoltaic solar energy conversion. Academic Press, Inc., New York

Falcão AFO (2010) Wave energy utilization: a review of the technologies. Renew Sustain Energy Rev 14:899–918

Falnes J (2002) Ocean waves and oscillating systems. Linear interactions including wave-energy extraction. Cambridge University Press, Cambridge

Gato LMC & Falcao AFdeO (1988) Aerodynamics of the Wells Turbine. Int J Mech Sci 30:383–395

Gato LMC & Falcao AFdeO (1990) Performance of the Wells Turbine with a Double Row of Guide Vanes. JSME Int J II 33(2):265–271

Gato LMC, Eca LRC, Falcao AFO (1991) Performance of the Wells Turbine with Variable Pitch Rotor Blades. J Energy Res Technol 113:141–146

Janssen PAEM (2004) The interaction of ocean waves and wind. Cambridge University Press, Cambridge

Kinoue Y, Setoguchi T, Kim TH, Kaneko K (2002) Air turbine using self-pitch-controlled blades for wave energy conversion. In: Proceedings of 12th International Offshore and Polar Engineering Conference, Kitakyushu, Japan, May 26–31, 2002, 668–673

Komen GJ, Cavaleri L, Donelan M, Hasselmann K, Hasselmann S, Janssen PAEM (1996) Dynamics and modelling of ocean waves. Cambridge University Press, Cambridge

Lamb H (1932) Hydrodynamics, 6th edn, Dover Publications, USA

Mazer JA (1997) Solar cells: an introduction to crystalline photovoltaic technology. Kluwer Academic Publishers, Boston

Mulligan WP, Rose DH, Cudzinovic MJ et al (2004) Manufacture of solar cells with 21% efficiency. In: Proceedings of 19th European Photovoltaic Solar Energy Conference (EU PVSEC '04), Paris, France, 387

Newman JN (1977) Marine Hydrodynamics. MIT press, Cambridge

Newman JN and Lee C-H (2002) Boundary-element methods in offshore structure analysis. In: 20th Int'l Conference on Offshore Mech. and Arctic Engineering, ASME: New York, 2001; J Offshore Mechanics and Arctic Engineering, 124:81-89

Raghunathan S (1995) The wells air turbine for wave energy conversion. Prog Aerospace Sci 31:335–386

Raghunathan S, Tan CP, Wells NA (1982) Theory and Performance of a Wells Turbine. AIAA TN 6:157–160

Salter SH (1974) Wave power. Nature 249:720–724

Salter SH (1993) Variable pitch air turbines. In: Proceedings of 1993 european wave energy symposium, 21–24th July, Edinburgh, pp 435–442

Sarmento AJNA, Gato LMC, Falcao AFdeO (1990) Turbine-controlled wave energy absorption by oscillating water column devices. Ocean Eng 17:481–497

Setoguchi T, Takao M, Kaneko K, Inoue M (1998) Effect of guide vanes on the performance of a Wells turbine for wave energy conversion. Int J Offshore Polar Engrg 8(2):155–160

Streetman BG (1995) Solid state electronic devices, 5th edn. Prentice Hall

Sze SM (2011) Physics of semiconductor devices, 3rd edn. Wiley, New York

Weider A, Levy H, Regev I, Ankri L, Goldenberg T, Ehrlich Y, Vladimirsky A, Yosef Z and Cohen M. (2006) SunSailor: solar powered UAV. Technical report, Students Project Report, Faculty of Aerospace Engineering, Technion IIT, Haifa, Israel

MIX
Papier aus verantwortungsvollen Quellen
Paper from responsible sources
FSC® C105338

If you have any concerns about our products,
you can contact us on
ProductSafety@springernature.com

In case Publisher is established outside the EU,
the EU authorized representative is:
Springer Nature Customer Service Center GmbH
Europaplatz 3, 69115 Heidelberg, Germany

Printed by Libri Plureos GmbH
in Hamburg, Germany